Springer Series in Operations Research and Financial Engineering

Editors:
Thomas V. Mikosch   Sidney I. Resnick   Stephen M. Robinson

Springer Series in Operation Research and Financial Engineering

Yves Pochet
Laurence A. Wolsey

Production Planning
by Mixed Integer
Programming

With 77 Illustrations

Yves Pochet
Laurence A. Wolsey
CORE / IAG and INMA
Université Catholique de Louvain
34 Voie du Roman Pays
B-1348 Louvain-la-Neuve
Belgium
pochet@core.ucl.ac.be
wolsey@core.ucl.ac.be

Series Editors:

Thomas V. Mikosch
University of Copenhagen
Laboratory of Actuarial Mathematics
DK-1017 Copenhagen
Denmark
mikosch@act.ku.dk

Stephen M. Robinson
University of Wisconsin-Madison
Department of Industrial Engineering
Madison, WI 53706
U.S.A.
smrobins@facstaff.wise.edu

Sidney I. Resnick
Cornell University
School of Operations Research and
    Industrial Engineering
Ithaca, NY 14853
U.S.A.
sir1@cornell.edu

Mathematics Subject Classification (2000): 90B30, 90C11, 90-01, 90-02

ISBN   978-1-4419-2132-1   e-ISBN   978-0-387-33477-6

Printed on acid-free paper.

©2006 Springer Science+Business Media, Inc.
Softcover reprint of the hardcover 1st edition 2006

9 8 7 6 5 4 3 2 1

springer.com

To our families

Brigitte
Florence
Christophe
Maxime

Marguerite
Jonathan
Julien

Preface

This book is about modeling and solving

multi-item,
single/multi-machine,
single/multi-level,
production planning problems
with time-varying demands
by mixed integer programming.

Since the beginnings of operations research and management science, models for production planning have been an important object of study with the Harris EOQ formula or Wilson's (Q, r) model, and Wagner–Whitin's dynamic lot-sizing model, the cornerstones for the treatment of stationary and time-varying (dynamic) demand, respectively.

The introduction of Materials Requirement Planning (MRP) systems in the 1970s was a major step forward in the standardization and control of production planning systems, but MRP and its successors were first and foremost information systems necessary but not sufficient for the efficient planning of the factory or enterprise. Much criticism was leveled at the inability of such systems to deal effectively with lead times and capacity constraints. Even in today's Enterprise Resource Planning (ERP) systems and Advanced Planning and Scheduling (APS) systems, the planning modules are still seen as unusable, or unable to handle the complexity of the underlying capacitated planning problems.

Starting in the 1960s and 1970s, the first serious efforts were made to describe mixed integer programming (MIP) models for single- and multi-stage planning problems of the type that arise regularly in practice, and that MRP and APS systems are designed to tackle.

Unfortunately MIP systems at the time were only able to solve "toy" instances, and so efforts were mainly concentrated on simple and rapid heuristics.

Starting in the early 1980s, motivated by the successes in tackling pure 0–1 and traveling salesman problems by strong cutting planes, a systematic study of the polyhedral structure of production planning models was initiated. The result is that today we know a considerable amount about the "right" way to formulate many simple production planning submodels as mixed integer programs, and this knowledge, combined with the remarkable progress of general MIP systems, enables us to solve or approximately solve many practical production planning problems that were considered far out of reach ten or so years ago.

The goal of this book is to enable a reader with a background in linear programming to use the knowledge and tools provided here to solve real-world production planning problems.

The book is addressed to practitioners with production planning or supply chain planning problems to be solved, and to students in management, industrial engineering, operations research, applied mathematics, computer science, and the like from final-year undergraduate up to Masters and PhD levels.

The book can be tackled on three levels.

Part I has material for everyone with an introduction to the modeling, formulation, and optimization approach to solving problems as MIPs, a chapter on classical Manufacturing Planning and Control models and systems (including MRP) and on more recent Advanced Planning and Scheduling systems, a chapter providing an introduction to MIP algorithms (including branch-and-bound, cutting plane, branch-and-cut, and neighborhood search MIP-based heuristic algorithms) and to the key issue of the quality of formulations in solving MIPs. These three chapters provide background material to render the book accessible to the widest possible audience.

Then a central chapter (Chapter 4) presents the classification of simple (single-item) production planning subproblems that is used throughout the book, and a procedure to improve the formulation of real-life production planning problems. This procedure is based on the classification scheme and the identification of the available reformulation results adapted to each problem structure. In the final chapter of Part I (Chapter 5), the reader tackles his first case studies, and is already required to improve the formulation and solution times of several production planning problems encountered in practice by using a library of extended reformulations, cutting planes, and neighborhood search heuristic algorithms, in combination with a standard MIP solver.

Part II provides a second-level course that could be entitled "Basic Polyhedral Combinatorics for Production Planning." The first of three chapters (Chapter 6) consists of a more rigorous introduction to the decomposition and reformulation approach developed throughout the book. In particular, different types of decomposition algorithm (reformulation and branch-and-cut, column generation, Lagrangian relaxation) are described and discussed.

The second (Chapter 7) contains a more or less complete polyhedral tour of the uncapacitated lot-sizing problem (LS-U). This incredibly rich problem allows us to present a wide variety of reformulation results using cutting planes (the so-called (l, S) inequalities) or additional variables (facility location and shortest path reformulations) and dynamic programming optimization algorithms (forward and backward), and it is hoped, provides a guide on ways to try to tackle other combinatorial/structured MIP problems.

Finally Chapter 8 presents the reformulation results known for the simplest MIP models (simple mixed integer sets and mixing sets, knapsack sets with a continuous variable and the Gomory mixed integer set, and single node flow sets) that are necessary for an understanding of both the cutting planes generated by today's commercial MIP systems (mixed integer rounding (MIR) and Gomory inequalities, knapsack cover inequalities and flow cover inequalities), and the cutting planes and reformulations that arise later in the book when examining different variants of lot-sizing models.

Parts III and IV provide a modeling and reference manual for the user who has identified a particular production planning subproblem – using for instance the classification scheme from Chapter 4 – and wishes to know and understand what formulations are available to improve his MIP model, and as a course for doctoral students in "Advanced Polyhedral Techniques and Production Planning."

Part III contains the important practical extensions of the single-item lot-sizing model (including constant and varying capacities, backlogging, start-ups, sales, more complicated cost functions and time windows, as well as other classical variants).

Part IV has two chapters, one (Chapter 12) dealing with multi-item models, and in particular on the different ways to formulate the allocation of resources to products (production modes) and the associated restrictions on production quantities, and a second (Chapter 13) dealing with multi-level problems, and in particular a presentation of the echelon stock reformulation. This concept is essential for the application of the single-item reformulation results to multi-level production planning problems, and to other supply chain production and distribution problems. The classification scheme for production planning problems presented in Part I for single-item problems is extended in Part IV to multi-item and multi-level problems.

Finally Part V (Chapter 14) contains three comprehensive case studies where we illustrate the modeling, formulation, solution, and sensitivity analysis approach, and three more technical case studies where we illustrate the various reformulation models and algorithms presented in the book. All case studies are directly derived from or based on real-life planning problems.

Figure 0.1 summarizes these three course levels, where the most important chapters at each level are indicated in bold.

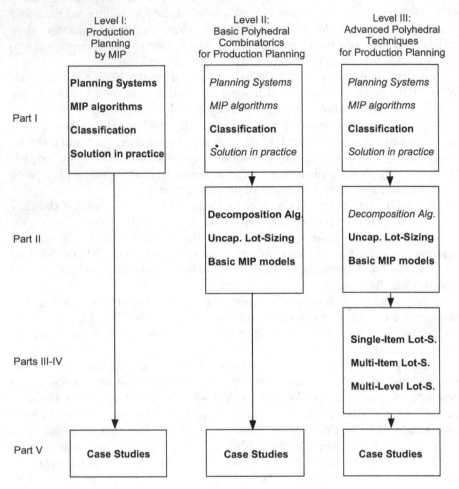

Figure 0.1. The three course levels.

Acknowledgments

Our first debt is to the numerous doctoral students and colleagues who have contributed willingly or unwillingly to the subject of this book. The doctoral students who have risen to the bait, excluding YP who was LAW's first student and swallowed hook, line, and sinker, include El-Houssaine Aghezzaf, Miguel Constantino, François Vanderbeck, Hugues Marchand, Cécile Cordier, Gaetan Belvaux, Marko Loparic, Francisco Ortega, Mathieu Van Vyve, and Quentin Louveaux. We have also had the opportunity and pleasure of collaborating with numerous colleagues, including Imre Barany, Tony Van Roy, Manfred Padberg, Ron Rardin, Stan van Hoesel, Albert Wagelmans, Oktay Günlük, Bram Verweij, Andrew Miller, Alper Atamturk, Michele Conforti, Agostinho Agra, and George Nemhauser. Special thanks are due to Mathieu Van Vyve for filling in several blanks in the original version of Tables 4.4–4.6 and for his important role in the conception and development of LS–LIB.

Another crucial factor has been the stimulating collaboration with different industrial partners in the course of three EU-financed projects PAMIPS, MEMIPS, and LISCOS, and in particular the practical production planning problems and mixed integer programs thrown at us before, during, and after. We are especially grateful to Bob Daniel, Richard Laundy, and Yves Colombani of Dash (the software company producing Mosel and Xpress-MP), Anna Schreieck, Beate Brockmüller, and Josef Kallrath of BASF, Ludwigshafen and Wim van de Velde of Procter and Gamble, Belgium.

The fine atmosphere and wonderful research environment at CORE, a research center in the Université catholique de Louvain, go without saying, so the support provided to CORE by the university, our respective departments of Management (IAG) and Mathematical Engineering (INMA), and the IAP Program of the Belgian Science Policy are most gratefully acknowledged. The authors YP and LAW have benefited from sabbatical leaves in 2003–2004 and 2004–2005, respectively, so again sincere thanks are due to the university and to the colleagues who replaced us.

Finally, but most of all, we would like to thank our families for their patience, love, and tolerance of a certain necessary addiction to work.

Université catholique de Louvain   Yves Pochet
December 2005   Laurence A. Wolsey

Case Studies and Web Site

This book contains problems and case studies in several chapters from Parts I and V. The corresponding versions of the data files and initial formulations in the Mosel modeling language, the modified formulations based on extended reformulations, cutting planes, and/or heuristic algorithms requiring both Mosel and LS–LIB, as well as the limited version of the LS–LIB[1] library, described in Chapter 5 and sufficient to treat all the problems and cases, are available on the Web site

`http://www.core.ucl.ac.be/LS-LIB/PPbyMIP`

Free restricted-size student versions of the modeling software Mosel and the MIP solver Xpress-MP can be downloaded from the Dash Web site

`http://www.dashoptimization.com`

The reader interested in knowing more about the Mosel modeling language may consult the documentation and reference manuals available at the same address.

Both Mosel and Xpress-MP are required in order to use the library LS–LIB. However the extended reformulation procedures ($XForm$) from LS–LIB can be used just with Mosel to generate input or matrix files in standard mathematical programming format. The corresponding reformulated problems can then be solved using any mixed integer programming system.

Apart from the black-box reformulation approach described in Chapter 5, which requires LS–LIB, all the formulations presented in this book can be implemented using the alternative classical reformulation approach described in Chapter 5 and other modeling and optimization software, such as AMPL, OPL/CPLEX, GAMS, and LINGO/LINDO.

---

[1] A full version of the LS–LIB library [135], as well as any modifications and updates, and instructions for use, can be obtained via the Web site `http://www.core.ucl.ac.be/LS-LIB`. This site also contains a variety of other lot-sizing test problems.

Contents

Part III   Single-Item Lot-Sizing

Part IV Multi-Item Lot-Sizing

Part I

Production Planning and MIP

Introduction

Production planning is viewed here as the planning of the acquisition of the resources and raw materials, as well as the planning of the production activities, required to transform raw materials into finished products meeting customer demand in the most efficient or economical way possible.

In industrial environments, the problems to be addressed in this field call for decisions about the size of the production lots of the different products to be manufactured or processed, about the time at which such lots have to be produced, and often about the machine or production facility on which the production must take place, or about the sequencing of the production lots. The usual objective is to meet forecast demand at minimum cost. These problems are typically short- to medium-term, or operational to tactical planning problems.

Supply chain planning is similar to production planning, but extends its scope by considering and integrating procurement and distribution decisions. Supply chain design problems cover a longer time horizon and include additional decisions such as the selection of suppliers, the location of production facilities, and the design of the distribution system.

The goal of production planning is thus to make planning decisions optimizing the trade-off between economic objectives such as cost minimization or maximization of contribution to profit and the less tangible objective of customer satisfaction. To achieve this goal, manufacturing planning systems are becoming more and more sophisticated in order to increase both the productivity and the flexibility of the production operations. For instance, to improve productivity, the current trend towards supply chain coordination implies the integration of production planning models and models involving procurement, production, distribution, and sales. Also, the need to be able to respond quickly to market or customer demand changes has created a need for refined production planning models better able to represent and exploit the flexibility of the production process, without losing in overall productivity.

In this general context of more integrated and sophisticated manufacturing systems, production planning models are very often mixed integer programming (MIP) models, because of problem features such as set-up costs and times, start-up costs and times, machine assignment decisions, and so on. A set-up occurs at the beginning of a new production lot. It typically implies additional production costs and times to prepare the machines and tooling. Such costs and times are fixed per batch and are not proportional to the batch size. Therefore binary or integer variables are required to model them. A start-up corresponds to the start of a sequence of batches of an item, following the production of a different item. Such MIP planning models may be difficult to solve for the large-size instances usually encountered in production planning systems.

However, sophisticated techniques can be used to improve or tighten the mathematical formulations of the models, or to design efficient optimization algorithms for solving them. Using an adequate or tight reformulation for a MIP model, or an efficient algorithm, may drastically reduce the running time needed to solve it. For more difficult instances, these techniques allow one also to increase the size of models solvable to optimality, or near optimality (i.e., giving solutions that are provably within a few percent of optimality).

It is the aim of Part I to *provide the reader with the expertise required to model and solve industrial production planning problems by using state-of-the-art optimization techniques and reformulation results.*

In order to be accessible to the widest possible audience, we follow a step-by-step approach, from building an initial model and a first mathematical formulation to solving industrial case studies using sophisticated reformulations and algorithms.

We take a modeling perspective. The objective in this part is to be able to use the reformulation results from the literature by analyzing the structure of the initial model, and by applying a *classification scheme* pointing to adequate reformulations. No mathematical analysis of these reformulations is necessary in Part I.

These reformulations – described in the subsequent Parts II to IV – can then be implemented either using a classical reformulation approach with any modeling language and MIP solver, or using a "black box" approach based on a library of reformulations and heuristics (LS–LIB). This library is designed in such a way that the user only needs to follow the classification scheme, without any knowledge of the mathematical description of the reformulations, but it requires the utilization of specific modeling and optimization software, namely Mosel and Xpress-MP.

We present two case studies showing how to use these reformulation approaches.

We assume some prior but elementary knowledge of optimization (linear programming models and the simplex algorithm). The knowledge required corresponds to the level of an introductory undergraduate course in optimization for engineering, business, or economics students.

In Parts II to IV of this book, we describe in detail all the reformulation results and techniques mentioned and used in Part I. This deeper mathematical description and understanding allows one to develop more sophisticated, and more efficient, tools for solving the same type of planning problems. The effectiveness of this second level of improvement is demonstrated on some of the case studies, but requires more taste for mathematical approaches.

We conclude this introduction by describing more precisely the contents of Part I.

- We start in Chapter 1 by describing, with an example, the *systematic modeling approach* that must be taken to build a correct initial model representing a production planning problem. This example is a variation of a well-known problem from the literature, namely the multi-item capacitated lot-sizing problem. The results obtained by applying standard MIP software to the example indicate the need for more sophisticated tools in order to be able to solve large-size instances to optimality.

- In Chapter 2 we study the *classical production planning models and systems*. We first review some production planning models considered in ERP (Enterprise Resources Planning) or MRP systems. We then present and criticize the typical heuristic solution approach implemented in such systems. Finally, we take a broader perspective and define the planning tasks in the general context of the more recent APS (Advanced Planning and Scheduling Systems), which include the well-known Manufacturing Planning and Control Systems, Material Requirements Planning (MRP-I), Manufacturing Resources Planning (MRP-II), and Hierarchical Production Planning (HPP). We also provide examples of procurement, production and distribution planning problems without mathematical models or formulations to illustrate the planning tasks and the planning process along the supply chain.

- Chapter 3 contains an introduction to *mixed integer programming* (MIP) models and algorithms for those whose background is limited to linear programming. This includes a description of the *branch-and-bound, cutting plane*, and *branch-and-cut* optimization algorithms used to find the optimal solution of a MIP problem. It also includes the description of some *neighborhood search MIP-based heuristic algorithms* used either to construct an initial feasible solution, or to improve the best known feasible solution, of a MIP problem. Some emphasis is also put on the impact of the quality of the mathematical formulation used on the running time of these algorithms. This chapter can be skipped by those with an appropriate background.

- In Chapter 4 we present the *algorithmic approach* used to solve production planning problems. First we briefly illustrate the reformulation approach on the *LS-U* (uncapacitated single-item lot-sizing) model. Then we explain informally how practical planning models that are almost always multi-item and multi-period can be decomposed into single-item subproblems, and solved using good formulations for appropriate single-item subproblems.

  We then introduce the *classification scheme for single-item production planning models* that is central to our methodology because most optimization approaches are based on the decomposition of the problem into single-item subproblems.

  The classification is then used to describe the reformulation results collected from the literature. We propose a *reformulation procedure* designed to identify and classify the structure of industrial applications, and to be able to apply the known reformulation results for related or embedded submodels to these applications.

  Finally, we come back to our starting example from Chapter 1, and analyze its structure using our classification scheme. We illustrate the use of the reformulation procedure by showing how it allows one to reformulate and solve this problem more efficiently.

- We conclude in Chapter 5 by showing how to use the reformulations, the cutting plane routines, and the primal heuristics available for solving production planning problems. In particular, we describe the library of procedures LS–LIB, and illustrate the proposed black-box reformulation approach on two industrial case studies.

1

The Modeling and Optimization Approach

Motivation

To cope with the increasing complexity of their business, many large- and medium-size companies have implemented computerized manufacturing planning systems in the past decades. Such systems are used to standardize the planning processes followed by the various plants or production departments. In most cases, they are pure transactional systems maintaining up-to-date procurement and production information on each item, recording and distributing planning decisions. This is of course crucial.

However, significantly superior results can be obtained by changing these tools into planning systems for coordination and optimization. For instance, the Kellogg Company has developed an optimization system to plan the production and distribution decisions for its cereal and convenience foods business. This planning system subsumes an operational short-term system to plan and optimize the flow of goods, as well as a tactical medium-term system to help in making budgets, in solving capacity expansion problems, and in the consolidation of decisions. Kellogg reports annual cost savings of 4 million dollars with the operational system, and projects annual savings of the order of 40 million dollars with the consolidated tactical system.

Till now the mathematical program behind the Kellogg Planning System (KPS) has been a linear program. This does not allow the KPS operational model to account for production and packaging set-up times (the time lost because of equipment adjustments in between batches of different products). This severe restriction also obliges the managers to review and modify manually the plans suggested by KPS in order to obtain feasible plans that take into account plant floor reality. This also means that the model used needs to be improved to allow Kellogg to fully optimize short-term productivity.

The extension of KPS to a mixed integer program, so that one is able to take into account set-up times, is under consideration. This is also the case for many other companies trying to develop planning systems able to optimize productivity. The necessary first step is often the development of new production planning models.

The resulting large-size mixed integer programs are typically much harder to solve to optimality, or near optimality, than linear programs. Nevertheless, it is often possible to (re)formulate them in such a way that the solution time is drastically reduced. Unfortunately some of these advanced or sophisticated reformulation techniques are not generic, in the sense that they depend on specific structure in the problem/model to be applicable. In other words, the identification of structure in the production planning problem is important during model construction, especially for the use of the reformulation techniques. Therefore a systematic modeling approach must be taken.

Objective

As a starting step towards the final objective of solving mixed integer production planning problems, the specific objective of this chapter is to learn how to systematically *transform a problem description into a mathematical model*. The problem description is usually unstructured and given as the minutes of a meeting, an internal memo, or a report.

This mathematical model, an abstraction from the real problem, should be

- *correct*, that is, planning decisions or solutions suggested by the model should represent reality with the desired level of detail, both in terms of feasibility (through model constraints) and optimality (objective function), and
- *structured*, that is, described using standard objects (products, machines, or resources,..,) and standard building blocks or generic constraints (flow conservation, capacity,...).

The structuring of the model will later play an essential role during the model classification and reformulation phases. The classification scheme will formally describe these generic constraints, and the combinations of generic constraints for which reformulation results are known.

Contents

In this chapter:

- We describe in Section 1.1 the concept of an *optimization problem* (indices or objects, variables, constraints, objective function) with a tiny example.
- We illustrate in Section 1.2 the *systematic modeling approach* that must be taken to build a correct and structured initial mathematical model of a production planning problem on a more realistic multi-item example, and we motivate the search for more sophisticated tools, by solving the initial mathematical program obtained in this example, using standard optimization software.

1.1 A Tiny Planning Model

1.1.1 Problem Description

A manufacturer produces a wide variety of bicycles. We are interested in the production plan of a single high-tech racing model whose production requires special materials and production equipment. At most one batch is produced per month, because of low demand and important economies of scale in the manufacturing costs. Because of the need to install special equipment and tools at the beginning of a batch, there is a high set-up cost, and thus it has been decided that it makes no sense to produce more frequently.

The batch manufacturing cost is best approximated by the fixed charge cost represented in Figure 1.1. The set-up cost represents the equipment and tool installation and preparation costs, and then the constant marginal cost corresponds to the constant time required to produce each bicycle. For the racing model, the set-up cost is 5000 euros, and the marginal cost is 100 euros. Hence, it costs 5100 euros to produce a batch of 1 bicycle, and 6000 euros for a batch of 10 bicycles.

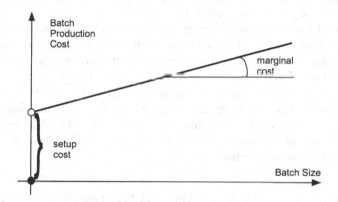

Figure 1.1. The fixed charge cost.

The capacity restrictions are ignored in planning this single-product variant because the work center and workers are shared by the many bicycle variants, and because capacity can be increased by hiring temporary workers, if necessary.

The company faces irregular or seasonal demand, sales being higher in spring and summer. Table 1.1 gives the sales forecasts in number of bicycles per month for the racing model in the coming year.

Moreover, there will be around 200 racing bicycles in stock at the end of the current year. This projected inventory is based on current production plans, sales forecasts, and customer orders up to the end of the year. To hold one bicycle in inventory during one month costs on average 5 euros, including

Table 1.1. Bicycle Manufacturer: Sales Forecasts for Next Year

Jan	Feb	Mar	Apr	May	Jun	Jul	Aug	Sep	Oct	Nov	Dec
400	400	800	800	1200	1200	1200	1200	800	800	400	400

the cost of capital and storage costs. Note that there is enough space available in the warehouse to store the bicycles.

The manufacturer wants to plan the production and inventory levels of this particular racing bicycle, in order to satisfy demand and minimize the corresponding manufacturing and inventory costs. He wants to plan production for next year up to the end of the peak demand period, that is, up to the end of August. (Why ?)

1.1.2 Some Solutions

Many people would start by trying to find and enumerate some good solutions of the above planning problem. And some people would be happy enough with such an approach. In this toy problem, we only seek the best compromise between inventory costs and production costs.

Because of the economies of scale, production costs are minimized by producing very large batches. This leads one to produce the whole demand (7200 units up to end of August, minus the initial inventory of 200 units) in January, but then to incur large inventory costs up to the end of August.

The opposite extreme is to minimize storage costs, which is one major concern in production planning. The minimum inventory plan consists of producing so as to just satisfy the demand in each month. But this requires one to set up the machines every month, resulting in small batches, and therefore significant manufacturing costs.

Table 1.2. Tiny Example: The Minimum Manufacturing Cost Solution

	Jan	Feb	Mar	Apr	May	Jun	Jul	Aug	Total
Demand	400	400	800	800	1,200	1,200	1,200	1,200	7,200
Production	7,000	0	0	0	0	0	0	0	7,000
Unit cost	700,000	0	0	0	0	0	0	0	700,000
Set-up cost	5,000	0	0	0	0	0	0	0	5,000
End inventory	6,800	6,400	5,600	4,800	3,600	2,400	1,200	0	
Inv. cost	34,000	32,000	28,000	24,000	18,000	12,000	6,000	0	154,000

These two extreme solutions, and their costs, are represented in Tables 1.2 and 1.3, respectively. To compute the inventory costs, we have assumed that, during each month, the inventory level evolves linearly over time from

Table 1.3. Tiny Example: The Minimum Inventory Cost Solution

	Jan	Feb	Mar	Apr	May	Jun	Jul	Aug	Total
Demand	400	400	800	800	1,200	1,200	1,200	1,200	7,200
Production	200	400	800	800	1,200	1,200	1,200	1,200	7,000
Unit cost	20,000	40,000	80,000	80,000	120,000	120,000	120,000	120,000	700,000
Set-up cost	5,000	5,000	5,000	5,000	5,000	5,000	5,000	5,000	40,000
End Invent.	0	0	0	0	0	0	0	0	
Invent. cost	0	0	0	0	0	0	0	0	0

the starting level to the ending level. If INV_t represents the inventory level at the end of month t, then the inventory cost can be represented by

$$\text{inventory cost} = \sum_{t=Jan}^{Aug} 5 * \frac{(INV_{t-1} + INV_t)}{2}$$

which is equivalent to

$$\text{inventory cost} = 2.5 * INV_0 + \sum_{t=Jan}^{Jul} 5 * INV_t + 2.5 * INV_{Aug} ,$$

where INV_0 is the initial stock (constant). Observe also that $INV_{Aug} = 0$ in these two solutions. In Tables 1.2 and 1.3, we have used this last and simpler expression to compute the inventory costs as a function of the monthly ending inventory levels.

We observe that the total cost of the minimum inventory solution is 740,000 euros which is significantly cheaper than the total cost of 859,000 euros of the minimum production cost solution. However, is it optimal? By using trial and error, it is often possible to improve such initial solutions. But how can one guarantee that an improved solution is indeed optimal? We need a more systematic approach able to provide a proof of optimality.

1.1.3 A First Model

Our approach is very different from the ad hoc approach suggested above. The essence of the modeling and optimization approach is to distinguish or separate the modeling and optimization phases, and to build a correct model, before going to the optimization phase.

- The objective of the *modeling phase* is to describe a mathematical abstraction (a model) of the problem to be solved. This model identifies
  - the *objects* to be manipulated (products, resources, time periods, etc.),
  - the *data* associated with the objects (demand for products during time periods, capacity of resources, etc.),

- the *decisions* (also called decision variables, or simply *variables*) to be taken relative to the objects in order to propose or define a solution to the problem,
- the *constraints* to be satisfied by the decisions in order to define feasible or acceptable solutions to the problem, and
- the *objective function* which provides a way to evaluate or compare feasible solutions, and to select the best or optimal solution among the feasible ones.

• The objective of the *optimization phase* is to find an optimal solution of the model.

Some claim that modeling is an art. Our goal here is to show that it is also a science. Thus, to build correct models with the right level of detail, it is necessary to follow a systematic approach in defining the objects, data, variables, constraints, and objective function corresponding to the problem description.

Moreover, to ease this translation from the real problem to the abstract model, it is often very helpful to define *higher-level structures*. This is usually achieved by defining appropriate concepts and notation. Such higher-level structures are also mandatory for our main classification and reformulation phases presented later in Chapter 4. For instance:

• Similar objects are grouped into object classes represented by *mathematical indices*, allowing one to use indexed notation for data, variables, and constraints.
  - The index t is used to represent an element in the set $\{1, \ldots, NT\}$ of time periods, where NT is defined as the number of time periods in the planning horizon.
  - The index i is used to represent an element in the set $\{1, \ldots, NI\}$ of products, where NI is defined as the number of products in the planning problem.
  - And so on.

• Similar data or variables are grouped into *data classes or variable classes*, allowing one to use generic definitions and naming conventions.
  - The data $demand(i, t)$ defines the demand for all products i and time periods t.
  - The variable $prod(i, t)$ defines the production-level decisions for all products i and time periods t.
  - And so on.

• Similar constraints are grouped into *constraint classes* or *global constraints*, allowing one to use generic definitions and common mathematical formulations across a constraint class.
  - The generic or global constraint $demand\_satisfaction(i, t)$ defines the demand satisfaction constraint for all products i and time periods t.
  - And so on.

To illustrate this systematic modeling approach, we now define the optimization model associated with our tiny example.

(i) First the *identification and naming of indices, data, variables, and constraints* is performed by scanning through the problem description, and systematically marking the objects, data, and so on, as they are encountered. This gives the following results.

Objects and Indices	Mathematical Notation
One manufacturer	
One product (racing bicycle)	Object: bike
Monthly time periods	Object: periods
	Index: $t = 1, \ldots, NT$ and $NT = 8$
Production resources: ignored	
Storage resources: ignored	

Data	Mathematical Notation
Production set-up cost	For bike [euro]: $q = 5\,000$
Production unit cost	For bike [euro/unit]: $p = 100$
Demand forecasts	For bike, period t [unit]: d_t
	$d = [400, 400, 800, 800, 1200,$
	$\qquad\quad 1200, 1200, 1200]$
Initial stock	For bike [unit]: $s\_ini = 200$
Inventory holding cost	For bike [euro/unit, period]: $h = 5$

Variables	Mathematical Notation
Production batch size	For bike, period t [unit]: $x_t \geq 0$
End inventory level	For bike, period t [unit]: $s_t \geq 0$

Constraints	Mathematical Notation
Demand satisfaction	For bike, period t [unit]: dem_sat_t

Objective function	Mathematical Notation
Minimize sum of production and inventory costs	[euro]: $cost$

(ii) Then, to complete the model it remains to define the *mathematical formulation* of the constraints and objective function.

The demand satisfaction generic constraint dem_sat_t, defined for all products (here, only bike) and all periods $t \in \{1, \ldots, NT\}$, simply states that the amount of product bike available in period t (which is defined by the ending inventory s_{t-1} from period $t - 1$ plus the production x_t in period t) must satisfy at least the demand d_t of period t; that is, $s_{t-1} + x_t \geq d_t$. The number of bikes in excess of the demand d_t defines the ending inventory s_t. This gives the constraint

$$dem\_sat_t := s_{t-1} + x_t = d_t + s_t \quad \text{for } t = 1, \ldots, NT \,,$$

where the variable s_0 occurring in dem_sat_1 represents the initial stock and is replaced by the constant s_ini.

This demand satisfaction constraint is common to almost all planning models though it takes a slightly different form depending on the existence of multiples machines, backlogging, or other variants. It is often called a *flow balance* or *flow conservation* constraint because its feasible (x, s) solutions correspond to the feasible flows in the network represented in Figure 1.2, where the nodes correspond to time periods.

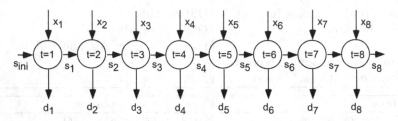

Figure 1.2. The flow conservation global constraint.

Modeling the objective function requires the addition of new binary variables, that take only the two values 1 or 0, to indicate the existence (1) or nonexistence (0) of a set-up in each period. Remember that a set-up, representing the installation and preparation of the equipment and tools, is required to produce a batch.

Variables (Revised)	Mathematical Notation
Production batch size	For bike, period t [unit]: $x_t \geq 0$
Production set-up	For bike, period t [-]: $y_t \in \{0, 1\}$
End inventory level	For bike, period t [unit]: $s_t \geq 0$

Assuming that y_t takes the value 1 when there is a set-up in period t, the objective function can be modeled as the sum of the production set-up costs, production unit or variable costs, and the inventory holding costs.

$$cost := \sum_{t=1}^{NT}(p\,x_t + q\,y_t) + \sum_{t=1}^{NT} h\,\frac{(s_{t-1} + s_t)}{2} ,$$

which is equivalent to

$$cost := \sum_{t=1}^{NT}(p\,x_t + q\,y_t) + \frac{h}{2}\,s\_ini + \sum_{t=1}^{NT-1} h\,s_t + \frac{h}{2}\,s_{NT} .$$

Finally, we must add the necessary constraints to ensure that the required set-ups are performed in every period.

Constraints (Revised)	Mathematical Notation
Demand satisfaction	For bike, period t [unit]: dem_sat_t
Set-up enforcement	For bike, period t [unit]: vub_t

The classical generic constraint used to enforce a set-up is the so-called variable upper bound (VUB) constraint defined as

$$vub_t := x_t \leq (\sum_{k=t}^{NT} d_k)\, y_t,$$

where $\sum_{k=t}^{NT} d_k$ is a true upper bound on x_t when there is a set-up in period t and no ending inventory (see Figure 1.2). Therefore, either there is no set-up in period t (i.e., $y_t = 0$) and the above constraint forces $x_t \leq 0$ (i.e., $x_t = 0$), or there is a set-up in period t (i.e., $y_t = 1$) at a cost of q and the above constraint imposes a valid upper bound on x_t, if we assume that the ending stock $s_{NT} = 0$. In this particular model instance, we can assume without loss of generality that $s_{NT} = 0$ because it holds for any optimal solution.

(iii) The structure of the *complete model* is identified by
  • the demand satisfaction global constraint for the single product *bike* over eight consecutive time periods,
  • the variable upper bound global constraint for the single product *bike* over eight consecutive time periods,
  • the initial inventory of product *bike*,
  • the nonnegativity and integrality restrictions on the variables, and
  • the cost function to be minimized.

$$\min \quad cost := \sum_{t=1}^{NT}(p\, x_t + q\, y_t) + \sum_{t=1}^{NT-1} h\, s_t + \frac{h}{2}\, s_{NT}$$

$$dem\_sat_t := \quad s_{t-1} + x_t = d_t + s_t \qquad \text{for all } t = 1,\ldots,NT$$
$$s_0 = s\_ini\, , \ s_{NT} = 0$$

$$vub_t := \quad x_t \leq (\sum_{k=t}^{NT} d_k)y_t \qquad \text{for all } t = 1,\ldots,NT$$

$$x_t, s_t \in \mathbb{R}_+, \ y_t \in \{0,1\} \qquad \text{for all } t = 1,\ldots,NT.$$

1.1.4 Optimizing the Model

Instances of optimization models involving only linear constraints and a linear objective function, and both continuous variables (s_t and x_t) and binary variables (y_t), are called *mixed integer or mixed binary programs (MIP)*. Such programs can be solved by general purpose branch-and-bound or branch-and-cut algorithms.

These general-purpose algorithms, as well as an introduction to the ideas of reformulation, are briefly described in Chapter 3.

For the more complex and larger size production planning applications that we consider in this book, we need to use or develop more sophisticated and specific optimization algorithms. We propose an approach based on model classification and reformulation in Chapter 4.

For completeness, feeding our first model and data into a MIP optimizer, we obtain the optimal solution of our tiny example given in Table 1.4. It has a total cost of 736,000 euros and is similar to the minimum inventory solution, but with two set-ups removed.

Table 1.4. Tiny Example: The Optimal Solution

	Jan	Feb	Mar	Apr	May	Jun	Jul	Aug	Total
Demand	400	400	800	800	1,200	1,200	1,200	1,200	7,200
Production	600	0	1,600	0	1,200	1,200	1,200	1,200	7,000
unit cost	60,000	0	160,000	0	120,000	120,000	120,000	120,000	700,000
set-up cost	5,000	0	5,000	0	5,000	5,000	5,000	5,000	30,000
End Inventory	400	0	800	0	0	0	0	0	
Inv. cost	2,000	0	4,000	0	0	0	0	0	6,000

1.2 A Production Planning Example

Here we take a more realistic example in order to further illustrate the systematic modeling approach that must be taken to build a mathematical model of a production planning problem. We start by describing the problem and its general context.

1.2.1 Problem Description

GW and the Global Supply Chain Department

GW is a large worldwide company in the fast-moving consumer goods industry, selling hundreds of brands to millions of consumers dispersed all over the world.

Bill Widge is the head of the Global Supply Chain Optimization (GSCO) Department. He is responsible for the development, implementation, and integration into the manufacturing information system (the well-known PASI-2 system) of new optimization approaches in order to improve capacity utilization and process flexibility.

The company has installed a common *manufacturing information, planning, and control system (MPCS)* in all its facilities. This system is an advanced information system that allows the company to plan and coordinate the procurement, production, and distribution activities. The implementation and customization of this system has been a major project for the company, spanning several years. It has led to major improvements in terms of supply chain coordination.

Unfortunately, because the same planning system is used in all facilities, the planning procedures used are generic procedures that have failed to improve the *productivity* (broadly speaking, the ratio of the quantity of outputs produced over the quantity of inputs utilized) and *flexibility* (ability to respond quickly to the perpetually changing requirements from the marketplace) of the manufacturing plants as much as their coordination.

Therefore, the Board has decided to create the GSCO department to remedy this weakness. The objective assigned to GSCO is the development of optimization based planning tools to support the planning tasks and improve the productivity of key processes. The ultimate goal is to integrate specific planning tools for these processes into the generic information system PASI-2.

The Problem

The problem we consider here is a successful GSCO project. It is aimed at productivity optimization for the largest plant in the food sector. This plant produces two families of products, designated Cereals and Fruits.

The following problem description is a summary of the information that was available at the start of the project.

The Production Process

The production process is composed of three major steps: preparation, mixing, and packaging.

- First, the raw materials, which are stored in huge tanks, need to be prepared (cleaned, heated, etc.) before they can be used in the process. There is only one preparation line.
- Next, the Cereals and Fruits are produced. This step is called mixing because the major operation consists of blending the different ingredients. Other operations at the mixing step include heating, crushing, and drying. There is only one mixing line.
- Finally, the products are packed on two dedicated packaging lines, one for Fruits and the other for Cereals.

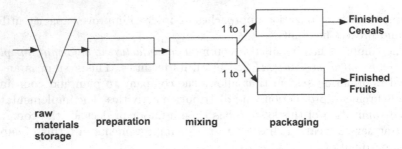

Figure 1.3. The Cereals and Fruits production process.

For each product obtained after mixing, there is only one packaging format available. In other words, there is a one-to-one correspondence between mixing products and finished products.

This production process is represented in Figure 1.3. Note that for building, and progressively fine-tuning a model, such a graphical flow representation of the problem elements helps to synthesize the information and is an important part of the modeling process.

The Bottleneck

The bottleneck or the major constraint of the whole production process is the mixing operation for the following reasons.

- There are very few raw materials. They have short and reliable procurement lead times, and their storage capacity is high relative to the needs.
- The preparation step is a very fast continuous process, and takes very little time. Thus a batch can always be prepared during the mixing of the previous batch.
- The pace of the mixing step is dictated by the mixing operation/machine. The other operations of the mixing step can be synchronized with the mixing operation without slowing down the process.
- Moreover, the mixing machine is inflexible in the sense that there are significant cleaning times at the end of each batch. These cleaning times must be respected to guarantee product quality, and do not depend on the sequence of products. They come from regulations imposed by the Food Administration. There are no cleaning or machine preparation times for the other operations of the mixing step. Apart from the cleaning times for mixing, the production rate can otherwise be assumed to be constant on every machine, with no economies of scale.
- Thanks to a recent investment in a second packing line, each line is now dedicated to a specific product family, and packaging can be carried out without switching times in between products. The joint packaging capacity exceeds the mixing capacity.

- The packaging lines are composed of flexible automated machines. Within a product family, they can switch from one product to another with almost no productivity loss.
- Finally, for quality reasons, the storage capacity of intermediate products between mixing and packaging is very limited, and intermediate products have to be packed almost directly after mixing.

In summary, the bottleneck of the process is the mixing operation, because there are large cleaning times at the end of each mixing batch. The preparation and all the subtasks of the mixing step can be synchronized with the mixing operation. The packaging operations involve no switching times and have a large enough aggregate capacity (the total packing capacity for the two product families) to absorb the output of the mixing operation. This packaging step must be synchronized with mixing because of limited storage capacity between these two production steps.

The Production Policy and Current Planning System

Because of limited product variety, and in order to reduce the global supply chain lead time, the company has imposed a make-to-stock (MTS) production policy at the plant level. This means that the plant production must be able to meet the demand coming from the distribution system directly from stock, that is, with zero delivery lead time at the plant level. To achieve this, the company establishes forecasts of weekly demand addressed to the plant, and computes safety stocks needed to cover the difference between actual and forecast demand. This process has been effective in the past. The forecasts are of good quality, and the safety stocks allow the company to achieve excellent customer service levels.

This bottleneck and production policy information is used in Figure 1.4 to update the flow representation of the process. The current planning system is typical of ERP/MRP type systems.

- Once a week, a Master Production Schedule (MPS) is generated for the next few weeks. This schedule plans the production at the finished product level (packaging level) in order to meet forecast demand and safety stock requirements.
- The MRP system determines when and how much to produce or order of each intermediate product (mixing and preparation level) or raw material over the schedule horizon. The MRP calculation is based on the packaging orders (batches) defined at the MPS level.
- Finally, detailed scheduling of the packaging and mixing operations is carried out a week in advance based also on the MPS.

The planning jargon used here (ERP, MRP, MPS, etc.) is briefly explained in Chapter 2.

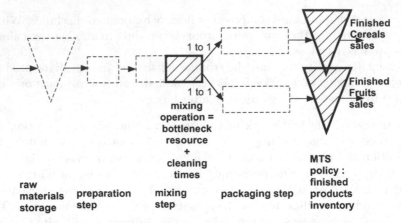

Figure 1.4. The Cereals and Fruits production process and policy.

In the current MPS/MRP system, weekly time buckets are used for the planning tools; that is, time is broken down into time periods of one week. The plant operates five days a week, and the mixing machine must be cleaned at the end of the week. Therefore no production batch runs over the weekend. Weekly time buckets are also chosen because the forecasting system is using weekly time buckets.

A time horizon of six weeks is currently used for the MPS and MRP. This is slightly longer than the total procurement and manufacturing lead time so as to allow GW to order and receive the raw materials on time. All raw material purchasing orders can be calculated from the master production plan for the finished products.

The Challenge

Unfortunately, the MPS process, which plays a central role in the planning system, does not take the limited capacity of the bottleneck operation (mixing) into account. Therefore the planner has to revise the MPS plan manually in order to get a feasible production plan. This looks similar to the Kellogg case mentioned at the beginning of the chapter, and has the same consequences: a slow and inefficient planning process, unable to optimize capacity utilization and to guarantee satisfaction of external demand.

The main difficulty in this planning problem is to optimize the trade-off between productivity (which requires large mixing batches to avoid losing capacity through frequent stoppages for cleaning at the bottleneck) and flexibility (which requires small batches to be able to produce as late as possible and to react quickly to market changes).

This difficulty arises because the MPS is not driven by the most scarce constraint in the process: the mixing capacity. Therefore, the goal of Bill Widge and the GSCO department is to design, develop, and install an efficient MPS tool giving feasible production plans, both with respect to demand satisfaction and capacity utilization. The aim is also to improve flexibility by producing as late as possible.

1.2.2 Modeling

In the application of our stepwise modeling approach, we need to identify the scope of the model, fix the boundaries of the model universe (which products to consider? which resources to model?), and decide on the general structure of the model. The level of detail of the model is also a major decision: enough detail is necessary to really optimize the productivity–flexibility trade-off, whereas unnecessary detail will make the problem impossible or harder to solve to optimality.

The role of the generic constraints becomes clear in this second example. We are able to reuse some of the constraints encountered earlier, and thus significantly simplify the modeling task.

(i) Identification and naming of indices, data, variables, and generic constraints by scanning through the problem description.

Objects and Indices	Mathematical Notation
One plant	$--$
Product families C and F	$--$
Individual finished products	Object: products
	Index: $i = 1,\ldots,NI$
Weekly time periods	Object: periods
	Index: $l = 1,\ldots,NT$ and $NT = 15$
Mixing line	Object: machine
	Index: $k = 1$
Packaging line for Cereals	Object: machine
	Index: $k = 2$
Packaging line for Fruits	Object: machine
	Index: $k = 3$
Other prod. resources: ignored	$--$
Storage resources: ignored	$--$

Remarks and Assumptions:
- The MPS model must consider finished products (or mixing products, because there is a one-to-one correspondence between them) individually, in order to be able to represent satisfaction of forecast demand.
- Weekly time buckets are used as in the current system, because there is no need to increase/decrease the level of detail.
- The time horizon needed to establish the MPS is much longer than the total procurement and manufacturing lead time (about six weeks), because it is necessary to anticipate the capacity requirements over

a long enough horizon in order to optimize capacity utilization. A horizon of 15 weeks was selected based on a deeper analysis of short-term variations of demand.

- The capacity of the mixing stage is the main or global bottleneck because of the cleaning times, but the capacity of each individual packaging line also needs to be taken into account because there is not enough packaging capacity to produce (mix and pack) only Fruits or only Cereals in a week.
- All operations other than mixing and packaging are neglected in the model because they can be synchronized with mixing and do not impose any additional capacity restrictions.

Data	Mathematical Notation
Demand forecast	For product i, period t [unit]: D_t^i
End period safety stock	For product i, period t [unit]: SS_t^i
Initial stock	For product i [unit]: SS_0^i
Cleaning time after mixing	For product i [hour]: β^i
Constant production rate	For product i, machine k [hour/unit]: α^{ik}
Machine capacity	For machine k [hour]: L^k
Product family Cereals	$=F^2$ (Subset of products)
Product family Fruits	$=F^3$ (Subset of products)

Remarks and Assumptions:
- We assume that the cleaning times at the end of the mixing batches are product-dependent, but not time-dependent.
- To be able to model mixing and packaging capacity utilization, we also need to know the number of working hours for each time period and each machine.
- Finally, the family (Cereal or Fruit) of each product must be known in order to assign the mixing batches to the packaging lines, and to model the packaging capacity restriction.

Variables	Mathematical Notation
Mixing batch size	For product i, period t [unit]: $x_t^i \geq 0$
Production set-up	For product i, period t [-]: $y_t^i \in \{0,1\}$
End period inventory level	For product i, period t [unit]: $s_t^i \geq 0$

Remarks and Assumptions:
- The main decisions in the model are the batch sizes for the mixing step for each finished product and each time bucket.
- To represent the machine cleaning times, we need to use production set-up variables for each product and period.
- In order to represent the trade-off between productivity and flexibility, we need to model the finished product inventory levels.

Constraints	Mathematical Notation
Demand satisfaction	For product i, period t [unit]: $dem\_sat_t^i$
Set-up enforcement	For product i, period t [unit]: vub_t^i
Mixing capacity restriction with cleaning times	fFor period t [hour]: mix_cap_t
Packaging capacity restriction without cleaning times	For product i, machine $k = 2,3$ [hour]: $pack\_cap_t^k$

Remarks and Assumptions:
- The generic set-up enforcement constraint is used as in the first example to assign correct values to the production set-up variables.
- It is also assumed that the packaging of a batch occurs in the same time bucket as the corresponding mixing batch, because there is no intermediate storage capacity.

Objective function	Mathematical Notation
Minimize total inventory	[euro]: $inventory$

Remarks and Assumptions:
- The objective of the model is to produce as late as possible, which can be expressed by minimizing the level of finished product inventory.

The structure of the optimization model is identified by
- the generic demand satisfaction constraints for each finished product over 15 consecutive weekly time periods, including initial stocks and safety stocks,
- the generic capacity utilization constraint for the mixing machine in each time period including cleaning times,
- the generic capacity utilization constraint for each packaging line in each time period, and
- the inventory minimization objective function.

This structure is apparent in the final graphical description of the MPS model in Figure 1.5.

(ii) Mathematical formulation of the generic constraints and objective function.

The demand satisfaction global constraint $dem\_sat_t^i$, defined for all products $i \in \{1, \ldots, NI\}$ and all periods $t \in \{1, \ldots, NT\}$, takes the same general form as in the first example:

$$dem\_sat_t^i := s_{t-1}^i + x_t^i = D_t^i + s_t^i \quad \text{for } i = 1, \ldots, NI \text{ and } t = 1, \ldots, NT,$$

where again the variable s_0^i occurring in $dem\_sat_1^i$ represents the initial stock and is replaced by the constant SS_0^i, and with the additional safety stock requirements:

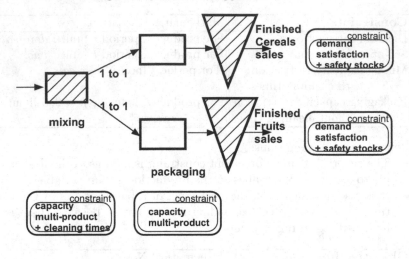

Figure 1.5. The structure of the Cereals and Fruits MPS model.

$$s_t^i \geq SS_t^i \qquad \text{for } i = 1, \ldots, NI \text{ and } t = 1, \ldots, NT.$$

Modeling the cleaning times for the mixing capacity global constraint requires the addition of the mixing binary set-up variables y_t^i for all products $i \in \{1, \ldots, NI\}$ and all periods $t \in \{1, \ldots, NT\}$. A set-up of product i in period t is defined here as the realization of the cleaning operation at the end of a batch of i in period t, and arises when the corresponding binary variable y_t^i takes the value 1. Again, a set-up is required when there is a batch of product i in period t (i.e., $x_t^i > 0$). This is modeled as before with the variable upper bound constraint or set-up enforcement constraint:

$$vub_t^i := x_t \leq \left(\sum_{k=t}^{NT} D_k^i + SS_{NT}^i \right) y_t,$$

where we again assume that the end-stocks $s_{NT}^i = SS_{NT}^i$ for all $i \in \{1, \ldots, NI\}$.

Now, using these set-up variables and remembering that $k = 1$ identifies the mixing machine, the mixing capacity constraint simply ensures that there is enough capacity in period t to produce all batches and perform all required cleaning operations:

$$mix\_cap_t := \sum_i \alpha^{i1} x_t^i + \sum_i \beta^i y_t^i \leq L^1 \qquad \text{for } t = 1, \ldots, NT.$$

The packaging capacity constraints are similar, but do not contain any cleaning times:

$$pack\_cap_t^k := \sum_{i \in F^k} \alpha^{ik} x_t^i \leq L^k \qquad \text{for } t = 1, \ldots, NT \text{ and } k = 2, 3.$$

Finally, the objective function simply corresponds to minimizing the sum of ending inventory levels over all products and periods:

$$inventory := \sum_{i=1}^{NI} \sum_{t=1}^{NT} s_t^i .$$

This concludes the modeling of this Cereal and Fruit mix-and-pack MPS problem.

1.2.3 Mathematical Formulation

The following optimization model (mixed integer linear program) summarizes the new MPS model designed by GSCO. It is based upon capacity utilization at the bottleneck and demand satisfaction for finished products in each time period as represented in Figure 1.5.

The indices identified are

- $i \in \{1, \ldots, NI\}$ representing one of the finished products whose production has to be planned,
- $t \in \{1, \ldots, NT\}$ representing one of the time periods (time buckets or weeks) of the planning horizon, and
- $k \in \{1, \ldots, 3\}$ representing one of the machines of the model ($k = 1$ corresponds to the mixing line, $k = 2$ corresponds to the Cereal packaging line, and $k = 3$ corresponds to the Fruit packaging line).

Using these indices, the following data have been defined.

- D_t^i represents the forecast demand for item i in period t in [units].
- SS_t^i represents the safety stock of item i needed at the end of period t in [units].
- SS_0^i represents the initial stock of item i at the beginning of the planning horizon in [units].
- α^{ik} represents the capacity consumed on machine k to produce one unit of product i in [hours/unit].
- β^i represents the mixing capacity consumed per cleaning operation at the end of a batch of product i in [hours/cleaning].
- L^k represents the capacity available on machine k in each time period in [hours].
- F^2 and F^3 form a partition of the NI products into the Cereal family and the Fruit family, respectively.

The decision variables are

- x_t^i to represent the amount of product i produced during time period t [units];
- y_t^i that takes the value 1 if there is a set-up of product i (i.e., a cleaning operation because of the production of a batch of product i) in period t, and 0 otherwise [0/1];
- s_t^i to represent the inventory level of product i at the end of time period t [units].

The final formulation obtained is written as follows.

$$\min \quad inventory := \sum_i \sum_t s_t^i \tag{1.1}$$

subject to

$$
\begin{array}{llr}
dem\_sat_t^i := & s_{t-1}^i + x_t^i = D_t^i + s_t^i & \text{for all } i,t & (1.2) \\[2mm]
& s_0^i = SS_0^i \, , \; s_t^i \geq SS_t^i & \text{for all } i,t & (1.3) \\[2mm]
vub_t^i := & x_t^i \leq M_t^i y_t^i & \text{for all } i,t & (1.4) \\[2mm]
mix\_cap_t := & \sum_i \alpha^{i1} x_t^i + \sum_i \beta^i y_t^i \leq L^1 & \text{for all } t & (1.5) \\[2mm]
pack\_cap_t^k := & \sum_{i \in F^k} \alpha^{ik} x_t^i \leq L^k & \text{for all } t \ \text{ and } k = 2,3 & (1.6) \\[2mm]
& x_t^i, s_t^i \in \mathbb{R}_+, \; y_t^i \in \{0,1\} & \text{for all } i,t, & (1.7)
\end{array}
$$

where $M_t^i = \sum_{k=t}^{NT} D_k^i + SS_{NT}^i$ is a valid upper bound on the production quantity x_t^i of item i in period t.

1.2.4 Implementation

To conclude this modeling chapter, we illustrate the resolution of this MPS optimization model on an instance involving 12 finished products, 6 cereal products and 6 fruit products, to be produced over a planning horizon of 15 time periods, using standard state-of-the-art optimization software. As you will observe, the quality of the solutions obtained, in reasonable computing time on such a small instance, is not good enough to use such a simple and direct approach in the new industrial planning system of GW. This motivates the development of more sophisticated and more efficient tools.

We implement and test the solution of this model using the Mosel algebraic modeling language and the default version of the Xpress-MP Optimizer MIP solver. This easy to read and straightforward Mosel implementation follows closely the mathematical formulation, and proceeds by defining the indices, data, variables, and constraints.

In the first part of the Mosel program or file, the indices are defined, and the data are declared and read from data files. Some auxiliary data such as M_t^i for all i and t are also computed using the Mosel programming language.

```
model GWGSCO
uses "mmetc","mmxprs","mmsystem"  ! Mosel libraries

! INDICES =================================================
declarations
   NI=12   !number of products
   NK=3    !number of machines
   NT=15   !number of time periods
end-declarations

! DATA ===================================================
declarations
   CAP:  array (1..NK) of real
   FAM:  array (1..NI) of integer !=2 for cereals ;
                                  !=3 for fruits
   SS: array (1..NI) of real      !constant over time
   SSINIT: array (1..NI) of real
   ALPHA: array (1..NI,1..NK) of real !=1 for all machines
   BETA: array (1..NI) of real    !defined only for k=1
   DEM: array (1..NI,1..NT) of real
end-declarations

!READ DATA FILES =========================================
initializations from 'gw_mps.dat'
   CAP
   [FAM, SS, SSINIT, BETA] as 'PRODUCT'
   DEM
end-initializations

!ASSIGN TRIVIAL DATA VALUES ==============================
forall (i in 1..NI, k in 1..NK)
   ALPHA(i,k) := 1

!COMPUTE AUXILIARY DATA FOR VUB CONSTRAINT ================
declarations
   DEMCUM: array (1..NI,1..NT) of real !residual demand till
                                ! the end of the planning horizon
   BIGM: array (1..NI,1..NT) of real   !variable upper bound
end-declarations
```

```
forall(i in 1..NI,t in 1..NT)
  DEMCUM(i,t):= SS(i) + sum(tt in t..NT) DEM(i,tt)
forall(i in 1..NI,t in 1..NT)
  BIGM(i,t):= DEMCUM(i,t)
forall(i in 1..NI,t in 1..NT, k in 2..NK | FAM(i)=k)
  BIGM(i,t):= minlist((CAP(1) - BETA(i))/ALPHA(i,1),
                      CAP(k)/ALPHA(i,k),
                      BIGM(i,t) )
```

The second part of the Mosel program consists in the definition of the variables and constraints. Observe the two special object types used by Mosel: mpvar for variables, and linctr for linear expressions and constraints. Observe also that the statement of the formulation in Mosel is very close to the algebraic notation used to describe the mathematical formulation. The same holds true for other algebraic modeling languages, such as AMPL, GAMS, or OPL.

```
!VARIABLES ======================================================
declarations
  x: array(1..NI,1..NT) of mpvar
  y: array(1..NI,1..NT) of mpvar
  s: array(1..NI,1..NT) of mpvar
end-declarations

forall(i in 1..NI,t in 1..NT)
  y(i,t) is_binary

!CONSTRAINTS ====================================================

declarations
  inventory : linctr
  dem_sat, inv_min, vub: array(1..NI,1..NT) of linctr
  mix_cap: array(1..NT) of linctr
  pack_cap: array(1..NK,1..NT) of linctr
end-declarations

inventory:= sum(i in 1..NI,t in 1..NT) s(i,t)

forall(i in 1..NI, t in 1..NT)
  dem_sat(i,t) := if(t>1,s(i,t-1),SSINIT(i)) + x(i,t) =
                  DEM(i,t) + s(i,t)

forall(i in 1..NI, t in 1..NT)
  inv_min(i,t)  := s(i,t) >= SS(i)

forall(i in 1..NI, t in 1..NT)
```

```
vub(i,t)        := x(i,t) <= BIGM(i,t)*y(i,t)

forall(t in 1..NT)   !k=1 for mixing
  mix_cap(t)    := sum(i in 1..NI)ALPHA(i,1)*x(i,t) +
                  sum(i in 1..NI)BETA(i)*y(i,t)<= CAP(1)

forall(k in 1..NK,t in 1..NT | k>1)
  pack_cap(k,t):= sum(i in 1..NI |FAM(i)=k)
                  ALPHA(i,k)*x(i,t) <= CAP(k)
```

The final part of the Mosel program controls the execution of the mixed integer programming algorithm (branch-and-cut), as well as the solution output. Here we ask for the best solution found, without cutting planes, with a time limit of 600 seconds, and then print out the solution obtained.

```
!SOLUTION ====================================================
setparam("XPRS_verbose",true) ! Enable message printing
setparam("XPRS_CUTSTRATEGY",0)   ! Disable automatic cuts
setparam("XPRS_MAXTIME",600)     ! Maximum run time
minimize (inventory)


!PRINT SOLUTION ==============================================
 forall (i in 1..NI, t in 1..NT) do
  writeln("ITEM ",i," and PERIOD ",t,
        ":     PROD= ",getsol(x(i,t)), " (", getsol(y(i,t)),
        ")    STOCK= ", getsol(s(i,t)) )
 end-do

!EXIT ========================================================
exit(0) end-model
```

The only addition with respect to the initial formulation consists in the computation of the big M_t^i parameter in constraint (1.4). For each i and t, the batch size cannot exceed

$$M_t^i = \min\{ SS^i + \sum_{l=t}^{NT} D_l^i , (L^1 - \beta^i)/\alpha^{i1} , L^k/\alpha^{ik} \}, \qquad (1.8)$$

where k is such that product i belongs to family F^k ($k = 2$ or 3).

The data are read from the single data file gw_mps.dat, which contains the values of the capacity, product data – family, safety stock, and initial stock, cleaning times β – and demand data.

```
CAP: [1400, 700,   700]
```

```
PRODUCT:[!FAM,SS   , SSINIT,BETA
(  1 ) [ 2, 10.00, 83.00, 30.]
(  2 ) [ 2, 10.00, 31.00, 20.]
(  3 ) [ 2, 10.00, 11.00, 30.]
(  4 ) [ 2, 10.00, 93.00, 40.]
(  5 ) [ 2, 10.00, 82.00, 40.]
(  6 ) [ 2, 10.00, 72.00, 10.]
(  7 ) [ 3, 20.00, 23.00, 30.]
(  8 ) [ 3, 20.00, 91.00, 20.]
(  9 ) [ 3, 20.00, 83.00, 10.]
( 10 ) [ 3, 20.00, 34.00, 50.]
( 11 ) [ 3, 20.00, 61.00, 30.]
( 12 ) [ 3, 20.00, 82.00, 20.]
]

DEM: [
  0, 95, 110, 96,  86,124,  83,108, 114,121, 110,124, 104, 86,  87,
 98, 96,  96, 98, 103,104, 122,101,  89,108, 101,109, 106,108,  76,
106,  0,  89,123,  96,105,  83, 82, 112,109, 119, 85,  99, 80, 123,
 98,121,   0,105,  98, 96, 101, 81, 117, 76, 103, 81,  95,105, 102,
  0,124, 113,123, 123, 79, 111, 98,  97, 80,  98,124,  78,108, 109,
103,102,   0, 95, 107,105, 107,105,  75, 93, 115,113, 111,105,  85,
110, 93,   0,112,  84,124,  98,101,  83, 87, 105,118, 115,106,  78,
 85, 92, 101,110,  93, 96, 120,109, 121, 87,  92, 85,  91, 93, 109,
122,116, 109,  0, 105,108,  88, 98,  77, 90, 110,102, 107, 99,  96,
120,124,  94,105,  92, 86, 101,106,  75,109,  83, 95,  79,108, 100,
117, 96,  78,  0, 108, 87, 114,107, 110, 94, 104,101, 108,110,  80,
125,112,  75,  0, 116,103, 122, 88,  85, 84,  76,102,  84, 88,  82
]
```

1.2.5 Optimization Results

The results in Table 1.5 have been obtained with the default version of the Xpress-MP Optimizer. The problem has been solved twice.

- In the first run, we have solved the problem by branch-and-bound, using Xpress-MP defaults, except that cut generation has been switched off.
- In the second run, we have solved the problem by branch-and-cut using Xpress-MP defaults including cut generation.
- In both cases Xpress-MP uses a branch-and-bound algorithm (see below and in Chapter 3), but in the second case it tightens the formulation by adding general cuts (MIR, knapsack and Gomory cuts; see Part II) so as to obtain improved bounds.

The general role of such cut generation procedures is explained in more detail in Chapter 3, devoted to the description of mixed integer programming algorithms, and is only roughly described here.

- The *branch-and-bound algorithm* is based on the solution of the *linear relaxation of the initial model*, which is the model obtained by replacing the integrality restrictions on the variables (in our example, $y_t^i \in \{0, 1\}$) by their (relaxed) bound restrictions (in our example, $0 \le y_t^i \le 1$).

- The relaxed problem is a pure linear program, and is thus easy to solve, but its optimal solution does not solve the initial problem if the relaxed integer variables take on fractional optimal values (y_t^i strictly between 0 and 1).

- In the latter case, the relaxed problem provides only a lower bound on the optimal solution value (in a minimization problem), simply because the relaxed problem is defined by adding feasible solutions to the original problem. This is the *lower bounding part* of the algorithm.

- The branch-and-bound algorithm proceeds by enumerating a sequence of linear relaxations, whose feasible solutions define some partition of the initial (i.e., nonrelaxed) model. Moreover, the best solution among all linear relaxations is (proved to be) the optimal solution of the initial model. This is the *branching part* of the algorithm.

- During this enumeration, some feasible solutions of the initial model are generated, that is, solutions where the relaxed integer variables take integer values. The objective value of each such feasible solution provides an upper bound on the optimal objective value. This is the *upper bounding part* of the algorithm.

- Finally, the branch-and-bound algorithm is exact if the enumeration is complete, and provides only an approximate or heuristic solution if the enumeration is truncated. In the latter case, the quality of the solution is usually measured by the so-called *duality gap* defined as

$$\text{Duality Gap} = \frac{\text{Best UB - Best LB}}{\text{Best UB}} \times 100\%,$$

where Best LB and Best UB are, respectively, the best values found for the lower bound and the upper bound when the enumeration is stopped. As the optimal solution value must lie somewhere in the range $[BestLB, BestUB]$, the duality gap measures the maximum relative deviation from optimality of the best feasible solution.

Note that in the solution reports from the Xpress-MP Optimizer, the duality gap is computed relative to the best lower bound (dividing by $BestLB$) instead of the best upper bound, and is therefore larger.

In general, the running time of the branch-and-bound algorithm (i.e., more precisely, the number of linear programs to be solved during the enumeration), as well as the quality of its approximate solutions (measured by the duality gap), depend heavily on the quality of the initial lower bound.

This is why the initial formulation of a model is so important in mixed integer programming. This is also why adding cuts or constraints, either automatically by Xpress-MP, or by any other reformulation technique, to improve the initial lower bound allows one to obtain better solutions.

Coming back to our illustrative example, Table 1.5 reports the results obtained within a a time limit of 600 seconds (industrial users are often interested in getting good solutions quickly). It compares the behavior of the branch-and-bound algorithm with and without Xpress-MP cuts.

Table 1.5. GW MPS Example (1.1)–(1.7)

Algorithm Formulation	Vars Cons	LP Val.	XLP Val. Ncuts	Best LB Best UB	Best UB t. (secs) Gap (%)
Basic form. B & B	540	2854	2854	3296	22
without Xpress-MP cuts	585		0	6295	47.64
Basic form. B & B	540	2854	5416	5620	147
with Xpress-MP cuts	585		280	5732	1.95

$NI = 12$, $NT = 15$. Maximum 600 second runs.

In Table 1.5, the column "Vars/Cons" shows the number of variables and constraints in the formulation, "LP Val." is the value of the initial linear relaxation of the formulation, "XLP Val." and "Ncuts" give the value of the linear relaxation after the addition of the Xpress-MP cuts and the number of cuts added, "Best LB" and "Best UB" give the value after 600 seconds of the best lower bound and best upper bound (best feasible solution), respectively, "Best UB t." indicates the time in seconds needed to obtain this best feasible solution, and "Gap" gives the final duality gap when the enumeration was stopped.

- We see that the addition of 280 constraints by Xpress-MP has increased the lower bound from 2854 to 5416. It means that the total inventory objective function value will be at least 5416 in the optimal solution, and this bound is known after a few seconds of computing time (after the addition of cuts and the solution of the corresponding linear relaxation, but before any enumeration). This is due to the automatic reformulation of some *low-level relaxation* or structure identified by the optimization system (see Chapter 3, Section 3.4 for the definition of this concept, and Chapter 8 for the study of reformulations for such low-level relaxations).
- In comparison, without reformulation it takes as much as 600 seconds to obtain a (weaker) best lower bound of 3296!

- Also, the cuts have allowed the branch-and-bound algorithm to find good feasible solutions in 600 seconds, whereas only bad solutions are found without cuts.
- Moreover, the best feasible solution found in 600 seconds (found after 147 seconds) with cuts is guaranteed to be less than 2% away from the optimal solution.
- Finally, it is impossible with this initial formulation and Xpress-MP cuts to obtain the optimal solution and prove its optimality after several hours of computing time.

It is not a simple matter to improve or tighten formulations. Our goal is to provide the necessary modeling and reformulation tools to allow the reader to perform this task. In particular, we illustrate this approach in Section 4.5 with the production planning example. We show how to use the classification and reformulation scheme developed in Chapter 4 to improve the results and obtain either good solutions quickly, or a provably optimal solution in reasonable time.

These improvements are based on the identification of the specific production planning structures contained in the model (called *high-level relaxations* in Section 3.4), and on their reformulations. The mathematical study of these reformulations is the main topic of Parts II to IV.

Exercises

Exercise 1.1 Consider the tiny example from Section 1.1, with a second type of bike (mountain bike) whose production has to be planned over the same planning horizon, from January to August.

The forecast demand for the mountain bikes is 200 bikes per month, except in July and August when the demand will increase (most likely) up to 500 bikes per month.

Initially (January 1st), there is no mountain bike in stock. The production set-up cost is 3000 euros, the unit production cost is 60 euros per bike, and the inventory holding cost is 3 euros per bike, per month in inventory.

i. By using the same MIP model as for the racing bikes and changing the data, determine the optimal production plan for the mountain bikes.

ii. In addition, there is a global production capacity restriction: at most 1500 bikes can be produced during each month. Change your MIP model to account for the joint production capacity limit. Build a model to plan simultaneously the production of both types of bike and optimize total production and inventory costs.

iii. Solve the corresponding MIP, and analyze the optimal solution obtained. In particular, what is the effect of the joint capacity restriction on the individual production plans?

Exercise 1.2 How would you change the GW–GSCO formulation from Section 1.2 if only one product could be packed on each packaging line in each time period?

Note that such constraints, called *production mode constraints*, are typically added in order to ease or simplify the organization of the mixing line, or to reduce the cleaning costs, but have an impact on the line flexibility. We assume implicitly here that the time periods are shorter than in the initial model (e.g., days instead of weeks).

Exercise 1.3 How would you change the GW–GSCO model from Section 1.2 if only one product could be mixed on the mixing line in each time period?

We assume implicitly here that the time periods are shorter than in the initial model (e.g., days instead of weeks). Given the current restriction on the stock of intermediate products (no stock after mixing, before packing), does this adapted model make sense?

Exercise 1.4 Consider the GW–GSCO model from Section 1.2, but under a scenario in which we can carry a set-up over from one week to another on the mixing line. That is, we do not have to clean the mixing line at the end of a week if the last lot produced in a week is of the same product as the first lot of the next week.
i. Change your MIP model to account for the set-up carryover possibility.
ii. Solve the corresponding MIP. Is it more difficult to solve than the initial model? In what way is it more difficult?
Hint: Although there are different ways to model this, additional variables are definitely needed, as well as new constraints to relate the new variables to the set-up variables.

Exercise 1.5 Consider the GW–GSCO model from Section 1.2. How would you change the model if the cleaning times on the mixing line were *sequence-dependent*, that is, if the cleaning time for a lot of a given product depended on what product was mixed immediately after it. The mixing line has to be cleaned at the end of each week. This can be modeled as a special sequence-dependent cleaning time, from the last product mixed to a dummy product representing the idleness of the line at the end of the week.
i. Formulate this modified problem as a mixed integer (MIP) program. Is your model correct?
ii. Create some data set for the sequence-dependent changeover times, and solve the corresponding MIP. Is it more difficult to solve than the initial model? In what way is it more difficult?
Hint: An additional set of variables is needed, as well as new constraints to link this new set of variables to the set-up variables. Also, we can assume that a product is never produced more than once in any time period.

Exercise 1.6 In the GW–GSCO model from Section 1.2, if it were possible to hold products in inventory immediately after mixing, and before packing, how would you change your model?

i. Assuming that the objective is to minimize total inventory, defined as the sum of stocks before and after packaging, change your MIP formulation to account for this possibility.

ii. Solve the corresponding MIP. Is it more difficult to solve than the initial problem? In what way is it more difficult?

Exercise 1.7 A company wants to plan the production of several finished products ($PRODUCTS = \{A, B, C, D\}$), over the next four months. It is using a make-to-stock production policy, and the estimated demands for each product for the next four months are known. The demands are given in Table 1.6 in units of product.

Table 1.6. Demand, Production Capacity Limit and Inventory Cost

Products	Demand Month 1 [units]	Demand Month 2 [units]	Demand Month 3 [units]	Demand Month 4 [units]	Production Capacity [units/month]	Inventory Cost [euros/ (unit,month)]
Product A	5,000	6,000	3,000	10,000	8,000	35
Product B	900	1,000	4,000	5,000	5,000	39
Product C	6,000	9,000	4,000	2,000	8,000	45
Product D	10,000	11,000	14,000	16,000	15,000	85

There is a monthly production capacity limit for each product, because each product has its own production facility, except for packaging. These monthly capacity limits are constant over time and given in Table 1.6 in units of product.

There is also a single packaging department in the company, transforming each product into a finished or packed product. So, there is also a global capacity limit on the total number of product units packed during a month. This limit is constant over time, and is estimated to be 28,000 units per month.

It is possible to store the finished products, as well as the products before packaging, in unlimited quantity because the warehouse is big enough. Nevertheless, there is a unit inventory cost for each product, corresponding to storage costs and the opportunity cost of capital. These costs are given for each finished product in Table 1.6, in euros per month and per finished product unit. They are time-independent. The inventory of a product before packaging is estimated to cost 4 euros less than after packaging, per month and unit.

The current or initial inventory is empty, there is no final inventory requirement, and there are no restrictions on the availability of raw materials. The planning objective is to minimize inventory costs, and demand has to be satisfied on time during the whole four-month horizon. The following steps need to be carried out to achieve this objective.

i. Formulate this problem as a Linear Program (LP).

ii. Develop a model, reading all the data from a file, and solving this LP, using Mosel or any algebraic modeling language.

iii. Solve the model, and print the optimal solution, using Mosel/Xpress-MP or any LP solver.

iv. If you were the chief operations officer (COO) of that company, would you try to increase the packing capacity? What sort of data would you collect, and what sort of computations would you perform in order to answer to this question?

Exercise 1.8 Minimizing the inventory costs in Exercise 1.7 leads to an unsatisfactory production plan in which each product is packed during each month. This is not satisfactory because the packaging line has to be cleaned between two campaigns of different products, and this cleaning or sterilization process consumes very expensive products. So, we must try to avoid packaging all products within the same month. We continue to work with and extend the model of Exercise 1.7 to remedy this situation.

The company now also wishes to take into account the cost of setting up or cleaning the packaging line. When a product is packed during a month, the set-up cost is incurred once (only once because, usually, a product is packed at most once in a month). These packing set-up costs are product-dependent, but time-independent, and are $(SUCOST_p) = (500,000,\ 900,000,\ 800,000,$ and $900,000)$ euros, respectively, for products $p = A,\ B,\ C,$ and D. Finally, there is no set-up cost for making the products because the production lines are dedicated to each product. Answer the following in order to optimize the production plan.

i. Formulate this problem as a Mixed Integer Program (MIP).

ii. Develop a model, reading all the data from a file, and solving this MIP, using Mosel or any algebraic modeling language.

iii. Solve the model, and print the optimal solution, using Mosel/Xpress-MP or any MIP solver.

Hint: To model the set-up costs, first introduce 0/1 or binary variables y_{pt} that indicate whether product p is packed during month t. Then add constraints to link the quantity x_{pt} of product p packed during month t to the set-up decision y_{pt}. Finally add the set-up costs in the objective function.

Notes

Introduction The description of the planning system and models developed by the Kellogg Company can be found in Brown et al. [31].

Section 1.1 We refer to Heipcke [88] for a more general introductory overview of optimization models, with a larger scope of applications than production planning.

Section 1.2 The GW company case, and its MPS story, are pure fiction. The case is inspired by several research projects in which the authors have been involved. The data used in Section 1.2.4 are derived from the standard test cases in Trigeiro et al. [161].

The model of Section 1.2.3 has been implemented and tested using the Mosel algebraic modeling language (version 1.4.1) and the default version of the Xpress-MP Optimizer MIP solver (version 15.30). More information about this software can be found at http://www.dashoptimization.com.

Here we have always used the default version of this commercial software. Similar results to those presented in this chapter are obtained using other modeling and optimization software.

All the tests reported here have been carried out on a 1.7 GHz PC (centrino) with 1 GB of RAM running under Windows XP.

Exercises Exercises 1.7 and 1.8 are adapted (and the data taken) from the syllabus and teaching material delivered with the Xpress-MP software.

2

Production Planning Models and Systems

Motivation

In the two industrial production planning systems, the Kellogg Company and GW, mentioned or analyzed in Chapter 1, we observed some important requirements for the new production planning model and tool.

- It was supposed to remedy important weaknesses of the current planning system (the inability to model and plan capacity utilization accurately, because of neglected machine preparation times in the case of the Kellogg Company, and because of neglected machine cleaning times in the MPS model for GW).
- It needed some coordination with the global planning system in charge of supporting all planning decisions from the strategic and long-term horizon level to the very detailed and short-term level. We heard about the tactical plan and the operational plan for Kellogg, and the MRP, ERP, and MPS for the GW Company case, and we observed that these planning levels are not independent (decisions at one level act as constraints at another level).
- It required a high level of integration in the decision processes in place, to avoid manual replanning and make sure that the decisions suggested by the model truly support and have an impact on the real planning decisions.

This need for improvement, coordination, and integration can be observed in almost all industrial projects. In order to develop an effective planning model, the modeler must be aware of the planning process and system used, of the limitations of the current system, of the architecture and structure of the existing system, and of the decision processes used by the planning teams.

This is our motivation for the inclusion of this chapter on production planning models and systems.

Objective

To do a useful job, the modeler must have sufficient knowledge about existing planning models, systems, and processes to be able to evaluate the current system, and in order to design improved, coordinated, and integrated solutions.

The general objective of this chapter is to provide this necessary knowledge. More specifically, the objective is to

- describe or survey the structure of the planning systems used by many – or most – companies,
- learn the general principles of the planning procedures, and
- study some generic classes of production planning models encountered in such systems.

We also provide some analysis and criticism of the planning models and methods used in these systems to help readers to develop some evaluation criteria to measure their performance, and to identify situations where the optimization approach may help to improve the productivity and flexibility of manufacturing systems.

Contents

In this chapter:

- In Section 2.1 we first give *mathematical formulations* of some of the classical *production planning models* considered in ERP (enterprise resource planning) or MRP systems;
- Then we analyze in detail in Section 2.2 the well-known *generic MRP planning procedure* used to solve these models by
  - describing its inputs and its structured data model,
  - presenting the single-item decomposition planning heuristic that forms the basis of most MRP planning systems, and
  - analyzing the limitations of the MRP decomposition approach;
- Next in Section 2.3 we take a broader view and define the planning tasks of *APS (Advanced Planning Systems)*, which subsume the well-known manufacturing, planning, and control systems; material requirements planning (MRP-I); manufacturing resource planning (MRP-II); and hierarchical production planning (HPP); and
- Finally, to illustrate the planning tasks and the planning process along the supply chain, we describe in Section 2.4, without mathematical models or formulations, the generic *strategic network design* and *supply chain master planning problems* as further examples of procurement, production, and distribution planning problems.

2.1 Some Production Planning Models

The purpose of this section is to provide further examples and mixed integer programming formulations of production planning models. The formulations described here correspond to classical models in ERP or MRP systems. The next section describes the global structure of such systems.

Modeling Elements

There are a number of modeling elements present in many or most production planning problems. Production planning deals mainly with the determination of *production lots or batches*, specifically the size of batches and the time of production, in order to meet some demand over a given finite horizon, called the *planning horizon*. Demand is usually generated from forecasts in a make-to-stock environment, or by customer orders in a make-to-order environment, or often by a combination of the two.

In order to define feasible and economical production plans, several other characteristics of the manufacturing system are usually taken into account: the availability of *resources* (machine hours, workforce, subcontracting, etc.), the *production and inventory costs*, and other performance measures such as *customer-service level*.

The simplest such production planning model is presented next. It is known as the *single-item uncapacitated lot-sizing model (LS-U)*. It corresponds to the planning of a single item to meet some dynamic demand over a discretized planning horizon. It contains all the modeling elements cited above, apart from the fact that there are no resource capacity restrictions. Our tiny economical example in Chapter 1 is one instance of this *LS-U* model.

There are also modeling elements that are present in some, but not all, models. Such elements usually make the models more complex and more difficult to solve.

- For instance, the products may compete for the allocation of capacity from some shared resources. This has been illustrated with the mixer or the packaging lines in our industrial example in Section 1.2, and is typical of the *Master Production Schedule (MPS)* model presented hereafter. This MPS approach is often used to plan the production of finished products.
- In some other cases, the products interact through multi-level product structures. In other words, a product can be an output of some production stage and also an input of some other production stage, or it may be delivered from an external supplier. This creates some precedence constraints between the supply and the consumption of that product. These restrictions are usually modeled through inventory balance constraints. Examples of such models are the *Material Requirements Planning (MRP)* model, or the MPS/MRP integrated model described later in this Section. This MRP model is used to integrate the production and procurement plans of all products and components.

- Finally, there are other elements needed to refine the model, or to model capacity utilization in a more precise way. For instance, the demand satisfaction process may allow demand for finished products to be backlogged. In this case, it is possible – but penalized because it has a negative impact on customer satisfaction – to deliver to a customer later than required. This occurs, for example, when a factory does not have enough capacity to deliver to all customers on time.

- In some other cases, it is necessary to model capacity utilization more precisely in order to guarantee to obtain feasible production plans. For instance, the capacity consumed when a machine starts or finishes a production batch, or when a machine switches from one product to another, may need to be considered. In these cases, we obtain models with set-up times, start-up times, changeover times, or models with sequencing restrictions. This was the case for the mixer in our industrial example in Chapter 1. On the other hand, such models may be too complex to be solved with set-up or start-up time restrictions, and then simpler models involving only set-up or start-up costs may be worth considering.

Uncapacitated Lot-Sizing Model

The first model is the *single-item, single-level, uncapacitated lot-sizing model*. This model is the core subproblem in production planning because it is the problem solved repeatedly for each item (from end products to raw materials) in the material requirements sequential planning system (see Section 2.2).

We use the index t, with $1 \leq t \leq n$, to represent the discrete time periods, and n is the final period at the end of the planning horizon. The purpose is to plan the production over the planning horizon (i.e., fix the lot size in each period) in order to satisfy demand, and to minimize the sum of production and inventory costs.

Classically, as in our tiny economical example in Chapter 1, the production costs exhibit some economies of scale that are modeled through a fixed charge cost function. That is, the production cost of a lot is decomposed into a fixed cost independent of the lot size, and a constant unit or marginal cost incurred for each unit produced in the lot. The inventory costs are modeled by charging an inventory cost per unit held in inventory at the end of each period. Any demand in a period can be satisfied by production or inventory, and backlogging is not allowed. The production capacity in each period is not considered in the model, and is therefore assumed to be infinite.

For each period t, with $1 \leq t \leq n$, the data p_t, q_t, h_t, and d_t model the unit production cost, the fixed production cost, the unit inventory cost, and the demand to be satisfied, respectively. For simplicity we suppose that $d_t \geq 0$ for all periods t. The decision variables are x_t, y_t, and s_t. They represent the production lot size in period t, the binary variable indicating whether there is a positive production in period t ($y_t = 1$ if $x_t > 0$), and the inventory at the end of period t, respectively.

The natural formulation of this uncapacitated lot-sizing problem can be written as follows, using the demand satisfaction and set-up enforcement (variable upper bound) generic constraints described in Chapter 1.

$$\min \quad \sum_{t=1}^{n}(p_t x_t + q_t y_t + h_t s_t) \tag{2.1}$$

subject to

$$s_{t-1} + x_t = d_t + s_t \qquad \text{for all } t \tag{2.2}$$
$$s_0 = s_n = 0 \tag{2.3}$$
$$x_t \le M_t y_t \qquad \text{for all } t \tag{2.4}$$
$$x \in \mathbb{R}_+^n, \; s \in \mathbb{R}_+^{n+1}, \; y \in \{0,1\}^n, \tag{2.5}$$

where M_t is a large positive number, expressing an upper bound on the maximum lot size in period t. Constraint (2.2) expresses the demand satisfaction in each period, and is also called the flow balance or flow conservation constraint. This is because every feasible solution of $LS\text{-}U$ corresponds to a flow in the network shown in Figure 2.1, where $d_{14} = \sum_{i=1}^{4} d_i$ is the total demand. Constraint (2.3) says there is no initial and no final inventory. Constraint (2.4) forces the set-up variable in period t to be 1 when there is positive production (i.e., $x_t > 0$) in period t. Constraint (2.5) imposes the nonnegativity and binary restrictions on the variables. The objective function defined by (2.1) is simply the sum of unit production, fixed production, and unit inventory costs.

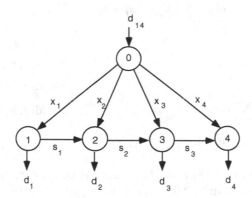

Figure 2.1. Uncapacitated lot-sizing network $(n = 4)$.

Master Production Scheduling Model

The next model is known as the *multi-item (single level) capacitated lot-sizing model*. It corresponds to the simplest Master Production Scheduling problem solved to plan the production of finished products in a Manufacturing

Planning and Control System (MPCS) (see Section 2.3). Our GW example in Section 1.2 is another example of such a MPS model.

The purpose is to plan the production of a set of items, usually finished products, over a short-term horizon corresponding at least to the total production cycle of these items. For each item, the model is the same as the *LS-U* model in terms of costs and demand satisfaction. In addition, the production plans of the different items are linked through capacity restrictions coming from the common resources used.

We define the indices i with $1 \leq i \leq m$ to represent the set of items to be produced, k with $1 \leq k \leq K$ to represent the set of shared resources with limited capacity, and t with $1 \leq t \leq n$ to represent the time periods. The variables x, y, s and the data p, q, h, d have the same meaning for each item i as in the model *LS-U*. A superscript i has been added to represent the item i for which they are each defined.

The data L_t^k represent the available capacity of resource k during period t. The data α^{ik} and β^{ik} represent the amount of capacity of resource k consumed per unit of item i produced, and for a set-up of item i, respectively. The coefficient β^{ik} is often called the set-up time of item i on resource k, and represents the time spent to prepare the resource k just before the production of a lot of item i. Together with α^{ik}, it may also be used to represent some economies of scale in the productivity factor of item i on resource k.

The natural formulation of this multi-item capacitated lot-sizing model, or basic MPS model, can be written as follows,

$$\min \quad \sum_i \sum_t (p_t^i x_t^i + q_t^i y_t^i + h_t^i s_t^i) \tag{2.6}$$

subject to

$$s_{t-1}^i + x_t^i = d_t^i + s_t^i \qquad \text{for all } i,t \tag{2.7}$$

$$x_t^i \leq M_t^i y_t^i \qquad \text{for all } i,t \tag{2.8}$$

$$\sum_i \alpha^{ik} x_t^i + \sum_i \beta^{ik} y_t^i \leq L_t^k \qquad \text{for all } t,k \tag{2.9}$$

$$x \in \mathbb{R}_+^{mn}, \ s \in \mathbb{R}_+^{m(n+1)}, \ y \in \{0,1\}^{mn}, \tag{2.10}$$

where constraints (2.6)–(2.8) and (2.10) are the same as for the *LS-U* model, and the generic constraint (2.9) expresses the capacity restriction on each resource k in each period t.

Material Requirements Planning Model

As a last example model, we describe the *multi-item multi-level capacitated lot-sizing model*, that can be seen as the integration of the previous MPS model for finished products, and the *LS-U* models for all intermediate products and raw materials, into a single monolithic model. It is often referred to as the Material Requirements Planning model, or the integrated MPS/MRP model.

The purpose of this model is to optimize simultaneously the production and purchase of all items, from raw materials to finished products, in order to satisfy for each item the external or independent demand coming from customers and the internal or dependent demand coming from the production of other items, over a short-term horizon.

The dependency between items is modeled through the definition of the product structure, also called the *bill of materials (BOM)*. The product structures are usually classified into Series, Assembly or General structures; see Figure 2.2.

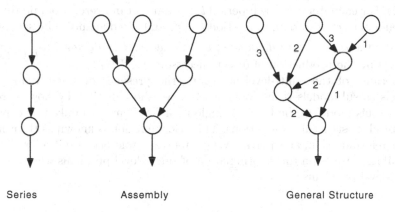

Series   Assembly   General Structure

Figure 2.2. Types of product structures in multi-level models.

The indices, variables, and data are the same as before, except that, for simplicity, we also use the index j with $1 \leq j \leq m$ to identify items. For item i, we use the additional notation $D(i)$ to represent the set of direct successors of i, that is, the items consuming directly some amount of item i when they are produced. Note that for series and assembly structures, these sets $D(i)$ are singletons for all items i, and for a finished product i, we always have $D(i) = \emptyset$. For $j \in D(i)$, we denote by r^{ij} the amount of item i required to make one unit of item j. These r^{ij} values are indicated along the edges (i,j) in Figure 2.2. This parameter r is used to identify the dependent demand, whereas d_t^i corresponds to the independent demand. For each item i, we denote by γ^i the lead-time to produce or deliver any lot of i. More precisely, x_t^i represents the size of a production or purchase order of item i launched in period t, and delivered in period $t + \gamma^i$.

The natural formulation for the general product structure capacitated multi-level lot-sizing model, or the monolithic MRP model, is

$$\min \quad \sum_i \sum_t (p_t^i x_t^i + q_t^i y_t^i + h_t^i s_t^i) \qquad (2.11)$$

subject to

$$s^i_{t-1} + x^i_{t-\gamma^i} = [d^i_t + \sum_{j \in D(i)} r^{ij} x^j_t] + s^i_t \qquad \text{for all } i, t \qquad (2.12)$$

$$x^i_t \le M^i_t y^i_t \qquad \text{for all } i, t \qquad (2.13)$$

$$\sum_i \alpha^{ik} x^i_t + \sum_i \beta^{ik} y^i_t \le L^k_t \qquad \text{for all } t, k \qquad (2.14)$$

$$x \in \mathbb{R}^{mn}_+, \; s \in \mathbb{R}^{m(n+1)}_+, \; y \in \{0,1\}^{mn}, \qquad (2.15)$$

where the only difference with respect to the previous MPS model resides in the form of the generic demand satisfaction or flow conservation constraint (2.12). For each item i in each period t, the amount delivered from production or vendors is $x^i_{t-\gamma^i}$ ordered in period $t - \gamma^i$, and the demand to be satisfied is the sum the *independent demand* d^i_t and the *dependent demand* $\sum_{j \in D(i)} r^{ij} x^j_t$ implied by the production of direct successors $j \in D(i)$.

Because of the multi-level structure, the presence of single item *LS-U* models as submodels is less obvious, but we show in Part IV how to reformulate this model in the form of single-item *LS-U* models linked by capacity and product structure restrictions. This reformulation is known as the echelon stock reformulation, and plays a very important role because it allows one to use all the results on the reformulation of single-level problems when treating multi-level problems.

2.2 The MRP Planning Model

Many industrial production planning models are variants or extensions of the the generic MRP model (2.11)–(2.15), described in Section 2.1, which is typical of discrete parts manufacturing systems. Provided that the BOM structure allows one to describe the product structure, which is usually the case for discrete parts manufacturing, this model potentially plans the procurement or production of all components needed to satisfy external customer demand over a medium-term horizon.

The numerous extensions or adaptations to this basic model correspond usually to better or refined models to include overtime, product or component substitutes in BOMs, alternate routings or machine selection to perform production operations, shipping and transportation to and from other sites, buying or subcontracting of some components, productivity and capacity utilization, and so on.

Nevertheless, the basic MRP model (2.11)–(2.15) is the kernel of many or most multi-item single-facility production planning models, and is solved in most integrated planning systems (see Section 2.3 for a general introduction to such systems). Moreover, most MRP and ERP planning systems use the same basic or trivial decomposition approach based on *LS-U* in order to solve this model or, at least, to provide feasible solutions.

In this section, we describe this simple but generic MRP model and its inputs, using the standard operations management terminology for production planning models. We also describe the traditional and heuristic MRP decomposition approach, and discuss its weaknesses.

> In such boxes, we establish the link between the generic MRP planning model and its inputs described here, and the mathematical programming formulation (2.11)–(2.15).

A major difference between the traditional MRP approach and the modeling/optimization approach is that the latter forces the user/modeler to make a clear distinction, and avoid some confusion, between the data required as input to the model and the model formulation itself (decisions, constraints, and objective), and also between the model formulation and the algorithm used to build a feasible or optimal production plan.

2.2.1 The Planning Model and Its Inputs

The data required to define and implement the MRP model are now described.

Independent Demand over the Planning Horizon

The main objective of production planning is to meet the so-called *independent demand*, which is defined for each facility as the demand coming from external sources. This comprises demand from customers for the main finished products, but also spare parts demand and demands from the distribution system or from other facilities.

> The independent or external demand for item i in period t is represented by d_t^i in Equation (2.12)

In a *make-to-stock (MTS)* production policy, this independent demand must be already in stock when the customer demand arrives at the facility. Therefore, all the procurement and production activities must be carried out in anticipation of this demand, and be based on demand forecasts. This policy is typically used for standard products, with little product variety or diversity, such as fast-moving consumer goods and many standard items of household equipment.

In a *make-to-order (MTO)* or *assemble-to-order (ATO)* production policy, some activities can still be performed after the external ordering of the products. The *delivery lead-time* is the time promised to customers for delivery. Therefore, at the time of ordering, the facility must hold enough raw materials or semi-finished products in inventory in such a way that the remaining production lead-time required to terminate the finished products ordered is less than (or equal to) the commercial lead-time. This implies that planning is decomposed in two phases or two separate problems. The upstream phase, also called anticipation or *"push" phase*, plans the procurement and production from raw materials up to some semi-finished products, and is based on demand forecasts for these semi-finished products. This is similar to MTS planning. The downstream phase, called the *final assembly*, on-order phase, or *"pull" phase*, schedules the production from the semi-finished products held in inventory up to the finished products, and is based on effective customer orders. This decomposition is illustrated in Figure 2.3. This approach is typical of production systems where there exists a large variety of finished product variants, based on a limited variety of raw materials or semi-finished products. This makes it more economical to hold these semi-products in inventory, but imposes a positive commercial lead-time to complete production. This is, for instance, the policy used by Dell to assemble its PCs.

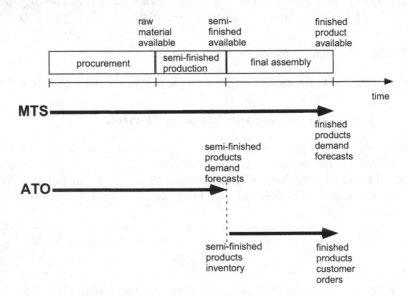

Figure 2.3. MTS and ATO production policies.

Formulation (2.11)–(2.15) is used to represent either a MTS policy, or the push phase of an ATO policy.

For all production policies, the *planning horizon* must be long enough to cover at least the total or cumulative lead-time, including procurement, production, and satisfaction of demand. This is necessary if one is to reach a high *customer-service level*, defined as the fraction of customer demands delivered on time, because we need to order the right materials now from our suppliers (i.e., the right quantity of each material) to be included into the finished products that will be delivered one lead-time from now. In other words, the total lead-time represents the required anticipation time in the planning process or, equivalently, the minimal planning horizon length. Then, the planning model will be solved and used in a *rolling horizon* manner. That is, the solution proposed for the early time periods will be implemented, the model data and parameters will be updated for the subsequent time periods, the model will be solved again, and so forth.

> In formulation (2.11)–(2.15) the number of time periods n is at least as large as the total cumulative lead-time from the ordering of raw materials to the completion of finished products, expressed in number of periods.

Bill of Materials (BOM) to Compute Dependent Demand

The *bill of materials* defines the product structure by specifying for each component (finished or semi-finished product) all of its direct predecessor components (raw materials or semi-finished products), as well as the number of each required per unit of the successor component. This BOM information allows one to transform the finished product or external time-phased demand – forecasts or orders – into detailed time-phased requirements for all components in the production system.

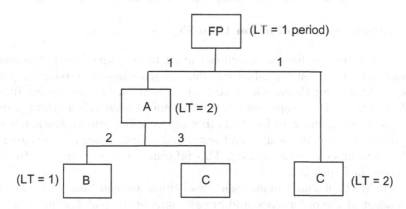

Figure 2.4. The bill of materials for finished product FP.

A BOM example is given in Figure 2.4, where

- each unit of finished product FP is obtained by assembling one unit of A with one unit of C,
- each unit of A is itself directly obtained from two units of B and three units of C, and

- items B and C are raw materials.

We use this simple example to illustrate the MRP planning process. In this example, a total of four units of raw material C are required to produce each unit of FP, three units of C per unit of item A are consumed when a production order of item A is performed, and one unit of C per unit of FP is used when an order of FP is released.

> The BOM structure is modeled in Equation (2.12) by r^{ij} for all
> items i and all $j \in D(i)$, that is, all direct successors j of i,
> where r^{ij} is the number of units of i required per unit of j.

The demand for intermediate products (such as A or C in the example) coming from the production orders of their successors is called *dependent demand*, as opposed to independent for finished products, because it depends entirely on the production plans of successor items. Such plans are controlled by the planner, whereas independent or external demand is not. For instance, item C will be consumed only when a production order of A or FP is started (a decision under the control of the planner), but not when a finished product FP is ordered or delivered. This distinction between dependent and independent demand is crucial and constitutes the basis of the MRP planning process.

> The dependent demand for item i in period t is represented by
> $\sum_{j \in D(i)} r^{ij} x_t^j$ in Equation (2.12).

Procurement and Production Lead-Times

Procurement and production activities cannot be performed instantaneously. In order to build realistic production plans, procurement or production lead-times, *lead-times* for short, are taken into account for all components in the BOM structure. They represent the total time needed to complete a procurement or production order, including preparation, administration, waiting, production, quality control and tests, and delivery, and are measured as an integer number of time periods. This information is written next to each component in Figure 2.4.

In the MRP planning model, such lead-times are constant over time, are independent of the order sizes, and are an input of the planning process.

> The constant procurement or production lead-time for item i is
> represented by γ^i in Equation (2.12).
> We can now rephrase our minimum length condition on the
> planning horizon. The number of time periods n is at least as
> large as the sum of γ^i values along any path in the BOM graph.

However, there is an important difference between the value of γ^i used in the planning optimization model (Equation (2.12)) and the value of the production lead-time used in the MRP planning process.

In the optimization model, γ^i is the *minimum lead-time* required for a batch of item i to be produced, minimum in the sense that no queue time (waiting time for the availability of machines or resources) is included. This holds because the explicit capacity constraints of the optimization model guarantee that there are enough capacity and resources to produce each lot without any delay, and therefore no safety queue time is required.

In contrast, the MRP planning process does not take capacity restrictions directly into account, and the constant production lead-time principle forces the planner to take a worst-case approach. The minimum production lead-time γ^i must be augmented by some *safety lead-time* to guarantee the feasibility of the production plans. For instance, in Figure 2.4, the lead-time for A has been fixed to two periods because item A is sometimes produced in large lots or on machines that are heavily loaded. Therefore a lead-time of two periods is reserved for all production orders, even though most of these orders are of small size, and will be released when there is enough capacity to complete them effectively after one time period. As a side effect, this also increases the level of work-in-progress inventory.

We come back later to the dramatic effects of the necessary inflation of production lead-times in the MRP planning process.

Routing of Components

In addition to the product structure defined by the BOM and to the production lead-times, the *routing* of products through different work centers, as well as the time and capacity consumed at each work center by a production order, are described in order to model and control capacity utilization.

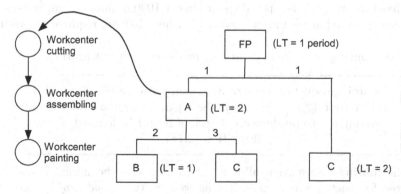

Figure 2.5. The routing of semi-finished product A.

The simplest routing model consists of the decomposition of the production order of each BOM component into a sequence of production operations. This is illustrated in Figure 2.5 where the production of component A is shown to require three successive operations (cutting, assembling, and painting).

The corresponding routing data for component A is given in Table 2.1. The sequence of operations is defined by numbering the operations. For each operation in the sequence, and for important or critical resources (manpower, machines, departments as a whole, etc.), the *unit production time* ([minutes per unit]) and the resource preparation or *set-up time* ([minutes per order or batch]) are defined. This set-up time is independent of the batch size. Usually, transportation and transfer times between operations are also modeled.

Table 2.1. Routing Data of Semi-Finished Product A

Routing of Component $i = A$				
Operation Number	Operation Description	Resource (k)	Unit Time $(\alpha^{ik}, [\text{min}])$	Set-up Time $(\beta^{ik}, [\text{min}])$
10	Cutting	Mach S100	1.5	25
15	Transfer	Forklift	–	20
20	Assembling	Mach ASS	0.5	10
30	Painting	Mach PPP	2.5	30

For instance, according to Table 2.1, a production order of 10 (resp., 20) units of component A requires 130 (resp., 175) minutes in total, assuming that 20 (resp., 40) units of component B and 30 (resp., 60) units of component C are available, and assuming that the resources are also available when needed. Even if the lot size of component A is almost always below 20 and requires thus less than 200 minutes, the production lead-time for component A has been fixed to two periods – two days or almost 1000 minutes – simply because the machines used are not always available when they are required to produce A.

These routing data are first used to model capacity utilization.

> The unit production time of item i on resource k is denoted α^{ik} in Equation (2.14). Similarly, the set-up or preparation time of resource k to produce one batch of item i is denoted β^{ik} in Equation (2.14).

The routing information allows one to compute the minimal lead-time required for each production order, as well as their *load profiles* (i.e., the evolution of the load over time) induced by the production plans in each work center and on each critical resource.

Capacity of Resources

To perform finite capacity planning, one needs additional information on the *actual or usable capacity* of each resource in each time period. The actual capacity is defined as the number of effective production hours that can be performed on the resource during the time period. This capacity will be compared with the load profiles computed from the production plans and routing data.

Usually, the available capacity is obtained as the product of the *gross capacity* (i.e., the office or worked hours), and the *productivity factor* (i.e., the fraction of worked hours that are effectively used for production). This productivity factor accounts for unavoidable breaks, interruptions, disturbances, or inefficiencies during the utilization of the resource.

Table 2.2. The Usable Capacity

Resource Description	Gross Capacity [hours/day]	Productivity Factor	Usable Capacity [hours/day]
Mach S100	8	0.95	7.6
Forklift	8	0.85	6.8
Mach ASS	16	0.85	13.6
Mach PPP	8	0.95	7.6

The only capacity information needed in production planning models is the net or usable capacity. These data are illustrated in Table 2.2 for the resources used in the routing of component A, where the productivity factors are higher for the automated cutting machine $S100$ and painting cell PPP than for the resources and operations requiring some manual intervention. There are 16 gross hours per day for assembly because two identical machines are available during one shift.

> The net capacity on resource k in time period t is represented by L_t^k in Equation (2.14).

Inventory Records

For all components, the independent and dependent time-phased demand define together the so-called *gross requirements*, corresponding to the total consumption, by external customers or internally by the production orders, of the components over time. This consumption requirement can be satisfied either from current inventory or from additional production or purchase orders. In order to compute the amounts that still need to be produced or purchased, the inventory status of each component must be known. This includes

- the *on-hand inventory*, which is the physical inventory in the warehouses;
- the *allocated or reserved inventory*, which is the part of the on-hand inventory that is reserved for production orders that have already been released, and is therefore not available any more to satisfy the gross requirements;
- the *back-orders*, which correspond to overdue or late component orders, and will be satisfied or delivered at the next reception; and
- the *on-order inventory*, which is the quantity of components already ordered (purchase or production) but not yet received, and for each such released order the *scheduled receipt* time period is known.

The *available inventory* is the inventory status used in production planning models, and is defined as the on-hand inventory minus the allocated inventory. It is often called inventory. The *inventory position* is defined as the available inventory augmented by the on-order inventory minus the back-orders. It is the most useful inventory status for inventory control, but it is rarely directly used in production planning models.

> The planned available inventory of item i at the end of period t is represented by the variable s_t^i in formulation (2.11)–(2.15). The on-order inventory of item i, scheduled to be received in period t, corresponds to the fixed quantity $x_{t-\gamma^i}^i$ released in the past (typically with $t - \gamma^i \leq 0$).
> The planned back-orders of item i at the end of period t will be represented by adding a new backlogging variable r_t^i in the formulation of the flow balance equation (2.12).

The *net requirements* of a component are the time-phased requirements obtained by subtracting the available inventory, and the on-order inventory when its reception is scheduled, from the gross requirements. They represent the amount still to be purchased or produced in order to satisfy the total or gross requirements.

The inventory status of each component is central and crucial information for the reliability of MRP systems. They are updated very regularly to incorporate the most recent events or transactions (order release, order reception, physical removal from stock, etc.) in order to reflect accurately the real situation on the shop floor and in the warehouses.

Planning Rules

Finally, the product database has to contain some more information relative to the definition and parameters of the planning rules used. Typically, it contains

- the rules and parameters for *safety stocks*, where the safety stock of a component is defined as the minimum stock to be held at the end of each planning period in order to be able to cover small variations of demand or consumption during the realization of the plan;

- the rules and parameters for *safety times*, where the safety time of a component is the time added to the component lead-time to cover unpredictable lead-time variations during the realization of the plan;
- the single-item *lot-sizing rules* and parameters for each component; such rules are used to transform the computed net requirements into economical procurement and production plans satisfying the requirements; we describe below the role of such single-item plans in the global MRP planning process; and
- component data required to use the lot-sizing rules: the procurement or production cost, the inventory holding cost, and so on.

The unit production cost, fixed set-up cost, and per unit and per period inventory holding cost are represented, respectively, by p_t^i, q_t^i, h_t^i in the objective function (2.11).
The safety times are part of the lead-time parameter γ^i in Equation (2.11).
There is no safety stock in formulation (2.11)–(2.15). Such safety stocks can be represented as simple lower bounds on the inventory variables s_t^i.

2.2.2 The Planning Process: Single Item Decomposition

So far we have studied the MRP model as defined by its inputs – products, BOM, routing, resources, capacity, inventory – and its mathematical representation. Now, the challenge is to design a solution approach for the mathematical programming problem (2.11)–(2.15).

Unfortunately, this model is usually too large to be solved directly, for the following reasons.

- Short time intervals/buckets are required to model demand satisfaction and capacity utilization accurately.
- Long planning horizons, and thus a large number of time periods, are required to cover the global procurement and production cycle.
- Capacity utilization needs to be tracked for all the critical resources.
- All the intermediate items need to be modeled in order to guarantee the feasibility of the planned flow of materials.

Therefore, decomposition approaches have been proposed to solve the planning model, leading to suboptimal production plans. The typical approach used in ERP/MRP planning systems is illustrated in Figure 2.6 for a MTS production policy and consists of the following steps.

(i) Master Production Scheduling (MPS)

The process starts with the computation of the Master Production Schedule, which, in a make-to-stock setting, is the production plan (lot or batch sizes

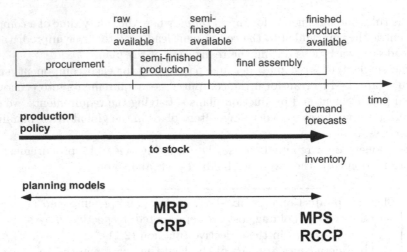

Figure 2.6. Planning models for an MTS policy.

per period) for finished products. This means that the MPS is only concerned with the plan of the last production operation yielding the finished product.

The MPS is built to satisfy the combination of firm customer orders – some are usually available for the very short term – and forecasts of customer orders throughout the planning horizon, as well as the required inventory levels at the end of the planning horizon. This last requirement is in anticipation of some future peak demand period, or simply to cover demand up to the next production batch for low-demand items. The MPS must take into account the existing inventory, the scheduled receipts of already released orders, as well as some safety stock requirements to cover forecasting errors.

The MPS mechanics are illustrated in Table 2.3 for the finished product FP from the BOM Figure 2.4, where the planning horizon has been fixed to six time periods; we are currently at the end of period 0, and all inputs to the MPS process are indicated in italics.
In this example:

- The gross requirements are defined, by convention, as the maximum of firm orders and forecasts in each time period, and correspond to updated forecasts.
- The required ending inventory plays the same role as an additional demand forecast for period 6.
- The net requirements are the minimal additional production quantities needed to satisfy the gross requirements, or equivalently the minimal quantities needed for the projected inventory to reach the safety stock level.
- The MPS is chosen to correspond to the net requirements and, therefore, the projected inventory corresponds to the safety stock after the consumption of the initial stock (and where the projected inventory in each period

Table 2.3. MPS Planning Process for Product *FP*

Planning Parameters	Time Periods	1	2	3	4	5	6
	Firm customer orders	*17*	*9*	*2*			
Ending inventory = *10*	Demand forecasts	*15*	*25*	*40*	*40*	*20*	*20*
	Gross requirements	17	25	40	40	20	30
	Scheduled receipts	*20*					
Safety stock = *5*	Net requirements		20	40	40	20	30
Current inventory = *7*	Projected inventory	10	5	5	5	5	5
	MPS planned orders (end)		20	40	40	20	30
Lead-time = *1*	MPS planned orders (start)	20	40	40	20	30	
	Available to promise	10	11	38	40	20	30

is equal to initial inventory plus scheduled receipt plus finished MPS orders minus gross requirements).

- The planned MPS orders have to start one period (the lead-time) before their completion.
- The *available to promise (ATP)* row gives the basic information needed to accept new customer orders; it indicates how many units of *FP* become available to satisfy new customer orders in each period.

In this example, we have just tried to minimize the finished product inventory by producing as little as possible. More economical plans can be built by minimizing production and inventory costs, but this would remain a single-item single-stage production plan.

(ii) Rough Cut Capacity Planning (RCCP)

The above approach can be used to determine, or even optimize, the MPS for each finished product individually. However, such finished products usually share some critical scarce resources, and some consolidation of the MPS plans is needed. This is carried out in parallel to the MPS process and is known as *Rough Cut Capacity Planning*. Its role is to check globally or "roughly" the feasibility of the MPS with respect to capacity utilization.

In the simplest case, the MPS is established without considering the capacity restrictions and RCCP consists of the computation of approximate load profiles implied by the MPS for some critical resources or for some aggregate view of the capacity (e.g., by department) using historical capacity utilization factors or simplified BOM structures. If the load exceeds the capacity, the planner has to adapt the MPS or increase the capacity manually.

In more sophisticated systems, the consolidation and modification of the MPS or the increase of capacity are suggested by the system.

In all cases, this approach remains approximate or rough because this capacity planning process does not take into account production stages other than the final one, and in particular does not consider

- the current inventory and in-progress orders at various stages, as if net requirements were equal to gross requirements at all stages but the final one, and
- the size and timing of production orders required at various production stages to produce the components consumed by the MPS.

Therefore, a detailed finite capacity verification step can only take place once the detailed production plans for all components are known.

(iii) Final Assembly Scheduling (FAS)

In the case of an assemble-to-order (ATO) production policy, a similar approach is used but the MPS is established for the decoupling items. The *decoupling items* are the semi-finished products at the interface between the push and pull planning phases; they are thus the last items produced to stock.

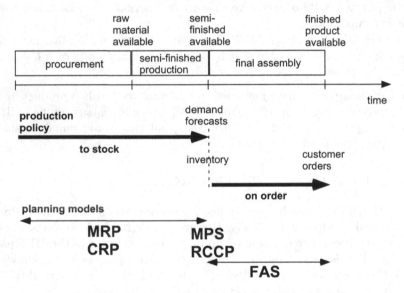

Figure 2.7. Planning models for an ATO policy.

For the MPS, the only modification with respect to the MTS policy is the need to compute customer demand forecasts at the level of the intermediate decoupling items, rather than at the finished product level. In other words, to behave as if customers were ordering directly the semi-finished products to be assembled.

Then, assuming that these decoupling items are available in stock when needed, the *Final Assembly Schedule (FAS)* determines when to realize the operations required to transform the intermediate items into the finished products, in order to meet firm customer orders on time. This approach is illustrated in Figure 2.7.

(iv) Material Requirements Planning (MRP)

The MPS and RCCP fix the production plan for all finished products, or decoupling items. Using a similar approach, that is, planning the production to meet uncertain forecasts, does not make sense for the other items in the BOM structure. One can do much better.

Once the production plan for finished products is fixed, one knows exactly when and in what quantity the components entering in the final production stage are required. This information has been called the dependent demand. So, we can replace uncertain forecasts by certain dependent demands, computed using the BOM structure. This eliminates the major source of uncertainty from the planning process, and hence the major reason to hold huge safety stocks. Then, we can plan the production of these components to meet their dependent demand. These production plans determine in turn the dependent demand of their immediate predecessors.

This process can be repeated, level by level in the BOM structure, all the way through, from the finished products back to the raw materials. It is known as the *Material Requirements Planning* process. Its sequential aspect is illustrated in Figure 2.8 on the BOM structure from Figure 2.4, assuming a MTS policy. Observe for instance that the total dependent demand and the production plan of item C can only be computed after the production plans of both FP and A have been fixed.

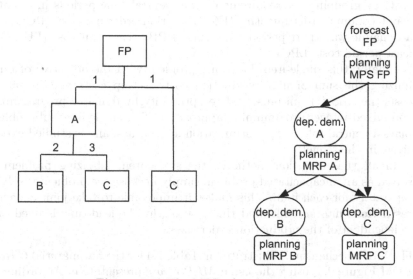

Figure 2.8. The MRP planning process.

For each item in the BOM structure subject to dependent demand, this sequential MRP planning process involves the following steps.

MRP Process: Step 1. Computation of the gross requirements.
These are time-phased requirements equal to the sum of dependent and independent demand. For some items such as spare parts, there can be a mix of a dependent and independent demand. In this case, forecasts must be computed for the independent part of the demand. The dependent demand is derived directly from the production plans of the direct successors in the BOM.

MRP Process: Step 2. Netting or computation of net requirements.
The net requirements are time-phased requirements. They correspond to the minimal additional (i.e., in addition to available stock and scheduled receipts) production quantities needed to satisfy the gross requirements.

MRP Process: Step 3. Planning or uncapacitated lot-sizing.
This last step consists in solving the single-item planning subproblem (LS-U) to determine the production plan meeting the net requirements, and satisfying some criterion. A production batch of an item in a period is called a *suggested production order*, or a *suggested procurement order*, or simply a *suggested order*.

Production plans or suggested orders are computed in MRP systems by using so-called *lot-sizing planning rules*. For instance the *lot for lot (LFL)* planning rule consists in taking the suggested orders equal to the net requirements, in every time period. This means that one produces exactly the demand, and therefore one minimizes the inventory level or cost. Other heuristic planning rules try to balance the set-up and inventory costs by grouping net requirements over several time periods in a static way (economic order quantity (EOQ), period order quantity (POQ)) or a dynamic way (part period balancing (PPB), least unit cost (LUC), or least period cost (LPC)).

Finally, this single-item lot-sizing problem with the objective of minimizing the sum of unit production costs, set-up costs and inventory costs (i.e., LS-U) can be solved to optimality by dynamic programming and mixed-integer programming approaches. This single-item subproblem plays a central role in our optimization approaches, and is studied extensively in the sequel.

In all these solution methods, the single-item lot-sizing problem is solved as an uncapacitated problem, simply because the problem is solved separately for each item. This makes it impossible to take joint capacity restrictions into account, and this is also why the lead-time is fixed and independent of the production order sizes.

The MRP mechanism is illustrated in Table 2.4 for the raw material C from the BOM Figure 2.4, using the usual *MRP record* presentation. According to the sequential MRP process, we assume that production plans are available for FP and A, and all data available prior to the computation of the MRP record are given in italics.

Table 2.4. MRP Record for Raw Material C

Planning Parameters	Time Periods	1	2	3	4	5
	Orders for FP (start)	*20*	*40*	*40*	*20*	*30*
	Orders for A (start)	*20*	*20*	*10*	*20*	*20*
	Gross requirements	80	100	70	80	90
	Scheduled receipts		*120*			
Safety stock = *0*	Net requirements				60	90
Current inventory = *150*	Projected inventory	70	90	20	70	110
Plan. rule: EOQ=*130*	Suggested orders (end)				130	130
Lead-time = *2*	Suggested orders (start)		130	130		

In the MRP record in Table 2.4:

- There is no independent demand for item C.
- The dependent demand for item C is three times the suggested orders of A plus the planned orders of FP (see the BOM structure in Figure 2.4).
- The initial inventory is large enough to cover the gross requirements up to period 3, and there are only net requirements in periods 4 and 5.
- The planning rule used is the fixed order size rule (FOQ), and an order of size 130 (this order size is computed using the EOQ formula as the best compromise between inventory and set-up costs for the average net requirement observed) is suggested each time the projected inventory becomes negative.
- The suggested MRP orders have to start two periods (the lead-time duration) before their completion.

This MRP planning process automatically computes suggested orders for all components in the product structure. The MRP records are updated regularly to take into account all transactions that have occurred and have modified the status of the production system, such as new customer orders, new order releases, order reception from suppliers, and so on.

In this dynamic context, the role of the planner (i.e., the user of the MRP system) is first to check the availability of the components and of the resources to perform the orders suggested in the coming or next few periods, and then to release the corresponding orders to the shop floor or to the supplier. In some cases, the MRP system makes infeasible or inadequate suggestions, mainly because it does not take capacity into account during the MRP process, and the planner has to adapt or improve the suggested plan manually. In such cases, the modified orders are transformed into *firm suggested orders* or blocked orders to prevent the MRP system changing them on the next run or automatic update.

(v) Capacity Requirements Planning (CRP)

The above approach determines the production plan for each component individually. As for MPS and RCCP, some consolidation of the MRP plans is needed. This is done after the MRP computations and it is known as *Capacity Requirements Planning*. Its role is to check the feasibility of the orders suggested by the MRP with respect to capacity utilization.

As for RCCP, there are several versions of CRP. In the simplest case, the CRP consists in the computation of detailed load profiles implied by the MRP orders. This is done by starting each MRP suggested order at its earliest start date, or at its latest finish date, and loading each work center or each resource according to the detailed description of the sequence of operations in the routing data. Once this is done for all suggested and in-progress orders, if the load exceeds the capacity in a work center, the planner has to adapt the suggested orders – start earlier or later to smooth the load – and to create firm suggested orders, or has to increase the capacity, manually. Hence, CRP identifies capacity problems, but does not resolve them.

In more sophisticated systems, the modification of the proposed orders or an increase of capacity are automatically suggested by the system.

In all cases, this approach remains very heuristic and suboptimal. Uncapacitated production plans are first generated, and then locally adapted to become feasible, by moving orders backward and forward in time or by increasing the capacity (overtime, alternate routing, etc.).

2.2.3 Limitations of MRP and the Optimization Answer

Although MRP systems are very powerful integrated production management and information systems, their planning modules implementing the myopic decomposition approach described above suffer from very severe limitations.

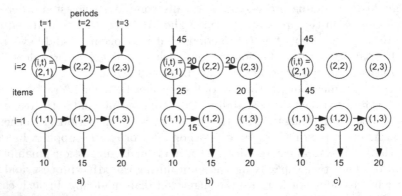

Figure 2.9. A two-level serial production planning example: (a) the minimum cost flow model; (b) its MRP solution with cost = 675; (c) an optimal solution with cost = 575.

This decomposition approach is called myopic because it does not exploit any knowledge about the model in the decomposition.

In other words, the decomposition is carried out in the same naive way for all models. The planning problem is decomposed into uncapacitated single-item subproblems, that are solved independently and sequentially, without backtracking, from finished products to raw materials. Capacity restrictions are taken into account only after the calculation of the production plan, and mainly to compute capacity requirements or to adapt the production plans locally (i.e., using minor modifications) with the hope of making them feasible.

In particular, in terms of productivity optimization, the major drawbacks of this myopic approach are the following.

Drawback 1:
Single-Level Decomposition $\Rightarrow$ Suboptimal Productivity
(Inventory and Production Costs)

The level-by-level decomposition of the product structure leads to suboptimal solutions with respect to the global minimum cost objective function. This is illustrated in the following simple example, the simplest one can imagine, involving only two items.

Suppose that we have an instance of the MRP model (2.11)–(2.15) with three periods and a serial BOM structure with 2 levels, one item at each level, where one unit of the raw material $(i = 2)$ is required to produce one unit of the finished product$(i = 1)$. For simplicity, we assume that the lead-time γ^i is zero for each item.

The external or independent demand for the finished product is $d^1 = (10, 15, 20)$. There is no external demand for the raw material. There is a fixed ordering cost of $q_t^2 = 200$ for the raw material, and fixed production cost of $q_t^1 = 100$ for the finished product for all t. The unit production cost is constant over time, and thus constant in all solutions, and therefore not considered. The inventory cost is $h_t^i = 5$ for all i, t. There are no capacity restrictions. This planning problem can be viewed as the fixed charge minimum cost network flow problem represented in Figure 2.9a.

In the MRP approach, we first determine the MPS for the finished product $i = 1$ in order to satisfy its external demand. The optimal solution, minimizing the set-up and inventory costs, is to produce 25 units in period 1, stock 15 from period 1 to period 2, and produce 20 units in period 3 with a total cost of 275. This production plan defines the internal or dependent demand for the raw material: 25 units of raw material have to be available in period 1, and 20 units have to be available in period 3. We then solve the MRP subproblem for the raw material, and find that it is optimal to order 45 units in period 1, and stock 20 units from period 1 to period 3 at a cost of 400. Note that an alternate optimal solution for the raw material is to order twice (25 units in period 1 and 20 units in period 3), and avoid the inventory costs. So the

MRP process has produced a global production plan with a cost of 675. This MRP solution is represented in Figure 2.9b.

The optimal solution with a cost of 575 is represented in Figure 2.9c.

Furthermore, when determining the MPS (which corresponds to solving the single-level minimum cost flow subproblem for item $i = 1$), the worst possible solution for item 1 (which is to produce the 45 units in period 1, and satisfy the demands in periods 2 and 3 from stock) forms part of the globally optimal solution shown in Figure 2.9c. This holds because it avoids the very costly procurement of the raw material. Such interactions between the items are simply ignored in the MRP decomposition process.

This example illustrates the difficulty of optimizing the production plans by solving independent single-level subproblems sequentially.

Drawback 2:
Single-Item Decomposition $\Rightarrow$ Infinite Capacity Planning $\Rightarrow$ Suboptimal Productivity (Capacity Utilization Plans)

The main characteristic of the MRP process is the decomposition into independent single-item planning subproblems. Because the resources are usually shared by several or many items, this decomposition scheme does not allow one to take capacity restrictions directly into consideration, that is, into consideration when the production plan is drawn up. In other words, the capacity available for item i depends on the production plans of some other items, and is therefore not known when planning item i.

Therefore finite capacity planning in MRP is carried out as follows. First, *infinite capacity* production plans (i.e., production plans defined as if capacity were infinite) are determined for all components (MPS and MRP). Next, these plans are translated into capacity requirements (RCCP at the MPS level, and CRP at the MRP level). Finally the plans are heuristically, and often manually, adjusted when some resources are overloaded. This clearly defines suboptimal capacity utilization plans. There is no reason to believe that the best or even good plans can be obtained in this way. The bottleneck (i.e., the most heavily loaded resource) capacity should be accounted for initially in the planning procedure, and exploited optimally, in order to optimize the global productivity.

Drawback 3:
Infinite Capacity Planning $\Rightarrow$ Constant Lead-Times $\Rightarrow$ Increased Inventory, Decreased Flexibility

Another consequence of infinite capacity planning is the impossibility of determining the production cycle and production lead-times as part of the output of the planning process.

As already explained in Section 2.2.1, in a finite capacity planning process it is possible to build realistic or feasible production plans without adding safety waiting times to the minimum production lead-times (γ^i in the MRP optimization model). This is done by taking work center capacity and routing data explicitly into account, and by only releasing orders for which enough capacity is available.

Unfortunately, in an infinite capacity planning approach, the load of the resources cannot be estimated or anticipated. Therefore, the effective production lead-time for each operation is the sum of the technical or minimum production lead-time γ^i and the waiting time for the availability of the resources. This waiting or queue time clearly depends on the resource load, and consequently varies over time for each resource. Because these waiting times cannot be anticipated, a worst-case approach has to be taken, and the constant lead-time used in MRP is inflated by a large enough safety time to guarantee that the lead-time can be met in all cases. This safety time is useful in the rare cases when the resources are heavily loaded, and useless in all the other cases.

A first consequence is that production orders are most often completed well in advance of the due-date or requirement date. Thus the safety times translate into increased work-in-progress inventory. A second and indirect consequence is that the total production cycle is augmented by the safety times at all production stages, the MPS time horizon is augmented accordingly, and the whole MRP planning process is based on longer-term forecasts. As long-term forecasts are usually much worse, larger end-product safety stocks are needed to protect the system against larger forecast errors. Finally, this longer MPS horizon requires more anticipation, and reduces the flexibility of the production system.

Summary

In summary, the myopic MRP decomposition scheme leads to important productivity and flexibility losses, two of the key levers in all manufacturing strategies, which is exactly the opposite of what is expected from a good planning system, and the opposite of what was initially expected from MRP systems. Indeed, the starting idea of MRP was to distinguish the dependent demand, which is computable, from the uncertain independent demand, for which forecasts are needed, with the objective of knowing when and how much is needed of each component, and thereby opening the way to a reduction of the global inventory levels.

The Optimization Approach

The observed limitations all relate to the MRP decomposition approach and planning process, and not to the MRP model itself. The MRP model formulated and discussed above adequately represents the planning problem faced by many companies, but a global solution and optimization approach is needed in order to reach the desired goal of improving the productivity and flexibility simultaneously.

This global optimization approach depends on the two main modeling ingredients to which we hope to contribute: the expertise needed to build correct and adequate mathematical models, and the expertise required to improve the initial problem formulations and to design optimization software allowing one to solve larger instances globally, without resorting to myopic decomposition.

2.3 Advanced Planning Systems

The purpose of this section is to describe the general context of production planning and supply chain planning models and systems.

2.3.1 Supply Chain Planning

A *supply chain (SC)* consists of a set of organizations, often legally separated, linked by materials, information, and financial flows, that produce value in the form of products and services for the ultimate customer. It can also consist of the geographically dispersed sites of a single and large company. Along this supply chain, raw materials have to be purchased, intermediate and finished products have to be produced or transformed, and finished products have to be sold and distributed.

Therefore a SC is usually modeled as a network composed of vendor nodes; plant nodes where products are produced or transformed; distribution center nodes where products are received, stored, and dispatched but not transformed; market nodes where products are sold or consumed; and transportation arcs connecting the nodes and supporting both the physical and information flow.

Supply Chain Planning (SCP) is defined as an integrated planning approach used to organize the SC activities.

- This multi-dimensional integration is concerned with the *functional integration* of the primary activities – purchasing, manufacturing, warehousing, transportation – and support activities that constitute the value chain of the SC.
- It is also concerned with the *inter-temporal integration* – often called hierarchical planning – of these activities over strategic, tactical, and operational planning horizons. Strategic problems deal with the management of change in the production process and the acquisition of the resources over long-term horizons based on aggregated data. Tactical problems analyze the resource allocation and utilization problems over a medium-term planning horizon using aggregate information. This consists in making decisions about, for instance, materials flow, inventory, capacity utilization,

and maintenance planning. Operational problems aim at planning and controlling the execution of the production tasks. For instance, production sequencing and input/output analysis models fit into this category. This integration is critical to success because the design of the SC must take into account the operations performed under this design, and because a company cannot maintain competitive operations and position with poor strategic decisions regarding its technology or the location of its plants and facilities.

• Finally it is concerned with the *spatial integration* of these activities.

Integrated planning is made possible because of the recent advances in information technology (IT). Focusing only on the procurement and manufacturing or production functions of the supply chain, *Manufacturing Planning and Control (MPC)* systems are developed to cope with these complex planning environments, and integrate these planning problems into a single integrated management system.

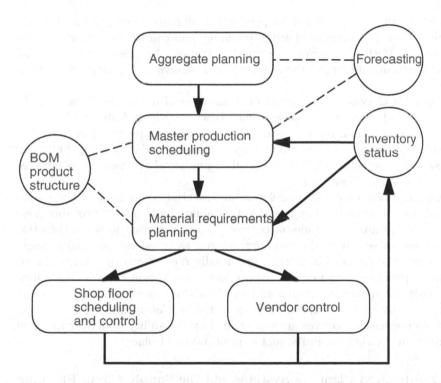

Figure 2.10. An MRP-II system.

For instance, Figure 2.10 describes how the tactical and operational planning problems are integrated in *Manufacturing Resources Planning (MRP-II)* systems, an example of an MPC system. In these systems, medium-term *aggregate or master planning* consists in deciding about capacity utilization, and aggregate inventory levels to meet the forecast demand over a medium-term horizon of about one year. A medium-term horizon is usually needed to be able to take into account some seasonal pattern in demand. MPS consists of planning the detailed short-term production of end-products in order to meet forecast demand and firm customer orders, taking into account the capacity utilization and aggregate inventory levels decided at the master planning stage. Here the time horizon is usually expressed in weeks and corresponds to the duration of the production cycle. MRP-I establishes the short-term production plans for all components (intermediate products and raw materials) from the production plan of end-products decided at the MPS stage, and from the product structure database (bills of materials). Then, *shop-floor control systems* (for manufactured components) and *vendor follow-up systems* (for purchased components) control the very short-term execution of the plans decided at the MRP-I stage. The time horizon at this stage is usually of a few days.

Other well-known integrated production planning concepts and systems fit into this general manufacturing, planning, and control framework. For instance, the MRP-II system represented in Figure 2.10 subsumes the original MRP-I system, and follows the *Hierarchical Production Planning (HPP)* principles.

Such MPC systems are based on transactional databases. However, the existence and storage of transactional data, as well as faster and cheaper data communication, do not automatically lead to improved decisions. The effective application of IT in SC management requires the building of effective decision-support systems. These are called *analytical IT systems*, as opposed to *transactional IT systems*.

Optimization planning models are an essential component of these analytical systems because they are able to evaluate and identify provably good plans and optimize the trade-off between financial and customer satisfaction objectives. In supply chain planning, as well as in operations management in general, the financial objectives are usually represented by transportation costs for purchasing and delivering products, production costs for machines, materials, manpower, start-ups and overheads, inventory holding costs, opportunity costs of the capital tied up in the stocks, insurance, and so on. Customer-service objectives are represented by the ability to deliver the right product, in the right quantity, at the right date and place.

2.3.2 Advanced Planning Systems and the Supply Chain Planning Matrix

The analytical IT or "computerized" planning systems, based on the transactional data gathered from an Enterprise Resource Planning transactional System, are called *Advanced Planning Systems (APS)*. The structure of the planning tasks of such APS is described in Figure 2.11, and is known as the *Supply Chain Planning Matrix (SCPM)*.

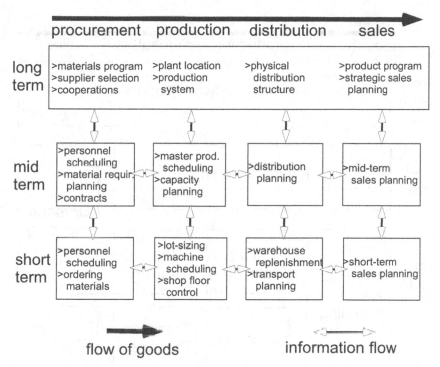

Figure 2.11. The Supply Chain Planning Matrix ([70]).

The main characteristics of an APS are the following:

- *Integral or global planning*: coordination of the planning of the entire supply chain;
- *Optimization focus*: the definition of alternatives, objectives, and constraints for all the planning tasks; and
- *Hierarchical approach*: the decomposition into planning modules, and their vertical and horizontal coordination by information flows.

These characteristics are reflected in Figure 2.11. In most of the applications, traditional MRP or ERP systems do not share these characteristics. They are restricted to the production function, and they do not optimize. Moreover, when they consider various planning horizons, they use essentially a sequential or independent approach for the different planning tasks. In other words, with respect to the APS as in Figure 2.11, there is no real bottom-up coordination with respect to planning horizons, and no left–right coordination with respect to planning functions.

This SCP matrix is also used by APS software providers to offer a set of software modules covering the matrix as much as possible. The typical module architecture of such systems is depicted in Figure 2.12.

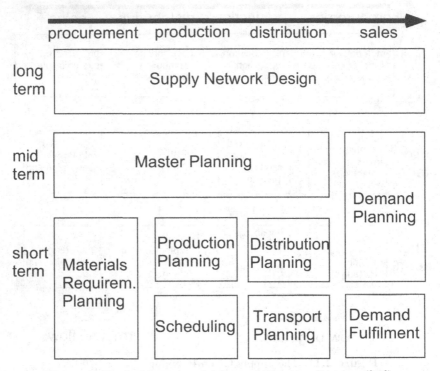

Figure 2.12. Architecture of Advanced Planning Systems ([70]).

Of course, the SCPM defines the general structure of an APS, but a single APS consisting of a fixed combination of software modules cannot respond to the management requirements of all supply chains. Typologies of supply chains are defined in the literature in order to identify supply chains having or sharing the same major characteristics, and therefore sharing the same planning requirements and tasks. One such typology is based on supply chain functional attributes – related to the functions of procurement, production, distribution, and sales – and structural attributes – related to the topography, integration, and coordination of the SC. For instance, this typology has been used to design APSs for specific industries, such as the computer assembly and consumer goods industries.

In the context of APS and its planning matrix, the objective of this book is to give a state-of-the-art description of the modeling and reformulation theory needed to design efficient optimization- or mathematical programming-based algorithms to support the supply chain planning tasks. In other words, we focus on one of the major characteristics of the APS approach, namely optimization.

2.4 Some Supply Chain Planning Problems

We describe here briefly, without mathematical formulations, two generic classes of supply chain planning problems. They extend the scope of production planning models presented so far by considering the entire supply chain rather than a single production facility or plant.

2.4.1 Strategic Network Design Problems

The purpose of supply chain strategic network design problems is to configure the supply chain network as a whole, from suppliers through production, warehousing, and distribution facilities, down to end customers, which can be downstream subsidiaries.

Decisions

The main decisions to be taken at this stage are the status of the nodes and arcs in the supply chain network. For each node, the decision is usually whether to install a facility at a specific location or site, and also the amount of product processed (bought, produced, transformed) at the node. For the arcs, the decision is whether to use a specific route for a given product to link some nodes in the network using a given transportation mode, and also the amount of product flowing through that route. These major decisions are either considered as static, requiring single-period decisions and a single-period model, or dynamic, involving a multiple-period model. In the case of a multiple-period model, similar decisions can be taken in each time period, and inventory arcs are added in the network at some specific nodes, typically modeling production and storage facilities.

Restrictions

These decisions have to be taken in order to satisfy the forecast demand of the customers. Therefore, the main constraint in this problem is the flow conservation constraint for each product at each location or node in each time period. This means that during each time period, the amount of product received from the local suppliers plus the amount available from initial inventory, the input flows, cover exactly the amount shipped to local customers or to other facilities and the amount put into final inventory, output flows, at each node.

Moreover, there are usually capacity restrictions attached to the various activities. These can be supply capacity, production, processing or transportation capacity, or storage capacity restrictions. Capacity installation (the

amount to install) and expansion (the increase of capacity) become decisions
to be taken in these problems.

Objective

The objective of the problem is to design a network able to satisfy customer
demand and to maximize the after-tax discounted yearly profit of the cor-
porations involved in the supply chain. This includes all costs and revenues
in the supply chain, namely revenues from sales and supply, manufacturing,
warehousing, inventory, and transportation costs. When the supply chain is
composed of different legal entities or covers several countries, this also requires decisions to be taken regarding the product transfer prices between
these entities, and the modeling of the legal restrictions on the pricing mechanisms.

Model Type

The manufacturing costs and investment costs often exhibit economies of scale
with respect to the amount produced or the capacity installed. This is usually
approximated or modeled using fixed charge cost functions, that is, cost functions with a fixed component to be paid if there is production/investment and
a linear component directly proportional to the amount produced or capacity
installed. In such cases, the resulting model is a mixed integer programming
model, very often linear.

The general structure of these problems is of the multi-period, multi-
product, multi-echelon or level, capacitated fixed charge network flow type.
This comprehensive modeling and optimization approach was used to design
the supply chain of Digital Equipment Corporation.

Challenges

Solving such complex and often large-scale models to optimality is still challenging. This is particularly true when transfer prices have to be incorporated
in these problems because this feature often makes the model nonlinear.

2.4.2 Supply Chain Master Planning Problems

The purpose of supply chain master planning problems is to optimize and syn-
chronize the materials flow along the complete supply chain over a medium-
term horizon. The main purpose is to adapt supply and production levels to
demand for aggregated products, taking the capacity of bottleneck resources
into account, with a centralized view considering all relevant costs and constraints. This global supply chain perspective for the mid-term decisions allows one to reduce inventory levels by improved coordination and by removing
redundant buffers between supply chain entities.

This problem takes the design of the supply chain network as fixed by a higher-level, longer horizon, planning module. The results of master planning impose restrictions on lower level detailed planning modules, which are very often functionally decomposed into short-term procurement, production, distribution, and transportation modules. Feedback mechanisms have to be implemented in order to coordinate these three planning levels.

Decisions

The main decisions are the aggregate production and distribution plan for all supply chain entities. In particular, production quantities are decided for each product group, each time period and location (plant or warehouse). Similarly, transportation quantities are decided for each link in the supply chain network, each product group and each time period.

These problems are always dynamic, multiple-period problems because their major objective is to optimize the trade-off between variations in processing levels and variations in inventory levels over time, in order to minimize the cost of satisfying the global supply chain demand for all products. Inventory levels over time are a consequence of the production and transportation decisions.

The main difference with respect to the strategic network design problem is the level of detail for the decisions modeled. Usually, the length of the planning periods is shorter, more detailed product groups are modeled by incorporating intermediate and storable products, and production and storage facilities are modeled in more detail. For example, set-up times and changeover times, the time or capacity consumed when a machine starts a production batch or when a machine switches from one product to another, are incorporated in master planning when they have a significant impact on capacity utilization. This level of detail is required in order to exploit the flexibility of the supply, production, and distribution processes in satisfying demand. To facilitate the coordination with short-term planning, it is often the case that the short-term horizon (the first few days or weeks) within the medium-term horizon is modeled using smaller time period intervals.

Restrictions

As for the strategic problem, the decisions are taken in order to satisfy the forecast demand. Therefore, the main constraint is again the flow conservation constraint for each product at each location in each time period.

The other main constraints are the capacity restrictions on supply, production, transportation, and inventory levels.

Objective

The objective of the problem is to optimize the trade-off among inventory costs, production, and transportation costs.

Model Type

Again, the manufacturing and transportation costs exhibit economies of scale with respect to the amount produced, and are modeled using fixed charge cost functions.

The general structure of these problems is of the multi-period, multi-product, multi-echelon or level, capacitated fixed charge network flow type. For instance, impressive returns with this master planning optimization approach have been obtained at the Kellogg Company.

Notes

Sections 2.1 and 2.2 In addition to the detailed planning case studies provided in this book, and to the generic planning models described here, we refer the reader to Voss and Woodruff [186] for an introduction to the modeling and solution of MRP optimization problems. For another general survey on production planning, we refer to Graves et al. [78].

Section 2.3 Our general definitions of supply chains and supply chain management are adapted from Christopher [38], Shapiro [149], and Stadtler and Kilger [155].

The presentation of the structure of Manufacturing Planning and Control Systems is derived from Vollmann et al. [185], integrating original characterizations of MRP-I systems by Orlicky [127] and Hierarchical Production Planning approaches by Hax and Meal [87]. The reader should refer to Vollman et al. [185] and Browne et al. [32] for a general description of MRP systems, to Hopp and Spearman [91] for a critical analysis of MRP systems, to standard operations management texts such as Silver et al. [151] and Johnson and Montgomery [93] for a more extensive treatment of the heuristic lot-sizing rules, and to Chopra and Meindl [37] for a modern textbook on Supply Chain Management.

The important distinction between analytical and transactional IT systems is emphasized in Shapiro [149] and Fleischmann et al. [70].

The description of the general architecture of Advanced Planning Systems comes from Fleischmann et al. [70], where a complete description of the Supply Chain Planning Matrix and its planning modules can be found. A similar structure focusing on the difference between transactional and analytical IT systems can be found in Shapiro [149].

The typology of supply chains we refer to is defined by Meyr at al. [120]. They illustrate how to use this typology to design an APS for the computer assembly and the consumer goods industries (see also Fleischmann and Meyr [69]) .

Section 2.4 Our description of the generic supply network design problem is inspired by the more complete review on the subject by Goetschalckx [76].

Its application to Digital Equipment Corporation can be found in Arntzen et al. [12].

A more complete introduction to the required coordination between the master planning SC module and the other SC modules, through disaggregation and feedback mechanisms, can be found in Rohde and Wagner [146].

The application of master planning to the Kellogg Company, and its impressive returns, can be found in Brown et al. [31].

3

Mixed Integer Programming Algorithms

Motivation

Our approach to help in solving industrial production planning problems is based on the solution of mixed integer programs by optimization methods. This means that we want either to find provably optimal solutions to these programs, or to find near-optimal solutions with a performance guarantee, expressed usually in terms of a percentage deviation of the objective value from the optimal value (duality gap; see Sections 1.2.5 and 3.3.4).

For readers that are not familiar with mixed integer programming, and in order to make this book accessible to a wide audience, we provide a – not too technical – introduction to mixed integer programming algorithms and reformulation techniques. This introduction contains all the material necessary to understand and to use the reformulation approaches and results presented in later chapters.

Our motivation is also to help the reader to develop a less myopic understanding of the reformulation approach by carefully defining the concepts and describing the main steps of the reformulation methods and the main questions one needs to answer in order to use these methods.

Objective

The general objective of this chapter is to present

- the general optimization methods used to solve mixed-integer programming models, namely the branch-and-bound and branch-and-cut methods, and
- the different reformulation techniques used to improve the mathematical formulations of these models.

Contents

More specifically:

- In Section 3.1 we define a *mixed integer program.*
- In Section 3.2 we provide an intuitive introduction to the analysis of running times of algorithms,
- In Section 3.3 we formalize the *branch-and-bound* algorithm used to solve general MIPs.
- In Section 3.4 we define the main steps of the *a priori reformulation* approach taken to tighten the initial mathematical formulation, including the important concepts of valid inequalities, good and bad formulations, and the ideal convex hull or tight reformulation.
- In Section 3.5 we formalize the *branch-and-cut* algorithm, using separation algorithms and cutting plane algorithms as building blocks.
- Finally in Section 3.6 we describe basic *construction and improvement heuristics* designed to find and improve feasible solutions quickly, and to be used in combination with a branch-and-bound algorithm.

3.1 Mixed Integer Linear Programs

All the example models that we have formulated so far, and that we consider in this book, belong to the general class of mixed integer linear programs.

Definition 3.1 *A* mixed integer linear program (MIP) *is an optimization program involving continuous and integer variables, and linear constraints. Any MIP can be written as*

$$(MIP) \quad Z(X) \ = \ \min_{(x,y)} \{ \ cx + fy \ : \ (x,y) \in X \ \},$$

where the set X is called the set of feasible solutions *and is described by m linear constraints, nonnegativity constraints on the x, y variables, and integrality restrictions on the y variables. In matrix notation*

$$X \ = \ \{ \ (x,y) \in \mathbb{R}^n_+ \times \mathbb{Z}^p_+ \ : \ Ax + By \geq b \ \},$$

where

- $Z(X)$ *denotes the optimal objective value when the optimization is performed over the feasible set X.*
- x *and y denote, respectively, the n-dimensional (column) vector of nonnegative continuous variables and the p-dimensional (column) vector of nonnegative integer variables.*
- $c \in \mathbb{R}^n$ *and $f \in \mathbb{R}^p$ are the (row) vectors of objective coefficients.*
- $b \in \mathbb{R}^m$ *is the (column) vector of right-hand side coefficients of the m constraints.*
- A *and B are the matrices of constraints with real coefficients of dimensions $(m \times n)$ and $(m \times p)$, respectively.*

Definition 3.2 *A mixed binary linear program* (MBP), *or mixed 0–1 program, is a MIP (according to Definition 3.1) in which the integer variables y are further restricted to take binary values. This means that the feasible set X of a MBP is defined by*

$$X = \{ (x, y) \in \mathbb{R}_+^n \times \{0, 1\}^p : Ax + By \geq b \} .$$

In these definitions, we have indicated the dimensions of all vectors for completeness. When these dimensions are clear from the context, or not important, they are usually omitted. Note also that the nonnegativity of variables is not essential in this definition. We have included this restriction because it is usually present in practice.

As we show, the linear relaxation of a MIP plays a very important role in the optimization algorithm used to solve it. It is obtained by removing the integrality restrictions on the y variables. Specifically if

$$P_X = \{ (x, y) \in \mathbb{R}_+^n \times [0, 1]^p : Ax + By \geq b \} ,$$

then X is the set of points in P_X with y integer; that is, $X = P_X \cap (\mathbb{R}^n \times \mathbb{Z}^p)$.

Definition 3.3 *The* linear relaxation (LR) *of the MIP* $\min\{cx + fy : (x, y) \in X\}$ *with* $X = P_X \cap (\mathbb{R}^n \times \mathbb{Z}^p)$ *is the linear program*

$$(LR) \qquad Z(P_X) = \min_{(x,y)} \{ cx + fy : (x, y) \in P_X \},$$

where the feasible set is P_X. We call P_X a formulation *for X.*

It is important to note that the linear programming relaxation of a MIP with feasible set X depends not just on X, but on the set of linear constraints used in describing P_X.

The set P_X is a larger set than X. Indeed, to get P_X we have just added to X all the points $(x, y) \in \mathbb{R}_+^n \times \mathbb{R}_+^p$ satisfying the linear constraints $Ax + By \geq b$, and with some non-integer y-coordinate. Because we minimize the same objective function over a larger set (i.e., $P_X \supseteq X$), the optimal objective values satisfy the following property.

Observation 3.1 *For any MIP with a minimization objective function, the linear relaxation defines a* lower bound *on the optimal objective value,*

$$Z(P_X) \leq Z(X).$$

On the other hand, upper bounds are obtained from feasible solutions.

Observation 3.2 *For any MIP with a minimization objective function, the objective value $Z = cx + fy$ achieved by any feasible solution $(x, y) \in X$ provides an* upper bound *on the optimal objective value,*

$$Z(X) \leq Z.$$

These lower and upper bounds will play an essential role in the solution of MIP using a branch-and-bound algorithm. This is explained in Section 3.3

To illustrate the notation of MIP, and the behavior of the branch-and-bound and branch-and-cut algorithms, we use the following small example involving only two integer variables ($p = 2$), and no continuous variables ($n = 0$). Such MIP programs with only integer variables are called *pure integer programs (PIP)*.

$$Z(X) \;=\; \min_{y} \{\; -y_1 - 2y_2 \;:\; y = (y_1, y_2) \in X \;\}, \qquad (3.1)$$

where the feasible set X is defined by

$$X \;=\; \{y = (y_1, y_2) \in \mathbb{Z}_+^2 \;:\; \begin{array}{rrr} y_1 & & \geq \;\;\; 1 \\ -y_1 & & \geq \;\; -5 \\ -y_1 & -0.8y_2 & \geq \; -5.8 \\ y_1 & -0.8y_2 & \geq \;\;\; 0.2 \\ -y_1 & -8y_2 & \geq \; -26 \;\}. \end{array}$$

This is a MIP as it fits Definition 3.1 with $n = A = c = 0$,

$$m = 5, \; p = 2, \; f = (-1, -2), \; B = \begin{pmatrix} 1 & 0 \\ -1 & 0 \\ -1 & -0.8 \\ 1 & -0.8 \\ -1 & -8 \end{pmatrix} \; \text{and} \; b = \begin{pmatrix} 1 \\ -5 \\ -5.8 \\ 0.2 \\ -26 \end{pmatrix}.$$

The MIP instance problem (3.1) is illustrated in Figure 3.1, where the feasible set P_X of the linear relaxation is represented by the shaded area. The feasible set X is the set of integer points satisfying the constraints defining P_X. Therefore X corresponds to the black dots inside P_X. The line orthogonal to the direction of minimization simply indicates points with the same objective value.

Graphically, the objective is to translate this line, as far as possible in the direction of minimization, and still find a feasible integer point lying on the translated line. Clearly, the line in Figure 3.1 has been translated too far out of the feasible set, and does not contain any feasible solution.

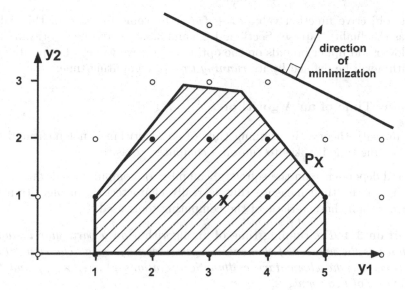

Figure 3.1. The MIP instance (3.1), its formulation P_X, and its feasible set X.

3.2 Running Time of Algorithms

Before describing the algorithms used to solve MIPs, we give now a very brief and intuitive introduction to the classical way of representing and analyzing the running time of (optimization) algorithms.

3.2.1 Performance of an Algorithm

Suppose that we want to solve a given optimization problem P, such as the PIP defined in (3.1), or the single-item LS-U instance formulated in Section 2.1. In addition we are given an algorithm A that solves P, either exactly or heuristically.

We suppose first that our intuitive understanding of the notion of an *algorithm* – as a sequence of computations starting with a mathematical program and its associated data, and producing a solution to the program – is precise enough.

The basic question to ask in studying the optimization problem P and algorithm A is the following.

Question 3.1 *Is algorithm A a fast and good algorithm for the program P?*

The two natural criteria or performance measures needed to analyze or compare the behavior of algorithms are the quality of the solutions obtained (good algorithm) and the time required (fast algorithm) to obtain these solutions. In the case of a MIP, the *quality of a solution* (x, y) will be defined by the objective function value $cx + fy$, or by some function of this value such as the duality gap (see Sections 1.2.5 and 3.3.4) when the algorithm outputs lower and upper bounds on the optimal objective value. The speed of an algorithm will be defined by its *running time* or computing time.

Running Time of an Algorithm

In complexity theory, the running time of an algorithm is not measured directly by the time it takes to run on a computer, so as to

- avoid dependence on the software, compiler, or computer used, and
- define a run-time measure that is really characteristic of algorithm A, rather than characteristic of the computing tools used.

Definition 3.4 *The* running time of an algorithm *on a particular instance is defined as the number of elementary operations performed. The elementary operations are the elementary arithmetic operations (+, −, ×, /) and the comparison of two numbers.*

The assumption behind this definition is that the running time needed to perform any elementary operation is a constant, and thus that the numbers encountered are not enormous. For example, summing n real numbers, or finding the minimum of n numbers, requires n elementary operations. Of course this constant will depend on the computing tools used, but it is not a characteristic of the algorithm itself.

Classes of Problems

In many cases, we are interested in designing algorithms to solve a more general class of problems of which P is a member. We adopt the following simple definition.

Definition 3.5 *A class of problems* is defined as a set of problems sharing a common mathematical structure or model. Particular problems in a class, often referred to as instances, differ in their size and their data.*

Thus Definition 3.1 defines the class MIP and each problem instance $P \in$ MIP is defined by the size parameters n, p, m, and by the data A, B, b, c, f. Subclasses of the MIP class are the MBP class and the PIP class. The example (3.1) is an instance from the class MIP and its subclass PIP.

We analyze algorithms and formulations for smaller subclasses of MIP. For instance, the *LS-U* problem defined in Section 2.1 is such a subclass. The problem classes we consider here are defined and classified in Chapter 4.

Running Time of an Algorithm for a Problem Class

Now we are interested in the running time of algorithms for solving all problem instances P belonging to a given class C. It is usual and natural to express the running time for solving $P \in C$ as a function of its size. For example, the running time of an algorithm A on any instance of LS-U with n time periods in the planning horizon is defined as the number $N(A, n)$ of elementary operations required, and could be equal to

$$N(A, n) = 3\frac{n(n-1)}{2} + n^2. \tag{3.2}$$

Remark: More precisely, note that the running time of an algorithm also depends on the size of the data (and not only the number of data items), because elementary operations on very large integer numbers take more time than on small integers. In complexity theory, it is usual to capture both the problem size and data size factors by expressing the running time for solving $P \in C$ as a function of the length of the data (string) required to define problem instance P.

For example, consider the LS-U model with n time periods. There are n demands to store, and if each demand is an integer between 0 and $31 = 2^5 - 1$, then it takes a string of at most $5n$ bits to store the demand data with the usual binary encoding scheme for integer numbers.

We do not enter into these details here, and express the running time as a function of the problem size only, assuming implicitly that all the elementary operations take constant time.

Run-Time Order of an Algorithm for a Problem Class

Finally, the running time required to solve small-size instances P of the class C is often not very relevant. To analyze the running time of an algorithm, one usually prefers to take an asymptotic perspective.

Definition 3.6 *The run-time order of algorithm A for problem class C is $f(n)$ if there exists a constant μ such that the running time required to solve any instance P is bounded from above by $\mu\, f(n)$, where n is the problem size parameter and n is sufficiently large.*

Typical values that one encounters for the run-time order $f(n)$ are n, n^2, n^3, ..., $\log n$, $\log^2 n$, ..., 2^n, or combinations thereof.

The mathematical notation for the run-time order is $O(n)$, $O(n^2 \log n)$, ...

In other words, the run-time order of an algorithm is the dominating term of an upper bound on the running time when the problem size goes to infinity. For example, the complexity measured in (3.2) corresponds to a run-time order of $O(n^2)$ because $N(A, n) \leq \frac{5}{2}n^2$ (for large enough n).

Definition 3.7 *An algorithm with polynomial run-time order $O(n^p)$ for some fixed value p is called a* polynomial algorithm.

For the problem classes that we consider here, the non-polynomial algorithms are called exponential algorithms.

The distinction between polynomial and exponential algorithms is crucial for the running time and the solution of large-size instances, as illustrated in Table 3.1. Observe also the difference between $log_2\, n$ and n.

Table 3.1. Illustration of Polynomial and Exponential Complexity

$n =$	10	10^2	10^3	10^4	10^5	10^6
$log_2\, n =$	3.32	6.64	9.97	13.29	16.61	19.93
$log_2^2\, n =$	11.04	44.14	99.32	176.56	275.88	397.27
$n^2 =$	100	10^4	10^6	10^8	10^{10}	10^{12}
$2^n =$	1024	$1.27\ 10^{30}$	$10.72\ 10^{300}$	$19.95\ 10^{3009}$	$9.99\ 10^{30102}$	

Definition 3.8 *Problems for which there exists a polynomial algorithm are considered to be "easy" problems, and belong to the "complexity" class of* polynomial problems.

Problem *LS-U* belongs to the class of polynomial problems. Also, the time required to solve a linear program using the interior point (barrier) algorithm is polynomial in its size. Therefore linear programs also belong to the class of polynomial problems. Note that the time to solve a linear program using the simplex algorithm is not polynomial, but is very fast in practice.

On the contrary, this does not hold for mixed-integer programs. The only guaranteed bound on the running time required is in general exponential in the size! Moreover, if a MIP is solved by a branch-and-bound approach, the time required depends heavily on the way in which the problem is formulated. This is the topic of the following sections.

3.2.2 The Size of a Formulation

By extension, we use the same *asymptotic notation* to define and characterize the size of a mathematical program as for the running time of algorithms.

For instance, the *LS-U* formulation given in Sections 1.1 and 2.1 contains $3n$ variables and $2n$ constraints, on top of the nonnegativity and integrality restrictions on the variables. This formulation is said to have $O(n)$ (i.e., of the order of n) variables and $O(n)$ constraints. We often write that this *LS-U* formulation is of size

$$O(n) \times O(n),$$

where the first factor characterizes the number of constraints, and the second the number of variables.

This notation does not characterize the running time needed to solve a mathematical program, but just the formulation size.

3.3 Branch-and-Bound Algorithm

We now describe the branch-and-bound algorithm which is the basic or general algorithm used for solving mixed-integer programming programs as defined in Definition 3.1. We illustrate the behavior of this algorithm on the simple two-dimensional PIP example (3.1).

For simplicity, we repeat and generalize the description of the optimization program MIP and its linear relaxation LR. Let

$$Z(V) \;=\; \min_{(x,y)} \{ \; cx + fy \; : \; (x,y) \in V \}$$

be the optimal value of the optimization problem defined over the feasible set V and with objective function $cx + fy$. By convention, we write $Z(V) = +\infty$ when the feasible set V is empty. Program MIP and its optimal value $Z(X)$ depend on the feasible set X, where

$$X \;=\; \{ \; (x,y) \in \mathbb{R}_+^n \times \mathbb{Z}_+^p \; : \; Ax + By \geq b \; \} \, .$$

Similarly, the linear program LR and its optimal value $Z(P_X)$ are defined by the feasible set P_X, where

$$P_X \;=\; \{ \; (x,y) \in \mathbb{R}_+^n \times \mathbb{R}_+^p \; : \; Ax + By \geq b \; \} \, .$$

We also recall our basic observation regarding the lower and upper bounds.

$$Z(P_X) \;\leq\; Z(X) \;\leq\; \bar{Z},$$

where $\bar{Z}$ is the objective value of any feasible solution found.

3.3.1 The Enumeration Principle

We describe first the general *"divide-and-conquer" principle* of the branch-and-bound algorithm for solving MIP.

(i) The initial lower bound on $Z(X)$ is provided by the optimal value $Z(P_X)$ of the linear relaxation LR. This program is easy to solve (in the complexity sense and in practice) because it is a linear program. Let $(x^\star, y^\star)$ be an optimal solution to LR.

   Assumption. We assume here without loss of generality that LR is bounded. Otherwise, when $Z(P_X) = -\infty$, problem MIP is either unbounded too, or infeasible. To distinguish between these two cases, it suffices to impose arbitrarily large bounds on the variables to obtain a bounded MIP, and to run the branch-and-bound algorithm on this modified problem. If it produces an optimal solution, then the original problem was unbounded, and otherwise infeasible.

Observation. Solving MIP can be rephrased as finding the best (with respect to the objective function) solution (x, y) in the set P_X with $y \in \mathbb{Z}^p$.

Principle. We try to solve MIP by solving a sequence of linear programs.

(ii) If $y^\star \in \mathbb{Z}^p$, then it is feasible for MIP as $(x^\star, y^\star) \in X$ and it also provides an upper bound on $Z(X)$. Therefore $(x^\star, y^\star)$ is an optimal solution to MIP, because the lower and upper bounds are equal ($cx^\star + fy^\star = Z(P_X) \le Z(X) \le \bar{Z} = cx^\star + fy^\star$ implies $Z(X) = cx^\star + fy^\star$).

(iii) Otherwise, $y^* \notin \mathbb{Z}^p$ and the solution $(x^\star, y^\star)$ is not feasible for MIP. We try to eliminate this useless solution from LR by adding linear constraints so as to keep a linear program.

  Let y_j with $j \in \{1, \ldots, p\}$ be some variable taking a fractional (non-integral) value $y_j^\star$ in the solution $(x^\star, y^\star)$ to LR.

Observation. In any feasible solution $(x, y) \in X$, we must have either $y_j \le \lfloor y_j^\star \rfloor$ or $y_j \ge \lceil y_j^\star \rceil$, where $\lfloor y_j^\star \rfloor$ and $\lceil y_j^\star \rceil$ denote the value of $y_j^\star$ rounded down and up to the nearest integer respectively.

  For instance, with $y_j^\star = \frac{32}{9}$, we must have either $y_j \le 3 = \lfloor \frac{32}{9} \rfloor$ or $y_j \ge 4 = \lceil \frac{32}{9} \rceil$ for all $(x, y) \in X$.

Branching Step. To eliminate the solution $(x^\star, y^\star)$, as well as all solutions with $\lfloor y_j^\star \rfloor < y_j < \lceil y_j^\star \rceil$, we replace the set P_X by the union of two disjoint sets P_X^0 and P_X^1, where

$$P_X^0 = P_X \cap \{(x, y) \in \mathbb{R}_+^n \times \mathbb{R}_+^p \ : \ y_j \le \lfloor y_j^\star \rfloor \} \quad \text{and}$$

$$P_X^1 = P_X \cap \{(x, y) \in \mathbb{R}_+^n \times \mathbb{R}_+^p \ : \ y_j \ge \lceil y_j^\star \rceil \} \, .$$

The variable y_j is called the *branching variable*, and the constraints $y_j \le \lfloor y_j^\star \rfloor$ and $y_j \ge \lceil y_j^\star \rceil$ are called the *branching constraints*.

Observation. We can now replace the search for the best integer solution in P_X by the search for the best integer solution in $P_X^0 \cup P_X^1$. Unfortunately, the price to pay in order to keep linear programs is that we have replaced a single linear program by two linear programs defined over two disjoint sets.

Example. The initial decomposition of P_X into P_X^0 and P_X^1 by branching on variable y_1 is illustrated in Figure 3.2 for the MIP example (3.1). The point $a = (\frac{32}{9}, \frac{101}{36})$ is the optimal fractional solution (x^*, y^*) of the linear relaxation (LR).

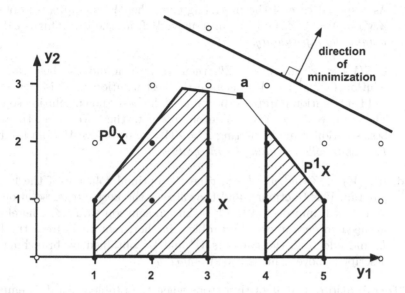

Figure 3.2. The MIP program (3.1) and its decomposition in the branch-and-bound algorithm.

(iv) We look now for the best integer solution lying in one of the formulations on the list $L = \{P_X^0, P_X^1\}$. We can continue the decomposition approach in the same way. This requires us to analyze separately each formulation in the list L.

Main iteration. We are given a list L of formulations, and the value $\bar{Z}$ of the best integer solution found so far. As long as no feasible solution is known, we set $\bar{Z} = +\infty$.

Selection and Solution Step. We select one formulation V from the list L, and solve the corresponding linear program (LP) to obtain $Z(V)$ and an optimal solution (x^V, y^V). This value $Z(V)$ is a lower bound on the value of the best integer solution in the set V.

Pruning Step. As in the first iteration detailed above, several cases may arise in examining the set V.

a. If $Z(V) \geq \bar{Z}$, then the best solution in V cannot be strictly better than $\bar{Z}$, because $Z(V)$ is a lower bound on the value of the best solution is V. Therefore we do not need to consider the integer solutions in V, and we simply remove V from the list L. This is called *pruning by bound*.

b. As a special case of the preceding one, when V is empty, we obtain $Z(V) = +\infty \geq \bar{Z}$, and we can remove V from the list. This is called *pruning by infeasibility*.

c. If $Z(V) < \bar{Z}$ and $y^V \in \mathbb{Z}^p$, then we have found the best integer solution (x^V, y^V) in V (because the best solution in V is integral), and this solution improves the value of the best known solution so far. Therefore we do not need to decompose V further. We record the new best solution value by setting $\bar{Z} = Z(V)$, and remove V from the list L. This is called *pruning by integrality*.

d. If $Z(V) < \bar{Z}$ and $y^V \notin \mathbb{Z}^p$, then the optimal solution of the linear program V is fractional, and the value of the best integer solution in V may still improve on the best known solution value $\bar{Z}$. Therefore, we need to decompose the problem further, remove V from the list L, and add to L the two sets V^0 and V^1 obtained by branching as described above. This is called *branching*.

(v) **Termination.** The algorithm stops when the problem list L is empty. This is guaranteed to occur in finitely many steps, if the integer variables y are bounded. However, the number of formulations to consider in the list L can grow exponentially with the number p of integer variables.

Running Time. Theoretically, the branch-and-bound algorithm requires a number of iterations that is exponential in the number of integer variables (p). Each iteration consists of the solution and treatment (pruning or branching) of one linear program from the list L, which can be carried out in polynomial time if an appropriate interior point algorithm is used.

3.3.2 The Branch-and-Bound Algorithm

We now summarize the branch-and-bound algorithm.

Branch – and – Bound

---

1. *Initialization*

    $L = \{P_X\}$

    $\bar{Z} := +\infty$

    Assume that LR is bounded $(Z(P_X) > -\infty)$

2. *Termination*

    If $L = \emptyset$ Then

    { If $\bar{Z} = +\infty$ Then $X = \emptyset$ (*infeasible* problem)

    If $\bar{Z} < +\infty$ Then the solution $(x, y) \in X$ with

    $$\bar{Z} = cx + fy \text{ is } optimal$$

    STOP

    }

3. *Node Selection and Solution*

    Select $V \in L$ and let $L := L \setminus \{V\}$

    Compute the optimal LP-value $Z(V)$ and solution (x^V, y^V) of V

4. *Pruning*

    If $Z(V) \geq \bar{Z}$ Then GO TO 2. (V is either *infeasible* or *dominated*

    by the best solution (upper bound) found so far)

    If $Z(V) < \bar{Z}$ Then

    { If $y_j^V \in \mathbb{Z}$ (i.e., is integral) for all $j = 1, \ldots, p$ Then

    { (a *better feasible solution* (upper bound) is found)

    Update the upper bound by setting $\bar{Z} := Z(V)$

    Update the list L by removing dominated programs

    (for each $W \in L$: If $Z(W) \geq \bar{Z}$, then $L := L \setminus \{W\}$)

    GO TO 2.

    }

    }

5. *Branching*

    (occurs only when program V has not been pruned)

    (i.e., when $Z(V) < \bar{Z}$ and $y_j^V \notin \mathbb{Z}$ for some $j \in \{1, \ldots, p\}$)

    Select j for which $y_j^V \notin \mathbb{Z}$ (y_j is the *branching variable*)

    Update list L by adding programs with restricted y_j values

    Set $L := L \cup \{V^0, V^1\}$ where

    $$V^0 = V \cap \{(x, y) \in \mathbb{R}_+^n \times \mathbb{R}_+^p \;:\; y_j \leq \lfloor y_j^V \rfloor \}$$

    $$V^1 = V \cap \{(x, y) \in \mathbb{R}_+^n \times \mathbb{R}_+^p \;:\; y_j \geq \lceil y_j^V \rceil \}$$

    ($y_j \leq \lfloor y_j^V \rfloor$ and $y_j \geq \lceil y_j^V \rceil$ are the *branching constraints*)

    GO TO 2.

It is standard to represent the sequence of problems or sets obtained as elements of the list L during the branch-and-bound algorithm by an *enumeration tree*. The first *node* of the tree, called the *root node*, always represents the initial linear relaxation (LR). Then, the two sets V^0 and V^1 obtained by branching from a set V are represented as child nodes of the node V.

A Branch-and-Bound Example

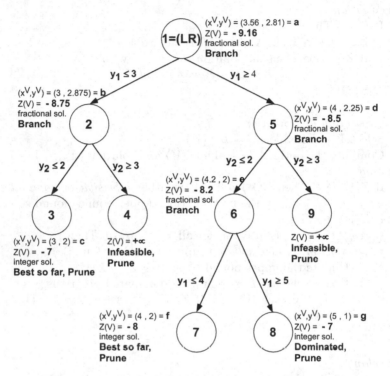

Figure 3.3. The branch-and-bound tree for Example (3.1).

The branch-and-bound enumeration tree corresponding to the simple example (3.1) is represented in Figure 3.3. In the figure:

- The nodes are numbered according to the selection order in Step 3.
- The linear program corresponding to node n is defined by the initial linear relaxation (LR) (node 1 or root node) augmented by the branching constraints on the path from the root node to node n.
- The solutions (x^V, y^V) found in Step 3 are illustrated in Figure 3.4 by letters from a to g in the graphical representation of program (3.1).
- The pruning status in Step 4 is indicated next to the nodes (not pruning means going to branching in Step 5).
- The branching variables and constraints in Step 5 are indicated on the branches (arcs) of the tree.

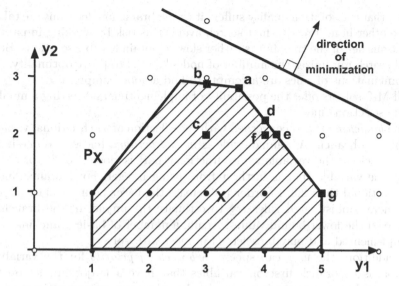

Figure 3.4. The MIP program (3.1) and points (x^V, y^V) found in Step 3 of the branch-and-bound algorithm.

3.3.3 Node Selection and Branching Rules

Finally, note that to implement a branch-and-bound algorithm, one needs to specify a node selection rule and a branching rule.

The *node selection rule* specifies which linear set from the list L to select at each iteration of Step 3. This rule has an impact on the order in which the nodes are treated, and therefore on the evolution of the lower and upper bounds during the execution of the algorithm. Some standard rules are:

- *Depth-first search* in which one selects a child of the preceding node after branching, and backtracks (i.e., moves back in the enumeration tree to select the next node) as few nodes up as possible after pruning a node;
- *Breadth-first search* in which one selects nodes that are closest to the top of the tree;
- *Best-bound search* in which one selects the node with the best (lowest) lower bound;
- *Combinations* of these, such as
  - breadth-first search for a certain number of nodes followed by depth-first search, or
  - depth-first search as long as branching is performed, and then best-bound after pruning a node.

The choice of the rule is often heuristic. Depth-first search allows one to obtain solutions quickly (because there are more branching constraints deep in the tree, and such bound constraints help to obtain integer solutions), but with the risk that the solution quality suffers if wrong branching decisions are taken. On the other hand breadth-first search avoids this risk by working in parallel on all branches of the tree, but is rather slow to obtain feasible solutions. Best-bound search minimizes the number of nodes treated to prove optimality, and the combinations of rules are attempts to find a good compromise.

All MIP solvers offer the possibility of implementing and testing a number of node selection rules.

The *branching rule* is concerned with the selection of the fractional variable on which to branch. A common rule is *most fractional* (closest to one-half). Here one selects the variable whose fractional part is closest to 0.5, or equivalently the variable that is farthest from being integer. For instance, 3.6 is more fractional than 2.3. In order to increase the lower bound as fast as possible, more sophisticated rules try to estimate the impact of the branching variable on the lower bound values. Again, a number of basic branching rules are implemented and may be selected in most MIP solvers.

In addition, the user can specify a *branching priority* for the variables. The goal is to branch first on variables that have a major impact on the solution quality. For instance, in production planning, it is sometimes useful to branch first on the variables corresponding to the initial periods of the planning horizon (if they take fractional values).

3.3.4 Solution Quality and Duality Gap

For large-size industrial applications, the branch-and-bound tree is often truncated because solving the problem to optimality takes too much computing time. In such cases, the node selection and branching rules also have an important impact on the quality of the solution obtained. The quality of the solution is usually measured by the so-called duality gap.

Definition 3.9 *When the branch-and-bound enumeration is unfinished, the duality gap measures the maximum relative deviation from optimality of the best feasible solution found. It is defined as*

$$Duality\ Gap = \frac{Best\ UB - Best\ LB}{Best\ UB} \times 100\ [\%]\ ,$$

where Best LB is the minimum lower bound among all outstanding problems or nodes in the list L (i.e., Best LB = $\min_{V \in L} Z(V)$), and Best UB is the value of the best solution found, when the enumeration is stopped.

Note that in the solution reports from the Xpress-MP Optimizer, the duality gap is computed relative to the best lower bound (dividing by *Best LB*) instead of the best upper bound, and is therefore larger.

3.4 Reformulation

The branch-and-bound algorithm is a general-purpose algorithm able in theory to solve any MIP as described in Definition 3.1. However, in practice the number of nodes may grow exponentially with the number p of integer variables, and the computing time may become too large for the production planning instances typically encountered in industrial applications.

3.4.1 The Quality of a Formulation

Good and Bad Formulations

However, there is still hope to solve large size instances, based partly on the following observation.

Observation 3.3 *The number of branch-and-bound nodes needed to solve a MIP depends heavily on the formulation used. It is therefore crucial to identify a good formulation.*

The importance of the initial formulation is now illustrated with our two-dimensional PIP example. We have already seen in Figure 3.3 that nine nodes were needed to complete the branch-and-bound enumeration using the initial formulation given by

$$X = P_X \cap \mathbb{Z}_+^2 \text{ and } P_X = \{y = (y_1, y_2) \subset \mathbb{R}_+^2 \ : \begin{array}{rcr} y_1 & \geq & 1 \\ -y_1 & \geq & -5 \\ -y_1 -0.8y_2 & \geq & -5.8 \\ y_1 -0.8y_2 & \geq & 0.2 \\ -y_1 \ -8y_2 & \geq & -26 \ \}. \end{array}$$

On the other hand, we can check in Figure 3.1 that the following inequalities are satisfied by all feasible points $(x, y) \in X$. Such inequalities are called valid inequalities

$$\begin{array}{rcr} -y_1 -y_2 & \geq & -6 \ , \\ y_1 -y_2 & \geq & 0 \ , \\ -y_2 & \geq & -2 \ . \end{array}$$

By adding all these valid inequalities to the initial formulation, one gets the formulation and linear relaxation represented in Figure 3.5. Observe that this reformulation does not change the set of feasible solutions X (because we have added valid inequalities only), but the linear formulation is now a smaller set.

Using this alternative and tightened formulation of X, one can solve the same PIP problem, using the same branch-and-bound algorithm, at the root node, as a pure linear program. This holds because the optimal solution of the new linear relaxation is an integral solution, represented by point f in Figure 3.5, and therefore no branching is required.

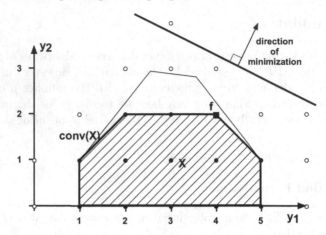

Figure 3.5. The MIP program (3.1) with an improved formulation.

Tight or Convex Hull Formulation

Moreover, this new formulation enables us to solve this PIP as a pure LP (i.e., without any branching step) for any linear objective function because our new linear relaxation coincides with the *convex hull* of the points in X.

Without giving a formal definition before Part II, the convex hull of X, denoted conv(X), is a formulation (i.e., its feasible set is defined by a set of linear inequalities), and it is characterized by the fact that all its extreme points (vertices in Figure 3.5) belong to X. In this sense, the convex hull formulation is the smallest or tightest valid formulation or valid linear programming relaxation for X. It is the best formulation (i.e., conv(X) $\subseteq P_X$ for any formulation P_X of X).

As all extreme points of conv(X) belong to X, all extreme optimal solutions to the linear relaxation belong to X, whatever the direction of minimization. Therefore it suffices to solve the linear relaxation to get an extreme optimal solution to LR, which is also an optimal solution to MIP.

This can be generalized to any MIP with rational data.

Observation 3.4 *Any MIP problem can be solved as a pure linear program by using as formulation the description by linear inequalities (or polyhedral description) of the convex hull of its feasible set X. Such a formulation is also called a* tight *formulation. A feasible set, such as* conv(X), *described by linear inequalities is called a* polyhedron.

In practice, using this reformulation approach (replacing the formulation by the polyhedral description of conv(X)) encounters two major difficulties.

1. The polyhedral description of conv(X) is not known, and finding this description is, in general, at least as hard (in the complexity sense) as

solving problem MIP. Thus, we have just moved the challenge from solving a hard MIP to finding the adequate formulation as an LP.

2. For most MIP instances, the polyhedral description of conv(X) involves a number of linear inequalities that is exponential in the number of variables. Solving this linear relaxation may not be possible, or may require a very long computing time.

Therefore, one cannot expect to use this ideal or tight formulation (conv(X)) in a direct reformulation and solution approach. We postpone the discussion of the second difficulty to Section 3.5 devoted to the branch-and-cut algorithm.

3.4.2 Valid Inequalities

Adding valid inequalities or constraints a priori to the initial formulation, that is, before the optimization starts, is a first level of improvement. The objective is

- to obtain a tightened formulation (tightened linear relaxation, i.e., closer to conv(X)),
- in order to improve (increase) the lower bounds provided by the linear relaxations solved at each node, and so
- to reduce the number of branch-and-bound nodes needed to solve the MIP program, and finally
- to reduce the total branch-and-bound computing time.

Definition 3.10 *A valid inequality (VI) for the feasible set X of MIP,*

$$X = \{ (x,y) \in \mathbb{R}^n_+ \times \mathbb{Z}^p_+ : Ax + By \geq b \},$$

is a constraint or inequality $\alpha x + \beta y \geq \gamma$ (with $\alpha \in \mathbb{R}^n$, $\beta \in \mathbb{R}^p$ and $\gamma \in \mathbb{R}$) satisfied by all points in X; that is,

$$\alpha x^\star + \beta y^\star \geq \gamma \text{ for all } (x^\star, y^\star) \in X.$$

Of course, we are not interested in any VI, but only in the VIs tightening the initial formulation, that is, eliminating part of the initial formulation or linear relaxation. Otherwise the reformulation will have no impact on the number of branch-and-bound nodes.

For instance, in the PIP represented in Figure 3.5, the inequality $y_1 \leq 6$ is valid, but it does not tighten the formulation. On the contrary, the valid inequality $y_1 + y_2 \leq 6$ is very useful because it is one of the inequalities defining conv(X).

Definition 3.11 *A facet-defining valid inequality for X is a valid inequality that is necessary in a description of the polyhedron conv(X).*

As we have already observed, facet-defining VIs are strong inequalities in the sense that they must be part of the formulation if one wishes to solve MIP (optimize over X for all objective functions $cx + fy$) without any branching.

Unfortunately, the polyhedron $\text{conv}(X)$ and its facet-defining VIs are generally unknown. Therefore one performs a *partial reformulation*, where only a subset of the (facet-defining) valid inequalities defining $\text{conv}(X)$ are added to the formulation. Moreover, the VIs added are not systematically facet-defining for $\text{conv}(X)$, because it is too difficult or time-consuming to check, or simply unknown.

As a consequence, we expect such a partial reformulation to reduce the number of nodes enumerated, but not to eliminate branching completely.

Analyzing Relaxations to Find Valid Inequalities

There are several ways to identify VIs, or classes of VIs, for a given formulation or feasible set X. It is beyond the scope of this introduction to explain methods of identifying classes of VIs. Nevertheless, we illustrate the general approach in order to help the reader to use the available reformulation results and software adequately (which is our main objective in Part I), to understand our reformulation and decomposition approach, and to prepare the transition to Part II.

Usually, finding facet-defining valid inequalities for $\text{conv}(X)$ requires one to take into account the complete structure of the set X, which is itself a very hard problem.

Therefore, one uses a simpler approach which is to identify facet-defining VIs for various relaxations of X, where a *relaxation Y of X* is any superset of X (i.e., $X \subset Y$). The following observation derives directly from the notion of relaxation and the definition of valid inequality.

Observation 3.5 *Any valid inequality for a relaxation Y of X is also a valid inequality for X.*

One typical *low-level relaxation Y* consists of a single constraint from the inequalities defining X and the bounds on the variables. This clearly defines a relaxation of X.

For instance, the initial inequality $y_1 + 0.8y_2 \leq 5.8$ and the bounds $1 \leq y_1 \leq 5$, $0 \leq y_2 \leq 3$ from the PIP defined in (3.1) and represented in Figure 3.5, define the relaxation

$$Y = \{(y_1, y_2) \in \mathbb{Z}_+^2 \ : \ y_1 + 0.8y_2 \leq 5.8$$
$$1 \leq y_1 \leq 5$$
$$0 \leq y_2 \leq 3 \qquad \},$$

where the bound $y_2 \leq 3$ is derived directly from the initial constraint $y_1 + 8y_2 \leq 26$ and the nonnegativity of y_1 ($26 \geq y_1 + 8y_2 \geq 8y_2$, which implies $y_2 \leq \lfloor \frac{26}{8} \rfloor = 3$ as y_2 must be an integer). Now, it is easy to verify that the inequality $y_1 + y_2 \leq 6$ is valid for Y, and thus for X. This inequality is facet-defining for conv(Y) and also by chance for conv(X).

The low-level relaxation approach can also be used to derive the valid inequality $y_1 - y_2 \geq 0$ from the initial inequality $y_1 - 0.8y_2 \geq 0.2$ and the trivial bounds on y_1 and y_2.

Such *valid inequalities based on low-level relaxations* or structures are derived automatically by the most advanced MIP solvers, that is, without any prior knowledge of the structure of the model other than the matrix of constraint coefficients. Examples of such inequalities are the the Knapsack, the Mixed Integer Rounding (MIR), and the Flow Cover VIs that we study in Part II. They are all based on single constraints and simple or variable bounds on the variables.

It is sometimes possible to derive valid inequalities and facets for more global or *high-level relaxations* Y. This is of interest because more global relaxations should lead to better or stronger reformulations of conv(X).

For instance, the single-item uncapacitated lot-sizing model (LS-U) is a high-level structure or relaxation present in many production planning problems. This relaxation is defined by the demand satisfaction and set-up forcing (variable upper bound) constraints described in Chapter 1.

Using the same approach, one can derive valid inequalities and facets for conv(X^{LS-U}) in order to improve the formulation of production planning models.

We illustrate the reformulation results known for model LS-U in Sections 4.1.1 and 4.1.2. It is the specific objective of Chapter 4 to classify the high-level relaxations encountered frequently in production planning models, and their known reformulation results that can be exploited in branch-and-cut algorithms.

The automatic reformulation approach implemented in MIP solvers, as well as the reformulations based on higher-level relaxations lead generally to a very large number of valid inequalities. This is the reason why they are sometimes added as cuts in a branch-and-cut approach (see Section 3.5), rather than a priori to the initial formulation.

3.4.3 A Priori Reformulation

Suppose that the analysis of some relaxations of the feasible set X yields a set or a family of valid inequalities. The inequalities can be added a priori to the formulation (i.e., before the optimization starts), and used directly in the branch-and-bound algorithm.

Let us first introduce some new concepts and notation. We are given a MIP to solve over the feasible set X, defined by the formulation P_X.

$$P_X = \{ (x,y) \in \mathbb{R}^n_+ \times \mathbb{R}^p_+ : Ax + By \geq b \}$$

and
$$X = P_X \cap (\mathbb{R}^n \times \mathbb{Z}^p) \ .$$

We have already seen that several formulations or reformulations exist for the same feasible set X. For example, we have described three alternative formulations of the set X in our small two-dimensional example defined in (3.1).

The usual way to define valid or correct reformulations of X consists of adding to the formulation a set or a family $\mathcal{C}$ of valid inequalities for X. For simplicity of notation we assume that the family $\mathcal{C}$ of valid inequalities for X is given explicitly by a list of constraints

$$\mathcal{C} = \{\alpha^j x + \beta^j y \geq \gamma^j \text{ for all } j = 1, \cdots, |\mathcal{C}| \ \} \ .$$

Definition 3.12 *The reformulation of X by the family $\mathcal{C}$ of valid inequalities is the formulation defined by the initial constraints $Ax + By \geq b$ and the valid inequalities in $\mathcal{C}$. It does not modify the set of feasible solutions X,*

$$
\begin{aligned}
X &= \{ (x, y) \in \mathbb{R}_+^n \times \mathbb{Z}_+^p : Ax + By \geq c \} \\
&= \{ (x, y) \in \mathbb{R}_+^n \times \mathbb{Z}_+^p : Ax + By \geq c \\
& \qquad\qquad\qquad \alpha^j x + \beta^j y \geq \gamma^j \text{ for all } j = 1, \cdots, |\mathcal{C}| \ \} \ ,
\end{aligned}
$$

but the reformulation allows one to obtain a tighter formulation $\tilde{P}_X$, where

$$\tilde{P}_X = P_X \cap C \subseteq P_X$$

and C is the set of points satisfying all valid inequalities in the family $\mathcal{C}$,

$$C = \{ (x, y) \in \mathbb{R}^n \times \mathbb{R}^p : \alpha^j x + \beta^j y \geq \gamma^j \text{ for all } j = 1, \cdots, |\mathcal{C}| \ \}.$$

The *a priori reformulation approach* consists of using the reformulation and its linear relaxation $\tilde{P}_X$, instead of P_X, in the branch-and-bound algorithm. It is practical only if the number $|\mathcal{C}|$ of inequalities added is not too large.

A Priori Reformulation Example

As an example of this approach, we come back to our two-dimensional pure integer program defined in (3.1).

To illustrate the a priori reformulation approach, we suppose that we have found two valid inequalities by analyzing the single constraint relaxations of X. They define a partial description of conv(X):

$$C = \left\{ (y_1, y_2) \in \mathbb{R}_+^2 : \begin{array}{r} -y_1 - y_2 \geq -6 \\ y_1 - y_2 \geq 0 \end{array} \right\} \ .$$

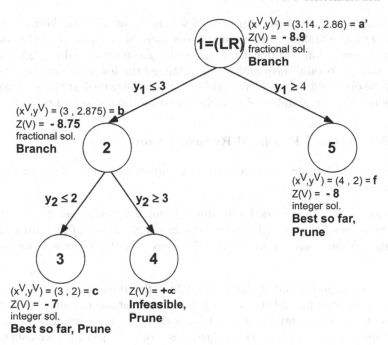

Figure 3.6. The branch-and-bound tree for the a priori reformulation $\tilde{P}_X$ of Example (3.1).

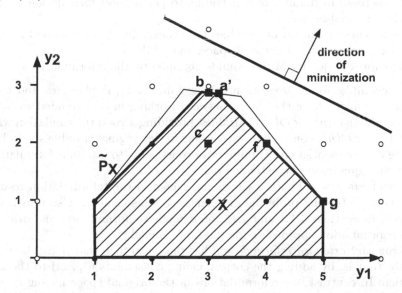

Figure 3.7. The formulation $\tilde{P}_X$ of Example (3.1) and points (x^V, y^V) found in Step 3 of the branch-and-bound algorithm.

Adding these inequalities a priori and solving by branch-and-bound allows one to obtain the branch-and-bound tree represented in Figure 3.6. The corresponding formulation and linear relaxation $\tilde{P}_X$ are represented in Figure 3.7. Adding these two valid inequalities has tightened the linear relaxation, which can also be observed by the increased lower bound obtained at the root node, and has reduced the number of branch-and-bound nodes from 9 to 5.

3.4.4 A Priori and Extended Reformulation

Another way to obtain an a priori reformulation is based on an extended reformulation.

Definition 3.13 *An* extended reformulation *for the feasible set X of a MIP is a formulation (defined by linear constraints in the sense of Definition 3.3) involving new (and usually more) variables (see Chapter 6 for a formal definition).*

By using the additional modeling flexibility offered by an extended space of variables, it is often possible to obtain good reformulations of (some relaxation of) conv(X) requiring far fewer constraints than good reformulations in the original space of variables. Of course, there is a price to pay for increasing the number of variables, and one has to find the best compromise among

- the increase in number of constraints to get a good formulation in the original variable space,
- the increase in number of variables (and constraints) to get a good formulation in an extended variable space, and
- the improved quality of the bounds obtained by the reformulations.

For example, for a program involving initially $O(n)$ variables and constraints, we might have the choice between working in an extended variable space involving $O(n^2)$ or $O(n^3)$ variables, providing a good reformulation with only $O(n^2)$ or $O(n^3)$ constraints, or working in the original variable space, but needing an exponential $O(2^n)$ number of constraints to obtain a formulation of the same quality as the extended one.

Several examples of such a choice are shown in the reformulation results for standard lot-sizing problems in Chapter 4. This choice is also illustrated explicitly in Sections 4.1.1 and 4.1.2 on extended reformulation and original space reformulation approaches for the *LS-U* model.

A *compact extended reformulation* has the advantage that it can be used directly, that is, by adding the corresponding constraints a priori to the formulation. In contrast, for reformulations in the original space having an exponential number of inequalities, the indirect cutting plane or branch-and-cut approach described in Section 3.5 has to be used.

3.5 Branch-and-Cut Algorithm

Motivation

Once valid inequalities have been derived to improve a formulation, we often face the second difficulty mentioned for tight reformulations. In many cases, the number of valid inequalities identified by analyzing X (or its relaxations) grows exponentially with the number of variables. This occurs particularly when the reformulation is carried out in the original space of variables.

In such a situation, it is not practical to add all the valid inequalities a priori in the formulation. We look thus for another way to benefit from the improved formulation (tighter formulation $\tilde{P}_X$, increased lower bounds, reduced number of branch-and-bound nodes), while keeping the time to solve the linear relaxation reasonable.

The following observation gives us some indication on how to solve the linear relaxation defined by the reformulation $\tilde{P}_X$ efficiently.

Observation 3.6 *For a given MIP, let $\tilde{P}_X$ be the reformulation obtained by adding an exponential number of valid inequalities to the initial formulation P_X. For a fixed objective function $cx + fy$, only a limited number (equal to at most $n + p$, the number of variables) of these valid inequalities is necessary to solve the linear relaxation over the set $\tilde{P}_X$ to optimality.*

This observation holds because, when solving the linear relaxation over $\tilde{P}_X$, the valid inequalities that are inactive at the optimal solution can be dropped from the formulation, and the optimal solution (a vertex of the polyhedron $\tilde{P}_X$ in the geometrical interpretation) is obtained as the intersection of at most $n + p$ active constraints, where $n + p$ is the dimension of the variable space.

For instance, in the pure integer program represented in Figure 3.5, the optimal solution f of the linear relaxation defined over $\tilde{P}_X = \text{conv}(X)$ is the intersection of two facet-defining inequalities for $\text{conv}(X)$, where $n + p = 2$ is the dimension of the variable space and is independent of the number of facets describing $\text{conv}(X)$.

3.5.1 Separation Algorithm

Observation 3.6 suggests also the idea of the cutting plane algorithm and separation algorithm used to solve the linear program over the set $\tilde{P}_X$. Instead of adding initially all valid inequalities from $\mathcal{C}$ to the formulation (in which case most of these exponentially many inequalities will be inactive and useless), the initial formulation P_X is used and the linear relaxation LR is first solved. Then the VIs from $\mathcal{C}$ are added only when they are needed, that is, only when they are not satisfied at the optimal solution of LR.

The separation problem for valid inequality family $\mathcal{C}$ and a given point (x, y) is the problem of deciding whether $(x, y) \in C$, that is, satisfies the constraints in $\mathcal{C}$.

Definition 3.14 *Given a feasible set X of a MIP, and a family C of valid inequalities for X, the* separation problem $SEP((x^\star, y^\star)|C)$ *for a given* $(x^\star, y^\star) \in P_X$ *is*

- *either to prove that $(x^\star, y^\star) \in C$, that is, to prove that $(x^\star, y^\star)$ satisfies all valid inequalities from C,*
- *or to find a valid inequality $(\alpha^j x + \beta^j y \geq \gamma^j) \in C$ that is violated (not satisfied) at $(x^\star, y^\star)$, that is, such that $\alpha^j x^\star + \beta^j y^\star < \gamma^j$.*

In other words, solving the separation problem for a given point $(x^\star, y^\star)$ means finding an inequality from C that is violated at $(x^\star, y^\star)$, or proving that no such inequality exists.

An algorithm for solving the separation problem is called a *separation algorithm*. It is of course often challenging to find an efficient separation algorithm for a family involving exponentially many valid inequalities. This is an important step in assessing the complexity of a cutting plane algorithm.

The separation algorithm is called *exact* if it guarantees to find a violated inequality in the family C when one exists. Otherwise the separation algorithm is called *heuristic*. Heuristic separation algorithms are often used to speed up the solution of the separation problem.

3.5.2 Cutting Plane Algorithm

First, we recall the notation used to represent the optimal solution values of the optimization problem MIP and its tightened linear relaxation $\tilde{P}_X = P_X \cap C$. Let

$$Z(\tilde{P}_X) = Z(P_X \cap C) \;=\; \min_{(x,y)} \{\, cx + fy \,:\, (x,y) \in P_X \cap C\}$$

be the optimal value of the optimization problem with objective function $cx + fy$ defined over the points in the sets P_X and C, where C is the set of points satisfying the valid inequalities in C. As before, $Z(P_X)$ denotes the optimal value of the linear relaxation without reformulation, that is, without the valid inequalities from C.

The tightened reformulation improves the lower bound value (i.e., $Z(\tilde{P}_X) \geq Z(P_X)$), but the optimal value of program MIP is unchanged (i.e., $Z(X \cap C) = Z(X)$) because the inequalities in C are valid for X.

The *cutting plane algorithm* solves the improved linear relaxation over $\tilde{P}_X$ and computes $Z(\tilde{P}_X)$, without adding a priori the valid inequalities in C, but by calling instead the separation algorithm repeatedly to generate the valid inequalities needed from the family C.

We describe directly the cutting plane algorithm in a more general setting. In the branch-and-cut algorithm, the cutting plane algorithm will be called for each formulation V (defined by the initial formulation P_X plus branching constraints) in the list L; that is, it is called at each node, to compute the improved lower bound $Z(V \cap C) \geq Z(V)$, where

$$Z(V \cap C) = \min_{(x,y)} \{ cx + fy : (x,y) \in V \cap C \}.$$

Note that $Z(P_X \cap C)$ will be the lower bound computed at the root node.

So, let V be a formulation with $V \subseteq P_X$, that is, a subset of P_X defined by linear constraints (P_X and branching constraints). The cutting plane algorithm computes $Z(V \cap C)$.

Cutting Plane Algorithm to compute $Z(V \cap C)$

a.   Initialize $W := V$, where W will contain the final formulation at the end of the cutting plane algorithm
Compute $Z(W)$ the optimal LP-value without VIs from C
Let $(x^\star, y^\star)$ be an optimal LP-solution

b.   Solve the separation problem $SEP((x^\star, y^\star)|C)$

Either all inequalities from C are satisfied at $(x^\star, y^\star)$ Then
$\{\ (x^\star, y^\star) \in V \cap C$ is an optimal solution with value $Z(V \cap C)$
    STOP
$\}$

Or SEP returns some violated inequality $(\alpha^j x + \beta^j y \geq \gamma^j)$ Then
$\{$ Add $(\alpha^j x + \beta^j y \geq \gamma^j)$ to the formulation:
        $W := W \cap \{(x,y) \in \mathbb{R}^n \times \mathbb{R}^p \mid \alpha^j x + \beta^j y \geq \gamma^j\}$
    Compute the optimal LP-value $Z(W)$
    Let $(x^\star, y^\star)$ be an optimal LP-solution
    GO TO b.
$\}$

During this algorithm, we always have $V \cap C \subseteq W \subseteq V$ because W is initialized as the set V and is updated only by adding valid inequalities from C. Therefore, when the optimal solution $(x^\star, y^\star)$ over W belongs to C, it is also the optimal solution over the set $V \cap C$. Finally, the algorithm terminates because $|C|$ is finite. Hence, the algorithm is correct.

The constraints returned by the separation problem are called *cuts* or *cutting planes* because they are hyperplanes used to eliminate or cut off the current solution $(x^\star, y^\star)$.

By Observation 3.6, we can hope that the separation problem does not need to be called too many times before the cutting plane algorithm terminates. Experience shows that only a fraction of the total exponential number of valid inequalities in C is indeed added. So, we have replaced the solution of a large or huge LP, by a (it is hoped, short) sequence of smaller LPs.

In practice, the total running time of the cutting plane algorithm is reduced by generating and adding several cuts at each iteration. This is called the *multiple cut* variant.

3.5.3 Branch-and-Cut Algorithm

Putting all the pieces together, the *branch-and-cut algorithm* is

- the branch-and-bound algorithm,
- applied to a reformulation defined by a family C of valid inequalities (i.e., each linear set V solved during branch-and-bound is tightened and replaced by $V \cap C$), and
- where the solution of the linear program to compute $Z(V \cap C)$ at Step 3 of branch-and-bound is performed by a cutting plane algorithm.

The *cut-and-branch* variant is the branch-and-cut algorithm in which the cutting plane algorithm is only called at the root node (first iteration of Step 3) to compute $Z(P_X \cap C)$. So it consists of the computation of $Z(P_X \cap C)$ at the root node using a cutting plane algorithm, followed by pure branch-and-bound applied to the resulting formulation (initial formulation P_X plus the cuts added at the root node). Compared to branch-and-cut, this variant allows one to save processing time for the nodes in the enumeration tree, but works with a weaker formulation.

In order to save some computing time in the separation algorithm, it is usual to store the cuts added to a linear relaxation feasible set V for the successor nodes. This is because cuts active at one node are likely to be active at some later node. This is done either by keeping the cuts in the formulation (i.e., keeping W instead of V at the end of the cutting plane algorithm in Step 3, and branching on the set W in Step 5 of branch-and-bound), or by storing the cuts in a *"cut pool"* containing a list of existing cuts. Then, before calling the regular separation algorithm, one starts by checking if one of the inequalities in the cut pool is violated. This is called *cut pool separation*.

The complete branch-and-cut algorithm, in which all the cuts generated are kept in the formulation, is summarized next.

Branch-and-Cut for a family C of valid inequalities
1. *Initialization* $\qquad L = \{P_X\}$ $\qquad \bar{Z} := +\infty$ $\qquad$ Assume that LR is bounded $(Z(P_X) > -\infty)$ 2. *Termination* $\qquad$ If $L = \emptyset$ Then $\qquad$ { If $\bar{Z} = +\infty$ Then $X = \emptyset$ (*infeasible* problem) $\qquad\quad$ If $\bar{Z} < +\infty$ Then the solution $(x, y) \in X$ with $\qquad\qquad\qquad\qquad \bar{Z} = cx + fy$ is *optimal* $\qquad$ STOP $\qquad$ }

Branch-and-Cut for a family $\mathcal{C}$ of valid inequalities (continued)

3. *Node Selection and Solution with the Cutting Plane Algorithm*
   Select $V \in L$ and let $L := L \setminus \{V\}$
   3a. Compute $Z(V)$ and an optimal *LP*-solution (x^V, y^V) over V
   3b. Solve the separation problem $SEP((x^V, y^V)|\mathcal{C})$
       Either all inequalities from $\mathcal{C}$ are satisfied at (x^V, y^V) Then
       $\{ (x^V, y^V) \in V \cap C$ is an optimal solution with value $Z(V \cap C)$
           GO TO 4.
       $\}$
       Or SEP returns a violated inequality $(\alpha^j x + \beta^j y \geq \gamma^j)$ Then
       $\{$ Add $(\alpha^j x + \beta^j y \geq \gamma^j)$ to the formulation:
           $$V := V \cap \{(x,y) \in \mathbb{R}^n_+ \times \mathbb{R}^p_+ \mid \alpha^j x + \beta^j y \geq \gamma^j\}$$
           Compute the optimal LP-value $Z(V)$ and solution (x^V, y^V)
           Let $(x^\star, y^\star)$ be an optimal LP-solution
           GO TO 3b.
       $\}$
4. *Pruning*
   If $Z(V) \geq \bar{Z}$ Then GO TO 2. (V is either *infeasible* or *dominated*
                                        by the best solution (upper bound) found so far)
   If $Z(V) < \bar{Z}$ Then
   $\{$ If $y_j^V \in \mathbb{Z}$ (i.e., is integral) for all $j = 1, \ldots, p$ Then
       $\{$ (a *better feasible solution* (upper bound) is found)
       Update upper bound by setting $\bar{Z} := Z(V)$
       Update list L by removing dominated programs
                 (for each $W \in L$: If $Z(W) \geq \bar{Z}$ Then $L := L \setminus \{W\}$)
       GO TO 2.
       $\}$
   $\}$
5. *Branching*
   (occurs only when program V has not been pruned)
   (i.e., when $Z(V) < \bar{Z}$ and $y_j^V \notin \mathbb{Z}$ for some $j \in \{1, \cdots, p\}$)
   Select j for which $y_j^V \notin \mathbb{Z}$ (y_j is the *branching variable*)
   Update list L by adding programs with restricted y_j values
       Set $L := L \cup \{V^0, V^1\}$ where
       $V^0 = V \cap \{(x,y) \in \mathbb{R}^n_+ \times \mathbb{R}^p_+ \mid y_j \leq \lfloor y_j^V \rfloor \}$
       $V^1 = V \cap \{(x,y) \in \mathbb{R}^n_+ \times \mathbb{R}^p_+ \mid y_j \geq \lceil y_j^V \rceil \}$
           ($y_j \leq \lfloor y_j^V \rfloor$ and $y_j \geq \lceil y_j^V \rceil$ are the *branching constraints*)
   GO TO 2.

Note finally that the branch-and-cut algorithm for solving program MIP is an alternative to the a priori reformulation approach. If it takes as input the same reformulation by the family $\mathcal{C}$ of valid inequalities, if the separation algorithm is exact and if the same node selection and branching rules are used, then the branch-and-bound algorithm will enumerate the same sequence of nodes with both the a priori reformulation and the branch-and-cut approaches.

Therefore, the choice between these two approaches depends mainly on the running time required to solve each node of the enumeration tree.

Cutting Plane and Branch-and-Cut Example

For the two-dimensional pure integer program defined in (3.1), the branch-and-cut enumeration tree based on exact separation of the two valid inequalities

$$C = \left\{ \begin{array}{l} -y_1 - y_2 \geq -6 \\ y_1 - y_2 \geq 0 \end{array} \right\}$$

is exactly the same as the tree obtained with a priori reformulation. This tree is given in Figure 3.6. The difference with a priori reformulation is in the solution of the linear relaxations at each node of the tree.

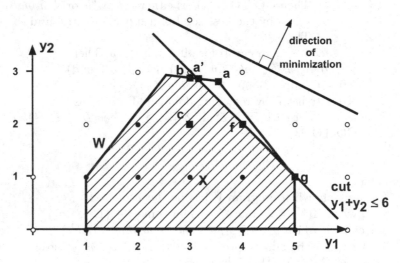

Figure 3.8. The MIP program (3.1) and the cut generated in the cutting plane algorithm.

The cutting plane algorithm at the root node will execute the following steps:

- First solve the initial linear relaxation (with $W = P_X$) to obtain point a in Figure 3.8;
- Solve the separation problem $SEP(a|C)$ and identify the violated inequality or cut $y_1 + y_2 \leq 6$;

- Update W by adding this cut, and resolve the linear relaxation over W to obtain point a' in Figure 3.8;
- Solve the separation problem $SEP(a'|\mathcal{C})$ and prove (because no cut is found in the family $\mathcal{C}$) that a' is the optimal solution of the improved linear relaxation $\tilde{P}_X$.

The shaded feasible set W in Figure 3.8 represents the set of solutions satisfying the initial constraints with the single cut added. The cut added has cut off the triangle aga' from the linear relaxation set P_X. The set W is the feasible set of the linear program solved at the root node, and branching constraints are added to this root node formulation. It is easy to check that the other constraint from $\mathcal{C}$ is never violated and will not be generated at the other nodes of the branch-and-cut enumeration.

We have described or surveyed here the generic algorithms, and their algorithmic components, used to solve mixed integer programs. We specialize this optimization approach in the next chapter to the solution of multi-item production planning models. In particular, we describe how these models can be decomposed into simpler single-item planning problems, and how to derive, use, and integrate a priori reformulation, valid inequalities, and separation algorithms for this class of models.

3.6 Primal Heuristics – Finding Feasible Solutions

So far we have worked mainly on improving the formulation of a specific problem instance. There the main goal was to improve the quality of the linear programming lower bound, but there was also the hope that the y variables in the solution of the linear program would be closer to being integral so that finding a good or optimal feasible integer solution would be easier.

Here we look directly at ways to find (it is hoped) good feasible integer solutions. These may be useful when the branch-and-cut algorithm is too slow, or takes a long time to find good feasible solutions. Most of the simple heuristics we present can be implemented at each node of the branch-and-cut tree, but for simplicity we just consider that we are at the root node in the branch-and-cut tree. Thus we have the optimal LP solution $(\hat{x}, \hat{y})$ obtained using the initial formulation P_X or some improved formulation $\tilde{P}_X$, and possibly the best known feasible solution $(\bar{x}, \bar{y})$.

Typically there are two types of heuristics: construction heuristics that produce a feasible solution from scratch, and improvement heuristics that try to improve a given feasible solution $(\bar{x}, \bar{y})$. We assume here that $y \in \{0, 1\}^p$ for simplicity.

Truncated MIP

The first heuristic is the trivial heuristic MIP consisting of running a branch-and-bound or branch-and-cut algorithm for a fixed amount of time. The best solution obtained when the time limit is reached is the MIP heuristic solution. This heuristic is thus both a construction and an improvement heuristic.

Diving

A next class of heuristics is known as *diving*, and diving is in fact a strategy for carrying out depth-first search in the branch-and-cut tree. At each node all the y variables that take value 0 or 1 in the linear programming solution are frozen at that value, and we need to create one branch by fixing one of the y variables that is fractional to an integer value.

LP-driven dives traditionally use the latest LP solution $(\hat{x}, \hat{y})$, and fix a variable that is closest to integer. Namely, with $F = \{j : \hat{y}_j \notin \mathbb{Z}^1\}$, they find $g_k = \min_{j \in F} g_j$, where $g_j = min[\hat{y}_j, 1 - \hat{y}_j]$ for $j \in F$, and set $y_k = 0$ if $\hat{y}_k \leq 0.5$ and $y_k = 1$ otherwise.

IP-driven or guided dives use the incumbent solution $(\bar{x}, \bar{y})$. Having chosen the variable y_k to be fixed next, the branching direction is fixed by setting $y_k = \bar{y}_k$, that is, equal to its value in the incumbent solution.

LP-driven diving is a construction heuristic, whereas guided diving is an improvement heuristic. We do not emphasize such heuristics here because they are typically implemented efficiently within the MIP systems, and their implementation requires some minimal knowledge about how the tree is constructed, and so on. However, we already see that either the LP or the incumbent solution is usually the guide.

Below we concentrate on heuristics that just require us to solve one, two, or a small number of modified LP or MIP versions of the original MIP

$$\min\{cx + fy \ : \ Ax + By \geq b, \ x \in \mathbb{R}_+^n, \ y \in \{0,1\}^p\}.$$

The modifications consist of either the fixing of some variables, the addition of a constraint, or the relaxation of some integrality constraints. They are chosen in such a way that the modified LPs or MIPs are easier to solve than the original MIP, and lead to a heuristic solution of the original MIP guided by the information contained either in the linear relaxation solution or in the incumbent solution, or both.

3.6.1 Construction Heuristics

We describe two procedures for building an initial solution for MIP.

Let $Q = \{1, \ldots, p\}$ be the index set of the y variables.

LP-and-Fix or Cut-and-Fix

This is a very simple heuristic closely related to diving. We just fix whatever is integral in the LP solution $(\hat{x}, \hat{y})$, and solve the resulting MIP^{LP-FIX}

(MIP^{LP-FIX}) $\min\{cx + fy:\ Ax + By \geq b,\ x \in \mathbb{R}^n_+,\ y \in \{0,1\}^p,$

$$y_j = \hat{y}_j \text{ for all } j \in Q \text{ with } \hat{y}_j \in \{0,1\} \ \}.$$

Either the problem is infeasible, and the heuristic has failed, or else it provides an LP-and-fix heuristic solution.

In general, this heuristic produces better solutions when a tighter formulation is used, and when the corresponding LP solution has fewer fractional y variables. The cut-and-fix variant is the same heuristic applied to a formulation that has been tightened either with cuts or with an extended reformulation.

Relax-and-Fix

Here we suppose that the 0–1 variables y can be partitioned into R disjoint sets $Q^1, \ldots, Q^R$ of decreasing importance. We also generalize a little by choosing subsets U^r with $U^r \subseteq \cup_{u=r+1}^{R} Q^u$ for $r = 1, \ldots, R-1$. We then solve sequentially R MIPs, denoted MIP^r with $1 \leq r \leq R$, to find a heuristic solution to the original MIP.

For instance, in a production planning problem, Q^1 might be all the y variables associated with time periods in $\{1, \ldots, t_1\}$, Q^2 those associated with periods in $\{t_1 + 1, \ldots, t_2\}$, and so on, whereas U^1 would be the y variables associated with the periods in some set $\{t_1 + 1, \ldots, u_1\}$, and so on.

In the first MIP^1, we only impose the integrality of the important variables in $Q^1 \cup U^1$ and relax the integrality restriction on all the other variables in Q.

(MIP^1) $\min$ $\{cx + fy:$ $Ax + By \geq b$

$$x \in \mathbb{R}^n_+$$

$$y_j \in \{0,1\} \qquad \text{for all } j \in Q^1 \cup U^1$$

$$y_j \in [0,1] \qquad \text{for all } j \in Q \setminus (Q^1 \cup U^1) \ \}.$$

Let (x^1, y^1) be an optimal solution of MIP^1. Then we fix the variables in Q^1 at their values in y^1, and move to MIP^2.

In the subsequent MIP^r, for $2 \leq r \leq R$, we additionally fix the values of the y variables with index in Q^{r-1} at their optimal values from MIP^{r-1}, and add the integrality restriction for the variables in $Q^r \cup U^r$.

(MIP^r) $\min$ $\{$ $cx + fy:$

$$Ax + By \geq b$$

$$x \in \mathbb{R}^n_+$$

$$y_j = y_j^{r-1} \in \{0,1\} \quad \text{for all } j \in Q^1 \cup \cdots \cup Q^{r-1}$$

$$y_j \in \{0,1\} \qquad\qquad \text{for all } j \in Q^r \cup U^r$$

$$y_j \in [0,1] \qquad\qquad \text{for all } j \in Q \setminus (Q^1 \cup \cdots \cup Q^r \cup U^r) \ \}.$$

Let (x^r, y^r) be an optimal solution of MIP^r for $2 \leq r \leq R$.

Either MIP^r is infeasible for some $r \in \{1, \ldots, R\}$, and the heuristic has failed, or else (x^R, y^R) is a relax-and-fix heuristic solution of the original MIP.

As an example to illustrate the notation, consider a production planning problem defined over a planning horizon of 20 time periods. Using the current formulation, assume that it is only possible to obtain good solutions in reasonable computing time for problems with up to about 10 time periods. Therefore a relax-and-fix heuristic is used with the following sets Q^r and U^r, with $R = 4$.

- Q^1 contains all the y variables associated with periods in $\{1, \ldots, 5\}$.
- $Q^2 = U^1$ contains all the y variables associated with periods in $\{6, \ldots, 10\}$.
- $Q^3 = U^2$ contains all the y variables associated with periods in $\{11, \ldots, 15\}$.
- $Q^4 = U^3$ contains all the y variables associated with periods in $\{16, \ldots, 20\}$.

We describe now the iterations of the relax-and-fix heuristic.

- In the first MIP^1, the y variables associated with periods 1 up to 10 (i.e., in $Q^1 \cup U^1$) are restricted to be integer, the other y variables being relaxed.
- From the solution of MIP^1, we fix the y variables corresponding to periods 1 to 5 (i.e., Q^1), and solve MIP^2 where the y variables associated to periods 6 to 15 (i.e., in $Q^2 \cup U^2$) are now integer, and y variables associated with periods 16 to 20 (i.e., $Q \setminus (Q^1 \cup Q^2 \cup U^2)$) are relaxed.
- From the solution of MIP^2, we additionally fix the y variables corresponding to periods 6 to 10 (i.e., Q^2), and solve MIP^3 where the y variables associated with periods 11 to 20 (i.e., in $Q^3 \cup U^3$) are now integer, and there are no remaining variables to be relaxed because $Q \setminus (Q^1 \cup Q^2 \cup Q^3 \cup U^3) = \emptyset$.
- The optimal solution to MIP^3 – if any – gives the relax-and-fix heuristic solution, because of our choice of U^3. In this case, there is no need to proceed to MIP^4.

The status of the variables over the iterations is summarized in Table 3.2, where we assume that the integer variables of the production planning model are the binary set-up variables y_t^i for each item i in each period t.

Table 3.2. Iterations of the Relax-and-Fix Heuristic

Iteration	MIP	Fixed Variables $y_t^i = (y_t^i)^{r-1}$ for all i and	Binary Variables $y_t^i \in \{0,1\}$ for all i and	Relaxed Variables $0 \leq y_t^i \leq 1$ for all i and	Solution
$r = 1$	MIP^1	-	$1 \leq t \leq 10$	$11 \leq t \leq 20$	(x^1, y^1)
$r = 2$	MIP^2	$1 \leq t \leq 5$	$6 \leq t \leq 15$	$16 \leq t \leq 20$	(x^2, y^2)
$r = 3$	MIP^3	$1 \leq t \leq 10$	$11 \leq t \leq 20$	-	(x^3, y^3)

The basic idea of the relax-and-fix heuristic is clear in this example. At each iteration, we solve a MIP problem involving ten periods of binary variables, and to avoid being too myopic we then only fix the values of the variables corresponding to the first five of these periods. Thus the sets U^r allow us to smooth the heuristic solution by creating some overlap between the successive planning intervals.

Observe that the optimal objective value of MIP^1 provides a valid lower bound on the optimal value of MIP, because MIP^1 is a relaxation of MIP. This does not hold for the subsequent iterations r, with $r \geq 2$, because some integer variables have been heuristically fixed.

Finally, note that there are many variants of relax-and-fix, also called *time decomposition* or *time partitioning*, where some problem-specific relaxations or approximations are used to define the problems MIP^r.

For instance, in our production planning example, assume that the problem involves 200 time periods. The first relaxation MIP^1 could be constructed using the same sets Q^1 and U^1 as above, but with the difference that all variables and constraints for periods after period 20 are completely ignored. This defines a smaller size relaxation $\overline{MIP^1}$ than the original MIP^1, and is thus easier and faster to solve. Then $\overline{MIP^1}$ is solved, and the x and y variables corresponding to periods $1, \ldots, 5$ become fixed at their optimal values obtained in $\overline{MIP^1}$. Next, problem $\overline{MIP^2}$ covering periods 6 to 25 is solved (with binary set-up variables in periods $6, \ldots, 15$, and linearly relaxed set-up variables in periods $16, \ldots, 25$), the solution obtained for periods $6, \ldots, 10$ is fixed, and so on. In this way, all problems $\overline{MIP^r}$ are defined over a planning horizon involving only 20 time periods.

However, this procedure may fail to produce feasible solutions because of capacity restrictions in later periods. If this occurs, one possibility is to add lower bound constraints on the final inventory level (period 20 in $\overline{MIP^1}$) that guarantee feasibility for the complete problem without eliminating any feasible solutions. However such bounds can only be calculated fast for relatively simple problems without set-up times. Otherwise one can guess such bounds, in which case one is working with approximations instead of relaxations, and the optimal value of $\overline{MIP^1}$ no longer provides a valid lower bound on the optimal value of MIP.

Finally observe that in certain cases relax-and-fix heuristics can also be based on decomposition by machine, product family, or geographical location.

3.6.2 Improvement Heuristics

First, we consider two recent approaches. The information available is the linear programming solution at the top node $(\hat{x}, \hat{y})$, and the best known feasible solution $(\bar{x}, \bar{y})$. In both cases the idea is to use MIP to explore some promising neighborhood for a limited amount of time. Then if a better (or even worse) feasible solution is found, the step can be iterated.

Relaxation Induced Neighborhood Search (RINS)

The idea here is to explore the neighborhood between the LP solution $(\hat{x}, \hat{y})$ and the IP solution $(\bar{x}, \bar{y})$. If a y_j variable has the same value in both solutions, that value is fixed. Thus we solve the MIP^{RINS}

$$(MIP^{RINS}) \quad \min \ \{cx + fy : \ Ax + By \geq b$$
$$x \in \mathbb{R}^n_+, \ y \in \{0,1\}^p$$
$$y_j = \bar{y}_j \text{ for all } j \in Q \text{ with } \bar{y}_j = \hat{y}_j \ \}.$$

Either MIP^{RINS} is infeasible or does not find a feasible solution in the allotted time, so the heuristic has failed, or else the best solution found is a relaxation induced neighborhood search or RINS heuristic solution.

This heuristic can be seen as the improvement version of the cut-and-fix construction heuristic.

Local Branching (LB)

Here the neighborhood is just constructed using the integer solution. An integer k is chosen, and the neighborhood consists of those y vectors that do not differ from $\bar{y}$ in more than k coordinates. So the MIP^{LB} to be solved is

$$(MIP^{LB}) \quad \min \ \{cx + fy : \ Ax + By \geq b$$
$$x \in \mathbb{R}^n_+, \ y \in \{0,1\}^p$$
$$\sum_{j \in Q: \bar{y}_j = 0} y_j + \sum_{j \in Q: \bar{y}_j = 1} (1 - y_j) \leq k \ \}.$$

Either MIP^{LB} is infeasible or does not find a feasible solution in the allotted time, so the heuristic has failed, or else the best solution found is a local branching heuristic solution.

It is easy to see that there are many possible variants of these heuristics. One may just consider a subset of the variables in Q, or one may distinguish in importance between variables at value 0, and variables at value 1, and so on. There are also some obvious ways to combine the heuristics.

Observation 3.7 *Fixing constraints $y_j = \bar{y}_j$ for $j \in Q^*$, and for some $Q^* \subseteq Q$ (as in relax-and-fix or in RINS), can be represented using a single local branching constraint*

$$\sum_{j \in Q^*: \bar{y}_j = 0} y_j + \sum_{j \in Q^*: \bar{y}_j = 1} (1 - y_j) \leq 0.$$

This immediately suggests the possibility of generalizing (and relaxing) the relax-and-fix or RINS fixing constraints by using constraints of the form

$$\sum_{j \in Q^* : \bar{y}_j = 0} y_j + \sum_{j \in Q^* : \bar{y}_j = 1} (1 - y_j) \le k,$$

with some value of $k \ge 1$.

Exchange (EXCH)

Finally, we describe briefly an improvement version of the relax-and-fix heuristic, called exchange.

We keep the same decomposition of integer variables in sets Q^r and U^r, with $1 \le r \le R$. At each step r with $1 \le r \le R$, all integer variables are fixed at their value in the best solution $(\bar{x}, \bar{y})$ found so far (or in the last solution encountered), except the variables in the set Q^r (or $Q^r \cup U^r$) which are restricted to take integer values. So the problem $MIP^{EXCH,r}$ solved at step r is defined by

$$(MIP^{EXCH,r}) \quad \min \{cx + fy : \quad Ax + By \ge b$$
$$x \in \mathbb{R}^n_+$$
$$y_j = \bar{y}_j \qquad \text{for all } j \in Q \setminus Q^r$$
$$y_j \in \{0, 1\} \qquad \text{for all } j \in Q^r \qquad \}.$$

Then, if a better solution is found, this exchange procedure can be repeated. Note that the different steps r with $1 \le r \le R$ are independent of one another, and any subset of steps can be performed in any order.

These construction and improvement heuristics are illustrated in some applications and case studies described in this book. In particular, they are applied at the end of Chapter 4 to the GW–GSCO master production scheduling example from Section 1.2.

Notes

We refer to Wolsey [193] for a general introduction to integer programming models and techniques.

Section 3.6 For heuristic procedures using the relax-and-fix idea, see Stadtler [154] for an application to multi-level lot sizing with set-up times and capacity constraints, and Federgrün and Tzur [64] and Federgrün et al. [62] for problems with family set-up variables, among others. Local Branching has been proposed in Fischetti and Lodi [66]. The Relaxation Induced Neighborhood Search and Guided Diving procedures have been described in Danna et al. [52]. Several of these improvement heuristics are implemented in state-of-the-art MIP solvers.

4

Classification and Reformulation

Motivation

Given the diversity of planning functions in the supply chain planning matrix described in Chapter 2, and given the diversity of supply chains (each supply chain can be characterized by a combination of functional and structural attributes, implying a huge diversity in planning requirements; see Section 2.3), a single advanced planning system or a single monolithic mathematical programming planning model cannot represent all planning problems.

Therefore, in parallel to the supply chain typology, our approach for the construction of planning models is to decompose and classify them based on their main attributes: decisions, objectives, and constraints. This building block approach and classification helps us and allows us first to construct a model and an initial mathematical formulation for the planning problem to be addressed.

Beyond modeling, there is a second and major motivation for this classification. Most real-life production planning problems are complex because they involve many products and many resources, such as machines, storage facilities, and plants, and many restrictions have to be satisfied by acceptable production plans. This results in mixed integer programs of large size that are usually very difficult to solve.

In Chapter 3 we have surveyed the state-of-the-art generic branch-and-bound and branch-and-cut algorithms based on a priori reformulation, valid inequalities, and separation. In the literature many reformulation results are known and described for canonical production planning models, such as single-item and/or single-resource problems, which are much simpler than the complex real-life problems.

In order to be able to incorporate these reformulation results in specialized branch-and-bound/cut algorithms for solving production planning models, it is crucial to be able to identify which results to use, which requires one to identify which canonical submodels are present in a model. The classification scheme presented here pursues exactly this goal.

Objective

In the context of a decomposition approach, it is the specific objective of this chapter to

- describe and classify the *canonical production models* frequently occurring as relaxations or sub-models in real life production planning problems,
- identify and classify the *reformulation results* that are known for these canonical models in order to design efficient branch-and-bound/cut algorithms for solving practical production planning models.

For complex planning problems, the objective is also to present and illustrate the effectiveness of a systematic reformulation procedure allowing us to take advantage – through the classification scheme – of reformulation results for standard single-item subproblems to obtain improved formulations and to design branch-and-cut optimization algorithms.

In Chapter 5 we then demonstrate how to use the classification scheme and the reformulation procedure in practice, with appropriate software tools.

The (more) technical and detailed presentation and derivation of the reformulation results listed here, as well as some additional reformulation results and techniques useful for more complex models (but requiring a less automatic approach) are given in Parts II to IV, and illustrated in Part V.

Contents

Step by step:

- In Section 4.1 we illustrate on the *LS-U* (uncapacitated lot-sizing) production planning model the use and impact of reformulations on the performance of the branch-and-bound/cut algorithm, namely the effect of using the extended (or compact linear) reformulation technique and the cutting plane reformulation technique defined in Chapter 3.
- In Section 4.2 we describe the *decomposition approach* used to reformulate and solve complex planning models involving many items and resources, starting from available reformulations for simpler (i.e., single-item, single-machine) planning models.
- In Section 4.3 we describe our *classification scheme* for canonical single item production planning models in the form of a three-field identifier *PROB-CAP-VAR* for each model, and by giving a conceptual or verbal description as well as an initial mathematical formulation for each model.

- In Section 4.4 we describe a *systematic reformulation procedure* relying on tables of extended and cutting plane reformulation results for the most common single-item production planning models, including *LS-U*.
- In Section 4.5 we put together these ideas to illustrate the use and effectiveness of the systematic reformulation procedure on the Master Production Scheduling example from Section 1.2.

4.1 Using Reformulations for Lot-Sizing Models

In earlier chapters we have presented and illustrated the modeling and optimization approach, as well as the generic branch-and-bound and branch-and-cut algorithms used to solve the resulting models.

Model *LS-U* is the simplest high-level relaxation occurring in most production planning models. So, finding good reformulations for *LS-U* is an important first step. Here we use this model to illustrate the type of reformulation results available for canonical single-item planning models, namely a priori reformulations and cutting planes with separation.

The approach is illustrated on the first *LS-U* example described in Section 1.1. For simplicity, we recall here the initial formulation of this *LS-U* instance characterized by

- the demand satisfaction constraint for the single product *bike* over eight consecutive time periods,
- the variable upper bound constraint for the single product *bike* over eight consecutive time periods, and
- the initial inventory of product *bike*.

$$\min \quad cost := \sum_{t=1}^{NT}(p\ x_t + q\ y_t) + \sum_{t=1}^{NT-1} h\ s_t \tag{4.1}$$

$$dem\_sat_t := \quad s_{t-1} + x_t = d_t + s_t \qquad \text{for } 1 \le t \le NT \tag{4.2}$$

$$s_0 = s\_ini, \ s_{NT} = 0 \tag{4.3}$$

$$vub_t := \quad x_t \le (\sum_{k=t}^{NT} d_k)y_t \qquad \text{for } 1 \le t \le NT \tag{4.4}$$

$$x_t, s_t \in \mathbb{R}_+, \ y_t \in \{0,1\} \qquad \text{for } 1 \le t \le NT, \tag{4.5}$$

where the variables are x_t for production, s_t for inventory, and y_t for setup in period t, and the data are $NT = 8$, $p = 100$, $q = 5000$, $h = 5$, $d = [400, 400, 800, 800, 1200, 1200, 1200, 1200]$ and $s\_ini = 200$. This formulation is $O(NT) \times O(NT)$; that is, it involves on the order of NT constraints and NT variables (see Section 3.2), where NT represents the number of time periods in the planning horizon.

The performance of the branch-and-bound algorithm (using the default Xpress-MP optimizer, but without preprocessing and without using the cuts generated by the solver) on this initial formulation (4.1)–(4.5) is reported in Table 4.1.

Table 4.1. B&B Solution of the *LS-U* Example from Section 1.1

Formulation	LP Val.	CPLP Val.	OPT Val.
Size	Vars	CPLP Time	OPT Time
Algorithm	Cons	CPLP Cuts	OPT Nodes
(4.1)–(4.5)	712,189	–	736,000
$O(NT) \times O(NT)$	24	–	0
B & B	16	–	29

In Table 4.1, "Vars" and "Cons" represent the number of variables and constraints in the formulation, "LP Val." is the value of the initial linear relaxation of the formulation. "CPLP Val.", "CPLP Time" and "CPLP Cuts" give, respectively, the value of the lower bound at the root node after the addition of cutting planes, the cutting plane time at the root node, and the number of cuts in the formulation at the end of the root node. "CPLP" values are only reported for branch-and-cut algorithms (the Xpress-MP cuts are not used in this toy example). "OPT Val.", "OPT Time" and "OPT Nodes" are the value of the optimal solution, the total run-time and total number of nodes in the enumeration tree. Times are given in seconds, rounded to the nearest integer.

In our analysis, we concentrate on the lower bound value at the root node and on the total number of nodes in the enumeration tree. Both indicators measure the quality of the formulation used. The run-time (rounded to 0 second) and the gap (always 0 when an optimal solution is found) do not give much information in this tiny example.

Observation 4.1 *There are 29 branch-and-bound nodes with the initial formulation. Observe that this formulation is already using some tightening for the variable upper bound constraint (4.4). If this constraint is replaced by the simpler but usual big-M type constraint $x_t \leq My_t$, with $M = 10,000$, that is, if we do not introduce the tightest upper bound on x_t in (4.4), then the LP lower bound at the root node ("LP Val.") is reduced to 703,500 and the number of nodes needed to solve the model to optimality increases to 51 nodes. So, some straightforward a priori formulation tightening is already included in the initial model.*

4.1.1 Using A Priori Extended Reformulations

As explained in Section 3.4, we look now for tight reformulations of *LS-U*, or tight reformulations of high-level relaxations of *LS-U*.

LS-U is polynomially solvable as it can be solved by dynamic programming. Given the complexity equivalence between optimization and separation discussed in Part II, it is natural to look for a compact (i.e., polynomial in the number of variables and constraints) linear reformulation for LS-U. As an example, we describe and test here a well-known extended reformulation for LS-U.

Multi-Commodity Extended Reformulation

A classical way to tighten the formulation of fixed charge network flow problems is to decompose the flow along each arc of the network as a function of its destination. This defines a so-called multi-commodity formulation by assigning a different commodity to each destination node. The decomposition by commodity allows one to tighten the formulation by decreasing the upper bounds in the variable upper bound constraints, which is important as illustrated in Observation 4.1.

We have already given the network flow interpretation of LS-U in Figure 1.2 in Section 1.1. So we can apply the multi-commodity idea, and decompose the flow (production) x_t as a function of its destination node (demand period) t, $t + 1$, ..., NT. Similarly, we can decompose the flow (inventory) s_t as a function of its destination node (demand period) $t + 1$, $t + 2$, ..., NT.

So, we consider as one specific commodity the demand to be satisfied in each time period, and do not mix the commodities. Commodity t corresponds to the demand delivered in period t. We define the new variables x_{it} $(i \leq t)$ as the production in period i of commodity t, and the new variable s_{it} $(i < t)$ as the inventory at the end of period i of commodity t.

In this reformulation, we further constrain the initial inventory s_ini to be consumed in the first period; that is, $s_{01} = s\_ini$ and $s_{0t} = 0$ for $t = 2, \ldots, NT$. This can be done without loss of optimality, because $s\_ini \leq d_1$ and there is always an optimal solution where earlier production is delivered first (this is called FIFO [= First In First Out] or FPFD [= First Produced First Delivered] ordering).

Also, the variables s_{tt}, for $t = 1, \ldots, NT$, do not exist because commodity t must be delivered in period t, therefore no inventory of commodity t may exist at the end of period t.

If needed, for instance, in case of positive minimal stock at the end of the planning horizon, an additional commodity can be created to correspond to the end horizon inventory. This is not necessary here, as we assume that the stock at the end of the horizon is zero.

The flow conservation constraint obtained for commodity $t = 5$ is illustrated in Figure 4.1.

By modeling separately the demand satisfaction (flow conservation) for each commodity, the LS-U model (4.1)–(4.5) can be reformulated as

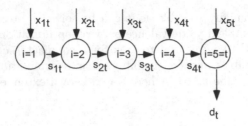

Figure 4.1. The flow conservation global constraint for commodity $t = 5$.

$$\min \quad cost := \sum_{i=1}^{NT} \sum_{t=i}^{NT} (p\, x_{it} + h\, s_{it}) \; + \; \sum_{i=1}^{NT} q\, y_i \tag{4.6}$$

$$
\begin{aligned}
dem\_sat_{it} := \quad & s_{i-1,t} + x_{it} = \delta_{it} d_t + s_{it} && \text{for } 1 \le i \le t \le NT && (4.7)\\
& s_{01} = s\_ini, \; s_{0t} = 0 && \text{for } 2 \le t \le NT && (4.8)\\
& s_{tt} = 0 && \text{for } 1 \le t \le NT && (4.9)\\
vub_{it} := \quad & x_{it} \le \hat{d}_t y_i && \text{for } 1 \le i \le t \le NT && (4.10)\\
& s_{it}, x_{it} \in \mathbb{R}_+, \; y_i \in \{0,1\} && \text{for } 1 \le i \le t \le NT, && (4.11)
\end{aligned}
$$

where the notation δ_{it} denotes 1 if $i = t$, and 0 otherwise. Constraint (4.7) is the flow conservation constraint of commodity t in all periods $i = 1, \ldots, t$, where the only period i with a demand for commodity t is $i = t$. Constraints (4.8) and (4.9) impose that there is no initial and no final inventory (end of period t) of commodity t, except the initial inventory of commodity 1. Constraint (4.10) forces the set-up variable y_i to be 1 when there is production for commodity t in period i. Using the decomposition of the flow, the tightest upper bound on x_{it} is $\hat{d}_t$, where $\hat{d}_1 = d_1 - s\_ini$ and $\hat{d}_t = d_t$ for $t > 1$. Constraint (4.11) imposes the nonnegativity and binary restrictions on the variables. Finally, Constraint (4.6) expresses the cost of the production plan.

This extended reformulation does not contain the initial variables and constraints. But it is nevertheless a valid reformulation of *LS-U* in the sense of Definitions 3.3 and 3.13. This can observed because an equivalent reformulation to (4.6)–(4.11) would be obtained by adding constraints to define the initial variables as a function of the new decomposed variables (i.e., $x_i = \sum_{t=i}^{NT} x_{it}$ and $s_i = \sum_{t=i+1}^{NT} s_{it}$ for all i), and by keeping the original objective function (4.1). The reformulation has then the same feasible solutions in the original (x, s, y) space as the original model.

Testing the Multi-Commodity Extended Reformulation

Reformulation (4.6)–(4.11) is of size $O(NT^2) \times O(NT^2)$. We can now test the effectiveness of the decomposition by commodity and tightening of the variable upper-bound constraints. The performance of the branch-and-bound algorithm (again using the default Xpress-MP optimizer, but without preprocessing and without the cuts generated by the solver) on the initial formulation and on the multi-commodity reformulation are compared in Table 4.2.

Table 4.2. B&B Solution of the *LS-U* Example from Section 1.1, Comparison of Initial and Multi-Commodity Formulations

Formulation Size Algorithm	LP Val. Vars Cons	CPLP Val. CPLP Time CPLP Cuts	OPT Val. OPT Time OPT Nodes
(4.1)–(4.5) $O(NT) \times O(NT)$ B & B	712189 24 16	– – –	736,000 0 29
(4.6)–(4.11) $O(NT^2) \times O(NT^2)$ B & B	736,000 72 72	– – –	736,000 0 1

We observe in Table 4.2 that the multi-commodity reformulation solves *LS-U* without any branching. The LP value is the optimal value, and one node suffices. The following theorem shows that this is not chance. The multi-commodity reformulation solves all instances of *LS-U* without branching.

Theorem 4.1 *The linear relaxation of formulation ((4.6)–(4.11)) always has an optimal solution with y integer, and solves LS-U. In other words, formulation (4.7)–(4.11) is a tight extended formulation of the convex hull of feasible solutions to LS-U. This is also called a complete linear description of LS-U.*

The multi-commodity extended reformulation has been given here to illustrate the type of results one can obtain with reformulations. Other extended reformulations giving a complete linear description of *LS-U* are known, as well as extended reformulations for canonical models other than *LS-U*. Pointers to these extended formulations are defined in our systematic reformulation procedure in Section 4.4.

4.1.2 Using Cutting Planes

We have just described the multi-commodity reformulation for the single-item model *LS-U* with NT periods. Although this reformulation is as tight as possible, it has $O(NT^2)$ constraints and $O(NT^2)$ variables and a model with 32 time periods has over a thousand variables and a thousand constraints. This may be too large a reformulation if it has to be applied to all items in a large-size multi-item production planning model (see Section 4.2).

One way to overcome this difficulty is to look for a complete linear description of model *LS-U* in the initial variable space involving only $O(NT)$

variables. And if this complete linear description needs an exponential (in NT) number of constraints, we can use the cutting plane and separation approach described in Section 3.5 to avoid adding all these constraints a priori.

A Class of Valid Inequalities

A first class of valid inequalities can be easily identified from the fractional solution of the linear relaxation of the initial formulation. Figure 4.2 represents the optimal solution of the linear relaxation of (4.1)–(4.5), where missing arcs correspond to arcs with zero flow.

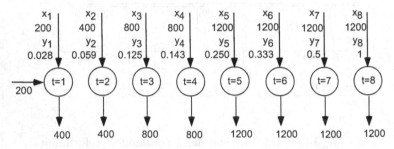

Figure 4.2. The solution of the linear relaxation of (4.1)–(4.5).

To eliminate or cut off this fractional solution we have to look at periods in which the corresponding y variable is fractional. Consider period 2 with $x_2 = 400$ but $y_2 = 0.059$. This value of y_2 is minimized because of the objective function, therefore y_2 is exactly the minimal value allowed by the set-up forcing constraint (4.4); that is, $y_2 \geq (=)\frac{x_2}{d_{2,8}}$ where the notation $d_{\alpha\beta}$ denotes $\sum_{i=\alpha}^{\beta} d_i$.

Observe that this reasoning also applies to y_8, but y_8 takes the value 1 because $y_8 \geq (=)x_8/d_{8,8} = 1$.

So, if period 2 was the last period of the horizon, we could also write $x_2 \leq d_2 y_2$ as set-up forcing constraint, and y_2 would take the value 1 with the current value of x_2.

But d_2 is a valid upper bound on x_2 only if period 2 is the last period or if there is no stock at the end of period 2 (i.e., $s_2 = 0$ and period 2 is separated from the later periods as if period 2 were the last one). Hence a valid upper bound on x_2 is $x_2 \leq d_2 + s_2$.

Therefore, a logical implication is

$$s_2 = 0 \;\Rightarrow\; x_2 \leq d_2 y_2.$$

This implication can be converted into the valid linear inequality

$$x_2 \leq d_2 y_2 + s_2 \,,$$

which is valid because in any feasible solution to *LS-U*:

- Either $y_2 = 0$ and the inequality is satisfied because $y_2 = 0$ implies $x_2 = 0$, and $0 \leq s_2$;
- Or $y_2 = 1$ and $x_2 \leq d_2 + s_2$ is a valid upper bound on x_2.

This inequality is violated by the current fractional point from Figure 4.2, and so we have simulated one pass of the separation problem. Using the same reasoning and starting from the upper bound $x_t \leq d_{tl} + s_l$ for any $l \geq t$ (remember that d_{tl} denotes $\sum_{i=t}^{l} d_i$), the above valid inequality can easily be generalized to

$$x_t \leq d_{tl} y_t + s_l \quad \text{for all } 1 \leq t \leq l \leq NT \quad (4.12)$$

for arbitrary demand data and time period.

Complete Linear Description

We denote the set of feasible solutions of model $LS\text{-}U$, that is, the solutions of (4.2)–(4.5), by X^{LS-U}. The class (4.12) of valid inequalities does not suffice to obtain a linear description of $conv(X^{LS-U})$ for any instance.

First we define the more general class of so-called (l, S) inequalities. It is shown in Chapter 7, Proposition 7.4, that the inequalities

$$\sum_{i \in S} x_i \leq \sum_{i \in S} d_{il} y_i + s_l \quad \text{for all } 1 \leq l \leq NT \text{ and } S \subseteq \{1, \ldots, l\} \quad (4.13)$$

are valid for X^{LS-U}. As an example, the valid inequality $x_2 + x_3 \leq d_{24} y_2 + d_{34} y_3 + s_4$ corresponds to the inequality (4.13) with $l = 4$ and $S = \{2, 3\} \subset \{1, \ldots, 4\}$.

The next theorem simply states that the (l, S) inequalities suffice to obtain the desired complete linear formulation. We can always eliminate the initial inventory, as we did in the multi-commodity formulation, by assuming a FPFD ordering and updating the (residual) demand vector accordingly. So, without loss of generality we assume that $s\_ini = 0$.

Theorem 4.2 *Assuming $d_t \geq 0$ for all t and $s\_ini = 0$, a complete linear description of $conv(X^{LS-U})$ is*

$$s_{t-1} + x_t = d_t + s_t \qquad \text{for } 1 \leq t \leq NT \qquad (4.14)$$

$$s_0 = 0 , \quad s_{NT} = 0 \qquad\qquad\qquad\qquad (4.15)$$

$$x_t \leq d_{t,NT} \, y_t \qquad\qquad \text{for } 1 \leq t \leq NT \qquad (4.16)$$

$$\sum_{i \in S} x_i \leq \sum_{i \in S} d_{il} y_i + s_l \qquad \text{for } 1 \leq l \leq NT, \, S \subseteq \{1, \ldots, l\} \qquad (4.17)$$

$$x_t, s_t, y_t \in \mathbb{R}_+, \, y_t \leq 1 \qquad \text{for } 1 \leq t \leq NT . \qquad (4.18)$$

Note that if $d_1 > 0$, the (l, S) inequality $x_1 \leq d_1 y_1 + s_1$ together with the initial equation $x_1 = d_1 + s_1$ and $y_1 \leq 1$ imply the equality $y_1 = 1$.

Separation Algorithm

We have obtained a complete linear programming formulation of $LS\text{-}U$ in the original variables (x, s, y). However, this formulation contains an exponential number of (l, S) inequalities (4.13), and a cutting plane approach must be used to avoid adding all these inequalities a priori to the formulation.

In order to use a class of valid inequalities in a cutting plane algorithm, the associated separation problem must be solved. Given a solution to the linear relaxation, it consists of either finding an inequality from the class violated by the solution, or proving that all inequalities from the class are satisfied by the given solution; see Chapter 3.

We denote by $P^{LS\text{-}U}$ the initial (linear) formulation (4.2)–(4.4) of $LS\text{-}U$ together with $x_t, s_t \geq 0$ and $0 \leq y_t \leq 1$ for all t.

Separation Given $(x^*, s^*, y^*) \in P^{LS\text{-}U}$:

- Either we find an (l, S) inequality violated by (x^*, s^*, y^*);
- Or we prove that all (l, S) inequalities are satisfied by (x^*, s^*, y^*).

As the (l, S) inequality may be rewritten as $\sum_{i \in S}(x_i - d_{il}y_i) \leq s_l$, to find the most violated (l, S) inequality for fixed $l \in \{1, \ldots, n\}$, it suffices to set

$$S^* = \{i \in \{1, \ldots, l\} : (x_i^* - d_{il}y_i^*) > 0\}$$

and test whether $\sum_{i \in S^*}(x_i^* - d_{il}y_i^*) > s_l^*$.

- If this holds, then the (l, S^*) inequality is the most violated inequality for the given value of l.
- Otherwise, there is no violated (l, S) inequality for the given value of l.

By enumerating over all possible values of l, we obtain a separation algorithm for the (l, S) inequalities whose running time is $O(NT^2)$; see Section 3.2.

Testing the Cutting Plane Reformulation

Table 4.3 compares the performance of the three formulations (again using the default Xpress-MP optimizer, without preprocessing or cuts generated by the solver) proposed to solve our bike production planning example from Section 1.1:

- The initial formulation (4.1)–(4.5) solved by branch-and-bound.
- The multi-commodity a priori reformulation (4.6)–(4.11) solved by branch-and-bound.
- The reformulation in the original space of variables using the initial formulation (4.1)–(4.5) and the separation algorithm for the (l, S) inequalities (4.13) in a branch-and-cut or cutting plane algorithm.

Table 4.3. B&B and B&C Solution of the *LS-U* Example from Section 1.1, Comparison of Reformulations

Formulation / Size / Algorithm	LP Val. / Vars / Cons	CPLP Val. / CPLP Time / CPLP Cuts	OPT Val. / OPT Time / OPT Nodes
(4.1)–(4.5) $O(NT) \times O(NT)$ B & B	712,189 24 16	– – –	736,000 0 29
(4.6)–(4.11) $O(NT^2) \times O(NT^2)$ B & B	736,000 72 72	- – –	736,000 0 1
(4.1)–(4.5) and (4.13) $O(NT) \times O(2^{NT})$ B & C	712,189 24 16	736,000 0 21	736,000 0 1

Our cutting plane algorithm requires six passes (and 21 cuts in total) to solve this instance of *LS-U* without branching, where one pass is defined as one iteration of cut generation for each l with $1 \leq l \leq NT$, followed by a single reoptimization.

4.1.3 Using Approximate Reformulations

The *LS-U* model is an ideal case. We know complete and compact (i.e., polynomial in size) extended linear reformulations, as well as a complete linear description in the original space of the convex hull of solutions $conv(X^{LS-U})$ with a fast separation algorithm. So, when a practical production planning problem involves *LS-U* as a submodel for an item, these reformulations are very effective in improving the formulation.

In many other cases, we only have partial reformulation results for the single-item submodels, say X^{LS}. That is, we have an initial formulation P^{LS}, some extended reformulation, or a class of valid inequalities in the original space that defines only an approximation $\overline{conv}(X^{LS})$ of the convex hull of solutions, but is significantly smaller than the initial formulation; that is,

$$conv(X^{LS}) \subset \overline{conv}(X^{LS}) \subset P^{LS}.$$

These approximate or partial reformulations can be used in the same way – a priori reformulations or cutting planes – as complete reformulations.

In all cases, the objective of the reformulation phase is to be able to use the best known results for submodels embedded in the planning model to be solved. This is the essence of the decomposition approach that we formalize next.

4.2 The Decomposition Approach for Complex Models

As we have already seen in the examples of Chapter 2, and in the master production scheduling example from Section 1.2.3, the structure of many, or most, multi-item production planning problems looks very similar when represented as mixed integer programs.

To be specific, the MPS example is more or less of the form

$$(MIPP^{item}) \qquad W^\star \; = \; \min \sum_i \sum_t (p_t^i x_t^i + h_t^i s_t^i + q_t^i y_t^i)$$

$$[\; s_{t-1}^i + x_t^i = d_t^i + s_t^i, \; x_t^i \le C_t^i y_t^i, \; y_t^i \le 1 \qquad \text{for all } t\;], \qquad \text{for all } i \quad (4.19)$$

$$[\; \sum_i a_t^{ik} x_t^i + \sum_i b_t^{ik} y_t^i \le L_t^k \qquad\qquad \text{for all } k\;], \qquad \text{for all } t \quad (4.20)$$

$$[\; x_t^i \le C_t^i y_t^i, \; y_t^i \le 1 \qquad\qquad\qquad\quad \text{for all } i\;], \qquad \text{for all } t \quad (4.21)$$

$$x_t^i \in \mathbb{R}_+^1, \; s_t^i \in \mathbb{R}_+^1, \; y_t^i \in \mathbb{Z}_+^1 \qquad\qquad\quad \text{for all } i,t.$$

This can be written more compactly as

$$(MIPP^{item}) \quad W^\star \; = \; \min \quad \sum_i \sum_t (p_t^i x_t^i + h_t^i s_t^i + q_t^i y_t^i)$$

$$(x^i, s^i, y^i) \in Y^i \qquad\qquad \text{for all } i \; ,$$

$$(x, s, y) \in Z \; ,$$

where Y^i represents the set of feasible solutions to the item i lot-sizing problem (i.e., lot sizes x^i, set-ups y^i, and inventory levels s^i defined for all time periods and satisfying the constraints (4.19) for item i), such as LS-U or some of its variants. On the other hand Z represents the solutions satisfying the set of linear constraints (4.20)–(4.21). The constraints defining Z are often called coupling or linking constraints because they link together the items that have to share the joint capacity.

This representation or scheme is not totally general, and certainly not unique. For instance, we can also view the linking set Z as the intersection of independent single-period sets. Now we can write the problem as the intersection of the time and period submodels as in formulation

$$(MIPP_{time}^{item}) \quad W^\star \; = \; \min \quad \sum_i \sum_t (p_t^i x_t^i + h_t^i s_t^i + q_t^i y_t^i)$$

$$(x^i, s^i, y^i) \in Y^i \qquad\qquad \text{for all } i$$

$$(x_t, s_t, y_t) \in Z_t \qquad\qquad \text{for all } t \; ,$$

where Z_t represents the set of feasible solutions to the period t submodel, that is, the lot sizes x_t, set-ups y_t defined for all items, and satisfying the constraints (4.20)–(4.21) for time period t.

The branch-and-bound/cut methods studied in Chapter 3, like most optimization methods, are based on easy-to-solve relaxations of the initial problem. For example, the above problem can be solved by some standard MIP software using a branch-and-bound algorithm based on the linear programming relaxation LR of the initial formulation. We suppose that the initial formulation for the lot-sizing sets Y^i is P^{Y^i}, and the initial formulation for the period t linking constraints in Z_t is P^{Z_t}. So, LR is defined by

$$LR = \min \quad \sum_i \sum_t (p_t^i x_t^i + h_t^i s_t^i + q_t^i y_t^i)$$

$$(x^i, s^i, y^i) \in P^{Y^i} \qquad \text{for all } i$$

$$(x_t, s_t, y_t) \in P^{Z_t} \qquad \text{for all } t.$$

Unfortunately, this direct branch-and-bound approach can only be used for the solution of small-size problems. In order to solve, or to find good solutions, for more realistic or real-size problems, one has to work with better or tighter relaxations or formulations providing improved lower bounds. Because of the multi-item structure of the initial problem, most efficient solution approaches are based on the following reformulation.

$$LB^{item} = \min \quad \sum_i \sum_t (p_t^i x_t^i + h_t^i s_t^i + q_t^i y_t^i)$$

$$(x^i, s^i, y^i) \in \overline{conv}(Y^i) \qquad \text{for all } i$$

$$(x_t, s_t, y_t) \in P^{Z_t} \qquad \text{for all } t,$$

where $\overline{conv}(Y^i)$ represents a partial (or complete) reformulation of the convex hull of the solutions of the single-item model Y^i. This bound LB^{item} can be obtained in several ways:

- Either by branch-and-bound using an a priori and compact linear reformulation of $\overline{conv}(Y^i)$;
- Or by branch-and-cut using a reformulation of $\overline{conv}(Y^i)$ involving many constraints, combined with a separation algorithm; see Chapter 3.

In some cases, we may also know good (or complete) linear reformulations for the single-period submodel. This in turn leads us to a stronger linear programming relaxation

$$LB^{item}_{time} = \min \quad \sum_i \sum_t (p_t^i x_t^i + h_t^i s_t^i + q_t^i y_t^i)$$

$$(x^i, s^i, y^i) \in \overline{conv}(Y^i) \qquad \text{for all } i$$

$$(x_t, s_t, y_t) \in \overline{conv}(Z_t) \qquad \text{for all } t,$$

where $\overline{conv}(Z_t)$ represents an approximate (or complete) linear description of the convex hull of the solutions of the single-period model Z_t.

These new lower bounds LB^{item} and LB^{item}_{time} are never worse, and typically much tighter than the linear relaxation bound LR. The following relations always hold between these bounds.

$$LR \leq LB^{item} \leq LB^{item}_{time} \leq W^\star .$$

Better lower bounds LB usually allow one to reduce the number of nodes needed to prove optimality, or to obtain good quality solutions. But obtaining these bounds requires more computing time than the time needed to obtain LR because of larger models or more cuts to be added in the cutting plane phase.

For any complex multi-item production planning problem to be solved by an optimization approach, the best reformulation thus depends on

- the existence of reformulation results (approximate or tight compact extended reformulations, valid inequalities, efficient separation algorithms) for the corresponding single-item and/or single-period submodels, and
- the impact of the reformulations on the computing time through a decreased number of branch-and-bound nodes but increased computing time at each node.

The model classification scheme presented next is crucial for an implementation of the decomposition approach. It forces us to present the description, analysis, and structuring of models in a way that facilitates the identification of structured submodels. Then, the systematic reformulation procedure of Section 4.4 identifies the submodels for which reformulation results are available.

Note finally that other optimization methods such as Lagrangian relaxation, Lagrangian decomposition, and Dantzig–Wolfe or column generation, exploit the same decomposition properties of the models. Instead of compact reformulations, these methods require the repeated solution of optimization problems defined over the single-item lot-sizing sets Y^i and the single-period sets Z_t. So to implement these algorithms, it is important to find efficient algorithms to optimize over the single-item/period feasible sets.

The links with these other methods are discussed further in Chapter 6.

4.3 Model Classification

Most practical supply chain planning problems are multi-item, multi-machine, and multi-level, but there exist very few reformulation results concerning such models. Therefore, the main optimization approach in solving such problems has been to integrate existing algorithms and known reformulation results for single-item problems, using a decomposition approach.

We describe here a classification scheme for single-item production planning models that allows one to benefit from this knowledge. Based on this scheme, the procedure to systematically reformulate and solve production planning models is described in Section 4.4, and illustrated in Section 4.5 on the GW master production scheduling example from Sections 1.2.2 to 1.2.4.

Parts II and III of this book describe the reformulation results according to our scheme for single-item models. Thus for each problem appearing in our classification, we need to describe in detail what results are known and can be used to implement the optimization/decomposition approach for these models. In Part IV we extend our classification to multi-item and multi-level production planning problems, and again present the useful reformulation results that are available. This structured knowledge is then exploited in Part V in solving several industrial cases.

In this section, we describe the basic single-item classification, its notational conventions, and the corresponding mathematical formulations.

4.3.1 Single-Item Classification

Planning problems deal with sizing and timing decisions for purchasing, production, or distribution of lots or batches. An item represents a physical product. The finite planning horizon is divided into time periods, indexed by t, $1 \le t \le n$, where n is the given number of time periods.

When considering canonical single-item models, for compactness of notations we use n to represent the length of the planning horizon. This notation is used throughout Sections 4.3 and 4.4, and in Parts II and III of the book. Alternatively, when considering specific production planning instances, we use NT to represent the number of time periods. Similarly, to represent the number of items in the multi-item models studied in Part IV, we use m in canonical models and NI in any particular planning instance.

We start by defining the basic single-item lot-sizing problem (LS). For a single item, we represent by

- d_t the demand to be satisfied in period t, that is forecast demand or customer orders due in period t;
- p'_t the variable or unit production cost in period t;
- h'_t the unit cost for holding one unit in inventory at the end of period t;
- q_t the fixed set-up cost to be paid if there is a positive production in period t;
- C_t the upper bound on production or capacity in period t.

The fixed charge production cost function in period t is characterized by the set-up cost q_t and the unit production cost p'_t.

Problem LS is the problem of finding the production plan for the single item, meeting the demand in every period, and satisfying the capacity restrictions; that is, the production is less than or equal to C_t in every period t, that minimizes the inventory and production costs. Note that in principle a variable amount of initial stock is allowed, at a cost of h'_0 per unit.

Our classification is dictated by the difficulty of solving single-item planning problems, or more precisely by the optimization and reformulation results presented in the literature. There are three fields $PROB$-CAP-VAR. In each field, we use $[x, y, z]^1$ to denote the selection of exactly *one* element from the set $\{x, y, z\}$, and $[x, y, z]^*$ to denote *any* subset of $\{x, y, z\}$. We simply use x, y, z to denote the selection of *all* the elements in the set $\{x, y, z\}$. Fields that are empty are dropped.

4.3.2 Description of the Field $PROB$

In the first field $PROB$, there is a choice of four problem versions $PROB = [LS, WW, DLSI, DLS]^1$.

LS (Lot-Sizing): This is the general problem defined above.

WW (Wagner–Whitin): This is problem LS, except that the variable production and storage costs satisfy $h'_t + p'_t - p'_{t+1} \geq 0$ for $0 \leq t \leq n$, where $p'_0 = p'_{n+1} = 0$. This condition means that, if set-ups occur in both periods t and $t + 1$, then it is more costly to produce in period t and stock till period $t+1$, than to produce directly in period $t+1$. In other words, given the set-ups it always pays to produce as late as possible. This condition is often referred to as the *absence of speculative motive for early production*. We define a new inventory cost as $h_t = h'_t + p'_t - p'_{t+1} \geq 0$ for $0 \leq t \leq n$ (see formulations below).

We name this cost condition WW because it was first introduced in the seminal paper of Wagner–Whitin. It is a little technical, but we show in Part II that it allows one to reduce the running time of the optimization algorithms, and to simplify the reformulation of the planning models. Moreover this condition is very often satisfied by the cost coefficients encountered in practice.

$DLSI$ (Discrete Lot-Sizing with Variable Initial Stock): This is problem LS with the restriction that there is either no production or production at full capacity C_t in each period t.

DLS (Discrete Lot-Sizing): This is problem $DLSI$ without an initial stock variable.

4.3.3 Description of the Field CAP

The second field CAP concerns the production limits or capacities $CAP = [C, CC, U]^1$. The three CAP variants of problem $PROB$ are

$PROB$-C (Capacitated): Here the capacities C_t vary over time.

PROB-CC (Constant Capacity): This is the case where $C_t = C$, a constant, for all periods t.

PROB-U (Uncapacitated): This is the case when there is no limit on the amount of the item produced in each period. In the absence of other constraints limiting the total amount produced over all items, this case means that the capacity C_t in each period t suffices to satisfy all the demands up to the end of the horizon.

Before presenting the third field VAR containing the many possible extensions, we present mixed integer programming formulations of the four basic variants with varying capacities *PROB-C*.

4.3.4 Mathematical Formulations for *PROB-CAP*

The standard formulation of LS as a mixed integer program involves the variables

- x_t the amount produced in period t for $1 \leq t \leq n$,
- s_t the stock at the end of period t for $0 \leq t \leq n$, and
- $y_t = 1$ if the machine is set up to produce in period t, and $y_t = 0$ otherwise, for $1 \leq t \leq n$.

We also use the notation $d_{kt} \equiv \sum_{u=k}^{t} d_u$ throughout.

LS-C can be formulated as

$$\min \quad \sum_{t=1}^{n} p_t' x_t + \sum_{t=0}^{n} h_t' s_t + \sum_{t=1}^{n} q_t y_t \qquad (4.22)$$

$$s_{t-1} + x_t = d_t + s_t \qquad \text{for } 1 \leq t \leq n \qquad (4.23)$$

$$x_t \leq C_t y_t \qquad \text{for } 1 \leq t \leq n \qquad (4.24)$$

$$x \in R_+^n, \; s \in R_+^{n+1}, \; y \in \{0,1\}^n , \qquad (4.25)$$

and X^{LS-C} denotes the set of feasible solutions to (4.23)–(4.25). Constraint (4.23) represents the flow balance constraint in every period t, the inflows are the initial inventory s_{t-1} and the production x_t, the outflows are the demand d_t and the ending inventory s_t. Constraint (4.24) represents the capacity restriction and also fixes the set-up variable y_t to 1 whenever there is positive production (i.e., $x_t > 0$). This constraint is also called a variable upper bound (VUB) constraint. The objective (4.22) is simply the sum of the set-up, inventory, and variable production costs.

WW-C can be formulated just in the space of the s, y variables as

$$\min \quad \sum_{t=0}^{n} h_t s_t + \sum_{t=1}^{n} q_t y_t \tag{4.26}$$

$$s_{k-1} + \sum_{u=k}^{t} C_u y_u \geq d_{kt} \qquad \text{for } 1 \leq k \leq t \leq n \tag{4.27}$$

$$s \in R_+^{n+1}, \; y \in \{0,1\}^n , \tag{4.28}$$

and X^{WW-C} denotes the set of feasible solutions to (4.27)–(4.28). To derive this formulation, the constraint (4.23) is used to eliminate x_t from the objective function (4.22). Specifically,

$$\sum_{t=1}^{n} p'_t x_t + \sum_{t=0}^{n} h'_t s_t = \sum_{t=1}^{n} p'_t(s_t - s_{t-1} + d_t) + \sum_{t=0}^{n} h'_t s_t$$

$$= \sum_{t=0}^{n} (h'_t + p'_t - p'_{t+1}) s_t + \sum_{t=1}^{n} p'_t d_t$$

$$= \sum_{t=0}^{n} h_t s_t + \sum_{t=1}^{n} p'_t d_t ,$$

where $p'_0 = p'_{n+1} = 0$. So, by defining $h_t = h'_t + p'_t - p'_{t+1}$, the objective function (4.22) becomes (4.26) to within the constant $\sum_{t=1}^{n} p'_t d_t$.

Now as $h_t \geq 0$ for all t, it follows that once the set-up periods are fixed (the periods t in which $y_t = 1$), the stocks will be as low as possible compatible with satisfying the demand and respecting the capacity restrictions. Based on this argument it is possible to prove that it suffices to find a minimum cost stock minimal solution in order to solve WW-C, where a *stock minimal solution* is a solution satisfying

$$s_{k-1} = \max(0, \max_{t=k,\ldots,n} [d_{kt} - \sum_{u=k}^{t} C_u y_u]). \tag{4.29}$$

In the proposed formulation (4.26)–(4.28) for WW-C, because of the presence of the initial stock s_0, any combination of set-up periods is feasible, and constraint (4.27) imposes a lower bound on the stock variables. The objective function (4.26), together with $h_t \geq 0$, guarantees that there exists an optimal solution to (4.26)–(4.28) that satisfies (4.29). It follows that the proposed formulation is valid, though its (s, y) feasible region is not the same as that of LS-C. Specifically (s, y) is feasible in (4.27)–(4.28) if and only if there exists (x, s', y) feasible in (4.23)–(4.25) with $s' \leq s$.

Remark 1. Whether the Wagner–Whitin cost condition is satisfied or not, the WW relaxation consisting of the constraints (4.27) is valid for problem LS, and often provides a very good approximation to the convex hull of solutions for problem LS.

Remark 2. Even though each single item subproblem may have WW costs, the existence of other constraints such as multi-item budget (production capacity) constraints or multi-item storage capacity constraints (PQ in the multi-item classification of Section 12.1) destroys the stock minimal solution property for individual items, and thus the items are more correctly classified as LS, rather than WW.

Remark 3. On the other hand, if in a multi-item problem the constraints linking together the different items involve only the set-up or start-up variables (PM in the multi-item classification of Section 12.1), then the stock minimal property of solutions is preserved, and the single items can be classified as WW if their costs satisfy the WW condition.

$DLSI$-C can be formulated by adding $x_t = C_t y_t$ in the formulation of LS-C. By summing constraints (4.23) from 1 up to t, one gets $s_t = s_0 + \sum_{u=1}^{t} x_u - d_{1t}$. Then, after elimination of the variables $s_t \geq 0$ and $x_t = C_t y_t$, we obtain an equivalent formulation of $DLSI$-C just in the space of the s_0 and the y variables, and X^{DLSI_0-C} is used to denote the set of feasible solutions to (4.31)–(4.32),

$$\min \quad h_0 s_0 + \sum_{t=1}^{n} q'_t y_t \tag{4.30}$$

$$s_0 + \sum_{u=1}^{t} C_u y_u \geq d_{1t} \qquad \text{for } 1 \leq t \leq n \tag{4.31}$$

$$s_0 \in R_+^1, \ y \in \{0,1\}^n , \tag{4.32}$$

where h_0 and q'_t are the new objective coefficients of variables s_0 and y_t obtained after eliminating the variables s_t and x_t by substitution. Specifically, the objective function (4.22) can be rewritten as

$$\sum_{t=0}^{n} h'_t s_t + \sum_{t=1}^{n} p'_t x_t + \sum_{t=1}^{n} q_t y_t$$

$$= h'_0 s_0 + \sum_{t=1}^{n} h'_t (s_0 + \sum_{u=1}^{t} C_u y_u - d_{1t}) + \sum_{t=1}^{n} p'_t C_t y_t + \sum_{t=1}^{n} q_t y_t$$

$$= (h'_0 + \sum_{t=1}^{n} h'_t) s_0 - \sum_{t=1}^{n} h'_t d_{1t} + \sum_{t=1}^{n} (q_t + (p'_t + \sum_{u=t}^{n} h'_u) C_t) y_t .$$

Then defining $h_0 = h'_0 + \sum_{t=1}^{n} h'_t$ and $q'_t = q_t + (p'_t + \sum_{u=t}^{n} h'_u) C_t$, it reduces to (4.30) except for the constant $-\sum_{t=1}^{n} h'_t d_{1t}$.

We also use the notation X^{DLSI_k-C} with $0 \leq k \leq n-1$ to denote the set of solutions of problem $DLSI_k$-C, which is problem $DLSI$-C except that the initial inventory is located in period k, and production can occur in periods $k + 1$ up to n. Problem $DLSI_k$-C involves thus variables $s_k \in R^1_+$ and $y_{k+1}, \ldots, y_n \in \{0, 1\}$.

DLS-C can be formulated just in the space of the y variables by fixing $s_0 = 0$:

$$\min \quad \sum_{t=1}^{n} q'_t y_t \tag{4.33}$$

$$\sum_{u=1}^{t} C_u y_u \geq d_{1t} \qquad \text{for all } 1 \leq t \leq n \tag{4.34}$$

$$y \in \{0, 1\}^n. \tag{4.35}$$

The set X^{DLS-C} denotes the set of feasible solutions to (4.34)–(4.35). We say that DLS has *Wagner–Whitin costs* if $q'_t \geq q'_{t+1}$ for all t, and without introducing a new problem class we denote this special case as $DLS(WW)$-C.

Observation 4.2 *The constant or uncapacitated problems $PROB$-$[CC, U]$[1] are all polynomially solvable. There is a polynomial dynamic programming algorithm solving LS-CC and the other seven problems can all be seen as special cases.*

All four varying capacity instances $PROB$-C are NP-hard, because all four problems are polynomially reducible to the 0–1 knapsack problem. This means that there are no polynomial algorithms known for them and, from complexity theory, it is very unlikely that there exists a polynomial algorithm for any of them.

We come back to the implications of these observations, to the relationships between these different problems, and to the analysis of algorithms and reformulations for these problems in Part II. So far, we consider that we have different versions of the single-item lot-sizing problem, along with mixed integer programming formulations adapted to each problem class.

4.3.5 Description of the Field VAR

The third field VAR concerns extensions or variants to one of the twelve problems $PROB - CAP$ defined so far; that is, $VAR = [B, SC, ST, LB, SL, SS]^*$. Although such variants can be combined, for simplicity we describe these variants in turn, and give a typical formulation for each problem LS-C-$[B, SC, ST, LB, SL, SS]$[1].

B (Backlogging): Demand must still be satisfied, but it is possible to satisfy a demand later than required. This occurs, for example, when a factory does not have enough capacity to deliver to all customers on time in a given period. Usually, the backlog or shortfall implies a penalty cost proportional to the amount backlogged and to the duration of the backlog.

Note that this backlogging variant is limited to independent or external demand, as the quantity backlogged is only a virtual flow used to model shortfalls in the delivery process and not a physical flow.

SC (Start-Up Costs): It is necessary to accurately model capacity utilization to obtain feasible production plans. This often requires one to model the capacity consumed when a machine starts a production batch, or when a machine switches from one product to another. In these cases, we obtain so-called set-up or start-up time models, changeover time models, or models with sequencing restrictions. However, in many cases, less accurate models involving only set-up or start-up costs are considered. Such models can be seen as obtained by relaxing (in the Lagrangian sense; see Chapter 6) the set-up or start-up time restrictions.

The simplest single-item start-up cost model is the following. If a sequence of set-ups starts in period t, a start-up cost g_t is incurred, which can be seen as the direct start-up cost plus an estimate of the opportunity cost of the start-up time or capacity consumed.

ST (Start-Up Times): As already explained, start-up times are used to model capacity utilization more accurately. The resulting models are more precise, but often more difficult to solve than their start-up cost variant.

If a sequence of set-ups starts in period t, the capacity C_t is reduced by an amount ST_t. We use $ST(C)$ to indicate the start-up time ST is constant over time; that is, $ST_t = ST$ for all t.

LB (Minimum Production Levels): In some problems, in order to guarantee a minimum level of productivity, minimum batch sizes or production levels are imposed. For instance, this feature is often used in combination with start-up costs to approximate start-up time models and avoid small batches in the solutions. This constraint may also be imposed for technological reasons.

If production takes place in period t, a minimum amount LB_t must be produced. We use $LB(C)$ to denote constant lower bounds over time, i.e. $LB_t = LB$ for all t. Note that this leads to variable lower-bound constraints, and not simple lower bounds.

SL (Sales and Lost Sales): In some cases, the demands to be satisfied are not fixed in advance. This occurs, for instance, when capacity is too low to satisfy the total potential demand, or when the selling price does not always cover the marginal cost of production. The optimization problem becomes then a profit maximization problem, with additional sales variables.

In the single-item problem, we model this case in the following way. In addition to the demand d_t that must be satisfied in each period, an additional amount up to u_t can be sold at a unit price of c_t.

Note that this variant can also be used to model the Lost Sales variant in which, as opposed to backlogging, it is possible to not deliver part of the demand. In this case, the demand from period t that has to be satisfied is d_t, and the additional demand that may be lost or not delivered is u_t. The unit price c_t represents in this case the penalty cost that is avoided for each unit of the additional demand effectively delivered.

SS (Safety Stocks): The last variant that we consider is present in many practical applications, and absent from most scientific publications. When the demand is an output of a forecasting system, it is not known with certainty. Therefore, a minimum amount of planned inventory, called the safety stock, is required at the end of each period so as to handle this uncertainty, and to avoid delivery shortages when actual demand exceeds forecast demand.

The variants described here are common variants included in the field *VAR*. These plus additional variants concerning either changes in the demand model, production constraints/costs, or sales constraints, are described and analyzed in Chapter 11.

4.3.6 Mathematical Formulations for *PROB-CAP-VAR*

Backlogging

The standard formulation of *PROB-CAP-B* as a mixed integer program involves the additional variables

- r_t the backlog at the end of period t for $t = 1, \ldots, n$.

This cumulated shortfall r_t in satisfaction of the demand in period t is charged at a cost of b'_t per unit. It is assumed throughout that $r_0 = 0$.

This leads to the following formulation for problem *LS-C-B*.

$$\min \sum_{t=0}^{n} h'_t s_t + \sum_{t=1}^{n} b'_t r_t + \sum_{t=1}^{n} p'_t x_t + \sum_{t=1}^{n} q_t y_t \tag{4.36}$$

$$s_{t-1} - r_{t-1} + x_t = d_t + s_t - r_t \qquad \text{for } 1 \leq t \leq n \tag{4.37}$$

$$x_t \leq C_t y_t \qquad \text{for } 1 \leq t \leq n \tag{4.38}$$

$$x, r \in R_+^n, \; s \in R_+^{n+1}, \; y \in \{0,1\}^n , \tag{4.39}$$

and X^{LS-C-B} denotes the set of feasible solutions to the constraints (4.37)–(4.39).

Problem *WW-CAP-B* is problem *LS-CAP-B* except that the costs satisfy the *WW* cost condition. With backlogging, the costs are said to be *Wagner–Whitin* if both $h_t = p'_t + h'_t - p'_{t+1} \geq 0$ and $b_t = p'_{t+1} + b'_t - p'_t \geq 0$ for $1 \leq t \leq n-1$. This means that, with respect to backlogging, there are no speculative motives for late production.

As an extension of the simple formulation (4.26)–(4.28) for WW-C, it can be proved that the following formulation involving only the s, r, y variables, is a valid and sufficient formulation for WW-C-B.

$$\min \sum_{t=0}^{n} h_t s_t + \sum_{t=1}^{n} b_t r_t + \sum_{t=1}^{n} q_t y_t \tag{4.40}$$

$$s_{k-1} + r_l + \sum_{u=k}^{l} C_u y_u \geq d_{kl} \qquad \text{for } 1 \leq k \leq l \leq n \tag{4.41}$$

$$s \in \mathbb{R}_+^{n+1}, \ r \in \mathbb{R}_+^n, \ y \in \{0,1\}^n . \tag{4.42}$$

The notation X^{WW-C-B} is used to represent the set of feasible solutions to the constraints (4.41)–(4.42).

The validity and sufficiency of formulation (4.41)–(4.42) is based on the following nontrivial result. When the objective function (4.40) of WW-C-B satisfies $h_t, b_t \geq 0$ for all t, it suffices to find a minimum cost stock minimal and backlog minimal solution in order to solve WW-C-B, where a solution is called *stock minimal* (resp., *backlog minimal*) if $s_{k-1} = \max_{l \geq k}[d_{kl} - \sum_{u=k}^{l} C_u y_u - r_l]^+$ (resp. if $r_l = \max_{k \leq l}[d_{kl} - \sum_{u=k}^{l} C_u y_u - s_{k-1}]^+$).

After elimination of the $s_1, \ldots, s_n$ variables, $DLSI$-C-B has the following feasible region in the (s_0, r, y) space,

$$s_0 + r_t + \sum_{u=1}^{t} C_u y_u \geq d_{1t} \qquad \text{for } 1 \leq t \leq n \tag{4.43}$$

$$s_0 \in R_+^1, r \in R_+^n, y \in [0,1]^n. \tag{4.44}$$

and X^{DLSI_0-C-B} denotes the set of feasible solutions to (4.43)–(4.44).

Finally, DLS-C-B is obtained from $DLSI$-C-B by setting $s_0 = 0$.

Start-Up Costs

The basic formulation for LS-C-SC requires the introduction of new variables

- $z_t = 1$ if there is a start-up in period t; that is, there is a set-up in period t, but there was not in period $t-1$, and $z_t = 0$ otherwise.

The resulting formulation for LS-C-SC is

$$\min \sum_{t=1}^{n} p'_t x_t + \sum_{t=0}^{n} h'_t s_t + \sum_{t=1}^{n} q_t y_t + \sum_{t=1}^{n} g_t z_t \tag{4.45}$$

$$s_{t-1} + x_t = d_t + s_t \qquad\qquad \text{for } 1 \leq t \leq n \qquad (4.46)$$

$$x_t \leq C_t y_t \qquad\qquad\qquad \text{for } 1 \leq t \leq n \qquad (4.47)$$

$$z_t \geq y_t - y_{t-1} \qquad\qquad \text{for } 1 \leq t \leq n \qquad (4.48)$$

$$z_t \leq y_t \qquad\qquad\qquad \text{for } 1 \leq t \leq n \qquad (4.49)$$

$$z_t \leq 1 - y_{t-1} \qquad\qquad \text{for } 1 \leq t \leq n \qquad (4.50)$$

$$x \in R^n_+, \ s \in R^{n+1}_+, \ y, z \in \{0,1\}^n \ , \tag{4.51}$$

and the set of feasible solutions to (4.46)–(4.51) is denoted by $X^{LS-C-SC}$. We assume that y_0, the state of the machine at time 0, is given as data. The additional constraints (4.48)–(4.50) define the values of the additional start-up variables. These constraints are a linearization of the constraint $z_t = y_t(1 - y_{t-1})$, for all t. There can be a start-up in period t (i.e., $z_t = 1$) only if there is a start-up in period t (see (4.49)) and no start-up in period $t - 1$ (see (4.50)), and there must be a start-up in period t if both events occur simultaneously (see (4.48)).

The formulations of $[WW, DLSI, DLS]^1$-C-SC, as well as their corresponding feasible sets $X^{[WW,DLSI_0,DLS]^1-C-SC}$, are obtained by just adding the constraints (4.48)-(4.50) and the integrality restrictions $z \in \{0,1\}^n$ to the formulations $[WW,DLSI,DLS]^1$-C given above.

Start-Up Times

The basic formulation for LS-C-ST requires the same start-up variables z_t as the start-up cost model LS-C-SC. The formulation for LS-C-ST is the same as for LS-C-SC ((4.45)-(4.51)), except that the variable upper bound constraint (4.47) has to be replaced by the constraint

$$x_t \leq C_t y_t - ST_t z_t \quad \text{for } 1 \leq t \leq n.$$

Minimum Production Levels

The basic formulation for LS-C-LB requires no additional variables. The formulation for LS-C-LB is the same as for LS-C ((4.22)–(4.25)), augmented with the variable lower bound constraint

$$x_t \geq LB_t y_t \quad \text{for } 1 \leq t \leq n.$$

Sales

The standard formulation of LS-C-SL as a mixed integer program involves the additional variables

- v_t the amount sold in period t, on top of the fixed demand d_t, for $1 \leq t \leq n$,

and is given by

$$\max \sum_{t=1}^{n}(c_t v_t - p_t x_t) - \sum_{t=0}^{n} h'_t s_t - \sum_{t=1}^{n} q_t y_t \qquad (4.52)$$

$$s_{t-1} + x_t = d_t + v_t + s_t \qquad \text{for } 1 \leq t \leq n \qquad (4.53)$$

$$x_t \leq C_t y_t \qquad \text{for } 1 \leq t \leq n \qquad (4.54)$$

$$v_t \leq u_t \qquad \text{for } 1 \leq t \leq n \qquad (4.55)$$

$$x, v \in R^n_+, s \in R^{n+1}_+, y \in \{0,1\}^n , \qquad (4.56)$$

where the objective (4.52) maximizes the contribution to profit, and the flow balance constraint (4.53) is updated to take the sales outflow into account. Constraint (4.55) models the simple upper bound on sales.

Safety Stocks

To incorporate this requirement, we just need to add a simple lower bound SS_t on the stock level at the end of period t; that is, $s_t \geq SS_t$ for all periods t with $1 \leq t \leq n$

4.3.7 The Classification $PROB\text{-}CAP\text{-}VAR$

We have described the three fields $PROB\text{-}CAP\text{-}VAR$ of the single-item lot-sizing classification, namely,

$$[LS, WW, DLSI, DLS]^1 - [C, CC, U]^1 -$$
$$[B, SC, ST, ST(C), LB, LB(C), SL, SS]^*$$

where one entry is required from each of the first two fields, and any number of entries from the third.

For instance, $WW\text{-}U$ (in place of $WW\text{-}U\text{-}\emptyset$) denotes the uncapacitated Wagner–Whitin problem, whereas $DLSI\text{-}CC\text{-}B, ST$ denotes the constant capacity discrete lot-sizing problem with initial stock variable, backlogging, and start-up times.

Observation 4.3 *It turns out that almost all the variants $PROB\text{-}[CC, U]^1\text{-}VAR$ are still polynomially solvable if the start-up times or lower bounds, if any, are constant (versions $ST(C), LB(C)$).*

This terminates the description of the classification for single-item problems. It is clearly beyond the scope of this description to give a complete mathematical programming formulation of all possible variants from the classification. These different formulations are described in more detail in Parts II and III.

4.4 Reformulation Results: What and Where

In this section, we list first the reformulation results available (the "What") for the most common or standard single-item lot-sizing problems, classified according to the scheme described in Section 4.3.

More precisely, we give the results in the form of three reformulation tables for the uncapacitated and constant capacity single-item models:

- The basic models $[LS, WW, DLSI, DLS]^1$-$[U, CC]^1$ without variants.
- The models with backlogging $[LS, WW, DLSI, DLS]^1$-$[U, CC]^1$-B.
- The models with start-up costs $[LS, WW, DLS]^1$-$[U, CC]^1$-SC.

Note that we do not give reformulation tables for models with varying capacity (value C of the field CAP) because there are no complete reformulation results available for these high-level relaxations, due to the fact all these models define NP-hard optimization problems.

For variants other than backlogging or start-up costs, there are only a few results available. The partial reformulation results known for these models, and the reformulation results for lower-level relaxations contained in these models, are given in Parts II to IV of the book.

For each model in these tables, we indicate the reformulation results in three sections.

- **Formulation** reports on the existence and the size (order of the number of constraints and variables) of tight and compact linear a priori reformulations.
- **Separation** gives the complexity of the separation algorithms for the tight reformulations in the original variable space.
- **Optimization** contains the complexity of the best optimization algorithm known for the model.

In each case, we indicate a reference to the research paper or publication containing, to our knowledge, the original result, as well as a pointer to the section in this book where the result is described in detail.

The tables also indicate the missing results. An *asterisk* * indicates that the family of inequalities only gives a partial description of the convex hull of solutions. A *triple asterisk* *** indicates that we do not know of any result specific to the particular problem class.

Even if they are not used in a direct solution approach by branch-and-bound/cut, we have included results for the associated optimization problems because they are very much related to the other results, and because other optimization methods such as Lagrangian relaxation or Dantzig–Wolfe decomposition require the solution of the optimization version of these standard models.

Finally, we conclude this section by providing a *reformulation procedure* (the "Where") indicating how to use the results in the tables, and build improved formulations for complex production planning models. Note that this procedure requires the use of the classification scheme and reformulation tables, but does not require any knowledge about the mathematical description or analysis of the reformulations.

4.4.1 Results for $PROB$-$[U, CC]$

In Table 4.4 we present results for the models $[LS, WW, DLSI, DLS]$-$[U, CC]$. Note that the entries $[DLSI, DLS]$-U have been left blank as the results and algorithms are trivial. In the **Formulation** entries for LS-U, FL denotes a facility location reformulation, SP denotes a shortest path reformulation, and MC denotes the multi-commodity reformulation already presented in Section 4.1.1.

Table 4.4. Models $PROB$-$[U, CC]$

	LS	WW	$DLSI$	DLS
Formulation	$Cons \times Vars$	$Cons \times Vars$	$Cons \times Vars$	$Cons \times Vars$
U	$SP\ O(n) \times O(n^2)$ $FL\ O(n^2) \times O(n^2)$ $MC\ O(n^2) \times O(n^2)$ Section 7.4.2 [100, 61, 145]	$O(n^2) \times O(n)$ Section 7.5 [140]	$--$	$--$
CC	$O(n^3) \times O(n^3)$ Section 9.6.3 [176]	$O(n^2) \times O(n^2)$ Section 9.5.3 [140]	$O(n) \times O(n)$ Section 9.4.2 [125, 140]	$O(n) \times O(n)$ Section 9.3.1 Folklore
Separation				
U	$O(n \log n)$ Section 7.4.1 [23]	$O(n)$ Section 7.5 [140]	$--$	$--$
CC	$*$ Section 9.6.2/3 [139]	$O(n^2 \log n)$ Section 9.5.2 [140]	$O(n \log n)$ Section 9.4.1 [85, 125, 140]	$O(n)$ Section 9.3.1 Folklore
Optimization				
U	$O(n \log n)$ Section 7.3 [3, 63, 187]	$O(n)$ Section 7.3 [3, 63, 187]	$--$	$--$
CC	$O(n^3)$ Section 9.6.1 [71, 171]	$O(n^2 \log n)$ Section 9.5.1 [178]	$O(n^2 \log n)$ Section 9.4 [178]	$O(n \log n)$ Section 9.3.2 [178]

Reading these tables is straightforward. Looking at the WW–CC entry in the **Formulation** block, we see that, for the problem with Wagner-Whitin costs and constant capacities, there is an extended formulation with $O(n^2)$ constraints and $O(n^2)$ variables that gives the convex hull. Details are to be found in Section 9.5.3. We see also in the WW–CC entry in the **Separation** block that there is a separation algorithm for the same problem whose running time is $O(n^2 \log n)$. Finally we see from the WW–CC entry in the **Optimization** block that the fastest known algorithm to find an optimal solution for this problem runs in $O(n^2 \log n)$.

4.4.2 Results for Backlogging Models $PROB$-$[U, CC]$-B

Now we consider the same problems but with backlogging. The results are given in Table 4.5.

Table 4.5. Models with Backlogging $PROB$-$[U, CC]$-B

	LS	WW	DLSI	DLS
Formulation	$Cons \times Vars$	$Cons \times Vars$	$Cons \times Vars$	$Cons \times Vars$
U	$SP\ O(n) \times O(n^2)$ $FL\ O(n^2) \times O(n^2)$ Section 10.2.2 [22, 137]	$O(n^2) \times O(n)$ Section 10.2.3 [140]	$--$	$--$
CC	$O(n^3) \times O(n^3)$ Section 10.3.4 [178, 180]	$O(n^3) \times O(n^2)$ Section 10.3.3 [125, 178]	$O(n^2) \times O(n)$ Section 10.3.2 [125, 176]	$O(n) \times O(n)$ Section 10.3.1 [125]
Separation				
U	$*$ Section 10.2.2 [137]	$O(n^3)$ Section 10.2.3 [140]	$--$	$--$
CC	$*$ Section 10.3.4	$*$ Section 10.3.3 [134, 104, 125]	$O(n^3)$ Section 10.3.2 [125]	$O(n)$ Section 10.3.1 [125]
Optimization				
U	$O(n \log n)$ Section 10.2.1 [3, 63, 187]	$O(n)$ Section 10.2.3 [3, 63, 187]	$--$	$--$
CC	$O(n^3)$ Section 10.3.4 [176]	$O(n^3)$ Section 10.3.3 [176]	$O(n^2 \log n)$ Section 10.3.2 [176]	$O(n^2)$ Section 10.3.1 [176]

4.4.3 Results for Start-Up Cost Models $PROB$-$[U, CC]$-SC

Finally we list in Table 4.6 the results for problems with start-up costs.

$DLS(WW)$ refers to the special case of DLS-CC-SC with just set-up and start-up costs in which the set-up costs are non-increasing over time (i.e., $q_t \geq q_{t+1}$; see Section 10.5.1).

Table 4.6. Models with Start-Up Costs $PROB$-$[U, CC]$-SC

		LS	WW	DLS
Formulation		$Cons \times Vars$	$Cons \times Vars$	$Cons \times Vars$
U		$SP(SC)\ O(n^2) \times O(n^2)$ $FL(SC)\ O(n^3) \times O(n^2)$ Section 10.4.2 [170, 191]	$O(n^2) \times O(n)$ Section 10.4.3 [140]	$--$
CC		$* * *$	$* * *$	$O(n^2) \times O(n^2)$ $(WW)\ O(n^2) \times O(n)$ Section 10.5.1 [165, 163]
Separation				
U		$O(n^3)$ Section 10.4.2 [170, 191]	$Exercise 10.13$ Section 10.4.3 [140]	$--$
CC		$O(n^2) *$ Section 10.5 [46]	$* * *$	$*$ Section 10.5.1 [164]
Optimization				
U		$O(n \log n)$ Section 10.4.1 [3, 63, 187]	$O(n)$ [3, 63, 187]	$--$
CC		$O(n^4)$ Section 10.5 [71]	$* * *$	$O(n^2)$ $(WW)\ O(n \log n)$ Section 10.6 [67, 147, 164]

Finally there is a reformulation for WW-U-B, SC, described in Section 10.6, with $O(n^2)$ constraints and $O(n)$ variables.

4.4.4 The Reformulation Procedure

Here we present general guidelines on how to use the classification scheme and the reformulation tables in order to obtain good or state-of-the-art formulations for production planning models.

We demonstrate the approach in detail in the next section on the Master Production Scheduling Example from Chapter 1, and on elementary case studies in Chapter 5.

Rule 1. Construct an initial model and a mathematical formulation using the classification scheme from Section 4.3. In particular, characterize or classify the single-item models as $PROB$-CAP-VAR.

Rule 2. For each single-item model, select appropriate reformulations by identifying the closest cell or cells in the reformulation tables.

The choice of a reformulation depends often on a compromise between its quality or tightness and its size. Therefore, several reformulations can be selected. From a given cell identified from the classification, we can move to other cells in order to obtain valid or allowed reformulations of the model. The allowed moves are

- *move upwards* $CC \Rightarrow U$, usually performed to reduce the size of the reformulation or the number of cuts generated.
- *towards the right* $LS \Rightarrow WW$, usually to reduce the size of the reformulation or the number of cuts generated.
- *towards the right* $WW \Rightarrow \{DLSI_k\}_{k=0,\dots,n-1}$.
- *towards the left* $WW \Rightarrow LS$.

Rule 3. The different reformulations identified should then be implemented, tested, and compared in terms of solution quality and computing time.

The allowed moves from cell to cell given in Rule 2, as well as some other moves, are justified by the following relations that exist between the sets of feasible solutions associated with the problems in the classification; see Section 4.3.

$$X^{prob-cap-SC} \subseteq X^{prob-cap},$$

$$X^{prob-CC-var} \subseteq X^{prob-U-var},$$

$$X^{LS-cap-var} \subseteq X^{WW-cap-var} \subseteq \bigcap_{k=0}^{n-1} X^{DLSI_k-cap-var},$$

where in each relation *prob*, *cap*, and *var* represent any fixed value of the fields $PROB$, CAP, and VAR, respectively. For instance, as any solution of $X^{prob-CC-var}$ is included in the larger set $X^{prob-U-var}$, any valid constraint or formulation for the larger set $X^{prob-U-var}$ is also valid for $X^{prob-CC-var}$, and thus the move $CC \Rightarrow U$ is allowed.

The move $WW \Rightarrow LS$ is justified by the discussion and remarks in Section 4.3 relative to the choice between classification LS or WW for the field $PROB$. In a multi-item lot-sizing problem where the single items satisfy the WW cost condition, the classification and formulation LS are more appropriate when additional constraints (such as linking capacity constraints) destroy the stock minimal characteristic of optimal solutions.

As an illustration, consider a multi-item single-level single-machine problem. Suppose that the subproblem for each item is classified as WW-CC-B.

- We identify first the cell WW-CC-B in Table 4.5. A reformulation is proposed, but $O(n^3) \times O(n^2)$ appears very large, because this reformulation must be applied individually to all items.

- We can move upwards from CC to U in Table 4.5 to find a relaxation. The relaxation $WW\text{-}U\text{-}B$ is obtained for which a tight and more compact $O(n^2) \times O(n)$ reformulation is indicated.
- We can move towards the right in Table 4.5 to find another relaxation. We obtain the relaxations $DLSI_k\text{-}CC\text{-}B$, for $k = 0, \ldots, n-1$, for which a tight $O(n^2) \times O(n)$ reformulation is again known for each k. However, this leads to an $O(n^3) \times O(n^2)$ formulation, which is again large.

4.5 A Production Planning Example: Reformulation and Solution

We have already illustrated on a MPS example in Section 1.2.4 that the structure of a MIP formulation can be used in order to improve both the quality of the solution and the final duality gap (see Table 1.5). Such improvements were based on the reformulation of simple (low-level) structures embedded in the problem, such as single mixed integer constraints or single-node flow structures (see Chapter 8). Moreover, they are obtained automatically by using state-of-the-art branch-and-cut solvers.

Here we show how to profit from the classification scheme to recognize more global structures that are specific to production planning problems. It is then possible to use the known reformulation results for these canonical planning structures in order to obtain an even better formulation of the initial problem.

As a simple and basic illustration of this principle (more comes later in the case studies in Chapter 5 and in Part V), we analyze the initial formulation (1.1)–(1.7) of our MPS example and observe that the Wagner–Whitin cost condition is satisfied because there are no production costs and there are positive inventory costs. Moreover, constraints (1.2)–(1.4) define an uncapacitated lot-sizing structure for each product and constraint (1.3) defines safety stocks for each item. Therefore each single-item submodel is classified as

$$WW-U-SS.$$

Observe that the single-item problems could be classified as $LS\text{-}U\text{-}SS$ because the capacity constraints linking the different items are likely to destroy the stock minimal structure of optimal solutions (see the discussion and remarks in Section 4.3 relative to the choice between classification LS or WW for multi-item problems).

We illustrate here how to use some known a priori reformulation results for these single-item submodels. These reformulations are given here for completeness, but they are analyzed in depth in Parts II and III.

Removing the Safety Stocks

First, note that the reformulation Table 4.4 does not include the safety stock variant. So, before applying the $WW-U$ reformulation with $O(n^2)$ constraints and $O(n)$ variables from Table 4.4, we apply a standard linear programming trick to remove the simple lower bound on the inventory variables, that is, to remove the safety stocks.

The inequality $s_t^i \geq SS_{t-1}^i - D_t^i$ always holds because the entering stock of item i in period t that is not used to satisfy some demand in period t must be part of the inventory at the end of period t. Therefore, and without loss of generality, we can tighten the safety stock for each item i and for $t = 1, \cdots, NT$ by setting

$$SS_t^i := \max\{SS_{t-1}^i - D_t^i, \ SS_t^i\},$$

where SS_0^i is the initial inventory of item i.

Then, we can eliminate the lower bounds on inventory by defining net inventory variables $ns_t^i := s_t^i - SS_t^i \geq 0$ for all i and t. After replacing the inventory variables by the net inventory variables (i.e., replacing s_t^i everywhere by $ns_t^i + SS_t^i$), we obtain the following equivalent formulation.

$$\min \ \sum_i \sum_t ns_t^i + \sum_i \sum_t SS_t^i \tag{4.57}$$

$$ns_{t-1}^i + x_t^i = ND_t^i + ns_t^i \qquad \text{for all } i, t \tag{4.58}$$

$$x_t^i \leq M_t^i y_t^i \qquad \text{for all } i, t \tag{4.59}$$

$$\sum_i \alpha^{i1} x_t^i + \sum_i \beta^i y_t^i \leq C^1 \qquad \text{for all } t \tag{4.60}$$

$$\sum_{i \in F^k} \alpha^{ik} x_t^i \leq C^k \qquad \text{for all } t \text{ and } k = 2, 3 \tag{4.61}$$

$$ns_0^i = 0 \ , \ ns_t^i \in \mathbb{R}_+^1 \qquad \text{for all } i, t \tag{4.62}$$

$$x_t^i \in \mathbb{R}_+^1, \ y_t^i \in \{0, 1\} \qquad \text{for all } i, t, \tag{4.63}$$

where $ND_t^i := D_t^i + SS_t^i - SS_{t-1}^i \geq 0$ is the net demand of item i in period t, and where the upper bound M_t^i on the production of item $i \in F^k$ in period t in constraint (4.59) is taken as

$$M_t^i = \min\{\sum_{l=t}^{NT} ND_l^i, \ \frac{C^1 - \beta^i}{\alpha^{i1}}, \ \frac{C'^k}{\alpha^{ik}}\}.$$

Extended Reformulation WW-U

Each single-item model (4.57)–(4.59) and (4.62)–(4.63) in the above formulation is classified as WW-U. Table 4.4 indicates the existence of the following linear reformulation with $O(n^2)$ constraints and $O(n)$ variables for this WW-U model (written for item i, translated directly for the net demand data ND_t^i and the net inventory variables ns_t^i); see Chapter 7.

$$ns_{t-1}^i + x_t^i = ND_t^i + ns_t^i \qquad\qquad \text{for all } t \qquad (4.64)$$

$$ns_{t-1}^i \geq \sum_{j=t}^{l} ND_j^i \left(1 - \sum_{u=t}^{j} y_u\right) \qquad\qquad \text{for all } t, l \qquad (4.65)$$

$$ns_t^i, x_t^i \in \mathbb{R}_+^1, \ y_t^i \in [0,1] \qquad\qquad \text{for all } i, t \qquad (4.66)$$

The $O(n^2)$ constraints (4.65) impose that the stock at the end of period $t-1$ must contain the demand of period $j \geq t$ if there are no set-ups in periods t up to j (i.e., if $\sum_{u=t}^{j} y_u = 0$).

The first reformulation consists of constraints (4.57)–(4.63), plus the constraints (4.65) for all items instead of the constraints (4.59). It is easily implemented in Mosel. The results obtained with the Xpress-MP Optimizer using this a priori reformulation are compared in Table 4.7 with the results obtained using the initial or basic formulation (4.57)–(4.63), with and without the Xpress-MP system cuts. Column "LP Val." gives the initial linear relaxation or lower-bound value before the Xpress-MP cuts, and column "XLP Val." gives the lower bound obtained at the root node after the addition of of Xpress-MP cuts.

Table 4.7. Extended Reformulation WW-U for the GW MPS Example

Algorithm Formulation	Vars Cons	LP Val.	XLP Val. Ncuts	Best LB Best UB	Best UB t. (secs) Gap (%)
Basic form. B & B	540	2893	2893	3341	0
(w/o Xpress-MP cuts)	405		0	6415	47.92
Basic form. B & C	540	2893	5481	5614	56
(with Xpress-MP cuts)	405		239	5746	2.30
WW – U B & C	540	5395	5496	5652	269
(with Xpress-MP cuts)	1845		18	5732	1.40

$NI = 12$ and $NT = 15$. Maximum 600 second runs.

The optimization was stopped after 600 seconds. With the WW-U reformulation, we obtain a slightly better feasible solution (see column "Best UB"), and better initial (see column "XLP Val.") and final lower bounds (see column "Best LB"). The column "Best UB t." gives the time in seconds to find the best feasible solution. The duality gap is reduced to 1.40%. Note that these results are obtained with the combination of the generic Xpress-MP cuts (with default branch-and-cut parameter settings) and the specific production planning reformulations.

Other Extended Reformulations

As we already observed, the single-item problems can also be classified as LS-U because the capacity constraints linking the different items are likely to destroy the stock minimal structure of optimal solutions. Therefore the known reformulations for model $LS\text{-}U$ given in Table 4.4 (namely the facility location (FL), shortest path (SP), and multi-commodity (MC) reformulations; see Chapter 7) could also be used and tested (Rule 3 of the reformulation procedure).

We have described the multi-commodity reformulation at the beginning of this chapter. As another example, the facility location reformulation for the single-item $LS\text{-}U$ model (4.58)–(4.59), (4.62)–(4.63) (without lower bounds on the net inventory) is defined on the extended variable space x_{tl}^i, for all items i, periods t and $l \geq t$, where x_{tl}^i represents the amount of item i produced in period t to satisfy net demand in period $l \geq t$.

Using the facility location reformulation, and the substitutions $x_t^i = \sum_{l \geq t} x_{tl}^i$ and $ns_t^i = \sum_{k=1}^{t} \sum_{l=t+1}^{NT} x_{kl}^i$, the final facility location reformulation of (4.57)–(4.63) is

$$\min \quad \sum_i \sum_t \sum_{l \geq t} (l - t) x_{tl}^i + \sum_i \sum_t SS_t^i \tag{4.67}$$

$$\sum_{t=1}^{l} x_{tl}^i = ND_l^i \qquad\qquad \text{for all } i, l \tag{4.68}$$

$$x_{tl}^i \leq ND_l^i y_t^i \qquad\qquad \text{for all } i, t, l \text{ with } l \geq t \tag{4.69}$$

$$\sum_i \sum_{l \geq t} \alpha^{i1} x_{tl}^i + \sum_i \beta^i y_t^i \leq C^1 \qquad\qquad \text{for all } t \tag{4.70}$$

$$\sum_{i \in F^k} \sum_{l \geq t} \alpha^{ik} x_{tl}^i \leq C^k \qquad\qquad \text{for all } t \text{ and } k = 2, 3 \tag{4.71}$$

$$x_{tl}^i \in \mathbb{R}_+^1, \; y_t^i \in \{0, 1\} \qquad\qquad \text{for all } i, t, l \text{ with } l \geq t . \tag{4.72}$$

The shortest path reformulation is derived directly from the dynamic programming algorithm used to solve $LS\text{-}U$, and is described in Chapter 7.

As a last reformulation, we can also implement and test the $O(n^2) \times O(n^2)$ extended reformulation for the single-item constant capacity model $WW\text{-}CC$ referred to in Table 4.4 and described in Chapter 9. For some items, the total demand over the planning horizon is larger than the production capacity of one period. Therefore, for each item i, with $i \in F^k$, one can define a constant upper bound on production

$$U^i = \min\{\sum_{t=1}^{NT} ND_t^i, \; \frac{C^1 - \beta^i}{\alpha^{i1}}, \; \frac{C^k}{\alpha^{ik}} \},$$

such that $x_t^i \leq U^i y_t^i$ is valid for all t. In any case, model $WW\text{-}CC$ is larger than, but at least as strong, as model $WW\text{-}U$.

The results obtained using these extended reformulations with the Xpress-MP Optimizer are compared in Table 4.8 with the results obtained using the initial or basic formulation. All the results have been obtained with the default branch-and-cut system from Xpress-MP.

Table 4.8. Extended Reformulations for the GW MPS Example

Algorithm Formulation	Vars Cons	LP Val.	XLP Val. Ncuts	Best LB Best UB	Best UB t. (secs) Gap (%)
Basic form. B & B	540	2893	2893	3341	0
(w/o Xpress-MP cuts)	405		0	6415	47.92
Basic form. B & C	540	2893	5481	5614	56
(with Xpress-MP cuts)	405		239	5746	2.30
WW-U B & C	540	5395	5496	5652	269
(with Xpress-MP cuts)	1845		18	5732	1.40
LS-U (MC) B & C	2880	5395	5503	5667	88
(with Xpress-MP cuts)	2925		26	5732	1.13
LS-U (FL) B & C	1620	5395	5526	5702	534
(with Xpress-MP cuts)	1665		59	5730	0.49
LS-U (SP) B & C	1620	5395	5486	5672	419
(with Xpress-MP cuts)	417		22	5730	1.01
WW-CC B & C	2160	5395	5480	5651	319
(with Xpress-MP cuts)	2205		23	5732	1.41

$NI = 12$ and $NT = 15$. Maximum 600 second runs.

We observe in Table 4.8 that the results obtained with the different reformulations are similar. In 600 seconds, the best lower bound is achieved by the facility location reformulation, and the best feasible solution is obtained by the shortest path and the facility location reformulations. As expected, the LS-U reformulations tend to lead to (slightly) better lower bounds than the WW-U reformulation. The capacitated model WW-CC has no additional effect, probably because the capacity is always shared between items and the bound U^i on the individual production batches is not binding. The duality gap computed with the best lower and upper bounds among all the reformulations is 0.49%.

Given the good results obtained with the facility location reformulation, we solved the problem with this reformulation without any time limit, in order to obtain the optimal solution. The optimal solution is the solution of value 5730 found in less than 600 seconds, and it took 1195 seconds and 386,700 nodes in total to prove its optimality.

Reformulations in the Original Variable Space by Cutting Planes

We can observe in Table 4.8 that the better results (lower and upper bounds) have been obtained at the price of a large increase in the size of the formulation. This may slow down the solution of the linear relaxations, and reduce the number of branch-and-bound nodes evaluated within the time limit of 600 seconds.

An alternative leading to the same lower bound at the root node would be to reformulate the single-item models LS-U using the complete linear reformulation by valid inequalities in the original variable space (4.14)–(4.18) described in Section 4.1.2. It involves an exponential number of (l, S) constraints (4.17) that can be added using the separation algorithm described in Section 4.1.2.

We have tested this approach at the root node, starting from the basic formulation (4.57)–(4.63) where the safety stocks have been removed, by performing the following:

- Solving the linear relaxation;
- Solving the separation problem for each item i and each period l;
- Adding to the formulation each violated (l, S) inequality identified;
- Re-optimizing the new linear relaxation (only after the generation of cuts for all items i and all periods l);
- Solving again the separation problem for each item and period;
- Repeating this procedure until no more violated (l, S) inequalities are generated.

This can be easily implemented in the Mosel modeling language. On our MPS test problem, this procedure requires 14 passes (i.e., 14 iterations of cut generation for all items and periods with a single reoptimization) and generates 933 violated (l, S) cuts in total, in about 20 seconds. In order to reduce the size of the model, these cuts have been added as model cuts; that is, inactive cuts are removed from the model and put into a cut pool. In this way, only 458 of the cuts are kept in the final formulation at the top node.

Then the resulting formulation at the root node is passed to Xpress-MP, and the default MIP solver is used. The results of this cut-and-branch approach are given in Table 4.9.

Table 4.9. Cutting Plane Reformulation for the GW MPS Example

Algorithm Formulation	Vars Cons	LP Val.	CPLP Val. Ncuts	XLP Val. Ncuts	Best LB Best UB	Best UB t. Gap (%)
Basic form. B & C with (l, S) cuts and Xpress-MP cuts	540 405	2893	5395 458	5479 52	5672 5730	492 1.01

$NI = 12$ and $NT = 15$. Maximum 600 second runs.

We observe in Table 4.9 that the lower bound obtained with the 458 (l, S) inequalities generated as cuts at the root node before the addition of Xpress-MP cuts (see column "CPLP Val.") is effectively the same as the lower bound obtained with the extended reformulations (column "LP Val." in Table 4.8).

This holds because all reformulations define complete linear descriptions of the single item models.

Although this formulation is of the same quality as and of smaller size than the extended formulations, which allows one to evaluate more nodes in the same amount of time, the best lower bound obtained after 600 seconds is not better than with the extended reformulations. This may be due to the fact that we do not generate additional violated (l, S) inequalities in the branch-and-bound tree, and therefore the bounds in the tree may be worse than with the tight extended reformulations.

Note also that the optimal feasible solution is again found in less than 600 seconds.

Heuristic Primal Solutions

The reformulations used and tested so far are mainly aimed at improving the lower or dual bound on the objective function, but are not specifically designed to produce good feasible or primal solutions quickly.

So to obtain better upper bounds, we apply the relax-and-fix construction heuristic and the relaxation-induced neighborhood search improvement heuristic described in Section 3.6.

For relax-and-fix we have decomposed the planning horizon into three equal parts.

- In the first iteration, we relax the set-up variables for periods in $\{6, \ldots, 15\}$, solve the resulting MIP^1, and then fix the set-up decisions for periods in $\{1, \ldots, 5\}$.
- In the second iteration, with the fixed set-up decisions for periods in $\{1, \ldots, 5\}$, we relax the set-up variables for periods in $\{11, \ldots, 15\}$, solve the resulting MIP^2 and we additionally fix the set-up decisions for periods in $\{6, \ldots, 10\}$.
- In the third and last iteration, with the fixed set-up decisions for periods in $\{1, \ldots, 10\}$, we optimize the set-up decisions for periods in $\{11, \ldots, 15\}$.

This corresponds to $R = 3$, $Q^1 = \{1, \ldots, 5\}$, $Q^2 = \{6, \ldots, 10\}$, $Q^3 = \{11, \ldots, 15\}$, $U^1 = U^2 = \emptyset$ in the notation of Section 3.6.2.

To test the ability of the algorithm to generate good solutions quickly, we have limited the computation time of each iteration to maximum 40 seconds. So, we fix variables at their values in the best solution obtained after maximum 40 seconds, and the relax-and-fix algorithm takes maximum 120 seconds in total. Note that the only true lower bound produced by this relax-and-fix procedure is the best lower bound obtained at the end of the first iteration (solution of MIP^1) before any variable fixing.

We have implemented the relax-and-fix procedure in Mosel. This simply requires three successive runs of almost identical models. The only modifications are the status of the binary variables from relaxed to binary, and from binary to fixed. The results obtained are given in Table 4.10 using the WW-U and WW-CC reformulations.

First, the running times of the relax-and-fix heuristic are only 41 and 43 seconds, respectively, with formulations WW-U and WW-CC, because the time limit of 40 seconds is reached only for the second iteration MIP^2. Next, the relax-and-fix heuristic produces feasible solutions quickly, but of relatively moderate quality ("R&F Val.") compared to those obtained in 600 seconds without this procedure (see "Best UB" in Table 4.9). Also, the lower bounds obtained are very weak (see "Best LB" in Tables 4.10 and 4.9).

Table 4.10. Heuristic solution for the GW MPS Example

Formulation Algorithm	Vars Cons	LP Val.	Best LB	R&F Val. RINS Val.	R&F Time RINS Time
WW-U B&C/R&F/RINS	540	5395	5429	5928	41
(with Xpress-MP cuts)	1845			5743	2
WW-CC B&C/R&F/RINS	2160	5395	5429	5770	43
(with Xpress-MP cuts)	2205			5730	2

$NI = 12$ and $NT = 15$; Maximum 160 second runs.

We have also tested the relax-and-fix heuristic on the basic formulation (4.57)–(4.63). It failed to produce a feasible solution because the program obtained at iteration 2, after fixing the set-up decisions for periods $\{1, \ldots, 5\}$, was infeasible. Due to the weak relaxed model for periods $\{6, \ldots, 15\}$ (i.e., no reformulation is used), the set-up decisions obtained for the first periods do not anticipate the capacity problems in later periods and lead to an infeasible solution.

Therefore, it appears to be very important for the feasibility and quality of the relax-and-fix procedure to start with a good formulation of the problem, that is, with a good linear relaxation, or to find another way to anticipate the capacity restrictions in later periods.

Finally we have implemented and tested the relaxation-induced neighborhood search improvement heuristic described in Section 3.6.2. Specifically, we fix the set-up variables that have the same value (0 or 1) in the linear relaxation (root node solution) and in the solution obtained by relax-and-fix. We then solve the resulting MIP using the default Xpress-MP, with a time limit of 40 seconds (maximum 160 seconds, including relax-and-fix). The results in Table 4.10 show that the RINS procedure is able to improve the relax-and-fix solution, and even once to produce the optimal solution ("RINS Val."), in almost no additional running time.

The next chapter shows how to use the classification scheme and the reformulation procedure in practice, and includes two small case studies.

The objective of Parts II to IV is to present all the available reformulation approaches and results in a systematic way. Then, as in our illustrative example, Part V uses these results with the support of the classification scheme to solve industrial case studies.

Exercises

Applications and exercises relative to the classification scheme and the reformulation procedure are given in the case studies of Chapters 5 and 14.

Notes

Sections 4.1 The multi-commodity reformulation for fixed charge network flow problems, implemented and tested in Section 4.1.1, was proposed by Rardin and Choe [145].

Sections 4.3 and 4.4 The classification scheme and the reformulation tables are taken from Wolsey [194]. An earlier and somewhat different classification scheme has been proposed by Bitran and Yanasse [28], and these authors also prove that the four varying capacity problems $PROB\text{-}C$ are NP-hard, because these problems are polynomially reducible to the 0–1 knapsack problem.

Section 4.5 The formulations and results presented here (and in Section 4.1) have been implemented and obtained using the Mosel algebraic modeling language (version 1.4.1) and the default version of the Xpress-MP Optimizer MIP solver (version 15.30). In particular the separation algorithm used to generate the (l, S) inequalities (4.13) has been directly coded in Mosel. See http://www.dashoptimization.com for more information about this software. All the tests reported here have been carried out on a 1.7 GHz PC (centrino) with 1 GB of RAM running under Windows XP.

Apart from the multi-commodity reformulation, the reformulations of the single-item problems $WW\text{-}U$ and $LS\text{-}U$ used here are studied in detail in Chapter 7. The $WW\text{-}CC$ reformulation is studied in Chapter 9. Appropriate references to these results are given in these chapters.

An introduction to the techniques used to prove that some valid inequalities suffice to describe the convex hull of solutions to a model is given in Section 6.4. For a general presentation of the various techniques that can be used to prove that some valid inequalities are facet defining, and for related topics, we refer the reader to Nemhauser and Wolsey [126].

5

Reformulations in Practice

Motivation

When tackling a new production planning problem it is interesting to try out several algorithmic options rapidly to see which ideas work on the given problem. This is also true for reformulations, heuristics, and different branch-and-cut options. Modern modeling languages and MIP solvers are sophisticated tools that permit one to develop and test these algorithmic possibilities easily, but this approach requires high-level algorithmic and mathematical expertise.

In addition to this first and classical approach, we describe here a library LS–LIB of reformulations, cutting plane separation routines, and heuristics that considerably simplifies and speeds up this (prototyping) process. Using this library, the user just needs to follow the classification scheme, without any knowledge of the mathematical description of the reformulations, and modify his or her problem formulation and optimization calls by adding calls to the chosen library routines/procedures (where the names of the appropriate data and variables are passed to the routines).

The library LS–LIB requires the utilization of specific modeling and optimization software, namely, Mosel and Xpress-MP. However, the extended reformulation procedures ($XForm$) from LS–LIB can be used just with Mosel to generate input or matrix files of the tightened formulations that can be read by almost any MIP solver.

Thus for our MPS example from Section 1.2, the results obtained in Section 4.5 with reformulations and heuristics (see Tables 4.8 to 4.10) can be obtained easily with LS–LIB without knowing any description of the mathematical reformulations used.

Objectives

As indicated in Chapter 4 the classification scheme permits us to choose an appropriate formulation, or cutting planes, for a specific production planning problem. The resulting problem is then tackled using heuristics and/or branch-and-cut.

The objective of this chapter is to demonstrate and teach the reader how to use the classification scheme in practice, so as, it is hoped, to "better solve" certain production planning problems, either by using a classical approach based on a modeling language and a MIP solver, or by using the library of reformulations and heuristic procedures LS–LIB.

Content

- In Sections 5.1, 5.2, and 5.3, we show how to use the reformulations, the cutting plane routines, and the primal heuristics, respectively. In each case we consider two versions:
  i. The first *classical approach* in which the user studies the formulations/cuts presented later in this book and, after writing his initial formulation in some modeling language, either adds the extended formulation using the same modeling language, or writes an appropriate separation routine or heuristic using his favorite programming language;
  ii. The second *black-box approach* in which, once the user's initial formulation is written in Mosel, he uses LS–LIB, by calling the appropriate procedures and passing the appropriate parameters (names of variables, demand vectors, capacities, etc.) used in his model.
- Section 5.4 lists all the procedures for reformulation, cutting plane separation, and heuristics provided in LS–LIB as well as their calling parameters.
- The chapter terminates with two case studies, which are first described in Section 5.5, and then formulated and solved using the classification scheme of Chapter 4 and the library LS–LIB in Sections 5.6 and 5.7, respectively.

5.1 Extended Reformulations

We assume that we want to solve a production planning problem, for which we have identified some valid relaxations WW-U and LS-U for each single item. In this section we show how to use reformulations for WW-U and LS-U respectively. For each we present the classical and black-box LS–LIB approach. We also show how to use LS–LIB to reformulate with approximate or partial extended formulations that are smaller, but potentially weaker than the complete formulations.

5.1.1 The Classical Approach for WW-U

In Table 4.4 we see that an extended reformulation for WW-U is presented in Section 7.5. There we see that the reformulation is

$$s_{k-1} \geq \sum_{u=k}^{t} d_u(1 - y_k - \ldots - y_u) \quad \text{for } 1 \leq k \leq t \leq NT.$$

These inequalities have already appeared as (4.65) in Section 4.5, and are a subset of the (l, S) inequalities presented in Subsection 4.1.2. Collecting terms, they can be rewritten as

$$s_{k-1} + \sum_{u=k}^{t} d_{ut}y_u \geq d_{kt} \quad \text{for } 1 \leq k \leq t \leq NT. \tag{5.1}$$

It is an easy task to add these inequalities to the initial formulation. Specifically, for a multi-item model written in Mosel, we would typically add

```
!-------------------------------------------------------------
declarations
  ! NT,NI: integer ! Already declared
  ww: array(1..NI,1..NT,1..NT) of linctr
  CDEM: array(1..NI,1..NT,1..NT) of real !cumulative demand
end-declarations

forall(i in 1..NI,k in 1..NT,t in k..NT)
  CDEM(i,k,t):=sum(u in k..t) DEM(i,u)

forall(i in 1..NI,k in 1..NT,t in k..NT)
  ww(i,k,t):= s(i,k-1)+
          sum(u in k..t) CDEM(i,u,t)*y(i,u) >= CDEM(i,k,t)
!-------------------------------------------------------------
```

5.1.2 The Black-Box Approach for *WW-U*

Here we make use of a procedure provided in the LS–LIB library, namely,

$$\text{XFormWWU(S,Y,D,NT,TK,MC)},$$

where the parameters are as follows:
NT is the number of periods,
S is a stock vector for periods $0, 1, \ldots, NT$,
Y is a set-up vector for periods $1, \ldots, NT$,
D is the demand vector for periods $1, \ldots, NT$,
TK is an approximation parameter discussed below with $0 \leq TK \leq NT$, and
$MC \in \{0, 1\}$ is the Model-Cut parameter. If $MC = 0$, the reformulation constraints are added a priori to the original matrix and, if $MC = 1$, they are added as model cuts to the cutpool.

To call *XFormWWU* within the Mosel model, it suffices to add the following:

```
! -----------------------------------------------------------
! To include the reformulation library
uses 'lslib-PPbyMIP'
! -----------------------------------------------------------
! Declare the parameters that must be passed
declarations
    ! NT,NI: integer ! Already declared
    S: array(range) of linctr
    Y: array(range) of linctr
    D: array(range) of real
    NT: integer
    TK: integer
    MC: integer
end-declarations

! Loop through the items
! Indicate variable names and data used in your Mosel file
! Add reformulation
 TK:=NT
 MC:=0
 forall(i in 1..NI) do
     S(0):= 0
     forall(t in 1..NT) S(t):= s(i,t)
     forall(t in 1..NT) Y(t):= y(i,t)
     forall(t in 1..NT) D(t):= DEM(i,t)
     XFormWWU(S,Y,D,NT,TK,MC)
 end-do
! -----------------------------------------------------------
```

Note that if the $O(n^2)$ constraints of the extended formulation lead to a formulation that is too large, one has the option of using the approximate reformulation with the approximation parameter $TK < NT$. In this case the inequalities (5.1) will only be added for values $1 \leq k \leq t < NT$ with $t - k \leq TK$. This leads to a smaller formulation with only $O(NT \times TK)$ constraints. In addition, using a small value of TK, one often obtains as good a lower bound as with large values of TK.

For this simple case, the classical approach is as simple as the black-box approach. However as we show later, the changes needed to call a reformulation for $LS\text{-}U$, or any other reformulation, are trivial, whereas understanding and writing out the correct reformulations oneself is nontrivial, requires further reading, and is prone to error.

5.1.3 The Classical Approach for *LS-U*

Here we use the so-called shortest path reformulation for *LS-U*. From Table 4.4, we see that it has $O(n)$ constraints and $O(n^2)$ variables, and is described in Subsection 7.4.2.

We see that the reformulation is:

$$\min \quad \sum_{u=1}^{n} p_u x_u + \sum_{t=1}^{n} q_t y_t$$

$$-\sum_{t=1}^{n} \phi_{1t} = -1$$

$$\sum_{u=1}^{t-1} \phi_{u,t-1} - \sum_{\tau=t}^{n} \phi_{t\tau} = 0 \qquad \text{for } 2 \leq t \leq n$$

$$\sum_{u=1}^{n} \phi_{un} = 1$$

$$\sum_{\tau=t:d_{t\tau}>0}^{n} \phi_{t\tau} \leq y_t \qquad \text{for } 1 \leq t \leq n$$

$$\sum_{\tau=t}^{n} d_{t\tau} \phi_{t\tau} = x_t \qquad \text{for } 1 \leq t \leq n$$

$$\phi_{ut} \subset \mathbb{R}_+^1 , \ y_t \in [0,1] \qquad \text{for } 1 \leq u \leq t \leq n,$$

where the variable $\phi_{kt} = 1$ if production takes place in period k and the amount produced is d_{kt}; that is, in period k one produces to satisfy the demand for periods k up to t.

The resulting block to be added to the initial formulation is, in Mosel, as follows:

```
!-------------------------------------------------------------
declarations
  ! NT,NI: integer ! Already declared
  CDEM: array(1..NI,1..NT,1..NT) of real
end-declarations

forall (i in 1..NI,t in 1..NT,l in t..NT)
  CDEM(i,t,l):=sum(u in t..l) DEM(i,u)

declarations
  sp: dynamic array(1..NI,1..NT,1..NT) of mpvar
  node: array(1..NI,1..NT) of linctr
  nodf: array(1..NI) of linctr
  defy: array(1..NI,1..NT) of linctr
```

```
   defx: array(1..NI,1..NT) of linctr
end-declarations

forall(i in 1..NI,l in 1..NT,t in 1..NT)
   create(sp(i,l,t))

forall(i in 1..NI,t in 1..1)
   node(i,t):= -SUM(l in t..NT)sp(i,t,l)=-1
forall(i in 1..NI,t in 2..NT)
   node(i,t):= SUM(l in 1..t-1)sp(i,l,t-1) -
               SUM(l in t..NT)sp(i,t,l)=0
forall(i in 1..NI)
   nodf(i):= SUM(t in 1..NT)sp(i,t,NT)=1
forall(i in 1..NI,t in 1..NT)
   defy(i,t):= SUM(l in t..NT|CDEM(i,t,l) > 0) sp(i,t,l) <=
               y(i,t)
forall(i in 1..NI,t in 1..NT)
   defx(i,t):= x(i,t)=SUM(l in t..NT)CDEM(i,t,l)*sp(i,t,l)
!-----------------------------------------------------------
```

5.1.4 The Black-Box Approach for *LS-U*

Here we add the shortest path reformulation for *LS-U* by making use of the
LS–LIB procedure

$$\text{XFormLSU2(S,X,Y,D,NT,TK,MC)},$$

where again
NT is the number of time periods,
S is the stock vector,
X is production quantity vector,
Y is the set-up vector,
D the demand, and TK and MC are as before.

It suffices to add the following block to the initial Mosel program.

```
!---------------------------------
uses 'lslib-PPbyMIP'
!---------------------------------
declarations
   ! NT,NI: integer ! Already declared
   S: array(0..NT) of linctr
   X: array(1..NT) of linctr
   Y: array(1..NT) of linctr
   D: array(1..NT) of real
   NT: integer
```

```
  TK: integer
  MC: integer
end-declarations

TK:=NT
MC:=0
forall(i in 1..NI)do
    S(0):= 0
    forall(t in 1..NT) S(t):= s(i,t)
    forall(t in 1..NT) X(t):= x(i,t)
    forall(t in 1..NT) Y(t):= y(i,t)
    forall(t in 1..NT) D(t):= DEM(i,t)
    XFormLSU2(S,X,Y,D,NT,TK,MC)
end-do
!----------------------------------
```

Here again it is possible to use an approximate shortest path reformulation that may give a weaker lower bound, but is smaller. By setting $TK < NT$, one obtains a formulation with $O(NT)$ constraints and $O(NT \times TK)$ variables. This approximate reformulation can be found in Section 7.6. In this case the black-box approach is already significantly easier than the classical approach.

5.2 Cut Separation

Now suppose that we wish to solve a problem containing items classified as LS-U using cutting planes. We see from Table 4.4 that the (l, S) inequalities and their separation routine are described in Section 7.4.1. In fact these inequalities were introduced and demonstrated in Subsection 4.1.2.

5.2.1 The Classical Approach for LS-U

Here the user must program the separation routine mentioned above to test whether a given linear programming solution is cut off by one of the (l, S) inequalities.

5.2.2 The Black-Box Approach for LS-U

The following block can be added to the initial Mosel formulation just before the call to the optimizer. The declarations are as for the reformulation XFormLSU2(). The loop on "i in $1..NI$" permits all the data and variable names to be passed via the routine XCutLSU(), as well as preparation of the appropriate separation routine. Finally the call XCut_init actually activates the separation routine.

```
!---------------------------------
uses 'lslib-PPbyMIP'
!---------------------------------
declarations
   ! NT,NI: integer ! Already declared
   S:array(0..NT) of linctr
   X,Y:array(1..NT) of linctr
   D:array(1..NT) of real
end-declarations

forall(i in 1..NI) do
   S(0) := 0
   forall(t in 1..NT) do
     S(t):= s(i,t)
     X(t):= x(i,t)
     Y(t):= y(i,t)
     D(t):= DEM(i,t)
   end-do
   XCutLSU(S,X,Y,D,NT)
end-do

XCut_init
!---------------------------------
```

5.3 Heuristics in LS–LIB

In Section 3.6 we presented several heuristics that work by fixing or dropping the integrality constraints on some or all of the 0–1 variables. Here we suppose specifically that these are the set-up variables y_t^i.

5.3.1 Calling a Construction Heuristic

To call a simple version of the relax-and-fix heuristic described in Section 3.6.1, it suffices to add the following block to the Mosel formulation.

```
!---------------------------------------
uses 'lslib-PPbyMIP'
!---------------------------------------
declarations
   ! NT,NI: integer ! Already declared
   CY: array(1..NI,1..NT) of linctr
   HEURSOL:array(1..NI,1..NT) of integer
   MAXTIME: integer
   FIX,BIN: integer
```

```
end-declarations

forall(i in 1..NI,t in 1..NT)
   CY(i,t):= y(i,t)
MAXTIME:=30
FIX:= 4
BIN := 6


XHeurRF(CY,HEURSOL,COST,NI,NT,MAXTIME,FIX,BIN)
!----------------------------------------
```

where
FIX is the number of time periods in which the set-up variables are fixed in each relax-and-fix iteration (i.e., FIX is the constant value of $|Q^r|$ for $1 \leq r \leq R$ in the relax-and-fix algorithm described in Section 3.6.1);
BIN is the number of time periods in which the set-up variables are not relaxed (i.e., binary) in each iteration (i.e., BIN is the constant value of $|Q^r \cup U^r|$ for $1 \leq r \leq R$ in the relax-and-fix algorithm described in Section 3.6.1), with $FIX \leq BIN$;
$HEURSOL$ contains the 0–1 solution produced by the heuristic;
CY is the set of linear constraints indexed over $1..NI, 1..NT$ and defining the y_t^i variables as binary variables;
$MAXTIME$ is the number of seconds allowed for each partial MIP solved (overridden if no feasible solution has been found); and
$COST$ is the name of the expression containing the objective function (minimization is assumed to be the direction of optimization).

In our simple implementation with $R = \lceil \frac{NT}{FIX} \rceil$ iterations, and using the notation of Section 3.6.1, MIP^r optimizes over the 0–1 set-up variables y_t^i where the period t lies in the set $Q^r \cup U^r = \{(r-1)FIX + 1, \ldots, (r-1)FIX + BIN\}$, and the 0–1 set-up variables y_t^i where the period t lies in the set $Q^r = \{(r-1)FIX + 1, \ldots, rFIX\}$ are fixed at their optimal value in MIP^r at the end of the rth iteration.

5.3.2 Calling an Improvement Heuristic

To call local branching (see Section 3.6.2) once just at the top node, it suffices to replace the last line of the relax-and-fix Mosel block by

```
!-------------------------------------------------------------
    XHeurLB(CY,HEURSOL,COST,NI,NT,MAXTIME,PK)
!-------------------------------------------------------------
```

where
HEURSOL must be initialized with a feasible y_t^i solution, and on termination contains the new heuristic solution; and
PK is the local branching parameter k described in Section 3.6.2.

5.4 LS–LIB Procedures

In this section we present the procedures available in LS–LIB.

5.4.1 Reformulations – XForm

Each reformulation concerns a single-item subproblem. In Table 5.1, Columns 1 and 2 indicate the problem classification, and the first header row contains the possible procedure parameters explained below.

S denotes the stock vector $s_0, s_1, \ldots, s_{NT}$.
R denotes the backlog vector $r_1, \ldots, r_{NT}$.
X denotes the production vector $x_1, \ldots, x_{NT}$.
Y denotes the set-up vector $y_1, \ldots, y_{NT}$.
Z denotes the start-up vector $z_1, \ldots, z_{NT}$.
W denotes the switch-off vector $w_1, \ldots, w_{NT}$.
D denotes the demand vector $d_1, \ldots, d_{NT}$.
C denotes the constant capacity C.
NT denotes the number of periods $n = NT$.
TK is the approximation parameter.
MC indicates if constraints are added to the cut pool as Model Cuts ($MC = 1$) or are added a priori to the formulation ($MC = 0$).

A "Y" in the table indicates that the corresponding parameter is present.
A "-" indicates that the parameter is not present.
A "0" in the "S" column indicates that just s_0 is present.
A "1" in the "C" column indicates that the constant capacity is assumed to be $C = 1$.
An "L" in the "C" column indicates that the value of the constant lower bound on production is passed to the routine in place of the capacity parameter.

N.B. In all the LS–LIB procedures, it is assumed that the time horizon is represented in Mosel as the range $1..NT$ or $0..NT$, and the set of items/skus/products as the range $1..NI$. If the time periods are represented as a set of strings, or sets of integers, an appropriate translation is needed before calling the procedures; see, for example, the Powder Production case in Part V.

Example 5.1 *Examination of the row $WW - U - SC, B$ in Table 5.1 indicates that we need to declare the variables and constants marked with a "Y", and then call the reformulation for each item. We assume that the variables and data in the Mosel problem formulation are called "sname(i,t), dname(i,t), etc".*

```
! ----------------------------------------------------------
declarations
     ! NT, NI: integer ! Already declared
     S: array(0..NT) of linctr
     R: array(1..NT) of linctr
     Y: array(1..NT) of linctr
     Z: array(1..NT) of linctr
     D: array(1..NT) of real
     TK: integer
     MC: integer
end-declarations

TK:= NT
MC:= 0
forall(i in 1..NI) do
   S(0):= sname(i,0)
   forall (t in 1..NT) do
     S(t):= sname(i,t)
     R(t):= rname(i,t)
     Y(t):= yname(i,t)
     Z(t):= zname(i,t)
     W(t):= wname(i,t)
     D(t):= dname(i,t)
   end-do
   XFormWWUSCB(S,R,Y,Z,W,D,NT,TK,MC)
end-do
! ----------------------------------------------------------
```

5.4.2 Cutting Plane Separation – XCut

To call cutting plane separation routines, the procedure arguments are shown in Table 5.2, and are essentially identical to those in Table 5.1.

5.4.3 Heuristics – XHeur

In Table 5.3 we indicate the calling parameters for the heuristics. Remember that:

CY denotes the linear expressions indexed over $1..NI, 1..NT$ defining the y variables as binary variables.

SOL indexed over $1..NI, 1..NT$ contains as input an initial feasible solution if it is an improvement heuristic, and as output the heuristic solution found (if any).

$COST$ is the name of the expression containing the objective function (minimization is assumed to be the direction of optimization).

Table 5.1. XForm

Classification		S R X Y Z W D C NT TK MC	Cons × Vars	Reference
LS-U1		Y - Y Y - - Y - Y Y Y	$O(n^2)\times O(n^2)$	[100] multicom
LS-U2		Y - Y Y - - Y - Y Y Y	$O(n)\times O(n^2)$	[61] short path
LS-U	B	Y Y Y Y - - Y - Y Y Y	$O(n^2)\times O(n^2)$	[22, 137]
LS-U	SC	Y - Y Y Y - Y - Y - -	$O(n^2)\times O(n^2)$	[170, 192]
WW-U		Y - - Y - - Y - Y Y Y	$O(n^2)\times O(n)$	[140]
WW-U	B	Y Y - Y - - Y - Y Y Y	$O(n^2)\times O(n)$	[140]
WW-U	SC	Y - - Y Y - Y - Y Y Y	$O(n^2)\times O(n)$	[140]
WW-U	SC,B	Y Y - Y Y Y Y - Y Y Y	$O(n^2)\times O(n)$	[6]
WW-U	LB	Y - - Y - - Y L Y Y Y	$O(n^3)\times O(n^2)$	[177]
WW-CC		Y - - Y - - Y Y Y Y Y	$O(n^2)\times O(n^2)$	[140]
WW-CC	B	Y Y - Y - - Y Y Y Y Y	$O(n^3)\times O(n^2)$	[180]
DLSI-CC		0 - - Y - - Y Y Y Y Y	$O(n)\times O(n)$	[140, 125]
DLSI-CC	B	0 Y - Y - - Y Y Y Y Y	$O(n^2)\times O(n)$	[125, 179]
DLS-CC	B	- Y - Y - - Y Y Y Y Y	$O(n)\times O(n)$	[125]
DLS-CC	SC	Y - - Y Y - Y 1 Y Y Y	$O(n^2)\times O(n)$	[163]

Table 5.2. XCut

Classification		S R X Y Z W D C NT TK	Separation	Reference
LS-U		Y - Y Y - - Y - Y -	$O(n^2)$	[23]
LS-C		Y Y Y - - - Y Y Y -		[15]
WW-U		Y - - Y - Y - Y -	$O(n)$	[140]
WW-U	B	Y Y - Y - - Y - Y -	$O(n^3)$	[140]
WW-CC		Y - - Y - - Y Y Y -	$O(n^2\log n)$	[140]
DLSI-CC		0 - - Y - - Y Y Y Y	$O(n\log n)$	[125]
DLSI-CC	B	0 Y - Y - - Y Y Y Y	$O(n^3)$	[179]

Table 5.3. XHeur

Algorithm	CY SOL COST	NI NT MAXT	PAR1 PAR2	Reference
RF	Y Y Y	Y Y Y	FIX BIN	[153, 193]
MIP	Y Y Y	Y Y Y	– –	Section 3.6
CF	Y Y Y	Y Y Y	– –	Section 3.6.1
RINS	Y Y Y	Y Y Y	– –	[52]
LB	Y Y Y	Y Y Y	PK –	[66]
EXCH	Y Y Y	Y Y Y	FIX –	Section 3.6.2

The value returned by the function gives the value of the heuristic solution (BIG if no solution is found).

All these heuristics are described in Section 3.6. The construction heuristics are relax-and-fix (RF), truncated branch-and-cut (MIP), and cut-and-fix

(CF), respectively. The improvement heuristics are relaxation-induced neighborhood search (RINS), local branching (LB), and exchange (EXCH).

5.5 Two Practice Cases

Here we present two cases. The initial description of each problem consists of a verbal description of the problem, its context, and data. Our approach in the next two sections is to divide the solution and the report of each case into two main parts:

i. A classification, complete or partial, of the problem based on the description, and an initial problem formulation in some modeling language;
ii. A discussion of possible reformulation and solution strategies, and a report on computational results with two or more formulations or algorithms.

5.5.1 Consumer Goods Production Line: Problem Description

We consider a production line in the fast-moving consumer goods (FMCG) industry producing 30 different skus (stock keeping units), which belong to six different product families, using a make-to-stock production policy. Capacity is limited and is not far in excess of average demand. Day-to-day demand is fluctuating, and during the year there are two seasonal peaks.

Production is organized in batches of fixed duration corresponding to a full shift (8 hours), and a single product or sku is produced during each batch. Therefore, production of each sku is scheduled in multiples of full shifts. Capacity must be 100% utilized within each shift or batch to reduce backlogs and to build up stocks. The process is continuous (24 hours per day, seven days per week), with the exception of planned maintenance periods. Depending on the sku, a batch corresponds to a few days up to several months of shipments/demand.

There are six product families, one standard and five variants. The production process is such that one cannot switch directly from production of one nonstandard family to another, but must first switch to the standard family.

Safety stock levels have been defined based on the current forecasting and planning processes. These safety stock and initial stock restrictions have been removed by computing net demands; see Section 4.5. Therefore the minimum net stock level (i.e., stock above the safety stock) is zero at the end of each time period.

Due to the fluctuating demands, the planning horizon is 20 days (60 periods). The production plan is regenerated weekly. The objective is to minimize inventory holding and backlogging costs. It is suggested that the cost of holding stock be taken as 0.125 times the cost of backlog.

After looking at the production plans generated over several weeks, the shop floor wishes to impose additional constraints, namely, that the number of production batches of each sku, and the number of campaigns of each family, be as small as possible over the 20-day horizon. They claim that this will not change the objective by more than 3 to 5%. Specifically they suggest the following.

i. For all skus from a nonstandard family, if the minimum number of batches required to meet demand is one or two, then this minimum number of batches should be produced (i.e., there is no build-up of stocks for slow moving items).

ii. If there is some nonstandard family for which the maximum number of batches for any sku of the family over the 20-day horizon is two or less, then the number of campaigns of the family should be restricted to this maximum. A *campaign* of a given family is a set of consecutive batches or shifts during which only skus of this family are produced.

What is the effect of these additional restrictions on the production plans and objective function?

Data

There are $NI = 30$ items, $NT = 60$ shifts, and $NF = 6$ families. For any sku, the quantity produced in an eight-hour shift is $CAP = 20,000$. The families are numbered from 1 to 6, with the standard family as number 1. The NI-vector FAM indicates to which family each sku belongs.

$$FAM := [4, 1, 2, 1, 2, 1, 2, 5, 3, 1, 2, 4, 1, 2, 4, 5, 3, 6, 1, 2, 4, 5, 3, 1, 1, 2, 2, 4, 4, 1].$$

The cost of storage and backlog are 0.125 and 1 per unit of sku per period, respectively.

The net demand data (after removal of initial and safety stocks) is in the file cgpdemand.dat.

5.5.2 Cleaning Liquids Bottling Line: Problem Description

Here the items correspond to four product families produced on a single bottling line. The time intervals are days, and the time horizon for planning is 30 days. Only one item is produced per day, and production each day is for a maximum of two shifts (16 hours), with a minimum production time of 7 hours. When switching families, the line is modified during the night so as not to interfere with production, so no time is lost, and only start-up and switch-off costs are incurred. The production and storage costs per family are constant over time. In certain periods the line is scheduled to be shut down for maintenance. Such periods are indicated by a very high set-up cost.

Recently there has been considerable discussion about whether to allow backlogging, and if so, how high the backlogging cost should be relative to the storage cost. Factors of both 2 and 10 have been suggested, but some are convinced that even with a factor of 10 the solution will not change much.

Data

There are $NI = 4$ items, and $NT = 30$ periods.
The NT-vector q gives the item independent set-up cost per period.
The NI-vector a gives the number of units of item i produced per hour.
The NI-vector h gives the storage costs per item per period.

```
a:=[807,608,1559,1622]
h:=[0.0025,0.0030,0.0022,0.0022]
q:=[100,100,100,9999,100,100,100,100,100,9999,
    100,100,100,100,100,100,100,100,9999,100,
    100,100,100,100,9999,100,100,100,100,100]
```

Start-up and switch-off costs are both 50.
Lower and upper bounds on production in hours are 7 and 16, respectively.
The demands for each item and period contained in the file `cldemand.dat`
are measured in production hours.

5.6 The Consumer Goods Production Line Case

5.6.1 Initial Classification

Single-item: For each item, 100% capacity utilization means that it is a discrete lot-sizing model. The capacity is constant by item and there is backlogging. So, the classification is $DLS\text{-}CC\text{-}B$.
Multi-item resources: There is a single production line on which exactly one sku is produced in each period.
Multi-item sequencing: The sequencing constraints only involve families. These are problem specific, and are discussed further below.

5.6.2 Initial Formulation

$$\min \quad \sum_{i,t}(b_t^i r_t^i + h_t^i s_t^i)$$

$$s_{t-1}^i - r_{t-1}^i + C^i y_t^i = d_t^i + s_t^i - r_t^i \qquad \text{for all } i,t$$

$$\sum_i y_t^i = 1 \qquad \text{for all } t$$

$$s_t^i, r_t^i \in \mathbb{R}_+^1, \ y_t^i \in \{0,1\} \qquad \text{for all } i,t$$

$$+ \text{ constraints on the sequencing of families.}$$

We define $\phi_t^f = 1$ if an sku from family f is produced in period t, and let $I(f) = \{i : FAM(i) = f\}$ be the skus in family f. Now the sequencing constraints say that if an sku from some nonstandard family f is produced in period t, an sku from another nonstandard family g cannot be produced in period $t+1$. Recalling that Family 1 is the standard family, and the others $2,\ldots,6$ are nonstandard, one possible formulation for the sequencing of families is:

$$\phi_t^f = \sum_{i \in I(f)} y_t^i \qquad\qquad \text{for all } f,t$$

$$\sum_f \phi_t^f = 1 \qquad\qquad \text{for all } t$$

$$\phi_t^f + \sum_{g:g \neq 1,f} \phi_{t+1}^g \leq 1 \qquad\qquad \text{for } 2 \leq f \leq 6, \text{ and all } t$$

$$\phi_t^f \in \{0,1\} \qquad\qquad \text{for all } f,t .$$

5.6.3 Reformulation and Algorithms

In Table 4.5, we see that $DLS\text{-}CC\text{-}B$ is treated in Section 10.3.1.

The Classical Approach

In Section 10.3.1, we learn that it suffices to eliminate the backlog variables r_t^i from the flow conservation constraints, and to add the corresponding MIR inequalities, to obtain a tight formulation for $DLS\text{-}CC\text{-}B$. Here we work with the equivalent tight formulation obtained by eliminating the inventory variables. First we rewrite the flow conservation constraints as

$$r_t^i \geq d_{1t}^i - C^i \sum_{u=1}^{t} y_u^i \qquad \text{for all } i,t,$$

and then it suffices to add the MIR inequalities (see Section 8.1.1)

$$r_t^i \geq C^i f_t^i (\lceil \tfrac{d_{1t}^i}{C^i} \rceil - \sum_{u=1}^{t} y_u^i) \qquad \text{for } i = 1,\ldots,NI, \ \ t = 1,\ldots,NT,$$

when $f_t^i = \tfrac{d_{1t}^i}{C^i} - \lfloor \tfrac{d_{1t}^i}{C^i} \rfloor > 0$.

The Black-Box Approach

In Table 4.5 we also see that the reformulation of $DLS\text{-}CC\text{-}B$ is small ($O(n) \times O(n)$). We can therefore add the extended formulation with $TK = NT = 60$. Using the LS–LIB library, we add the following Mosel block.

```
!----------------------------------------
uses 'lslib-PPbyMIP'
!----------------------------------------
declarations
  ! NT, NI: integer ! Already declared
  C= CAP
  Y: array(1..NT) of linctr
  R: array(1..NT) of linctr
  D: array(1..NT) of real
  CDEM(1..NI,1..NT,1..NT) of real !cumulative demand
end-declarations

forall(i in 1..NI,k in 1..NT,t in k..NT)
  CDEM(i,k,t):=sum(u in k..t) DEM(i,u)

forall(i in 1..NI | CDEM(i,1,NT)>0) do
    forall(t in 1..NT) Y(t):=y(i,1,t)
    forall(t in 1..NT) R(t):=r(i,t)
    forall(t in 1..NT) D(t):=DEM(i,t)
    XFormDLSCCB(R,Y,D,C,NT,NT,0)
end-do
!----------------------------------------
```

5.6.4 Results

Based on the Mosel file `cgp.mos`, we obtain the results in Table 5.4 for the basic problem with a run-time limit of 600 seconds. The first line indicates the results with Xpress-MP default, and the second after addition of the Xform reformulation block. The columns represent the formulation; the number of columns, rows and 0–1 variables; followed by the initial LP value; the value XLP after the system has automatically added cuts at the root node; the best feasible solution obtained; the total time in seconds (a $*$ indicates that optimality is not proved in 600 seconds); the number of nodes in the branch-and-cut tree; and the gap (if any) after 600 seconds, respectively.

Table 5.4. *Consumer Goods* Before and After Reformulation

Formulation	Vars	Cons	Int	LP	XLP	Best UB	Time (secs)	Nodes	Gap (%)
cgp	4569	2275	1920	1627056	1779184	1987135	600*	98000	11.0
cgp-r	4569	3364	1920	1863547	1868012	1879048.5	78	939	0

It is standard to test whether setting priorities on the variables improves the performance of the enumeration. Here it seems natural to first branch on family variables ϕ_t^f with priority to earlier periods, followed by the item set-up variables again with priority to earlier periods. The additional Mosel code required is

```
forall(f in 1..NF,t in 1..NT)
    setmipdir(phi(f,t),XPRS_PR,50+t) !decide families first
forall(i in 1..NI,t in 1..NT|CDEM(i,1,NT)>0)
    setmipdir(y(i,t),XPRS_PR,200+t)
```

The results in Table 5.5 show that these branching directives do not improve the solution time and quality.

Table 5.5. *Consumer Goods* Before and After Reformulation with Priorities

Formulation	Vars	Cons	Int	LP	XLP	Best UB	Time (secs)	Nodes	Gap (%)
cgp	4569	2275	1920	1627056	1778278	2015702	600*	90300	12.1
cgp-r	4569	3364	1920	1863547	1868012	1887436	600*	27900	0.5

5.6.5 Sensitivity Analysis

Now we add the constraints suggested by the shop floor. First we define the minimum number of batches of sku i required to satisfy its demand as:

$$nb(i) = \lceil \frac{d_{1,NT}^i}{CAP} \rceil \, ,$$

where CAP is the output per shift, and the maximum number of campaigns of family f as

$$nc(f) = \max_{i \in I(f)} nb(i) \, ,$$

and then we model the restrictions as:

i. If there exists some sku i with $FAM(i) \neq 1$ with $nb(i) \in \{1, 2\}$, then add the constraint

$$\sum_t y_t^i \leq nb(i).$$

ii. If there exists some family $f > 1$ with $nc(f) \in \{1, 2\}$, then create new variables z_t^f for all t to model the start-up of campaigns, and add the constraints

$$z_t^f \geq \phi_t^f - \phi_{t-1}^f \qquad \qquad \text{for all } t$$
$$\sum_t z_t^f \leq nc(f)$$
$$z_t^f \in \{0, 1\} \qquad \qquad \text{for all } f, t.$$

Table 5.6. *Consumer Goods* with Shop Floor Recommendations

Formulation	Vars	Cons	Int	LP	XLP	Best UB	Time (secs)	Nodes	Gap (%)
cgp	4749	2472	2100	1627056	1744926	2147704	600*	44600	18.1
cgp-r	4749	3561	2100	1864253	1869864	1890699	291	3642	0

The results in Table 5.6 show first that the problem remains solvable to optimality using the single-item reformulations, and that the duality gap obtained without reformulations deteriorates. Next they show that there is an increase of only 0.6% of the optimal objective function value. This is due to the second restriction which forces the regrouping of batches of a same family into a limited number of campaigns, and therefore implies additional backlogs or stocks. The first restriction has no effect on global backlogging and inventory costs.

5.7 The Cleaning Liquids Bottling Line Case

5.7.1 Initial Classification

Single-item: The single item problems have constant capacity (16 hours) and constant lower bounds (7 hours), along with start-up and switch-off costs. As the unit production costs are constant over time, they can be ignored without loss of generality, and the Wagner-Whitin cost condition is satisfied because the inventory costs are constant over time. Thus we obtain the classification WW-CC-SC,LB.

Multi-item constraints and costs: There is only one machine, and at most one item can be produced per period.

5.7.2 Initial Formulation

$$\min \quad \sum_{i,t} h^i s_t^i + \sum_{i,t} q_t y_t^i + \sum_{i,t} g z_t^i + \sum_{i,t} \gamma w_t^i$$

$$s_{t-1}^i + x_t^i = d_t^i + s_t^i \qquad\qquad \text{for all } i,t$$

$$x_t^i \le C^i y_t^i \qquad\qquad \text{for all } i,t$$

$$x_t^i \ge L^i y_t^i \qquad\qquad \text{for all } i,t$$

$$z_t^i - w_{t-1}^i = y_t^i - y_{t-1}^i \qquad\qquad \text{for all } i,t$$

$$z_t^i \le y_t^i \qquad\qquad \text{for all } i,t$$

$$\sum_i y_t^i \le 1 \qquad\qquad \text{for all } t$$

$$x_t^i, s_t^i \in \mathbb{R}_+^1, \ y_t^i, z_t^i \in \{0,1\} \qquad\qquad \text{for all } i,t.$$

Here x_t^i denotes the production time of item i during t, and s_t^i, r_t^i, and d_t^i are also measured in production hours. The variable w_t^i models a switch-off of item i at the end of period t. Note that w_t^i is precisely the slack variable in the usual constraint $z_{t+1}^i \geq y_{t+1}^i - y_t^i$ used to define z_{t+1}^i; see Section 4.3.6.

5.7.3 Reformulation and Algorithms

We treat first the problem without backlogging, and consider the single-item aspects only. From Table 4.6, we see that valid inequalities are known for LS-CC-SC and thus for WW-CC-SC, but no extended formulation of reasonable size is known. However, to deal with the start-up variables a complete $O(n^2) \times O(n)$ formulation is available for WW-U-SC involving just the original (s, y, z) variables. This suggests the addition of the constraints

$$s_{t-1}^i \geq \sum_{u=t}^{l} d_u^i (1 - y_t^i - z_{t+1}^i - \cdots - z_u^i)$$

for all i, and for $1 \leq t \leq l \leq NT$.

On the other hand, to deal with the capacities, an $O(n^2) \times O(n^2)$ reformulation of WW-CC is available; see Section 9.5.3 as indicated in Table 4.4. This implies the addition of the following reformulation for each item i:

$$s_{k-1} \geq C \sum_{t \in [k,n]} f_t^k \delta_t^k + C\mu^k \qquad \text{for all } k$$

$$\sum_{u=k}^{t} y_u \geq \sum_{\tau \in \{0\} \cup [k,n]} \lceil \frac{d_{kt}}{C} - f_\tau^k \rceil \delta_\tau^k - \mu^k \qquad \text{for all } k, t, \ k \leq t$$

$$\sum_{t \in \{0\} \cup [k,n]} \delta_t^k = 1 \qquad \text{for all } k$$

$$\mu^k \geq 0, \ \delta_t^k \geq 0 \qquad \text{for all } t \in \{0\} \cup [k,n] \text{ and all } k$$

$$y \in [0,1]^n \ ,$$

where $f_0^k = 0$, $[k,n] = \{k, \ldots, n\}$ and $f_\tau^k = \frac{d_{k\tau}}{C} - \lfloor \frac{d_{k\tau}}{C} \rfloor$.

For the lower bound constraints, some valid inequalities are known, but for this we choose to stick to the original formulation.

5.7.4 Results

In Table 5.7 we present computational results showing the effects of the reformulations. Instance cl is the original formulation. Instance cl-$WWUSC$ is with the addition of the inequalities for WW-U-SC with $TK = 15$ and $MC = 0$. Instance cl-$WWCC$ is with the addition of the inequalities for WW-CC with $TK = 15$. Instance cl-$WWUSC$-CC has both the reformulations of WW-U-SC and WW-CC for each item. The nine columns represent the instance, the number of rows, columns, and 0–1 variables, followed by the initial LP value, the value XLP after the system has automatically added cuts, OPT the optimal value, the total number of seconds required to prove optimality, the number of nodes in the branch-and-cut tree, and the gap (if any) after 600 seconds.

Table 5.7. Results for Cleaning Liquids without Backlogging $TK = 15$

Instance	Cons	Vars	Int	LP	XLP	OPT	Time (secs)	Nodes	Gap (%)
cl	510	720	120	1509	3676	4404.5	171	68234	0
cl-WWUSC	1873	720	120	3775	4328	4404.5	12	203	0
cl-WWCC	2130	2220	120	3520	3956	4404.5	40	1384	0
cl-WWUSC-CC	3493	2220	120	4292	4330	4404.5	9	105	0

5.7.5 Sensitivity Analysis

If we introduce backlogging, it is necessary to refer to the classification Table 4.5, where we see that there exists a formulation for WW-CC-B. We also see below Table 4.6 that there is a compact extended formulation for WW-U-B, SC. As the formulations for WW-CC-B are large, we first try the formulation for WW-U-B, SC.

In Table 5.8, the results with backlogging are presented, with a time limit of 600 seconds. The parameter $\rho = 2$ is the ratio of unit backlogging cost over inventory unit cost, and the problems solved are the following.

clb2 is the original formulation with backlogging.

clb2-WWUBSC is with the WW-U-B, SC formulation and $TK = 15$.

clb2-WWUBSC-b is clb2-WWUBSC plus the WW-CC-B formulation with $TK = 5$.

clb2-WWUBSC-a is clb2-WWUBSC plus the WW-CC-B formulation with $TK = 15$.

Table 5.8. Results for Cleaning Liquids: Backlogging $\rho = 2$

Instance	Cons	Vars	Int	LP	XLP	OPT	Time (secs)	Nodes	Gap (%)
clb2	510	720	120	1507	2755	3386	600*	205100	4.0
clb2-WWUBSC	3327	868	120	3150	3176	3386	252	14064	0
clb2-WWUBSC-b	7479	5092	120	3236	3247	3386	597	3134	0
clb2-WWUBSC-a	37673	9892	120	3360	3360	3386	217	20	0

Next we try $\rho = 5$. One might expect this to give similar results to those of the original instance without backlogging because the backlogging cost is quite high.

Table 5.9. Results for Cleaning Liquids: Backlogging $\rho = 5$

Instance	Cons	Vars	Int	LP	XLP	OPT	Time (secs)	Nodes	Gap (%)
clb5	510	720	120	1509	3003	4074	600*	234800	11.8
clb5-WWUBSC	3327	868	120	3560	3564	4024	600*	38800	4.5
clb5-WWUBSC-b	7479	5092	120	3675	3702	4026	600*	4800	4.2
clb5-WWUBSC-a	37673	9892	120	3962	3962	4024	168	15	0

We observe that even with $\rho = 5$, there is a cost reduction of 380 which is close to 8%. Even with $\rho = 10$, the optimal value is 4241, a decrease of approximately 3% relative to the cost without backlogging.

5.7.6 Heuristics

Given that the instances with backlogging require 100 or more seconds to solve with the most effective reformulation, the user might be interested in finding good feasible solutions within 10–30 seconds. On the initial formulation clb, we tested the relax-and-fix heuristic followed by the exchange heuristic both with the parameters MAXTIME = 2 and FIX = BIN = 8, followed by the RINS heuristic for 30 seconds. On the first reformulation WW-U-B, SC with TK = 15, we tested the same heuristics but with MAXTIME = 5. The results are shown in Table 5.10 for $\rho = 2$, 5, and 10.

Table 5.10. Heuristics for Cleaning Liquids with Backlogging

Instance	RF	EXCH	EXCH	EXCH	EXCH	RINS	OPT
clb2	3386.0	3386.0	3386.0	3386.0	3386.0	3386.0	3386.0
clb5	4198.8	4198.8	4077.1	4047.5	4047.5	4024.2	4024.2
clb10	4456.2	4456.2	4378.2	4326.9	4326.9	4326.9	4241.1
clb2-WWUBSC-a	3386.0	3386.0	3386.0	3386.0	3386.0	3386.0	3386.0
clb5-WWUBSC-a	4153.1	4153.1	4056.6	4056.6	4056.6	4056.6	4024.2
clb10-WWUBSC-a	4241.1	4241.1	4241.1	4241.1	4241.1	4241.1	4241.1

In the last column, we have added the known optimal values to aid in assessing the effectiveness of the heuristics for this instance. We observe that relax-and-fix finds an optimal solution when $\rho = 2$ whether we start from either the weak or strengthened formulation. With $\rho = 5$, the optimal solution is found starting with the weak formulation with the RINS heuristic contributing to the quality. With $\rho = 10$, an optimal solution is found directly by relax-and-fix starting from the strong formulation.

Exercises

Exercise 5.1 The black-box approach
Consider the GW–GSCO problem from Section 1.2. The results in Section 4.5 (see Tables 4.8 to 4.10) have been obtained by the classical implementation approach for reformulations and heuristics.

Obtain similar results by using the black-box approach based on the library LS–LIB of reformulations and heuristics.

Exercise 5.2 Short-Term Formulation Planning in the Pharmaceutical Industry
The production process for pharmaceutics consists of three main stages.

1. *The bulk production stage.* The bulk is the active substance contained in the final product. Its production process is very long because of extensive quality control and tests. Therefore, the production is made to stock, and is based on long-term forecasts.
2. *The formulation stage.* The second step consists in the transformation of the bulk product into medicines, that is, products in their final form: pill, syrup, capsule, and so on. This step is also produced to stock with a horizon of several months. Formulation is done in large lots for productivity and quality control reasons, and formulation planning must take into account the bulk availability because it is impossible to adapt the bulk production plan in the short term.
3. *The packaging stage.* This final step can be produced very fast (and does not require quality tests) and is therefore made to order, just before the delivery to customers.

We consider the formulation planning problem for a single bulk used to formulate six products. One unit of bulk is used to formulate one unit of each formulated product. Each product is sold either to subsidiaries, or to tender contracts.

Given the bulk production plan and sales forecasts for each product and each sales order type, the purpose is to plan the formulation of products over a horizon of 42 weeks in order to satisfy the sales forecasts, without exceeding the availability of the bulk.

The other major constraint in this model consists of the time window constraint. For each type of sales order, subsidiary sales or tenders, each product must have a minimum remaining lifetime when packed and delivered to customers. Therefore, there exists for each product, and each sales order type, a maximum duration between the start of the formulation and the packaging/delivery to customers. This time duration is called the formulation time window.

Finally, among the formulation plans satisfying bulk availability and formulation time window restrictions, the operations manager first tries to minimize the number of formulation batches (maximum one batch per week per formulated product is allowed) in order to minimize the quality control workload, to simplify the traceability of the batches, and to increase the formulation productivity. In addition, for two formulation plans with the same total number of batches, the plan with the minimum overall stock levels is preferred. A batch of a given product has a maximum batch size fixed by the formulation equipment used.

The particular instance to be solved is defined by the following data.

a. The sales forecasts for each product-order type combination, and for each time period, are given in the file pharma.dat. Each line contains the sales forecasts over the planning horizon for one product-order type combination, in the order product1-subsid., product1-tender, product2-subsid., product2-tender, etc. All demands are net of initial inventory, and thus there is no initial inventory of formulated products, and the demands in the first periods are zero.

b. The initial bulk inventory allows one to formulate 3000 units of formulated products. In addition, there is a planned receipt of 20,000 units of bulk in the beginning of each week.

c. The formulation (including quality control) lead-time is 3 weeks for each product. Hence, a product whose formulation starts in period x is available to satisfy demands in period $x + 3$ and after.

d. The formulation time window is 5 weeks for subsidiaries, and 8 weeks for tenders. Hence, a formulation batch of a single product started in period x (and available in period $x + 3$) can only satisfy subsidiary demands up to period $x + 5$ and tender demands up to period $x + 8$.

e. The maximum batch sizes for the six products are $30,000$, 4000, 5000, 8000, 8000 and 8000 units, respectively.

Answer the following questions.

i. Build an initial formulation, and solve the corresponding MIP.

ii. Using the classification scheme and the reformulation tables from Chapter 4, identify valid relaxations for the problem. Reformulate the problem by using tight reformulations for these relaxations. Implement and test your reformulations using LS–LIB.

iii. Try to obtain the best possible solution in less than 600 seconds.

iv. Try to obtain the best possible duality gap in less than 600 seconds.

Exercise 5.3 Short-Term Planning for Glass Production Lines

We consider the production planning of a single-site glass plant, where three production lines or floats are used to produce six different product qualities or products. The daily demands of each product to be satisfied from production in this site are known and given for the next 15 days. For multi-site companies, such short-term and local demands are typically the output of a longer-term planning and demand allocation module. The goal is to find minimum cost production plans for the three floats satisfying the (allocated) demand for the next 15 days.

Each float produces a single product at a time, but is able to produce any product. A production campaign of a product on a float lasts for an integer number of days. When there is a changeover from a campaign of one product to a campaign of another product on a float at the beginning of a day, the float cannot be stopped but some production capacity is nevertheless lost. This lost capacity measures the quantity of waste produced during the changeover operation because of bad product quality. This lost capacity is machine- or float-dependent and is also sequence-dependent. The production rates are constant over all lines and all products, and cannot be changed or adapted to the demand.

The cost of a plan is the sum of the inventory holding costs for the products made in advance and the production costs. We assume here that each production day of a product on a line has a fixed cost, which is product-dependent.

The particular instance to be solved is defined by the following data.

a. The demands are given in the file float.dat. Each line contains the demands for all products in a given time period. Demands are net of initial inventory, and thus there is no initial inventory.

b. The production capacity per day is $CAP = 342$ units of product, for each product and each line.

c. The inventory holding costs per day are $h = (5, 5, 1, 3, 2, 1)$ for the different products.

d. The fixed production costs per line per day are $q = (800, 800, 600, 400, 400, 800)$ for the products.

e. The production capacity lost when switching from product i to product j on float line k, measured in units of product j, is given by $c_k^{ij} = st^j + 10(k-1) + 10|j-i|$, where the start-up capacities st are 40, 20, 20, 20, 40, 50, respectively, for the six products.

f. Initially, the float line i is ready (without losing capacity) to produce product i, for $1 \leq i \leq 3$.

Using the variables

- s_t^i to represent the inventory of product i at the end of period t,
- $y_{kt}^i \in \{0, 1\}$ to indicate whether line k produces product i in period t,
- $\chi_{kt}^{ij} \in \{0, 1\}$ to model whether there is a transition from product i in period $t-1$ to product j in period t on float line k,

an initial formulation for this planning problem is

$$\min \ \sum_i \sum_k \sum_t q^i y_{kt}^i + \sum_i \sum_t h^i s_t^i \qquad (5.2)$$

$$s_{t-1}^i + \sum_k [CAPy_{kt}^i - \sum_j c_k^{ji} \chi_{kt}^{ji}] = d_t^i + s_t^i \qquad \text{for all } i, t \qquad (5.3)$$

$$\sum_i y_{kt}^i = 1 \qquad \text{for all } k, t \qquad (5.4)$$

$$\chi_{kt}^{ij} \le y_{k,t-1}^i \qquad \text{for all } i, j, k, t \qquad (5.5)$$

$$\chi_{kt}^{ij} \le y_{kt}^j \qquad \text{for all } i, j, k, t \qquad (5.6)$$

$$\chi_{kt}^{ij} \ge y_{k,t-1}^i + y_{kt}^j - 1 \qquad \text{for all } i, j, k, t \qquad (5.7)$$

$$s_t^i \in \mathbb{R}_+^1, \ y_{kt}^i, \chi_{kt}^{ij} \in \{0,1\} \qquad \text{for all } i, j, k, t , \qquad (5.8)$$

where constraint (5.3) models demand satisfaction, constraint (5.4) imposes that every day exactly one product is set up on each line; constraints (5.5)–(5.7) define the changeover variables χ. Note that when $t = 1$ in constraint (5.5), $y_{k,t-1}^i$ must be replaced by the initial status of the production line i.

It is possible to tighten the initial formulation of the changeover variables by replacing constraints (5.5)–(5.7) by the unit flow formulation, described in Section 12.2.2, written as

$$\sum_i \chi_{kt}^{ij} = y_{kt}^j \qquad \text{for all } j, k, t \qquad (5.9)$$

$$\sum_j \chi_{kt}^{ij} = y_{k,t-1}^i \qquad \text{for all } i, k, t . \qquad (5.10)$$

where again when $t = 1$ in constraint (5.10), $y_{k,t-1}^i$ must be replaced by the initial status of the production line i.

Answer the following questions.

i. Implement the initial and tightened formulations, and test their performance by solving the corresponding MIPs. Compare the duality gaps obtained after 300 and 600 seconds.

ii. Using LS–LIB, implement relax-and-fix and RINS heuristics to solve this planning problem; see Section 3.6. Test your heuristics with the initial and tightened formulations, and try to obtain the best possible solution in less than 20 seconds.

Notes

Section 5.1 The explicit use of extended formulations and approximate extended formulations, as well as the first XForm routines, were developed by Van Vyve in his thesis [178]; see also Van Vyve and Wolsey [181] and Pochet et al. [135].

Section 5.2 Earlier attempts to automate the solution of production planning problems formulated as mixed integer programs include the MPSARX cut-and-branch system based on SCICONIC which recognized lot-sizing structure and generated (l, S) inequalities from its path separation routines (see Van Roy et Wolsey [175]), and BC-PROD based on EMOSL, a combined modeling and optimization language, in which a special syntax was imposed so as to recognize the embedded lot-sizing structures (variables, constraints, and data) and to add appropriate cutting planes; see Belvaux and Wolsey [25].

Section 5.3 The general heuristics implemented in LS–LIB are described in Section 3.6, and references are given in Section 3.6.2.

Section 5.4 LS–LIB and its global constraints are described in Pochet et al. [135].

Section 5.5 The Consumer Goods problem is a modified instance of a problem reported on earlier in Miller and Wolsey [125], that was studied as part of the EU-funded LISCOS project. The Cleaning Liquids Bottling line problem has been reported on in Belvaux and Wolsey [25].

Basic Polyhedral Combinatorics for Production Planning and MIP

6

Mixed Integer Programming Algorithms and Decomposition Approaches

In Parts II to IV we adopt a more rigorous approach. Our specific goals in Part II are both to provide a somewhat self-contained introduction to polyhedral combinatorics based on the uncapacitated lot-sizing problem and simple mixed integer sets, and to provide the background and results necessary to tackle the more complicated variants tackled in Parts III and IV.

First in this chapter we briefly give some basics of polyhedral theory allowing us to describe the general decomposition approaches for which finding good formulations and/or algorithms for the subproblems is crucial, and which allow us to distinguish between good and less good formulations. The uncapacitated lot-sizing problem studied in detail in Chapter 7 provides a remarkably rich model allowing us to use and demonstrate nearly all the important concepts of polyhedral combinatorics with many beautiful results in the form of simple algorithms and effective linear programming formulations. It also provides a point of comparison when tackling more complicated variants later. In Chapter 8 we present several fundamental results on valid inequalities, separation algorithms, and formulations for simple mixed integer sets that are either used later in the book to develop strong formulations for lot-sizing models or are used to generate strong cutting planes in several recent commercial mixed integer programming systems.

Turning now specifically to the contents of this chapter, the goal is to introduce more formally the important ideas about the formulation and decomposition of integer programs that motivate the results for single-item problems and multi-item problems presented in Parts II–IV. Some of these ideas were introduced less formally in Chapter 4.

- We start in Section 6.1 by collecting the polyhedral concepts that we use. We also present the theoretical equivalence between optimization and separation. This result is crucial – many variants of single-item lot-sizing problems with constant capacities are polynomially solvable; this equivalence tells us that for such problems finding the convex hull of the feasible region explicitly by inequalities or by an extended formulation is a possibility.

- In Section 6.2 we formalize the approach by decomposition and reformulation.
- In Section 6.3, we present various decomposition algorithms:
  1. Algorithms based on valid inequalities or extended formulations for the subproblems which are the topic of this book;
  2. Algorithms based on Lagrangian relaxation or column generation that are the algorithms of choice for many classes of problems in which it is relatively easy in practice to optimize over the "subproblems," but little is known about improved formulations; and
  3. Hybrid algorithms that combine both approaches.
- Having established the need to find improved formulations for a set X, and if possible formulations giving its convex hull, some of the arguments that can be used to show that a given formulation P for X is indeed the convex hull of X are listed in Section 6.4. Specific examples of such proofs are shown in later chapters.

6.1 Polyhedra, Formulations, Optimization, and Separation

We suppose that the problem has been formulated as an integer program

$$(IP) \qquad z = \min\{cx : Ax \geq b, x \in \mathbb{Z}_+^n\}$$

or as a mixed integer program

$$(MIP) \qquad z = \min\{cx + fy : Ax + By \geq b, x \in \mathbb{R}_+^n, y \in \mathbb{Z}_+^p\}.$$

For simplicity of notation we typically just discuss IP.

Definition 6.1 *Given the integer program IP, the set $X = \{x \in \mathbb{Z}_+^n : Ax \geq b\}$ is the set of feasible solutions.*

6.1.1 Formulations of an Integer Program

Definition 6.2 *The set of points $P = \{x \in \mathbb{R}^n : Ax \geq b\}$ satisfying a finite set of linear inequalities is a polyhedron.*

Definition 6.3 *A polyhedron P is called a formulation for X if $X = P \cap \mathbb{Z}^n$; that is X is precisely the set of integer points in P.*

We observe immediately that a set X can have an infinity of formulations. In Figure 6.1, we show two formulations for X.

Given two formulations P^1 and P^2 for X, P^1 is better than P^2 if $P^1 \subset P^2$. This makes sense because, for any objective function $c \in \mathbb{R}^n$,

$$z \geq \min\{cx : x \in P^1\} \geq \min\{cx : x \in P^2\}.$$

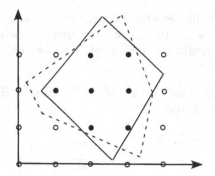

Figure 6.1. Two formulations for X.

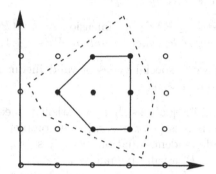

Figure 6.2. The convex hull of X.

Definition 6.4 *The* convex hull *of a set X, written* $\text{conv}(X)$*, is the set of points of the form* $x = \sum_{i=1}^{T} \lambda_i x^i$, $\sum_{i=1}^{T} \lambda_i = 1, \lambda_i \geq 0$ *for* $i = 1, \ldots, T$*, where* $\{x^1, \ldots, x^T\}$ *is any finite set of points of X. A point x expressed in this way is said to be a* convex combination *of the points* $x^1, \ldots, x^T$.

For IPs and rational MIPs, $\text{conv}(X)$ is a polyhedron. It follows that $\text{conv}(X)$ is the best of all possible formulations for X (see Figure 6.2), and

$$z = \min\{cx : x \in \text{conv}(X)\} \geq \min\{cx : x \in P\},$$

for all formulations P of X.

6.1.2 Valid Inequality Representation of Polyhedra

In this subsection we have in mind that P is a polyhedron, $X = P \cap \mathbb{Z}^n$, and in many cases $P = \text{conv}(X)$.

Definition 6.5 *An inequality $\pi x \geq \pi_0$ with $(\pi, \pi_0) \in \mathbb{R}^n \times \mathbb{R}^1$ is a* valid inequality *for $P \subseteq \mathbb{R}^n$ if it is satisfied by all points in P, that is, if $\pi x \geq \pi_0$ for all $x \in P$.*

One is particularly interested in "strong" or "strongest" valid inequalities. The natural candidates are the inequalities that are necessary to define the polyhedron P. To distinguish between valid inequalities, we need a couple of concepts.

Definition 6.6 *A set* $x^0, x^1, \ldots, x^k$ *of* $k + 1$ *points in* $\mathbb{R}^n$ *are affinely independent if the unique solution of*

$$\sum_{i=0}^{k} \alpha_i x^i = 0, \qquad \sum_{i=0}^{k} \alpha_i = 0$$

is $\alpha_i = 0$ *for* $i = 0, \ldots, k$, *or in other words if the* k *vectors of differences* $x^1 - x^0, \ldots, x^k - x^0$ *are linearly independent.*

Definition 6.7 *A set* $P \subseteq \mathbb{R}^n$ *is of* dimension k *(dim*$(P) = k$) *if the maximum number of affinely independent points in* P *is* $k + 1$.

Now we can single out special types of valid inequalities. Suppose that $\dim(P) = n - k$ for some $k \geq 0$.

i.   First there are valid inequalities that are satisfied at equality by all points of P, that is, hyperplanes $\{x : \pi x = \pi_0\}$ that completely contain P. There are k (linearly) independent equations with this property.
ii. Given a polyhedron P, a valid inequality $\pi x \geq \pi_0$
     – that contains $n - k$ (affinely) independent points of P satisfying the inequality at equality, and
     – such that $\pi x > \pi_0$ for some $x \in P$,
     is called a *facet-defining inequality*, and the corresponding set $F = P \cap \{x : \pi x = \pi_0\}$ is a *facet* of P. The dimension of F is $\dim(P) - 1$.
iii. Given a valid inequality $\pi x \geq \pi_0$ for P, the set $F = P \cap \{x : \pi x = \pi_0\}$ is called a *face* of P. It is a *proper* face if $\emptyset \subset F \subset P$.

In Figure 6.2, consider $P = \text{conv}(X)$. We have $n = 2$, $k = 0$, and $\dim(P) = \dim(X) = 2$. There is no equality $\pi_1 x_1 + \pi_2 x_2 = \pi_0$ that completely contains all points of X. $\text{conv}(X)$ is described by five facets, each of dimension $n - k - 1 = 1$ and containing two affinely independent points. The facet $F = \text{conv}(X) \cap \{x : x_2 = 1\}$ is generated by the facet-defining valid inequality $x_2 \geq 1$ and contains the affinely independent points $(2, 1)$ and $(3, 1)$.

Theorem 6.1 *If* $P = \{x \in \mathbb{R}^n_+ : Ax \geq b\}$ *with* $\dim(P) = n - k$, *the polyhedron* P *is completely described by* k *independent equalities of type i and one of the facet-defining inequalities of type ii for each facet of* P.

The interest of having a complete linear description of $P = \text{conv}(X)$ is that any linear integer program $\min\{cx : x \in X\}$ can be solved as a linear program $\min\{cx : x \in \text{conv}(X)\}$. Whether or not such a description is known and of manageable size, our *basic working hypothesis* is that

– the better the formulation of our integer program, the more effective will be the algorithm used to solve the problem, and

– adding "strong" valid inequalities, such as facet-defining inequalities for conv(X), to our initial formulation improves the formulation.

In particular if we are using a mathematical programming system based on linear programming and branch-and-bound, the better formulation will typically lead to improved linear programming bounds and less nodes in the branch-and-bound tree.

6.1.3 Extreme Point Representation of Polyhedra

Rather than being described by valid inequalities, polyhedra can also be described in terms of points and rays.

Definition 6.8 $x \in P$ is an extreme point *of a polyhedron P if there do not exist two points $x^1, x^2 \in P$, $x^1 \neq x^2$ with $x = \frac{1}{2}x^1 + \frac{1}{2}x^2$.*

In other words, an extreme point of P is a point of P that cannot be written as the convex combination of two other points in P.

Definition 6.9 $r \neq 0$ is a ray *of a polyhedron $P \neq \emptyset$ if $x \in P$ implies $x + \lambda r \in P$ for all $\lambda \geq 0$.*
A ray r of P is an extreme ray *if there do not exist two rays r^1, r^2 of P, $r^1 \neq \lambda r^2$ for some $\lambda > 0$, with $r = \frac{1}{2}r^1 + \frac{1}{2}r^2$.*

Theorem 6.2 *Every polyhedron $P = \{x \in \mathbb{R}^n : Ax \geq b\} \neq \emptyset$ with rank(A) = n can be represented as a convex combination of extreme points $\{x^t\}_{t=1}^T$ and a non-negative combination of extreme rays $\{r^s\}_{s=1}^S$:*

$$P = \{x : x = \sum_{t=1}^T \lambda_t x^t + \sum_{s=1}^S \mu_s r^s,$$
$$\sum_{t=1}^T \lambda_t = 1, \lambda \in \mathbb{R}_+^T, \mu \in \mathbb{R}_+^S\}.$$

The condition on the rank ensures that P has at least one extreme point, or equivalently that there is no ray r of P for which $-r$ is also a ray of P.

Example 6.1 *Consider the polyhedron P, shown in Figure 6.3:*

$$
\begin{aligned}
s + y &\geq 2.7 \\
2s + y &\geq 2 \\
3s + y &\geq 3 \\
s &\geq 0 \\
s + 0.7y &\geq 2.1
\end{aligned}
$$

The points $(6,0), (7,0), (0,5)$ in P are affinely independent, so dim(P) = 2. Removing the two inequalities $2s + y \geq 2$ and $3s + y \geq 3$ that do not define facets, and are thus not necessary to describe P, we obtain a minimal description of P:

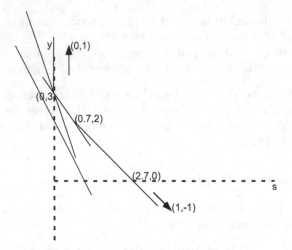

Figure 6.3. A polyhedron P.

$$
\begin{array}{rl}
s + & y \ \geq 2.7 \\
s & \ \geq \ 0 \\
s + & 0.7y \geq 2.1.
\end{array}
$$

To represent P in terms of extreme points and extreme rays, we see that there are two extreme points $(0.7, 2)$ and $(0, 3)$, and two extreme rays $(1, -1)$ and $(0, 1)$. Thus we obtain the alternative description:

$$
P = \{(s, y) \in \mathbb{R}^2 : \binom{s}{y} = \binom{0}{3}\lambda_1 + \binom{0.7}{2}\lambda_2 + \binom{0}{1}\mu_1 + \binom{1}{-1}\mu_2,
$$
$$
\lambda_1 + \lambda_2 = 1, \lambda \in \mathbb{R}_+^2, \mu \in \mathbb{R}_+^2\}.
$$

6.1.4 Cutting Planes and the Separation Problem

In principle one could add all the facet-defining inequalities a priori. However as there is an infinity of valid inequalities, and even the number of facet-defining inequalities can be incredibly large, it is not always possible or desirable to add all the inequalities to the formulation a priori.

Another possibility is to add valid inequalities as cuts or cutting planes thereby removing a point x^* that is not integral and that is part of the current formulation. Such points are typically obtained as the optimal solution of the linear program obtained by optimizing over the current formulation P of X; see Figure 6.4.

The problem of finding whether there is a valid inequality for X cutting off x^* is an important one.

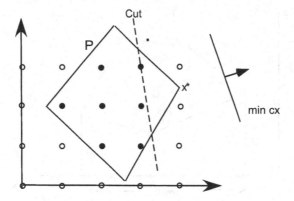

Figure 6.4. A cut removing x^*.

Definition 6.10 *Given* (X, x^*), *the* separation problem, *denoted* $SEP(X, x^*)$ *or* $SEP(\text{conv}(X), x^*)$, *is the problem of finding a valid inequality* (π, π_0) *for* $\text{conv}(X)$ *cutting off* x^*, *or deciding that there is no such inequality.*

Note that if no such inequality exists, $x^* \in \text{conv}(X)$.

Even if we do not have a complete description of $\text{conv}(X)$, we may have a family of valid inequalities $\mathcal{F}$. These give us implicitly the polyhedron

$$P_\mathcal{F} = \{x \in \mathbb{R}^n_+ : \pi x \geq \pi_0 \text{ for all } (\pi, \pi_0) \subset \mathcal{F}\},$$

for which we then wish to solve the separation problem $SEP(P_\mathcal{F}, x^*)$.

A generic cutting plane algorithm based on the separation problem has been described in Section 3.5.

6.1.5 Extended Formulations

Another way to strengthen a formulation is to look for an *extended formulation* involving *additional variables* that somehow leads to a more precise description of the problem. For $X = \{x \in \mathbb{Z}^n_+ : Ax \geq b\}$, suppose that it can be shown that

$$X = \{x \in \mathbb{Z}^n_+ : Bx + Gz \geq d \text{ for some } z \in \mathbb{R}^q\}.$$

Now let $Q = \{(x, z) \in \mathbb{R}^n_+ \times \mathbb{R}^q : Bx + Gz \geq d\}$.

The *projection* of Q into the x-space, denoted $\text{proj}_x Q$, is the polyhedron given by

$$\text{proj}_x Q = \{x \in \mathbb{R}^n : \text{ there exists } z \text{ for which } (x, z) \in Q\}.$$

Now $\tilde{P} = \text{proj}_x Q$ is a formulation for X as $X = \tilde{P} \cap \mathbb{Z}^n$. Such a projection is shown in Figure 6.5.

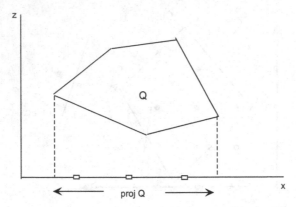

Figure 6.5. Extended formulation and projection.

Definition 6.11 *The polyhedron* $Q = \{(x,z) \in \mathbb{R}^n_+ \times \mathbb{R}^q : Bx + Gz \geq d\}$ *is an* extended formulation *for* $X = \{x \in \mathbb{Z}^n_+ : Ax \geq b\}$ *if* $\mathrm{proj}_x Q$ *is a formulation for* X.

An implicit formulation $\tilde{P}$ can be much stronger than the original formulation, and there are extended formulations, that we call *tight*, whose projection gives $\mathrm{conv}(X)$. What is particularly interesting is that the number of inequalities needed to describe $\mathrm{conv}(X)$ with an extended formulation may be small (perhaps polynomial) compared to the number of facet-defining inequalities (possibly exponential) needed to describe $\mathrm{conv}(X)$ in the original space.

6.1.6 Optimization and Separation: Polynomial Equivalence

The discussion so far has been motivated by the idea of finding an explicit description of $\mathrm{conv}(X)$, or a tight extended formulation. For which problem classes can we hope to find such descriptions of $\mathrm{conv}(X)$?

To get a partial answer to this question, we need to consider simultaneously the separation problem $SEP(X, x^*)$ and the corresponding optimization problem.

Definition 6.12 *Given* (X, c), *the* optimization problem, $OPT(X, c)$, *is to find an optimal solution* x^* *to the problem* $\min\{cx : x \in X\}$.

The following fundamental result tells us that these two problems are either both easy, or both difficult.

Theorem 6.3 *Subject to certain technical conditions,* OPT *is polynomially solvable if and only if* SEP *is polynomially solvable* .

The practical consequence of this is that there is only hope of finding a good description of $\mathrm{conv}(X)$ (i.e., all facet-defining inequalities are easily described, or there exists a tight extended formulation that is compact in the number of constraints and variables) if one of the problems $OPT(X, c)$ or $SEP(X, x^*)$, and thus both, are polynomially solvable.

On the other hand, for problems that are difficult (NP-hard problems in complexity theory), we can at best hope to find partial descriptions of $\text{conv}(X)$.

6.1.7 Optimization and Separation for Polynomially Solvable Problems

A variety of algorithms is used in Parts III and IV to establish that various OPT and SEP problems are polynomially solvable. However, throughout there is a certain emphasis on finding an explicit description of $\text{conv}(X)$ in the original variable space, and/or a tight extended formulation whose projection is $\text{conv}(X)$. Here we address a few points that are important when using such formulations.

Optimization by Linear Programming

Suppose that to solve $OPT(X, c)$, we solve

$$\max\{cx : x \in \text{conv}(X)\}$$

by linear programming. As X is typically a discrete set, we need to find an optimal extreme point solution of $\text{conv}(X)$ so as to be sure that the point obtained lies in X. Otherwise, if the face of optimal solutions consists of more than a single point, the linear program has nonextreme optimal solutions. Thus it is important to use an algorithm that terminates with an extreme point. Alternatively it is always possible to very slightly perturb the objective function vector c, so that the linear program has a unique (and hence extreme point) optimal solution.

Optimization Using a Tight Extended Formulation

Here, using the notation of Subsection 6.1.5, we wish to solve $OPT(X, c)$ by solving the linear program

$$\max\{cx + 0z : (x, z) \in Q\} = \max\{cx + 0z : Bx + Gz \geq d, x \in \mathbb{R}^n_+, z \in \mathbb{R}^q\},$$

where $\text{proj}_x(Q) = \text{conv}(X)$. Here there is a further potential difficulty. Even if (x^*, z^*) is an optimal extreme point of the linear program and thus an extreme point of Q, x^* may not be an extreme point of $\text{conv}(X)$. An example where this occurs is shown in Figure 6.6. The solution is again to slightly perturb the objective function vector c while keeping zero cost on the z variables. Now the projection of the face of optimal solutions must be a single point, and hence an extreme point of $\text{conv}(X)$.

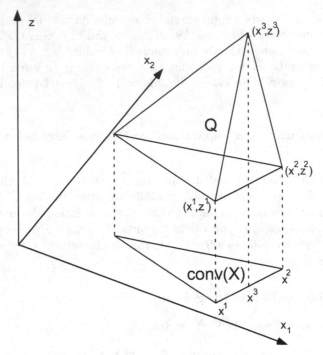

Figure 6.6. (x^3, z^3) is extreme in Q, but x^3 is not extreme in $\text{conv}(X)$.

Separation Using a Tight Extended Formulation

Suppose that we know a tight extended formulation Q with $\text{proj}_x(Q) = \text{conv}(X)$ and we wish to solve $SEP(X, x^*)$. We have that

$$x^* \notin \text{conv}(X) \text{ if and only if } \{z \in \mathbb{R}^q : Bx^* + Gz \geq d\} = \emptyset$$
$$\text{if and only if } \{z \in \mathbb{R}^q : Gz \geq d - Bx^*\} = \emptyset.$$

From Farkas' lemma, the latter holds if and only if there exists a dual vector $\pi \geq 0$ with $\pi G = 0$ and $\pi(d - Bx^*) > 0$. It follows that

$$\pi Bx \geq \pi d$$

is a valid inequality for $\text{conv}(X)$ cutting off x^* and that

$$\text{conv}(X) = \{x \in \mathbb{R}^n_+ : \pi Bx \geq \pi d \text{ for all } \pi \geq 0 \text{ with } \pi G = 0\}.$$

Solving a standard Phase 1 linear program with artificial variables w such as

$$\min\{\sum_i w_i : Gz + Iw \geq d - Bx^*, w \geq 0\}$$

provides a direct way to test if $\{z \in \mathbb{R}^q : Gz \geq d - Bx^*\} = \emptyset$, and, when $x^* \notin \text{conv}(X)$, the optimal dual variables of this linear program provide a vector

π satisfying the conditions of Farkas, and hence a valid inequality cutting off x^*. So when there exists a polynomial size tight extended formulation, there exists both a polynomial time LP optimization algorithm and a polynomial time LP separation algorithm.

6.2 Decomposition and Reformulation

Suppose that the optimization problem $OPT(X,c)$ that we are trying to solve is hard. It follows that $SEP(X,x^*)$ is hard, and and so there is no real chance of finding a good description of $\mathrm{conv}(X)$.

Suppose in addition that the feasible region X for which we have a formulation P_X can be broken up into two (or more) parts

$$X = Y \cap Z \qquad \text{with } Y, Z \subseteq \mathbb{Z}^n_+,$$

so the problem to be solved can be written equivalently as

$$\min\{cx : x \subset Y \cap Z\}.$$

Let P_Y, P_Z be the initial formulations for Y and Z, respectively.

The best possible case is that in which both problems $OPT(Y,c)$ and $OPT(Z,c)$ are easy. In this case both problems $SEP(Y,x^*)$ and $SEP(Z,x^*)$ are easy, and there is the possibility of describing $\mathrm{conv}(Y)$ and $\mathrm{conv}(Z)$ explicitly. If such descriptions can be found, we then have the improved formulation

$$\tilde{P}_X = \mathrm{conv}(Y) \cap \mathrm{conv}(Z).$$

The second possibility is that $OPT(Y,c)$ is easy, but $OPT(Z,c)$ is hard. In this case we can as a first step hope to obtain an improved formulation

$$\tilde{P}_X = \mathrm{conv}(Y) \cap P_Z,$$

and then possibly an even better formulation

$$\tilde{P}_X = \mathrm{conv}(Y) \cap \tilde{P}_Z,$$

where $\tilde{P}_Z$ is a good approximation to $\mathrm{conv}(Z)$.

In the case where both problems $OPT(Y,c)$ and $OPT(Z,c)$ are difficult, we can still aim for an improved formulation

$$\tilde{P}_X = \tilde{P}_Y \cap \tilde{P}_Z,$$

where $\tilde{P}_Y, \tilde{P}_Z$ are good approximations to $\mathrm{conv}(Y)$ and $\mathrm{conv}(Z)$, respectively.

Obviously these ideas can be repeated to break up the sets Y and Z. In particular suppose that

$$Y = \hat{Y}^1 \times \cdots \times \hat{Y}^m = \prod_{i=1}^{m} \hat{Y}^i, \quad \text{and} \quad Z = \hat{Z}^1 \times \cdots \times \hat{Z}^n = \prod_{j=1}^{n} \hat{Z}^j,$$

where each of the sets $\hat{Y}^i$ is of the form $\hat{Y}$, each of the sets $\hat{Z}^j$ is of the form $\hat{Z}$, and $Y^1 \times Y^2 = \{(y^1, y^2) : y^1 \in Y^1 \text{ and } y^2 \in Y^2\}$. Then if $\text{conv}(\hat{Y})$ and $\text{conv}(\hat{Z})$ are known, we have the reformulation

$$\begin{aligned}
\tilde{P}_X &= \text{conv}(Y) \cap \text{conv}(Z) \\
&= [\text{conv}(\hat{Y}^1) \times \cdots \times \text{conv}(\hat{Y}^m)] \cap [\text{conv}(\hat{Z}^1) \times \cdots \times \text{conv}(\hat{Z}^n)].
\end{aligned}$$

6.2.1 Decomposition of a Multi-Item Lot-Sizing Problem

We consider a typical multi-item lot-sizing problem with m items, n periods, demands d_t^i for item i in period t, individual production limits C_t^i, production, storage, and fixed costs p_t^i, h_t^i and q_t^i respectively, machine production rates a^i, set-up times b^i, and machine capacity B_t. Taking the following variables

x_t^i is the production of item i in period t,
s_t^i is the stock of item i at end of period t, and
$y_t^i = 1$ if item i is set up in period t, and $y_t^i = 0$ otherwise,

we obtain a first natural formulation

$$\min \quad \sum_{i,t} p_t^i x_t^i + \sum_{i,t} h_t^i s_t^i + \sum_{i,t} q_t^i y_t^i$$

$$s_{t-1}^i + x_t^i = d_t^i + s_t^i \qquad \text{for all } i, t \qquad (6.1)$$

$$x_t^i \leq C_t^i y_t^i \qquad \text{for all } i, t \qquad (6.2)$$

$$\sum_i a^i x_t^i + \sum_i b^i y_t^i \leq B_t \qquad \text{for all } t \qquad (6.3)$$

$$x_t^i, s_t^i \geq 0, y_t^i \in \{0,1\} \qquad \text{for all } i, t. \qquad (6.4)$$

Here (6.1) are the product conservation equations, (6.2) the variable upper bound capacity constraints imposing that a machine has to be set up for an item before it can be produced, and (6.3) the machine capacity constraints linking the production of different items.

Now we examine the structure of the feasible region X described by the constraints (6.1)–(6.4). We note that X can be written as

$$X = (\prod_{i=1}^{m} \hat{Y}^i) \cap (\prod_{t=1}^{n} \hat{Z}^t),$$

where

$$\hat{Y}^i = \{(x^i, s^i, y^i) \in \mathbb{R}^n \times \mathbb{R}^n \times \{0,1\}^n : \text{satisfying (6.1)-6.2) for all } t\}$$

is a single item lot-sizing region, and

$$\hat{Z}^t = \{(x_t, s_t, y_t) \in \mathbb{R}^m \times \mathbb{R}^m \times \{0,1\}^m : \text{satisfying (6.2)-(6.3) for all } i\}$$

is a generalized single-node flow region (see Chapter 8).

This is precisely the structure discussed above, so this provides the motivation to study the separation and optimization problems associated with $\hat{Y}$, the single-item lot-sizing set, and $\hat{Z}$, the single-node flow set, in order to obtain improved formulations for X.

6.3 Decomposition Algorithms

6.3.1 Decomposition Algorithms I: Valid Inequalities and Separation

Consider again the problem IP or $OPT(X, c)$ in the form

$$\min\{cx : x \in Y \cap Z\} \text{ where } X = Y \cap Z, Y = P_Y \cap \mathbb{Z}^n, Z = P_Z \cap \mathbb{Z}^n,$$

and P_Y and P_Z are formulations for Y and Z, respectively. We suppose that $OPT(Y, c)$ is easy.

The discussion so far suggests several different approaches that can be taken to solve this problem with a mixed integer programming system; see Chapter 3.

Algorithm 1. A Priori Reformulation. Find a formulation $\tilde{P}_Y$ that approximates or describes conv(Y). Use a standard MIP system to solve IP using as initial formulation

$$\tilde{P}_X^1 = \tilde{P}_Y \cap P_Z.$$

Algorithm 2. Use of an Extended Formulation. Find an extended formulation $Q_Y = \{(x, w) \in \mathbb{R}^n \times \mathbb{R}^p : Bx + Gw \geq d\}$ of Y such that $\text{proj}_x(Q_Y)$ provides a good approximation or an exact description of conv(Y). Use a standard MIP system to solve the mixed integer program

$$\min\{cx : (x, w) \in Q_Y, x \in Z\}$$

which is equivalent (in terms of linear relaxation or LP bound) to using the formulation

$$\tilde{P}_X^2 = \text{proj}_x(Q_Y) \cap P_Z.$$

Algorithm 3. Reformulation by Cutting Planes. Here one needs an exact or heuristic algorithm to solve $SEP(Y, x^*)$. One then solves linear programs starting with the initial formulation $P_Y \cap P_Z$. At each iteration one or more cutting planes $\pi^t x \geq \pi_0^t$ are generated, and added to the formulation. The iterations stop when no more cuts are generated. Suppose that T cuts are generated in all.

Now the resulting formulation

$$\tilde{P}_X^3 = P_Y \cap P_Z \cap \{x : \pi^t x \geq \pi_0^t \text{ for } t = 1, \ldots, T\}$$

can be input to a standard MIP system. This approach is also known as *cut-and-branch* when cuts are only added before the branch-and-bound enumeration.

Algorithm 4. Branch-and-Cut. If the cuts are also added at nodes of the branch-and-bound tree, we attempt to go one step further and generate an appropriate tight formulation at all (or many) nodes of the tree. As indicated in Chapter 3 this is now possible with several of the state-of-the-art MIP systems.

Should one of these methods be preferred to another? Let z_{LP}^i for $i = 1, 2, 3$ be the value of the LP solution $\max\{cx : x \in \tilde{P}_X^i\}$, measuring the strength of the first three formulations.

Proposition 6.4 *If the formulations giving* conv(Y) *are tight formulations, and the separation algorithm* SEP(Y, x^*) *for* Y *is exact,*

$$z_{LP}^1 = z_{LP}^2 = z_{LP}^3 = \min\{cx : x \in \text{conv}(Y) \cap P_Z\}.$$

Therefore the approaches are equivalent, and ease of implementation and running times should be the deciding factor in choosing between them.

6.3.2 Decomposition Algorithms II: Lagrangian Relaxation and Column Generation

There are cases in which $OPT(Y, c)$ is relatively easy to solve in practice, and next to nothing is known about a practical algorithm for $SEP(Y, x^*)$, or even about interesting classes of valid inequalities for Y. In such cases another class of well-known decomposition algorithms is more appropriate.

Consider again the problem

$$z = \min\{cx : x \in Y \cap Z\}.$$

Here we suppose either that $OPT(Y, c)$ is easy, or that small- to medium-sized instances are well solved in practice (even though it may theoretically be a hard problem). Also let $P_Z = \{x \in \mathbb{R}_+^n : Dx \geq d\}$ be the formulation for Z. We are interested in obtaining a strong lower bound by solving

$$w = \min\{cx : x \in \text{conv}(Y) \cap P_Z\} \leq z$$

with no knowledge about valid inequalities or strong formulations for Y.

Below we consider two classical approaches: (Dantzig–Wolfe) column generation and Lagrangian relaxation. Both algorithms iterate between two problems:

- A *master problem* in which, at iteration t, new "prices" π^t on the linking or complicating constraints $Dx \geq d$ (represented by a row vector, with one price for each constraint) are calculated based on the previous prices π^{t-1} and a set $x^1, \ldots, x^{t-1} \in Y$ of "proposals".
- A *subproblem* $OPT(Y, (c - \pi^t D))$

$$L(\pi^t) = \min\{(c - \pi^t D)x + \pi^t d : x \in Y\}$$

which provides a lower (or dual) bound on the optimal value w, and a new proposal x^t (x^t is an optimal solution to problem $OPT(Y, (c - \pi^t D))$ to be sent to the master problem.

Observe that $L(\pi^t) \leq w$ for any $\pi^t \geq 0$ because, for $x^\star \in Y \cap P_Z$ with $w = cx^\star$ (i.e., $x^\star$ is an optimal solution to $OPT(\text{conv}(Y) \cap P_Z, c))$, we have

$$w = cx^\star \geq cx^\star - \pi^t(Dx^\star - d) = (c - \pi^t D)x^\star + \pi^t d \geq L(\pi^t) ,$$

where the first inequality holds because $x^\star \in P_Z$ and $\pi^t \geq 0$ and the second inequality holds because $x^\star$ is feasible but not necessarily optimal for the subproblem $OPT(Y, (c - \pi^t D))$.

The final result of column generation and Lagrangian relaxation is a solution to the *Lagrangian dual problem*

$$w = \max_{\pi \geq 0} L(\pi)$$

with $w \leq z$.

Outline of the Column Generation Algorithm.

Let $x^1, \ldots, x^{t-1}$ be a set of points of Y.

The master problem at iteration t is the linear program:

$$
\begin{aligned}
z^t \;=\; &\min\ cx \\
&Dx \geq d \\
&x \in \text{conv}\{x^1, \ldots, x^{t-1}\}.
\end{aligned}
$$

After substitution of $x = \sum_{\tau=1}^{t-1} \lambda_\tau x^\tau, \sum_{\tau=1}^{t-1} \lambda_\tau = 1, \lambda_\tau \geq 0$ for $\tau = 1, \ldots, t-1$ (i.e., x is a convex combination of points $x^1, \ldots, x^{t-1}$), this problem can be rewritten in the form:

$$z^t = \min \sum_{\tau=1}^{t-1} \lambda_\tau (cx^\tau)$$

$$\sum_{\tau=1}^{t-1} \lambda_\tau (Dx^\tau) \geq d$$

$$\sum_{\tau=1}^{t-1} \lambda_\tau = 1$$

$$\lambda_\tau \geq 0 \qquad\qquad \text{for } \tau = 1, \ldots, t-1.$$

Let (π^t, π_0^t) be optimal dual variables for this linear program, so $\pi^t \geq 0$ are the dual variables on the constraints $Dx \geq d$ and π_0^t is the dual variable on the convexity constraint $\sum_{\tau=1}^{t-1} \lambda_\tau = 1$. Note that $z^t = \pi_0^t + \pi^t d \geq w$ because $\text{conv}\{x^1, \ldots, x^{t-1}\} \subseteq \text{conv}(Y)$.

The subproblem at iteration t is

$$\zeta^t = \min \ (c - \pi^t D)x - \pi_0^t$$
$$x \in Y.$$

It computes the minimum reduced cost of a new column $x \in Y$ to add to the master. If $\zeta^t < 0$, then the optimal solution x^t to the subproblem is the new "proposal" or column sent to the master problem for iteration $t+1$.

Note that the subproblem gives also a lower bound on the optimal value w because $w \geq L(\pi^t) = \zeta^t + \pi_0^t + \pi^t d$.

The algorithm terminates when $\zeta^t \geq 0$. Then $w = L(\pi^t) = c\tilde{x}^t$, where $\tilde{x}^t \in \{x : Dx \geq d\} \cap \text{conv}(Y)$ is a solution of the master problem.

Outline of the Lagrangian Relaxation Subgradient Algorithm.

Here the master problem at iteration t produces an update of the dual variables. This update is given by the simple formula

$$\pi^t = \max\{0, \pi^{t-1} + \mu_t(d - Dx^{t-1})\}$$

where the sequence of values μ_t is chosen appropriately and converges to zero.

The subproblem at iteration t is the optimization problem obtained by relaxing or dualizing the the constraints $Dx \geq d$

$$L(\pi^t) = \min \ (c - \pi^t D)x + \pi^t d$$
$$x \in Y,$$

with solution $x^t \in Y$. Observe that this is exactly the same subproblem as in the Column Generation Algorithm.

The algorithm terminates if $Dx^t \geq d$ and $\pi^t(Dx^t - d) = 0$. However, in general one only obtains that $\lim_{t \to \infty} L(\pi^t) \to w$, so the algorithm typically terminates with a lower bound $\max_{u=1,\ldots,t} L(\pi^u)$ on w without necessarily finding a good primal feasible solution x satisfying $Dx \geq d$.

Finally both these algorithms need to be embedded into a branch-and-bound scheme. Here the choice of branching variables or objects is not always obvious, especially as the convexity variables from the column generation procedure are not good objects on which to branch.

Should one of these two algorithms be preferred? How does their common bound w compare with the bound z_{LP} given by the algorithms of the previous section using cuts or reformulation?

Theorem 6.5 $w = \max_{\pi \geq 0} L(\pi) = z_{LP} = \min\{cx : x \in \text{conv}(Y) \cap P_Z\}$.

So the outcome of both these algorithms is to convexify the set Y while leaving the formulation of Z unchanged.

6.3.3 Decomposition Algorithms III: Hybrid Algorithms

It is possible to obtain the stronger lower bound

$$w^* = \min\{cx : x \in \text{conv}(Y) \cap \text{conv}(Z)\}$$

when one has either a practical algorithm for separation or optimization for both Y and Z. The case when one has a separation algorithm for both is obvious, and is a direct extension of algorithms from Section 6.3.1. We now consider briefly the other two possibilities.

Lagrangian Decomposition

$OPT(Y, c)$ and $OPT(Z, c)$ are both tractable.
The problem is reformulated as

$$\min \; \alpha \, cy + (1 - \alpha) \, cz$$
$$y - z = 0$$
$$y \in Y, z \in Z,$$

where $0 \leq \alpha \leq 1$. The reformulated problem is then solved by Lagrangian relaxation, dualizing the constraints $y - z = 0$. It follows immediately from Theorem 6.5 that the bound w^* is achieved.

Combined Column Generation and Cutting Planes

$SEP(Y, c)$ and $OPT(Z, c)$ are both tractable.

The master problem at iteration t.
This problem is constructed from a set $\{x^i\}_{i \in I^{t-1}}$ of points of Z and a set

$\{(\pi^j, \pi_0^j)\}_{j \in J^{t-1}}$ of valid inequalities for $\operatorname{conv}(Y)$ giving either the linear program:

$$z^t = \min cx$$

$$x - \sum_{i \in I^{t-1}} \lambda_i x^i = 0$$

$$\sum_{i \in I^{t-1}} \lambda_i = 1$$

$$\sum_{j \in J^{t-1}} \pi^j x \geq \pi_0^j \qquad \text{for } j \in J^{t-1}$$

$$\lambda_i \geq 0 \qquad \text{for } i \in I^{t-1},$$

or, if we eliminate the x variables, the linear program:

$$z^t = \min \sum_{i \in I^{t-1}} (cx^i)\lambda_i$$

$$\sum_{i \in I^{t-1}} (\pi^j x^i)\lambda_i \geq \pi_0^j \qquad \text{for } j \in J^{t-1}$$

$$\sum_{i \in I^{t-1}} \lambda_i = 1$$

$$\lambda_i \geq 0 \qquad \text{for } i \in I^{t-1}.$$

Let λ^t be a primal optimal solution and $(\mu^t, \mu_0^t) \in \mathbb{R}_+^{|J^{t-1}|} \times \mathbb{R}^1$ a dual optimal solution.

The order in which the two subproblems are solved below is arbitrary. We have chosen to look first for a violated inequality.

The Separation Subproblem – Adding Constraints.
Solve $SEP(Y, \sum_{i \in I_{t-1}} \lambda_i^t x^i)$.
If a valid inequality (π^t, π_0^t) is generated, cutting off the point $\sum_{i \in I^{t-1}} \lambda_i^t x^i$, set $J^t = J^{t-1} \cup \{t\}$, set $t \leftarrow t + 1$, and return to the Master.
Otherwise set $J^t = J^{t-1}$, and go to the optimization subproblem.

The Optimization Subproblem – Adding Columns.
Solve $OPT(Z, c - \sum_{j \in J^{t-1}} \mu_j^t \pi^j)$ with optimal value ζ^t and solution x^t.
If $\zeta^t < \mu_0^t$, the column corresponding to x^t has negative reduced cost. Set $I^t = I^{t-1} \cup \{t\}$, set $t \leftarrow t + 1$, and return to the Master.
Otherwise stop.

On termination $w^* = cx^*$, where $x^* = \sum_{i \in I^{t-1}} \lambda_i^t x^i \in \operatorname{conv}(Y) \cap \operatorname{conv}(Z)$.

6.4 Convex Hull Proofs

Suppose that $X \subset \mathbb{Z}^n$ is an integer set, and P is a formulation for X, so $P = \{x \in \mathbb{R}^n_+ : Ax \geq b\}$, $X = P \cap \mathbb{Z}^n$, and $\mathrm{conv}(X) \subseteq P$. Below we assume for simplicity that $\mathrm{conv}(X) \neq \emptyset$, that both $\mathrm{conv}(X)$ and P have extreme points, and that $\mathrm{conv}(X)$ is not just a single point.

We say that a nonempty polyhedron (with extreme points) is *integral* if all its extreme points are integral. The extension to mixed integer sets is straightforward. We say that an extreme point is integral if all the coordinates of the integer variables are integral. For simplicity, we consider here only integer sets.

How can one prove that $P = \mathrm{conv}(X)$, or equivalently that P is integral? Below we indicate several ways of proving such a result. Several of them are used later, particularly in Chapters 7 and 8.

First we list some of the ways that follow more or less from the definition

1. Show that all extreme points of P are integral.
   The contrapositive is:
2. Show that all points of P with $x \notin Z^n$ are not extreme points of P.
3. Show that all facets/faces of P have integral extreme points.
4. Show that the linear program: $\min\{cx : x \in P\}$ has an optimal solution in X for all $c \in \mathbb{R}^n$.
   One way to do this is:
5. Show that there exists a point $x^* \in X$ and point u^* feasible in the dual linear program: $w^D = \max\{ub : uA \leq c, u \in \mathbb{R}^m_+\}$ with $cx^* = u^*b$.
   Another is:
6. Assuming $b \in \mathbb{Z}^m$, show that, for all $c \in \mathbb{Z}^n$ for which the optimal dual value w^D is bounded, w^D is integer-valued.
7. Show that $\dim(\mathrm{conv}(X)) = \dim(P)$, and that if $\pi x \geq \pi_0$ is a facet-defining inequality for $\mathrm{conv}(X)$, then $\pi x \geq \pi_0$ must be identical to one of the inequalities $a^q x \geq b^q$ defining P.
8. Show that $\dim(\mathrm{conv}(X)) = \dim(P)$, and that for all $c \in \mathbb{R}^n$ for which $M(c) \neq X$ and the optimum value is finite, $M(c) \subseteq \{x : a^q x = b^q\}$ for some inequality $a^q x \geq b^q$ defining P, where $P \cap \{x : a^q x = b^q\}$ is a proper face of P. Here $M(c) = \arg\min\{cx : x \in X\}$ is the set of optimal solutions for a given cost vector $c \in \mathbb{R}^n$. This condition is sufficient because, if $\pi x \geq \pi_0$ defines a facet of $\mathrm{conv}(X)$, and we take $c = \pi$, then $M(\pi)$ can only lie in that facet. It follows that the faces of P contain all the facets of $\mathrm{conv}(X)$.
   Another possibility is to show that $\mathrm{conv}(X) = \mathrm{proj}_x(Q)$ where Q is some extended formulation for X.
9. Show that Q is integral.
10. Show that the linear program: $\max\{cx + 0w : (x, w) \in Q\}$ always has an optimal solution with $x \in \mathbb{Z}^n$.

We terminate with one very important way of showing that P is integral.

Definition 6.13 *A* $\{0, +1, -1\}$ *matrix* A *is* totally unimodular *(TU) if every square submatrix of* A *has determinant* $0, 1,$ *or* -1.

To recognize such matrices, we have the following very useful test.

Proposition 6.6 *A* $\{0, +1, -1\}$ *matrix* A *is totally unimodular if, for any subset* J *of the set* $N = \{1, \ldots, n\}$ *of columns, there is a partition* J_1, J_2 *of* J *such that for every row* $i = 1, \ldots, m$,

$$\left| \sum_{j \in J_1} a_{ij} - \sum_{j \in J_2} a_{ij} \right| \leq 1.$$

The importance of TU matrices resides in the following well-known result.

Theorem 6.7 *If* A *is totally unimodular,* $b \in \mathbb{Z}^m$, *and* $l, h \in \mathbb{Z}^n$, *then the polyhedron*

$$\{x \in \mathbb{R}^n : Ax \leq b, l \leq x \leq h\}$$

is integral, whenever it is nonempty.

Exercises

Exercise 6.1 Consider the set

$$X = \{y \in \mathbb{Z}_+^n : C \sum_{j=1}^n y_j \geq b\}.$$

i. Find $\text{conv}(X)$.
ii. Find $\text{conv}(X \cap \{0, 1\}^n)$.

Exercise 6.2 Consider the set

$$X = \{(x, y) \in [0, 1]^m \times \{0, 1\} : \sum_{j=1}^m x_j \leq my\}.$$

Show that

$$\text{conv}(X) = \{(x, y) \in [0, 1]^m \times [0, 1] : x_j \leq y \text{ for } j = 1, \ldots, m\}.$$

Exercise 6.3 Consider the set

$$X = \{(x, y) \in \mathbb{R}_+^n \times \{0, 1\}^n : \sum_{j=1}^n x_j \leq b, \ x_j \leq y_j \text{ for } j = 1, \ldots, n\}.$$

i. Show that, if $f = b - \lfloor b \rfloor > 0$, the inequality

$$\sum_{j \in S} x_j + f \sum_{j \in S} (1 - y_j) \leq b$$

is valid for X, and facet-defining if $|S| > b$.
ii. Given that all nontrivial facets of $\text{conv}(X)$ are of the above form, give a tight extended formulation for $\text{conv}(X)$.

Exercise 6.4 Show that if a 0–1 matrix A has the consecutive 1s property (i.e., for all rows $i = 1, \ldots, m$, if $a_{ij} = a_{ik} = 1$ for $k > j + 1$, then $a_{it} = 1$ for all $j < t < k$), then matrix A is totally unimodular.

Notes

Section 6.1 An introduction to polyhedra as they arise in integer programs can be found in several books, in particular Schrijver [148] and Nemhauser and Wolsey [126]. The equivalence of optimization and separation is due to Grötschel, Lovász, and Schrijver [79, 80]. A large majority of the remarkable developments in combinatorial optimization and polyhedral combinatorics over the past 25 to 30 years have been concentrated in the development of valid inequalities and in separation algorithms for these inequalities. The equal emphasis placed in this book on extended formulations is relatively new. However, such formulations have been known for some time; one of the earliest is the representation of the closed convex hull of the union of polyhedra of Balas [18, 21] that we present in Section 8.4.

Section 6.2 The decomposition approach described here is familiar to anyone who has attempted to use a decomposition algorithm. The idea of decomposing multi-item lot-sizing problems was already examined in the papers of Manne [115] and Dzielinski and Gomory [59].

Section 6.3 Details on different decomposition algorithms for integer programming can be found in textbooks on integer programming, such as Nemhauser and Wolsey [126], Martin [119], Parker and Rardin [131], and Wolsey [193]. Decomposition for linear programs originated with Dantzig and Wolfe [53], and column generation with Gilmore and Gomory [75]. Lagrangian relaxation was first applied by Held and Karp [89] to the traveling salesman problem. A fundamental paper showing its importance for integer programming is that of Geoffrion [73]. Lagrangian relaxation methods and Dantzig–Wolfe or column generation methods have been used to solve numerous production planning models. See, for instance, Afentakis et al. [2], Afentakis and Gavish [1], Chen and Thizy [36], Diaby et al. [56], Tempelmeier and Derstoff [158], Thizy and Van Wassenhove [160], Trigeiro et al. [161] for Lagrangian relaxation approaches, and Kang et al. [95] and Vanderbeck [182] among others for column generation approaches

Lagrangian decomposition was suggested in Jornsten and Nasberg [94] and Guignard and Kim [84], and the combination of cutting planes and column generation has been a challenge for several years; see, for instance, the recent work of Poggi and Uchoa [143] and Fukasawa et al. [72].

Section 6.4 A simple example demonstrating the different convex hull proof techniques can be found in Section 6.2 of Wolsey [193]. The classical primal–dual method (5) is due to Edmonds [60], and the very useful approach (8) is attributed to Lovász [109]. Total unimodularity is also treated in most of the books cited above. Proposition 6.6 is due to Ghoula-Houri [74].

7

Single-Item Uncapacitated Lot-Sizing

Here we consider the simplest possible production planning problem involving time-varying demand, and fixed set-up or order placement costs. This problem has been introduced and classified as *LS-U* in Part I. By studying the properties of this model and different ways to solve it, we hope to learn how to analyze and solve more complicated single-item variants in Part III, as well as to provide many of the results needed to tackle more complicated multi-item problems by the decomposition approach in Part IV.

- In Section 7.1 we formulate problem *LS-U*. This formulation is repeated here to make this chapter self-contained and independent of Part I.
- In Section 7.2 we characterize optimal solutions of *LS-U*.
- Then we describe and analyze dynamic programming algorithms for *LS-U* in Section 7.3.
- In Section 7.4 we describe linear programming reformulations, and in particular the convex hull formulation in the original space, the associated separation algorithm, and tight extended reformulations.
- In Section 7.5 we analyze the special case *WW-U*, where the cost function satisfies the nonspeculative Wagner–Whitin condition.
- Partial reformulation results are described in Section 7.6. These are useful in solving large problem instances.
- Finally in Section 7.7 we give proofs of some of the convex hull and tight reformulation results presented earlier in the chapter.

7.1 The Uncapacitated Lot-Sizing Problem (*LS-U*)

The *uncapacitated lot-sizing problem LS-U* can be described as follows. There is a planning horizon of $n = NT$ periods. The demand for the item in period t is $d_t \geq 0$ for $t = 1, \ldots, n$. For each period t, there are unit production costs p'_t, unit storage costs h'_t for stock remaining at the end of the period t, and a fixed set-up (or order placement) cost q_t which is incurred to allow production to take place in period t, but is independent of the amount produced.

A First Formulation of the Problem.
One natural way to view *LS-U* is as a minimum cost network flow problem with, in addition, fixed costs for the use of certain arcs. In Figure 7.1 we show such a network for an instance with $n = 4$ periods. The variable and fixed costs associated with each arc are shown.

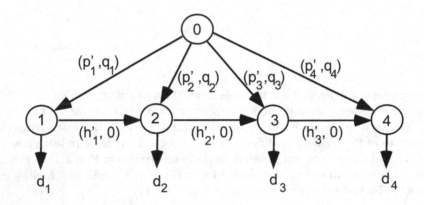

Figure 7.1. Formulation as a fixed charge network flow.

Note that the flow in arc $(0, t)$ represents the amount produced in period t, and the flow in arc $(t, t+1)$ represents the stock at the end of period t.

To formulate the problem as a mixed integer program, we first define variables. Specifically we use

x_t for the amount produced in period t,
s_t for the amount in stock at the end of period t, and
y_t for the 0–1 set-up variable which must have the value 1 if $x_t > 0$.

Now we can write a first mixed integer programming formulation:

$$\min \ \sum_{t=1}^{n} p'_t x_t + \sum_{t=0}^{n} h'_t s_t + \sum_{t=1}^{n} q_t y_t \qquad (7.1)$$

$$s_{t-1} + x_t = d_t + s_t \qquad \text{for } 1 \le t \le n \qquad (7.2)$$

$$x_t \le M y_t \qquad \text{for } 1 \le t \le n \qquad (7.3)$$

$$s \in \mathbb{R}_+^{n+1}, x \in \mathbb{R}^n, y \in [0,1]^n \qquad (7.4)$$

$$y \in \mathbb{Z}^n \qquad (7.5)$$

$$s_0 = s_0^*, s_n = s_n^*, \qquad (7.6)$$

where M is a large positive number.

Here the constraint (7.2) represents conservation of product (or flow conservation at node t in the network of Figure 7.1). The constraint (7.3) ensures that if $x_t > 0$, then $y_t > 0$ and so necessarily $y_t = 1$. The constraint (7.6) is optional in the sense that there are situations in which both the initial stock s_0 and the final stock s_n may be decision variables. However, in the majority of cases their values are fixed, and the values s_0^*, s_n^* are part of the data. Unless otherwise stated, we assume their values are fixed. What is more, we assume that $s_0^* = s_n^* = 0$, as in Figure 7.1. This assumption is justified by Observation 7.1 in the next section.

We denote by X^{LS-U} the set of feasible solutions to (7.2)–(7.6).

7.2 Structure of Optimal Solutions of *LS-U*

We start with an observation that allows us to simplify the presentation in most of this section, and can be viewed as a basic preprocessing step for problems with initial stocks and stock lower bounds (safety stocks).

Observation 7.1 *In LS-U, if the initial and final stocks are fixed and there are stock lower bounds, the problem can be reformulated with modified demands so that initial and final stocks, and the stock lower bounds, are all zero.*

Preprocessing step. Suppose that the initial fixed values are $s_0 = s_0^*, s_n = s_n^*$ and the lower bounds are $s_t \geq s_t^*$ for $1 \leq t \leq n - 1$. To treat the stock lower bounds, one first calculates updated lower bounds: $\underline{S}_0 = s_0^*$ and

$$\underline{S}_t = \max[\underline{S}_{t-1} - d_t, s_t^*] \quad \text{for } 1 \leq t \leq n.$$

Then one introduces net stock variables ns_t and makes the substitution $ns_t = s_t - \underline{S}_t \geq 0$ for $t = 0, \ldots, n$. The resulting balance equations are

$$ns_{t-1} + x_t = (d_t + \underline{S}_t - \underline{S}_{t-1}) + ns_t.$$

Thus we have obtained a modified problem with d_t replaced by $d_t + \underline{S}_t - \underline{S}_{t-1} \geq 0$ for $1 \leq t \leq n$, and the amount $\sum_{t=0}^{n} h_t' \underline{S}_t$ added to the objective function. Finally we set $ns_0 = ns_n = 0$.

Suppose that it has been decided in which periods there are set-ups, that is, $y \in \{0,1\}^n$ is known. The optimal production plan is now a minimum cost flow in the network of Figure 7.2 except that arcs $(0,t)$ are suppressed if $y_t = 0$. Note that this is the same network as above (Figure 7.1), but now we indicate the variables associated with each arc rather than the costs.

We now present a well-known and fundamental property of minimum cost network flow problems.

Observation 7.2 *In a basic or extreme feasible solution of a minimum cost network flow problem, the arcs corresponding to variables with flows strictly between their lower and upper bounds form an acyclic graph.*

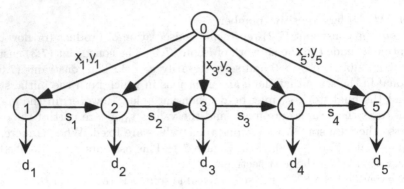

Figure 7.2. The fixed charge network flow variables.

This immediately tells us something important about the structure of optimal solutions to *LS-U*.

Proposition 7.1 *There exists an optimal solution to LS-U in which* $s_{t-1}x_t = 0$ *for all* t.

Proof. Given $y \in \{0,1\}^n$, consider an optimal basic feasible solution. Suppose $s_{t-1} > 0$. The stock must originate from production in some period k where $k < t$. So the flow on arcs $(0,k),(k,k+1),\ldots,(t-1,t)$ is positive. As these edges together with edge $(0,t)$ form a cycle, it follows from Observation 7.2 that $x_t = 0$. □

Now we can fully describe optimal solutions.

Proposition 7.2 *There exists an optimal solution to LS-U characterized by:*
i. A subset of periods $1 \le t_1 < \cdots t_r \le n$ *in which production takes place. The amount produced in* t_j *is* $d_{t_j} + \cdots + d_{t_{j+1}-1}$ *for* $j = 1,\ldots,r$ *with* $t_{r+1} = n+1$;
ii. A subset of periods $R \subseteq \{1,\ldots,n\}\setminus\{t_1,\ldots,t_r\}$. *There is a set-up in periods* $\{t_1,\ldots,t_r\} \cup R$.

The structure of an optimal solution is shown in Figure 7.3. The intervals $[t_1, t_2-1]$, $[t_2, t_3-1]$,... of a basic solution with no stock entering or leaving the interval, and production in the first period to satisfy demand in all periods of the interval, are called *regeneration intervals*. So, Proposition 7.2 shows that a basic optimal solution can be decomposed into a sequence of regeneration intervals, plus some additional set-ups without production (periods in R).

Note that if s_0 is a variable and $s_0 > 0$, then the first regeneration interval has a special form with $s_0 = \sum_{u=1}^{t} d_u$ for some t. Also in most practical instances, we will have nonnegative fixed costs $q_t \ge 0$ for all t, and in these cases, one can take $R = \emptyset$ in Proposition 7.2. However, if there are negative fixed costs, we have the following simple solution.

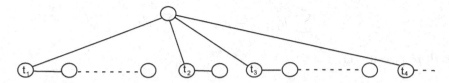

Figure 7.3. Structure of an optimal solution of *LS-U*.

Observation 7.3 *For instances of LS-U with negative fixed costs, it suffices to replace the fixed costs q_t by $q_t^+ = \max[q_t, 0]$ for $t = 1, \ldots, n$, and add the constant term $\sum_t \min[q_t, 0]$ to the objective function.*

Proof. If $q_t < 0$, $y_t = 1$ in any optimal solution. So replacing the term $q_t y_t$ by $0 y_t + q_t$ does not affect the optimal value of the problem. □

To terminate this section we introduce some notation, and then make a simple observation that allows us to simplify calculations later.

Notation. Throughout the text we use $d_{kl} \equiv \sum_{u=k}^{l} d_u$.

Observation 7.4 *The objective function (7.1) of LS-U can be alternatively written as either*

$$h_0 s_0 + \sum_{t=1}^{n} p_t x_t + \sum_{t=1}^{n} q_t y_t + K_1, \tag{7.7}$$

where $p_t = p_t' + \sum_{j=t}^{n} h_j'$ for $t = 1, \ldots, n$, $h_0 = \sum_{t=0}^{n} h_t'$, and $K_1 = -\sum_{t=1}^{n} h_t' d_{1t}$, or as

$$\sum_{t=0}^{n} h_t s_t + \sum_{t=1}^{n} q_t y_t + K_2, \tag{7.8}$$

where $h_t = h_t' + p_t' - p_{t+1}'$ for $t = 0, \ldots, n$ with $p_0' = p_{n+1}' = 0$ and $K_2 = \sum_{t=1}^{n} p_t' d_t$.

Proof. One can use the flow conservation constraints (7.2) to eliminate either the production variables $\{x_t\}$, or the stock variables $\{s_t\}$ from the objective function (7.1). Eliminating the stock variables using $s_t = s_0 + \sum_{u=1}^{t} x_u - d_{1t}$ gives

$$\sum_{t=1}^{n} p_t' x_t + \sum_{t=0}^{n} h_t' s_t = \sum_{t=1}^{n} p_t' x_t + (\sum_{t=0}^{n} h_t') s_0 + \sum_{t=1}^{n} h_t' \sum_{u=1}^{t} x_u - \sum_{t=1}^{n} h_t' d_{1t}$$

$$= (\sum_{t=0}^{n} h_t') s_0 + \sum_{t=1}^{n} (p_t' + h_t' + \cdots + h_n') x_t - \sum_{t=1}^{n} h_t' d_{1t}.$$

Similarly, eliminating the production variables, using $x_t = d_t + s_t - s_{t-1}$, gives

$$\sum_{t=1}^{n} p_t' x_t + \sum_{t=0}^{n} h_t' s_t = \sum_{t=1}^{n} p_t' (d_t + s_t - s_{t-1}) + \sum_{t=0}^{n} h_t' s_t$$

$$= \sum_{t=0}^{n} (h'_t + p'_t - p'_{t+1}) s_t + \sum_{t=1}^{n} p'_t d_t$$

with $p'_{n+1} = 0$. □

7.3 A Dynamic Programming Algorithm for *LS-U*

We now present an algorithm to solve *LS-U*. Using Observation 7.4, we take as objective function $\sum_{t=1}^{n} p_t x_t + \sum_{t=1}^{n} q_t y_t$. In addition by Observation 7.1 we take $s_0^* = s_n^* = 0$, and by Observation 7.3 we suppose that $q_t \geq 0$ for all t.

Let $G(t)$ be the minimum cost of solving the problem over the first t periods, that is, satisfying the demands $d_1, \ldots, d_t$, and ignoring the demands after period t, and let $\phi(k, t)$ be the minimum cost of solving the problem over the first t periods subject to the additional condition that the last set-up and production period is k for some $k \leq t$. Now it follows from the definition that

$$G(t) = \min_{k:k \leq t} \phi(k, t).$$

How can we calculate the values $\phi(k, t)$? Observe that given the structure of the optimal solutions described in Proposition 7.2 and the condition that the last production period is k (which implies that $s_{k-1} = 0$), the solution up till period $k - 1$ must be an optimal solution up to period $k - 1$, and so it necessarily has cost $G(k - 1)$. This is an instance of the so-called *Principle of Optimality*. So we have that

$$\phi(k, t) = G(k - 1) + q_k + p_k d_{kt}.$$

This is all that is needed to obtain a dynamic programming recursion.

Forward Dynamic Programming Recursion

$$G(0) = 0$$
$$G(t) = \min_{k:k \leq t} [G(k - 1) + q_k + p_k d_{kt}] \qquad \text{for } t = 1, \ldots, n.$$

By calculating $G(1), G(2), \ldots$, in order, one terminates with $G(n)$, the *value* of an optimal solution to *LS-U*.

To recover an *optimal solution*, a little more information must be kept. Let $\kappa_t = \arg \min_{k:k \leq t} [G(k - 1) + q_k + p_k d_{kt}]$. Then one can work backwards. κ_n gives the information that $y_{\kappa_n} = 1$ and $x_{\kappa_n} = d_{\kappa_n, n}$. Continuing, we need to find an optimal solution for periods $1, \ldots, \kappa_n - 1$ of cost $G(\kappa_n - 1)$, with the last production period before $\kappa_n - 1$ given by $\kappa_{(\kappa_n - 1)}$, and so on.

Note that in fact the calculations in the recursion are very simple because

$$\phi(k, t+1) = \phi(k, t) + p_k d_{t+1} \qquad \text{for all } k, t \text{ with } k \le t, \qquad \text{and}$$
$$\phi(k, k) = G(k-1) + q_k + p_k d_k \qquad \text{for all } k.$$

This calculation can clearly be carried out in $O(n^2)$ operations. We now demonstrate the algorithm.

Example 7.1 *Consider an instance of LS-U with $n = 5$, $d = (3, 2, 1, 2, 1)$, $q = (0, 12, 7, 4, 5)$, $p = (3, 0, 1, 0, 2)$, and $h = 0$.*

Applying the recursion, we obtain
$G(1) = \phi(1, 1) = q_1 + p_1 d_1 = 0 + 3 \times 3 = 9$, *and so* $\kappa_1 = 1$.

$G(2) = \min[\phi(1, 2), \phi(2, 2)] = \min[\phi(1, 1) + p_1 d_2, G(1) + q_2 + p_2 d_2]$
$= \min\{9 + 3 \times 2, \ G(1) + 12 + 0\} = \min\{15, 21\} = 15$, *and so* $\kappa_2 = 1$.

$G(3) = \min[\phi(1, 3), \phi(2, 3), \phi(3, 3)]$
$= \min[\phi(1, 2) + p_1 d_3, \phi(2, 2) + p_2 d_3, G(2) + q_3 + p_3 d_3]$
$= \min\{15 + 3 \times 1, \ 21 + 0 \times 1, \ G(2) + 7 + 1 \times 1\} = \min\{18, 21, 23\} = 18$, *so* $\kappa_3 = 1$.

$G(4) = \min\{18 + 3 \times 2, \ 21 + 0 \times 2, \ 23 + 1 \times 2, \ G(3) + 4 + 0 \times 2\}$
$= \min\{24, 21, 25, 22\} = 21$, *so* $\kappa_4 = 2$.

$G(5) = \min\{24 + 3 \times 1, \ 21 + 0 \times 1, \ 25 + 1 \times 1, \ 22 + 0 \times 1, \ G(4) + 5 + 2 \times 1\} = 21$,
so $\kappa_5 = 2$.

Thus the optimal value is $G(5) = 21$.

Working backwards to find an optimal solution, $\kappa_5 = 2$ so an optimal solution is obtained by setting $y_2 = 1$ and $x_2 = d_{25} = 6$, and completing with an optimal solution for the first $\kappa_5 - 1 = 1$ period of value $G(1)$. Here $\kappa_1 = 1$, and so $y_1 = 1, x_1 = d_{11} = 3$. Therefore a complete optimal solution is given by $y_1 = y_2 = 1$ with $x_1 = d_1 = 3$ and $x_2 = d_{25} = 6$.

Now we look at a backward variant and also consider whether it is possible to solve the problem faster.

A Faster Backward Dynamic Programming Recursion

Let $H(t)$ be the optimal value for problem *LS-U* with planning horizon $t, \dots, n$, ignoring periods $1, \dots, t-1$. We obtain a recursion by considering for which interval $[t, k]$ the demand is satisfied by production in t.

$$H(t) = \min_{k:k>t} [q_t + p_t d_{t,k-1} + H(k)] \text{ if } d_t > 0.$$

$$H(t) = \min\{H(t+1), \min_{k:k>t+1} [q_t + p_t d_{t,k-1} + H(k)]\} \text{ if } d_t = 0.$$

Starting from $H(n+1) = 0$, and calculating $H(n)$, $H(n-1)$, ... in order, one obtains the optimal solution value $H(1)$ of *LS-U*.

A straightforward implementation of this recursion again results in an $O(n^2)$ algorithm. Now we indicate how it is possible to carry out the calculations faster leading to an $O(n \log n)$ algorithm. For simplicity we suppose that $d_t > 0$ in each period.

Rewriting the recursion, we have

$$H(t) = \min_{k:k>t} [-p_t d_{kn} + H(k)] + q_t + p_t d_{tn}.$$

Thus the crucial calculation at each iteration is $k_t = \arg\min_{k:k>t}[-p_t d_{kn} + H(k)]$.

Consider now a plot of the points $(x^k, y^k) = (d_{1,k-1}, H(k))$ for $k = t+1, \ldots, n+1$ with $H(n+1) = 0$.

Observation 7.5 *The line through (x^k, y^k) with slope $-p_t$ intersects the vertical line $x = d_{1n}$ in $-p_t d_{kn} + H(k)$. Thus it suffices to find the lowest intercept. In Figure 7.4 the same graph is shown twice: on the left one sees the slopes, and on the right the resulting lower envelope.*

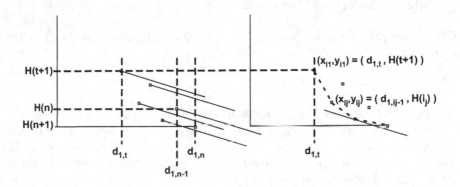

Figure 7.4. Backward DP and lower envelope.

Now consider the lower convex envelope of the points (x^k, y^k) for $k = t+1, \ldots, n+1$ with breakpoints (vertices of the lower convex envelope) $(x_{i_1}, y_{i_1}), \ldots, (x_{i_r}, y_{i_r})$ with $\{i_1, \ldots, i_r\} \subseteq \{t+1, \ldots, n+1\}$ and $i_r = n+1$.

Observation 7.6 *If (x^k, y^k) is not a breakpoint, there is some breakpoint (x_{i_j}, y_{i_j}) with $H(i_j) - p_t d_{i_j, n} \leq H(k) - p_t d_{kn}$. It follows that*

$$H(t) = \min_{j=1,\ldots,r} (H(i_j) - p_t d_{i_j, n}) + q_t + p_t d_{tn}.$$

Observation 7.7 *The segment of the lower convex envelope joining adjacent breakpoints $(x_{i_{j-1}}, y_{i_{j-1}})$ and (x_{i_j}, y_{i_j}) has slope*

$$m_{i_j} = \frac{H(i_j) - H(i_{j-1})}{d_{i_{j-1}, i_j - 1}} < 0$$

and the slopes are increasing in j.

Observation 7.8 *If $m_{i_j} \leq -p_t < m_{i_{j+1}}$, then*

$$H(t) = H(i_j) - p_t d_{i_j, n} + q_t + p_t d_{tn}.$$

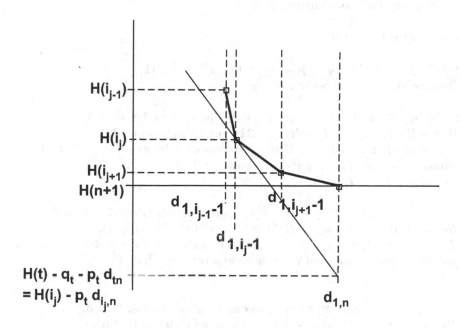

Figure 7.5. Backward DP, lower envelope, and computation of $H(t)$.

The computation of i_j such that $m_{i_j} \leq -p_t < m_{i_{j+1}}$, and of $H(t)$, is illustrated in Figure 7.5.

Observation 7.9 *Given the values m_{i_j} in increasing order, the value for which $m_{i_j} \leq -p_t < m_{i_{j+1}}$ can be found by bisection in $O(\log n)$.*

Finally having obtained the new point $(x^t, y^t) = (d_{1,t-1}, H(t))$, it is necessary to update the lower convex envelope by adding the point (x^t, y^t), and possibly removing some of the breakpoints from the beginning of the previous ordered list.

Observation 7.10 *For $j = 1, \ldots, r$, if the slope of the line joining (x^t, y^t) to (x_{i_j}, y_{i_j}) is greater than or equal to $m_{i_{j+1}}$, the points (x_{i_l}, y_{i_l}) for $l = 1, \ldots, j$ are no longer breakpoints in the convex envelope.*

As there are at most n breakpoints that can be removed from the lower convex envelope only once, the update of this lower envelope takes $O(n)$ in total. Therefore the running time of the backward algorithm is the following.

Proposition 7.3 *The backward DP recursion can be implemented to run in* $O(n \log n)$.

Example 7.2 *Consider an instance of LS-U with* $n = 5, d = (5, 4, 2, 3, 6), p = (3, 1, 3, 2, 1),$ *and* $f = (0, 30, 25, 15, 5)$.
Applying the above algorithm, we have

$$t = 6, (x^6, y^6) = (20, 0).$$

$t = 5, p_5 = 1, H(5) = q_5 + p_5 d_5 = 11, (x^5, y^5) = (14, 11)$.
Breakpoints z^5, z^6, slopes $m_5 = -\frac{11}{6}$.

$t = 4, p_4 = 2,$ *and* $-2 < -\frac{11}{6}$. *Thus by Observation 7.8,* $k_4 = 5$.
Hence $H(4) = q_4 + p_4 d_4 + H(5) = 32,$ *and* $(x^4, y^4) = (11, 32)$.
The slope of (z^4, z^5) *is* -7. *Using Observation 7.10, as* $-7 < -\frac{11}{6}$, *no breakpoints are eliminated, and the new breakpoints are* z^4, z^5, z^6.
The slopes are $(-7, -\frac{11}{6})$.

$t = 3, p_3 = 3,$ *and* $-7 \le -3 < -\frac{11}{6}$. *Thus by Observation 7.8,* $k_3 = 5$.
Hence $H(3) = q_3 + p_3 d_{34} + H(5) = 51,$ *and* $(x^3, y^3) = (9, 51)$.
The slope of (z^3, z^4) *is* $-\frac{19}{2}$. *Using Observation 7.10, as* $-\frac{19}{2} < -7$, *no breakpoints are eliminated, and the new breakpoints are* z^3, z^4, z^5, z^6.
The slopes are $(-\frac{19}{2}, -7, -\frac{11}{6})$.

$t = 2, p_2 = 1,$ *and* $-\frac{11}{6} < -1$. *Thus by Observation 7.8,* $k_2 = 6$.
Hence $H(2) = q_2 + p_2 d_{25} + H(6) = 45,$ *and* $(x^2, y^2) = (5, 45)$.
The slope of (z^2, z^3) *is* $\frac{6}{4}$. *Using Observation 7.10, as* $\frac{6}{4} > -\frac{19}{2}$, *the breakpoint* z^3 *is eliminated.*
The slope of (z^2, z^4) *is* $-\frac{13}{6}$. *As* $-\frac{13}{6} > -7$, *the breakpoint* z^4 *is eliminated.*
The slope of (z^2, z^5) *is* $-\frac{34}{9}$. *As* $-\frac{34}{9} < -\frac{11}{6}$, *the new set of breakpoints is* z^2, z^5, z^6.
The slopes are $(-\frac{34}{9}, -\frac{11}{6})$.

$t = 1, p_1 = 3,$ *and* $-\frac{34}{9} \le -3 < -\frac{11}{6}$. *Thus by Observation 7.8,* $k_1 = 5$.
Hence $H(1) = q_1 + p_1 d_{14} + H(5) = 53$.
An optimal solution is $y_1 = y_5 = 1, x_1 = 14, x_5 = 6$.

We finish this section by viewing the forward recursion as a shortest path problem.

The Shortest Path Algorithm.

Consider a directed graph with nodes $1, 2, \ldots, n+1$, where each arc (i, j) with $i < j$ corresponds to the regeneration interval $[i, j-1]$, that is, to a set-up in period i with production to satisfy demand for the interval $[i, j-1]$. Clearly any path from 1 to $n+1$ provides a feasible solution to *LS-U*. Furthermore this solution satisfies the structure of extreme solutions described in Proposition 7.2. If arc (i, j) has cost $c(i, j-1) = q_i + p_i d_{i,j-1}$, the length of a shortest path will necessarily be $G(n)$. Hence solving *LS-U* by the forward dynamic programming algorithm is equivalent to solving the shortest path problem on the graph described above. An instance of the directed graph for $n = 4$ is shown in Figure 7.6.

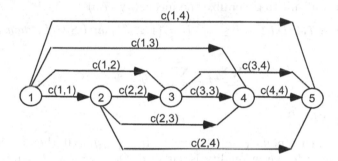

Figure 7.6. Shortest path to solve *LS-U* with $n = 4$.

7.4 Linear Programming Reformulations of *LS-U*

Here we use *LS-U* to demonstrate the classical polyhedral approach to the study of formulations for a combinatorial optimization problem. We exhibit valid inequalities for X^{LS-U}, a description of the convex hull of feasible solutions, extended formulations, and separation algorithms.

The first point to observe is that, because we have shown in the previous section that the optimization problem $OPT(X^{LS-U}, (p', h', q))$ is polynomially solvable, we know from Chapter 6 that the separation problem $SEP(X^{LS-U}, (x^*, s^*, y^*))$ is also polynomially solvable. Therefore it may be possible to provide a complete inequality description of $\text{conv}(X^{LS-U})$, and a fast combinatorial separation algorithm.

7.4.1 Valid Inequalities for *LS-U*

We want to find valid inequalities satisfied by all points in X^{LS-U}, that is, those satisfying (7.2)–(7.5). In particular we are looking for valid inequalities for X^{LS-U} that are not just obtainable as linear combinations of the constraints (7.2)–(7.4).

One such inequality already described in Chapter 4 is

$$x_t \leq d_t y_t + s_t. \tag{7.9}$$

We provide here a different argument for its validity. Using $s_{t-1} + x_t = d_t + s_t$, it can also be written as

$$s_{t-1} \geq d_t(1 - y_t).$$

In this form, it is easy to see that the inequality is valid. If $y_t = 0$, then there is no production in period t, and so the entering stock s_{t-1} must at least contain the demand for the period d_t. On the other hand, if $y_t = 1$, the inequality reduces to $s_{t-1} \geq 0$, which is always valid.

It is not difficult to generalize the inequality (7.9).

Proposition 7.4 *Let $1 \leq l \leq n$, $L = \{1, \ldots, l\}$ and $S \subseteq L$, then the (l, S) inequality*

$$\sum_{j \in S} x_j \leq \sum_{j \in S} d_{jl} y_j + s_l \tag{7.10}$$

is valid for X^{LS-U}.

Proof. Consider a point $(s, y) \in X^{LS-U}$. If $\sum_{j \in S} y_j = 0$, then as $x_j = 0$ for $j \in S$ and $s_l \geq 0$, the inequality is satisfied. Otherwise let $t = \min\{j \in S : y_j = 1\}$. Then $\sum_{j \in S} x_j \leq \sum_{j=t}^{l} x_j \leq d_{tl} + s_l \leq \sum_{j \in S} d_{jl} y_j + s_l$ where the first inequality follows from the definition of S and the nonnegativity of x_j, the second from the flow conservation equations, and the third using $y_t = 1$ and the nonnegativity of the y_j. □

What is more, repeating Theorem 4.2, it has been shown that these are the only inequalities needed.

Theorem 7.5 *When $s_0 = s_n = 0$, the original constraints (7.2)–(7.4) plus the (l, S) inequalities (7.10) give a complete linear inequality description of $conv(X^{LS-U})$.*

For later, using $\sum_{j \in L} x_j = d_{1l} + s_l$, we observe that the (l, S) inequality can be rewritten as either

$$\sum_{j \in L \setminus S} x_j + \sum_{j \in S} d_{jl} y_j \geq d_{1l}, \tag{7.11}$$

or, if $k = \min\{i \in S\}$, using $s_{k-1} + \sum_{j=k}^{l} x_j = d_{kl} + s_l$, as

$$s_{k-1} + \sum_{j \in \{k, \ldots, l\} \setminus S} x_j + \sum_{j \in S} d_{jl} y_j \geq d_{kl}, \tag{7.12}$$

or equivalently as

$$s_{k-1} + \sum_{j\in\{k,\dots,l\}\setminus S} x_j \geq \sum_{j=k}^{l} d_j(1 - \sum_{t\in S,t\leq j} y_t). \tag{7.13}$$

The complication from a linear programming point of view is that there is an exponential number of these inequalities. This means that it is impossible to add all of them to the formulation a priori. The alternative is either to choose a small subset to be added a priori (which inequalities?), or to generate them as cutting planes when they are needed. Taking the cutting plane approach, given a point (x^*, y^*, s^*) satisfying (7.2)–(7.4), nonintegral in y^*, we need to find one or more of the (l, S) inequalities cutting it off.

The Separation Problem for (l, S) Inequalities
Given (x^*, s^*, y^*) satisfying (7.2)–(7.4), either find an (l, S) inequality cutting off the point, or show that all the inequalities are satisfied, and so demonstrate that $(x^*, s^*, y^*) \in \mathrm{conv}(X^{LS-U})$.

It turns out that this separation problem is easily solved by inspection. We consider the inequalities in the form (7.11).

Algorithm for (l, S) Separation
1. For $l = 1, \dots, n$,
2. Calculate $\alpha_l \equiv \sum_{j=1}^{l} \min(x_j^*, d_{jl}y_j^*)$.
3. If $\alpha_l < d_{1l}$, output $L = \{1, \dots, l\}, S = \{j \in L : x_j^* > d_{jl}y_j^*\}$.
4. end For.

A straightforward implementation requires $O(n^2)$ comparisons in line 2. However, it is possible to rearrange the computations to obtain a running time of $O(n\log n)$. Observe that $0 \leq d_j y_j^* \leq d_{j,j+1}y_j^* \leq \dots \leq d_{jn}y_j^*$. So, using bisection, it is possible for each j to determine an integer $l(j) \in \{j, \dots, n\}$ such that $d_{j,l(j)-1}y_j^* < x_j^* \leq d_{j,l(j)}y_j^*$ in $O(\log n)$.
Then we can calculate α_l recursively as follows:

$$\alpha_l = \alpha_{l-1} + d_l(\sum_{j\in Y_l} y_j^*) + \sum_{j\in X_l}(x_j^* - d_{j,l-1}y_j^*),$$

where $Y_l = \{j \in \{1,\dots,l\} : l(j) > l\}$ and $X_l = \{j \in \{1,\dots,l\} : l(j) = l\}$. As $Y_l = (Y_{l-1} \cup \{l\}) \setminus X_l$, the additional time needed to calculate all the α_l is $O(n)$.

Example 7.3 *Consider an instance with $n = 4$ and $d = (7,2,6,4)$. For $l = 3$, the (l, S) inequalities are*

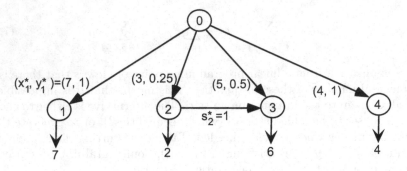

Figure 7.7. Fractional solution of *LS-U*.

$$
\begin{aligned}
x_1 & & \leq 15y_1 & & & & +s_3 \\
& x_2 & \leq & & 8y_2 & & +s_3 \\
& & x_3 \leq & & & 6y_3 & +s_3 \\
x_1 +x_2 & & \leq 15y_1 & +8y_2 & & & +s_3 \\
x_1 & +x_3 & \leq 15y_1 & & & +6y_3 & +s_3 \\
& x_2 +x_3 & \leq & & 8y_2 & +6y_3 & +s_3 \\
x_1 +x_2 & +x_3 & \leq 15y_1 & +8y_2 & & +6y_3 & +s_3.
\end{aligned}
$$

Suppose now that we are given the nonintegral solution $x^* = (7,3,5,4), s^* = (0,1,0,0), y^* = (1,\frac{1}{4},\frac{1}{2},1)$ *shown in Figure 7.7.*

Using the improved separation algorithm, we see that

$x_1^* \leq d_1 y_1^*$, *so* $l(1) = 1$.
$d_{23}y_2^* < x_2^* \leq d_{24}y_2^*$, *so* $l(2) = 4$.
$d_3 y_3^* < x_3^* \leq d_{34}y_3^*$, *so* $l(3) - 4$.
$x_4^* \leq d_4 y_4^*$, *so* $l(4) = 4$.

$l = 1$.   *Initially* $X_1 = \{1\}$, $Y_1 = \emptyset$ *and* $\alpha_1 = 7 \geq d_1$.
No violation.

$l = 2$. $X_2 = \emptyset$, $Y_2 = \{2\}$ *and* $\alpha_2 = 7 + 2y_2^* = 7.5 < d_{12} = 9$.
Violated inequality with $S = Y_2$.
$x_1 + d_2 y_2 \geq d_{12}$ *is violated by* $9 - 7.5 = 1.5$.

$l = 3$. $X_3 = \emptyset$, $Y_3 = \{2,3\}$ *and* $\alpha_3 = 7.5 + 6(y_2^* + y_3^*) = 12 < d_{13} = 15$.
Violated inequality with $S = Y_3$.
$x_1 + d_{23}y_2 + d_3 y_3 \geq d_{13}$ *is violated by* $15 - 12 = 3$.

$l = 4$. $X_4 = \{2,3,4\}$, $Y_4 = \emptyset$ *and* $\alpha_4 = 12 + (3-2) + (5-3) + 4 = 19 \geq d_{14}$.
No violation.

7.4.2 Extended Formulations for *LS-U*

Here we consider formulations for *LS-U* involving more informative variables in which the link between the continuous variables (x, s) and the 0–1 variables y is more precise. We present two such extended formulations. The first has a very natural choice of variables.

The Facility Location Extended Formulation

Let w_{ut} with $u \leq t$ be the amount produced in period u to satisfy demand in period t. This leads to the formulation

$$\min \quad \sum_{u=1}^{n} p_u x_u + \sum_{t=1}^{n} q_t y_t \tag{7.14}$$

$$\sum_{u=1}^{t} w_{ut} = d_t \qquad \text{for } 1 \leq t \leq n \tag{7.15}$$

$$w_{ut} \leq d_t y_u \qquad \text{for } 1 \leq u \leq t \leq n \tag{7.16}$$

$$x_u = \sum_{t=u}^{n} d_t w_{ut} \qquad \text{for } 1 \leq u \leq n \tag{7.17}$$

$$y \in [0,1]^n, w_{ut} \in \mathbb{R}_+^1 \qquad \text{for } 1 \leq u \leq t \leq n, \tag{7.18}$$

$$y \in \mathbb{Z}^n. \tag{7.19}$$

Here (7.15) ensures that the demand is satisfied in each period, and (7.16) that the set-up variable $y_u = 1$ whenever there is production in period u. The equation (7.17) provides the link between the original production variables x and the new variables w. Note that one can completely remove the x variables by eliminating them from the objective function using (7.17). Alternatively one could also add the constraints (7.2) to have the values of the original stock variables s.

Question How good is this new extended formulation $Q^{FL-U} = \{(x, y, w) :$ (7.15)- (7.18)\}?

The answer is that the formulation cannot be bettered.

Theorem 7.6 *The linear program*

$$\min\{px + qy : (x, y, w) \in Q^{FL-U}\} \tag{7.20}$$

has an optimal solution with y integer, and thus it solves LS-U.

We can also say what this means theoretically.

To compare formulations, we need to consider solutions in the same space of variables, so we need to look at the projection of Q^{FL-U} in the (x, y) or (x, y, s)-space; see Section 6.1.5.

Theorem 7.7 $proj_{x,y}Q^{FL-U} = proj_{x,y}conv(X^{LS-U})$.

Example 7.4 *We consider an instance with $n = 3$. Q^{FL-U} is of the form:*

$$
\begin{aligned}
w_{11} & & & = d_1 \\
w_{12} & & +w_{22} & = d_2 \\
& w_{13} & +w_{23}+w_{33} & = d_3 \\
w_{11} & & & \leq d_1 y_1 \\
w_{12} & & & \leq d_2 y_1 \\
& w_{13} & & \leq d_3 y_1 \\
& & w_{22} & \leq d_2 y_2 \\
& & w_{23} & \leq d_3 y_2 \\
& & w_{33} & \leq d_3 y_3 \\
w_{11}+w_{12} & +w_{13} & & = x_1 \\
& & w_{22}+w_{23} & = x_2 \\
& & w_{33} & = x_3
\end{aligned}
$$

$$w \in \mathbb{R}_+^6, 0 \leq y_1, y_2, y_3 \leq 1.$$

Formulation Q^{FL-U} has $O(n^2)$ constraints and $O(n^2)$ variables. We now present a second reformulation that is more compact as it has only $O(n)$ constraints, apart from the nonnegativity constraints.

Shortest Path Formulation of *LS-U*

We saw above in Proposition 7.2 that an optimal solution consists of a sequence of regeneration intervals $[i, j-1]$ in which an amount $d_{i,j-1}$ is produced in period i. In fact in the shortest path algorithm of Section 7.3, the arc (i, j) corresponds precisely to such an interval. This suggests the following choice of variable:

$\phi_{ut} = 1$ if an amount $d_{ut} > 0$ is produced in period u. In addition, if the first set-up period with production occurs in period t, then $\phi_{1,t-1} = 1$ and $d_{1,t-1} = 0$. Otherwise $\phi_{ut} = 0$. So essentially $\phi_{ut} = 1$ if the regeneration interval $[u, t]$ is part of the solution.

This leads to the following reformulation (assuming $s_0 = 0$ and $s_n = 0$).

$$\min \quad \sum_{u=1}^{n} p_u x_u + \sum_{t=1}^{n} q_t y_t \tag{7.21}$$

$$-\sum_{t=1}^{n} \phi_{1t} = -1 \tag{7.22}$$

$$\sum_{u=1}^{t-1} \phi_{u,t-1} - \sum_{\tau=t}^{n} \phi_{t\tau} = 0 \qquad \text{for } 2 \leq t \leq n \tag{7.23}$$

$$\sum_{u=1}^{n} \phi_{un} = 1 \tag{7.24}$$

$$\sum_{\tau=t:d_{t\tau}>0}^{n} \phi_{t\tau} \leq y_t \qquad \text{for } 1 \leq t \leq n \qquad (7.25)$$

$$\sum_{\tau=t}^{n} d_{t\tau}\phi_{t\tau} = x_t \qquad \text{for } 1 \leq t \leq n \qquad (7.26)$$

$$y \in [0,1]^n, \phi_{ut} \in \mathbb{R}^1_+ \qquad \text{for } 1 \leq u \leq t \leq n. \qquad (7.27)$$

The flow conservation constraints (7.22)–(7.24) model a solution as a sequence of regeneration intervals. Again the constraints (7.26) allow us to relate the new variables ϕ_{ut} to the original production variables x_t. Constraints (7.25) define the set-up variables y_t.

Let $Q^{SP-U} = \{(x, y, \phi) \in \mathbb{R}^n \times [0,1]^n \times \mathbb{R}_+^{\frac{n(n+1)}{2}} : (7.22)\text{-}(7.26)\}$.

To see why this extended reformulation is called a shortest path formulation, it is best to extend the network to include the y_t variables explicitly as bounds on certain arcs. An instance of the extended network for $n = 3$ is shown in Figure 7.8.

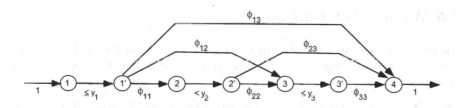

Figure 7.8. Shortest path reformulation of *LS-U*.

We see that the flow $\sum_i \phi_{i,t-1}$ enters node t, is bounded by y_t on arc (t, t'), and the flow $\sum_i \phi_{t,i}$ leaves from node t'. Thus this network incorporates the flow conservation constraints (7.22)–(7.24) and the set-up variable constraint (7.25). Therefore formulation Q^{SP-U} is equivalent to the shortest path (i.e., minimum cost unit flow) in the extended network with a cost q_t on arc (t, t') and a cost $p_u d_{ut}$ on arc $(u', t + 1)$. By comparing Figures 7.6 and 7.8, we can also observe that this shortest path formulation can be derived directly from the shortest path representation of the forward dynamic program solving *LS-U*.

Theorem 7.8 *The linear program*

$$\min\{px + qy : (x, y, \phi) \in Q^{SP-U}\} \qquad (7.28)$$

has an optimal solution with y integer, and thus solves LS-U.

This implies the same result as above concerning the strength of the formulation.

Theorem 7.9 $proj_{x,y}Q^{SP-U} = proj_{x,y}\text{conv}(X^{LS-U})$.

Example 7.5 *We consider an instance with $n = 3$. We present Q^{SP-U} and assume that $d_t > 0$ for $t = 1, 2, 3$.*

$$
\begin{aligned}
-\phi_{11} \quad -\phi_{12} \quad -\phi_{13} & & & & = -1 \\
\phi_{11} & & -\phi_{22} \quad -\phi_{23} & & = 0 \\
\phi_{12} & & +\phi_{22} & -\phi_{33} & = 0 \\
\phi_{13} & & +\phi_{23} & +\phi_{33} & = 1 \\
\phi_{11} \quad +\phi_{12} \quad +\phi_{13} & & & & \leq y_1 \\
\phi_{22} & +\phi_{23} & & & \leq y_2 \\
\phi_{33} & & & & \leq y_3 \\
d_{11}\phi_{11} \quad +d_{12}\phi_{12} \quad +d_{13}\phi_{13} & & & & = x_1 \\
+d_{22}\phi_{22} \quad +d_{23}\phi_{23} & & & & = x_2 \\
d_{33}\phi_{33} & & & & = x_3
\end{aligned}
$$

$$ x \in \mathbb{R}^3_+, \ y \in [0,1]^3, \ \phi \in \mathbb{R}^6_+. $$

7.5 Wagner–Whitin Costs

As we indicated when defining our initial classification in Part I, it turns out that in practice a large number of instances have a somewhat special cost structure.

Definition 7.1 *A lot-sizing problem has Wagner-Whitin costs if $p'_t + h'_t \geq p'_{t+1}$ for all t.*

Note that the modified objective function $\sum_t p_t x_t + \sum_t q_t y_t$ is Wagner–Whitin if $p_t \geq p_{t+1}$ for all t, and the modified objective function $\sum_t h_t s_t + \sum_t q_t y_t$ is Wagner–Whitin if $h_t \geq 0$ for all t.

What are the consequences of Wagner–Whitin costs? The first remarkable result is that for *WW-U* the faster backward dynamic programming recursion of Section 7.3 can be made to run in $O(n)$.

What about formulations and valid inequalities?

Observation 7.11 *With Wagner–Whitin costs, it is optimal to produce as late as possible. In other words, given the set-up periods $y \in \{0,1\}^n$, it is optimal to satisfy the demand in period t from the last set-up period before or equal to t. Alternatively s_{k-1} contains the demand d_j for period $j \geq k$ only if no set-up occurs in the time interval $[k, \ldots, j]$.*

This suggests a new mixed integer programming formulation for *WW-U*:

$$\min \quad \sum_{t=0}^{n} h_t s_t + \sum_{t=1}^{n} q_t y_t \qquad (7.29)$$

$$s_{k-1} \geq \sum_{j=k}^{l} d_j (1 - y_k - \ldots - y_j) \qquad \text{for } 1 \leq k \leq l \leq n \qquad (7.30)$$

$$s \in \mathbb{R}_+^{n+1}, y \in [0,1]^n \qquad (7.31)$$

$$y \in \mathbb{Z}^n. \qquad (7.32)$$

Because $h_t \geq 0$ for all t, there exists an optimal solution in which each s_{k-1} is as small as possible, and thus

$$s_{k-1} = \max[0, \max_{l \geq k} \sum_{j=k}^{l} d_j (1 - y_k - \ldots - y_j)].$$

So we see that $s_{k-1} = d_{kl}$ if and only if $y_j = 0$ for all $j = k, \ldots, l$ and $y_{l+1} = 1$. This is exactly Observation 7.11. This formulation is useless when the costs are not Wagner-Whitin, because if some $h_t < 0$, the objective function will be unbounded.

Observation 7.12 *The inequality (7.30), called the* (l, S, WW) *inequality*

$$s_{k-1} \geq \sum_{j-k}^{l} d_j (1 - y_k - \ldots - y_j),$$

can be rewritten as

$$s_{k-1} + \sum_{j=k}^{l} d_{jl} y_j \geq d_{kl},$$

or as an (l, S) *inequality*

$$\sum_{j=1}^{k-1} x_j + \sum_{j=k}^{l} d_{jl} y_j \geq d_{1l}.$$

What is really interesting in practice is that the above formulation is best possible. Let P^{WW-U}, X^{WW-U} be the set of points (s, y) satisfying (7.30)–(7.31), (7.30)–(7.32), respectively.

Theorem 7.10 *i. Any optimal extreme point solution of the linear program* $\min\{hs + qy : (s, y) \in P^{WW-U}\}$ *solves WW-U.*
ii. $P^{WW-U} = \text{conv}(X^{WW-U})$.

This result tells us that there is a linear program with just $2n$ variables and $n(n+1)/2$ constraints that solves $WW\text{-}U$. What is more, it tells us that the subset (l, S, WW) of the (l, S) inequalities of the form (7.30) suffice in the presence of Wagner–Whitin costs.

Given such a compact formulation, there appears to be little need for a separation algorithm for the (l, S, WW) inequalities. However, for large problems with many periods and items, we may wish to add them as cuts.

An $O(n)$ Algorithm to Separate the (l, S, WW) Inequalities.

To separate the (l, S, WW) inequalities, a most violated inequality for the point $(s^*, y^*) \in \mathbb{R}_+^{n+1} \times [0,1]^n$ involving s_{k-1} is the inequality

$$s_{k-1} \geq \sum_{j=k}^{l(k)} d_j (1 - y_k - \cdots - y_j)$$

where $l(k) = \max\{u : u \geq k, d_u > 0, \sum_{j=k}^{u} y_j^* < 1\}$.

As $1 \leq l(1) \leq l(2) \cdots \leq l(n) \leq n$, the values of $l(k)$ and the corresponding violations can be found in linear time.

In Table 7.1 we present a résumé of the results that we have seen in this chapter. They are identical to those in Table 4.4 for problems *LS-U* and *WW-U*.

Table 7.1. Reformulation results for *LS-U* and *WW-U*

	LS	WW
	$Cons \times Vars$	$Cons \times Vars$
Formulation		
U	SP $O(n) \times O(n^2)$	WW $O(n^2) \times O(n)$
	FL $O(n^2) \times O(n^2)$	
Separation		
U	(l, S)	(l, S, WW)
	$O(n \log n)$	$O(n)$
Optimization		
U	$O(n \log n)$	$O(n)$

7.6 Partial Formulations

When solving large-size instances involving many time periods, or many items for which extended reformulations need to be added to the initial formulation, the resulting linear relaxation may become too large to be solved directly in a branch-and-bound system. One option is then to use a cutting plane approach. An alternative is to reduce the size of the reformulations by using partial or approximate extended reformulations.

We illustrate this here with an approximate shortest path reformulation of the uncapacitated lot-sizing set X^{LS-U}.

We choose some parameter k ($\leq n$) that determines the size and strength of the formulation. For simplicity we assume that $d_t > 0$ for all $t = 1, \ldots, n$ and $s_0 = s_n = 0$. The variables are defined as follows:

$z_{it} = 1$ if production takes place in period i, and the amount produced is d_{it} satisfying all the demands from periods i up to t, with $t < i + k$.

$u_i = 1$ for $1 \leq i \leq n - k$ if production takes place in period i, and the amount produced is d_{it} for some $t \geq i + k$.

$v_t = 1$ for $k + 1 \leq t \leq n$ if exactly d_{it} is produced in some period $i \leq t - k$.

$w_t = 1$ for $t = 2, \ldots, n - k$ if demand for periods $t - 1, \ldots, t + k$ is all produced simultaneously in some period $i \leq t - 1$.

The resulting approximate formulation X_k^{SP} with $O(nk)$ variables and $O(n)$ constraints is

$$-\sum_{i=1}^{k} z_{1i} - u_1 = -1 \tag{7.33}$$

$$\sum_{i=\max[t-k+1,1]}^{t} z_{it} + v_t - \sum_{i=t+1}^{\min[t+k,n]} z_{t+1,i} - u_{t+1} = 0 \quad \text{for } 1 \leq t \leq n-1 \tag{7.34}$$

$$\sum_{i=n-k+1}^{n} z_{in} + v_n = 1 \tag{7.35}$$

$$u_t + w_t - v_{t+k} - w_{t+1} = 0 \qquad \text{for } 1 \leq t \leq n-k \tag{7.36}$$

$$\sum_{i=t}^{\min[t+k-1,n]} z_{ti} + u_t \leq y_t \qquad \text{for } 1 \leq t \leq n \tag{7.37}$$

$$x_t \geq \sum_{i=t}^{\min[t+k-1,n]} d_{ti} z_{ti} + d_{t,t+k} u_t \qquad \text{for } 1 \leq t \leq n \tag{7.38}$$

$$s_{t-1} \geq \sum_{i=1}^{t-1} \sum_{j=t}^{n} d_{tj} z_{ij} + \sum_{i=t}^{t+k-1} d_{ti} v_i + d_{t,t+k} w_t \qquad \text{for } 2 \leq t \leq n \tag{7.39}$$

$$z, u, v \geq 0, \ y \in [0,1]^n \tag{7.40}$$

$$v_t = 0 \text{ for } t \leq k \tag{7.41}$$

$$u_t = w_t = 0 \qquad \text{for } t \geq n-k+1 \tag{7.42}$$

$$w_1 = 0, z_{it} = 0 \qquad \text{for } t \geq i+k. \tag{7.43}$$

In Figure 7.9, we show a shortest path representation of X_k^{SP} for $n = 5$ and $k = 2$. In this example, u_1 approximates or aggregates all regeneration intervals $[1, t]$ with $t \geq 3$. In other words, u_1 represents the flow $\sum_{t=3}^{n} \phi_{1t}$ of the shortest path formulation. Similarly, v_5 approximates or aggregates all regeneration intervals of length strictly larger than k ending in period 5, that is, v_5 represents $\sum_{t=1}^{3} \phi_{t5}$.

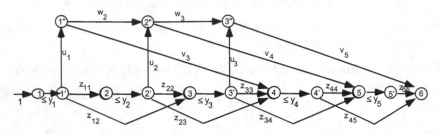

Figure 7.9. Approximate shortest path formulation for uncapacitated lot-sizing.

Proposition 7.11 *The polyhedron (7.33)–(7.43) is integral.*
If the linear program $\min\{px + hs + qy : (x, y, s, z, u, v, w)$ satisfying (7.33)-(7.43)\} has an optimal solution with $w_t = 0$ for all t, this solution is optimal for the uncapacitated lot-sizing problem $\min\{px + hs + qy : (x, y, s) \in X^{LS-U}\}$.

Something can be said about the strength of such an approximation in comparison with a cutting plane reformulation. For $\mathrm{conv}(X^{LS-U})$, we know from Section 7.4 that every facet-defining inequality is a (t, l, S) inequality of the form $\sum_{j \in [1, t-1] \cup S} x_j + \sum_{j \in [t, l] \setminus S} d_{jl} y_j \geq d_{tl}$ with $1 \leq t \leq l \leq n$ and $S \subseteq [t+1, l] = \{t+1, \dots, l\}$.

Proposition 7.12 *If $l \leq t + k$, the (t, l, S) inequality is valid for X_k^{SP}. If $d_t > 0$ for all t, the (t, l, S) inequality is valid for X_k^{SP} only if $l \leq t + k$.*

So reformulation X_k^{SP} is equivalent to the formulation obtained by adding all (t, l, S) inequalities with $l \leq t + k$ to the initial formulation (7.2)–(7.4) and (7.6).

7.7 Some Convex Hull Proofs

Proof of Theorem 7.5.

We use the proof technique number 8 from Section 6.4. We assume that

$s_0 = s_n = 0$, and also that $d_1 > 0$. Rewriting the set X^{LS-U} in the (x, y) space we have

$$\sum_{u=1}^{t} x_u \geq d_{1t} \qquad\qquad \text{for } t = 1, \ldots, n-1 \qquad (7.44)$$

$$\sum_{u=1}^{n} x_u = d_{1n} \qquad\qquad\qquad\qquad (7.45)$$

$$x_t \leq M y_t \qquad\qquad \text{for } t = 1, \ldots, n \qquad (7.46)$$

$$x \in \mathbb{R}_+^n, y \in \{0, 1\}^n, \qquad\qquad\qquad (7.47)$$

and P is described by the constraints (7.44), (7.45), $x \in \mathbb{R}_+^n, y \in [0, 1]^n$ and the (l, S) inequalities (7.11).

First we check that $\dim(\text{conv}(X)) = \dim(P)$. All points in X satisfy the equation (7.45) and $y_1 = 1$ as $d_1 > 0$. It is not hard to show that X contains $2n - 1$ affinely independent points, and so $\dim(\text{conv}(X)) = 2n - 2$. Now P also satisfies these equations as its description includes (7.45), $y_1 \leq 1$ and the (l, S) inequality with $l = 1, S = \emptyset$ which is the inequality $d_1 y_1 \geq d_1$.

Let $M(p, q)$ be the set of all optimal solutions to the problem $OPT(X^{LS-U}, (p, q))$ of minimizing the objective function (7.7) over the set X^{LS-U}.

First we observe that as $\sum_{t=1}^{n} x_t = d_{1n}$, we can add any multiple of this constraint to the objective function without modifying the set $M(p, q)$. Thus we suppose without loss of generality that $\min_t p_t = 0$. In addition, as we have that $y_1 = 1$, we can also assume that $q_1 = 0$.

If $q_t < 0$ for some $t \geq 2$, then $M(p, q) \subseteq \{(x, y) : y_t = 1\}$, so we suppose from now on that $q_t \geq 0$ for $t \geq 2$.

Now let $l = \arg\max_t \{p_t + q_t > 0\}$. As $(p, q) \neq (0, 0)$, we have that $1 \leq l \leq n$. Suppose that $p_k = q_k = 0$ for some $k < l$. Then $M(p, q) \subseteq \{(x, y) : x_l = 0\}$.

Otherwise we have that $p_i + q_i > 0$ for $i = 1, \ldots, l$ and $p_i = q_i = 0$ for $i = l+1, \ldots, n$. Let $L = \{1, \ldots, l\}$ and $S = \{t \in L : p_t > 0\}$ with $1 \in S$. Note that, from the definition of l, all demand after period l can be produced at zero cost. We show that

$$M(p, q) \subseteq \{(x, y) : \sum_{t \in S} x_t + \sum_{t \in L \setminus S} d_{tl} y_t = d_{1l}\}.$$

Consider an optimal solution (x^*, y^*, s^*). Suppose that $\tau = \min\{t \in L \setminus S : y_t^* = 1\}$.

As $p_\tau = 0$, one can produce an amount $d_{\tau l}$ or more at zero variable cost in τ. Also as $p_t + q_t > 0$ for all $t \in L$ with $t > \tau$, in an optimal solution $x_t^* = y_t^* = 0$ for all $t \in L$ with $t > \tau$. This holds because otherwise a strictly better solution would be obtained by reducing x_t^* and/or y_t^* and increasing x_τ^*, which is a contradiction to the optimality of (x^*, y^*, s^*). So $\sum_{t \in L \setminus S} d_{tl} y_t^* = d_{\tau l} y_\tau^* = d_{\tau l}$ and $\sum_{t \in S: t \geq \tau} x_t^* = 0$.

As $y_t^* = x_t^* = 0$ for all $t \in L \setminus S$ with $t < \tau$, $\sum_{t=1}^{\tau-1} x_t^* = \sum_{t \in S: t < \tau} x_t^* \geq d_{1, \tau-1}$. But as $p_t > 0$ for all periods $t \in S$ with $t < \tau$, a solution can only be optimal if $\sum_{t\in S:t<\tau} x_t^* = d_{1,\tau-1}$. This holds because otherwise a strictly better solution can be obtained by reducing $\sum_{t\in S:t<\tau} x_t^*$ and increasing x_τ^*, a contradiction.

So we have shown that $\sum_{t\in S:t<\tau} x_t^* + \sum_{t\in S:t\geq\tau} x_t^* + \sum_{t\in L\setminus S} d_{tl}y_t^* = d_{1,\tau-1} + 0 + d_{\tau l} = d_{1l}$.

In the special case where $y_t^* = 0$ for all $t \in L \setminus S$, we obtain similarly that $\sum_{t=1}^{l} x_t^* = \sum_{t\in S} x_t^* = d_{1l}$, because production in period $l+1$ has zero cost.

Finally note that when $l = n$, $L \setminus S \neq \emptyset$ because $\min_t p_t = 0$. Now it can be readily checked that all the faces used above in the proof are proper faces of P. □

Proof of Theorem 7.8.

Observe that there exists an optimal solution in which $y_t = 1$ for all $t \in \{1,\dots,n\}$ with $q_t \leq 0$. In this case, variable y_t can be eliminated from the linear program and the constant term $\sum_{t:q_t<0} q_t$ added to the objective function.

On the other hand, when $q_t > 0$, one must have y_t as small as possible and thus $y_t = \sum_{\tau=t:d_{t\tau}>0}^{n} \phi_{t\tau}$ from (7.25). Eliminating x_t using (7.26), the resulting equivalent linear program is

$$\min \ \sum_{u-1}^{n} p_u \sum_{\tau=u}^{n} d_{u\tau}\phi_{u\tau} + \sum_{t=1:q_t>0}^{n} q_t \sum_{\tau=t:d_{t\tau}>0}^{n} \phi_{t\tau} + \sum_{t=1:q_t<0}^{n} q_t$$

subject to

$$-\sum_{t=1}^{n} \phi_{1t} = -1$$

$$\sum_{u=1}^{t-1} \phi_{u,t-1} - \sum_{\tau=t}^{n} \phi_{t\tau} = 0 \qquad\qquad \text{for } 2 \leq t \leq n$$

$$\sum_{u=1}^{n} \phi_{un} = 1$$

$$\phi \in \mathbb{R}_{+}^{n(n+1)/2}.$$

This is a shortest path problem in an acyclic network, and the underlying matrix is TU, so using Theorem 6.7 it has an optimal solution with $\phi \in \{0,1\}^{n(n+1)/2}$, and thus $y \in \{0,1\}^n$. □

Proof of Theorem 7.10.

Here we pass by an extended formulation Q^{WW-U} for WW-U:

$$s_{t-1} = \sum_{j=t}^{n} d_j \delta_{t-1}^j \qquad \text{for } 1 \leq t \leq n \qquad (7.48)$$

$$\delta_t^j + \sum_{u=t+1}^{j} y_u \geq 1 \qquad \text{for } 0 \leq t < j \leq n \qquad (7.49)$$

$$\delta_t^j \in \mathbb{R}_+^1 \qquad \text{for } 0 \leq t < j \leq n, y \in [0,1]^n, s \in \mathbb{R}_+^{n+1}, \qquad (7.50)$$

where $\delta_t^j = 1$ if the stock s_t at the end of period t includes the demand d_j for period $j > t$. Q^{WW-U} is a correct formulation for the stock minimal solutions of WW-U because δ_t^j is forced to 1 when $\sum_{u=t+1}^{j} y_u = 0$.

The matrix corresponding to the constraints (7.49) consists of an identity matrix and a matrix with consecutive 1s in each row, which is known to be totally unimodular. Thus, by Theorem 6.7, Q^{WW-U} is an integral polyhedron.

Now consider its projection $\text{proj}_{s,y} Q^{WW-U}$ into the (s, y) space. In the projection, the only way to eliminate a δ_t^j variable is to replace it by either its lower bound $1 - \sum_{u=t+1}^{j} y_u$, or by its lower bound of 0. So, for each subset T_t of $\{t+1, \ldots, n\}$, we obtain an inequality of the form

$$s_t \geq \sum_{j \in T_t} d_j \left(1 - \sum_{u=t+1}^{j} y_u\right).$$

However, some of these inequalities are redundant and can be eliminated. Consider any point (s^*, y^*) that is cut off by one of the inequalities for some t and some $T_t \subseteq \{t+1, \ldots, n\}$. Suppose that $\sum_{u=t+1}^{k} y_u^* < 1 \leq \sum_{u=t+1}^{k+1} y_u^*$. This point is also cut off by the (at least as strong) inequality $s_t \geq \sum_{j=t+1}^{k} d_j (1 - \sum_{u=t+1}^{j} y_u)$. It follows that the latter inequalities suffice to describe $\text{conv}(X^{WW-U})$. □

Exercises

Exercise 7.1 Consider an instance of LS-U with $n = 4$ and $d = (8, 2, 6, 5)$ in which there is an initial stock $s_0 = 11$ and lower bounds on the stocks in each period $\underline{s} = (1, 2, 2, 6)$. In addition $q' = (17, 29, -12, 43)$, $p' = (2, 7, 8, 3)$, and $h' = (1, 2, 1, 2)$.
i. Convert into an instance of LS-U in standard form with no initial stock, and lower bounds of zero on the stocks.
ii. Convert the objective function into an equivalent form with $q \geq 0$ and $h = 0$.
iii. Convert the objective function into an equivalent form with $q \geq 0$ and $p = 0$.

Exercise 7.2 Consider an instance of LS-U with $n = 5$ periods, costs $p' = (7, 4, 2, 3, 4)$, $h' = (1, 3, 2, 1, 3)$, $q' = (25, 45, 26, 30, 20)$, and demands $d =$

(5, 3, 7, 4, 9) for which initial and final stock must be zero.

i. Rewrite the objective function in normalized form so that all the unit storage costs are zero.

ii. Rewrite the objective function in normalized form so that all the unit production costs are zero.

iii. Is the objective function Wagner–Whitin?

iv. Find an optimal solution using forward dynamic programming.

v. Solve using the backward dynamic programming algorithm.

Exercise 7.3 Consider an instance of *LS-U* with $n = 6$ periods and demands $d = (4, 2, 5, 1, 7, 1)$. Suppose that you are given a fractional solution

$$x^* = (10.99, 0.88, 0, 0.125, 7, 1), s^* = (6.99, 5.875, 0.875, 0, 0, 0),$$
$$y^* = (1, 0.0009, 0, 0.125, 1, 1).$$

i. Using the simple $O(n^2)$ algorithm, find an (l, S) inequality cutting off the point, or show that no such inequality exists.

ii. Repeat with the improved separation algorithm.

Exercise 7.4 In practice finding a most violated inequality is not always the best strategy. Consider the (l, S) inequality in the form

$$s_{k-1} + \sum_{j \in S} x_j + \sum_{j \in [k,l] \setminus S} d_{jl} y_j \geq d_{kl},$$

(see (7.13)), denoted (k, l, S). It is effective computationally to add violated inequalities with $l - k$ as small as possible.

i. Describe a separation algorithm to test if, for a fixed integer τ, there is a violated inequality with $l - k \leq \tau$.

ii. What is the complexity of your algorithm for fixed τ?

iii. Apply your algorithm to the instance of Exercise 7.3.

Exercise 7.5 Write out the linear programming dual of the facility location reformulation (7.14)–(7.18) of *LS-U*.

i. Interpret the dual variables and constraints.

ii. Find a dual feasible solution with the same value as that of the primal LP, thereby showing that the reformulation is tight.

Exercise 7.6 Consider a different extended formulation for *LS-U* with the additional variables:

π_{kl} is the amount produced in period k used to satisfy demands in the interval $[k, l]$. Thus $\pi_{kl} = \min[x_k, d_{kl} y_k]$.

i. Show that

$$\min px + qy$$
$$\sum_{u=1}^{t} \pi_{1u} \geq d_{1t} \text{ for } 1 \leq t \leq n$$
$$\pi_{kl} \leq x_k \text{ for } 1 \leq k \leq l \leq n$$
$$\pi_{kl} \leq d_{kl} y_k \text{ for } 1 \leq k \leq l \leq n$$
$$x \in \mathbb{R}^n_+, y \in \{0, 1\}^n, \pi \in \mathbb{R}^{n(n+1)/2}$$

is a valid formulation for *LS-U*.

ii. Show that this reformulation is tight.

Hint. Relate the formulation to the (l, S) inequalities.

Exercise 7.7 Solve an instance of *WW-U* with $n = 5$, $d = (7, 2, 6, 4, 11)$, $p = 0$, $h = (1, 2, 3, 1, 3)$, and $q = (34, 56, 21, 17, 39)$.

Exercise 7.8 Use the multicommodity reformulation of *LS-U*, introduced in Section 4.1.1, to construct an approximate reformulation with similar properties to those of the approximate shortest path formulation of Section 7.6. Show that with parameter τ, all (x, y, s) solutions of the approximate formulation satisfy the (k, l, S) inequalities with $l - k \leq \tau$.

Notes

Section 7.2 The use of regeneration intervals or points of regeneration to characterize the structure of optimal solutions is attributed to Manne [116]. The term regeneration point has been taken from the probability literature and the study of renewal processes, and its application to inventory management problems, is due to Karlin [97]. Regeneration intervals are crucial to the development of polynomial dynamic programming algorithms for different lot-sizing variants.

Section 7.3 In [188], Wagner and Whitin presented an $O(n^2)$ dynamic programming algorithm for *LS-U*. In the early 1990s, several ways of solving *LS-U* in $O(n \log n)$ were developed, see Aggarwal and Park [3], Federgrün and Tzur [63], and Wagelmans et al. [187].

Section 7.4 In Barany et al. [23], the (l, S) inequalities were introduced and shown to provide the convex hull. The first extended formulation proposed for *LS-U* was the facility location formulation of Krarup and Bilde [100]. The shortest path reformulation is due to Eppen and Martin [61].

Section 7.5 The Wagner–Whitin cost hypothesis was first introduced by Wagner and Whitin [188], even though the hypothesis was not needed for their dynamic programming algorithm. However, when a faster algorithm was developed for *LS-U*, it was also shown that *WW-U* could be solved in linear time, see Aggarwal and Park [3], Federgrün and Tzur [63], and Wagelmans et al. [187].

The expression for the stock minimal solutions under the Wagner-Whitin cost condition, and the polyhedral characterization showing that only $O(n^2)$ special (l, S) inequalities are needed to give the convex hull are from Pochet and Wolsey [140]. An alternative proof based on a characterization of the faces of the polyhedron can be found in Pereira and Wolsey [132].

Section 7.6 The approximate extended shortest path formulation is taken from Van Vyve and Wolsey [181], and is related to a formulation proposed earlier by Stadtler [153].

Section 7.7 The proof of Theorem 7.5 presented here is a minor variant of a proof due to Kolen [98] presented in detail in Pochet [133]. This variant has already appeared in Pochet and Wolsey [141]. Another interesting proof of Van Hoesel et al. [169] is based on a dual greedy algorithm. The first proof of Theorem 7.8 by Eppen and Martin [61] was based on an original technique of converting a dynamic formulation recursion into an extended formulation. The proof of Theorem 7.10 is essentially that of Pochet and Wolsey [140].

Basic MIP and Fixed Cost Flow Models

The aim in this chapter is to introduce some of the basic sets and valid inequalities that are useful in mixed integer programming, where useful means here that either the sets and inequalities arise as submodels (or as what are called low-level relaxations in Chapter 3) of lot-sizing problems studied in later chapters, and/or that the inequalities are being used as cutting planes in standard mixed integer programming solvers.

For each set X considered, we explore several possibilities:

- Characterize the optimal solutions of $OPT(X, c)$.
- Describe an algorithm for $OPT(X, c)$ and its complexity.
- Describe a family $\mathcal{F}$ of valid inequalities for X or describe conv(X).
- Describe a separation algorithm for conv(X), or for the polyhedron described by the family $\mathcal{F}$.
- Describe an extended formulation for X, if possible providing a tight formulation of conv(X).

It should be clear from Chapter 6 on algorithms and decomposition why such descriptions may be helpful when taking a decomposition approach. Our emphasis, however, is first on valid inequalities and reformulation, because if possible we would like to solve problems by inputting a strong formulation to a standard MIP solver. A second option, requiring more work, is to add one or more separation routines to such a system. We do not solve any problems by Lagrangian relaxation or column generation. However, as seen in Chapter 6, efficient optimization algorithms are of importance in solving the subproblems that arise when using these approaches.

We essentially study nine closely related sets.

- In Section 8.1 we study the simplest possible mixed integer set, *the Basic-MIP Set*.

$$X^{MI} = \{(s, y) \in \mathbb{R}^1_+ \times \mathbb{Z}^1 : s + y \geq b\}.$$

The inequality derived for this two-variable set, called the *simple mixed integer rounding (SMIR) inequality*, might equally well be called the basic disjunctive, split cut, or mixed integer Gomory inequality; it is fundamental.

- In Section 8.2 we consider the intersection of K basic-MIP sets with the same *integer* variable, called the *MIP Set*.

$$X_K^{MI} = \{(s,y) \in \mathbb{R}_+^K \times \mathbb{Z}^1 : s_k + y \geq b_k \text{ for } 1 \leq k \leq K\}.$$

The K SMIR inequalities suffice to obtain the convex hull description. This is used later to provide a tight formulation for *DLS-CC-B*.

- In Section 8.3 we consider the intersection of K basic-MIP sets with the same *continuous* variable, called the *Mixing Set*.

$$X_K^{MIX} = \{(s,y) \in \mathbb{R}_+^1 \times \mathbb{Z}^K : s + y_k \geq b_k \text{ for } 1 \leq k \leq K\}.$$

Here an exponential family of *mixing inequalities* is needed to describe the convex hull. This is used to model *DLSI-CC* and *WW-CC*.

- In Section 8.4 we review briefly three different methods that can be used to derive extended formulations for more general mixing sets. Such sets share the property that the fractional value of their "continuous" variable s can only take a small number of values. These techniques are used to reformulate other more complicated mixing sets in Sections 8.5 and 8.6.

- In Section 8.5 we consider a first generalization of the mixing set, called the *Continuous Mixing Set*.

$$X_K^{CMIX} = \{(s,r,y) \in \mathbb{R}_+^1 \times \mathbb{R}_+^K \times \mathbb{Z}^K : s + r_k + y_k \geq b_k \text{ for } 1 \leq k \leq K\}.$$

This is useful in modeling the lot-sizing problems with constant capacities and backlogging *DLSI-CC-B* and *WW-CC-B*.

- In Section 8.6 we consider two further closely related generalizations of the mixing set, the *Two-Capacity Mixing Set*:

$$X^{2DIV} = \{(s,y,z) \in \mathbb{R}_+^1 \times \mathbb{Z}_+^K \times \mathbb{Z}_+^K : s + y_k + Cz_k \geq b_k \text{ for } 1 \leq k \leq K\}$$

and the *Divisible Mixing Set*

$$X^{DMIX} = \{(s,y) \in \mathbb{R}_+^1 \times \mathbb{Z}^{|K_1|+|K_2|} : s + y_k \geq b_k \text{ for } k \in K_1,$$
$$s + Cy_k \geq b_k \text{ for } k \in K_2\},$$

where $C \geq 2$ is integer. These can be used to treat problems with a choice of production capacities, and with production lower bounds, respectively,

- In Section 8.7, we pass from "easy" sets to "hard" sets for which no polynomial description of the convex hull can be expected. We consider first the *Integer Continuous Knapsack Set*, that is, the integer knapsack set with a single continuous variable.

$$K^I = \{(s,y) \in \mathbb{R}^1_+ \times \mathbb{Z}^n_+ : \sum_{j=1}^{n} a_j y_j \leq b + s\}.$$

Here we derive the *mixed integer rounding (MIR) inequality*, and show that, when applied to a row of an optimal LP tableau, it gives precisely the *mixed integer Gomory (MIG) cut*. Both MIG and MIR inequalities are now generated automatically in the standard MIP solvers.

- In Section 8.8, we again consider a set with a single constraint, but now the integer variables are restricted to take 0–1 values. It is called the *Binary Continuous Knapsack Set*

$$K^B = \{(s,y) \in \mathbb{R}^1_+ \times \{0,1\}^n : \sum_{j=1}^{n} a_j y_j \leq b + s\}.$$

Various versions of the *cover inequalities* derived here are generated in the standard MIP systems.

- In Section 8.9 we consider a slightly more general mixed integer set, the *Binary Single-Node Flow Set*

$$K^F = \{(x,y) \in \mathbb{R}^n_+ \times \{0,1\}^n : \sum_{j \in N_1} x_j - \sum_{j \in N_2} x_j \leq b,$$
$$x_j \leq a_j y_j \text{ for } 1 \leq j \leq n\},$$

which appears as a simple relaxation of *LS-C* and *WW-C*. The *flow cover inequalities* that are derived here are also generated in some MIP solvers.

8.1 A Two-Variable Basic Mixed Integer Set

We start by examining the simplest possible mixed integer set with just two variables. We consider the basic-MIP set

$$X^{MI} = \{(s,y) \in \mathbb{R}^1_+ \times \mathbb{Z}^1 : s + y \geq b\}.$$

8.1.1 Valid Inequalities and Formulations

Proposition 8.1 *i. Let $f = b - \lfloor b \rfloor \geq 0$. The simple mixed integer rounding (SMIR) inequality*

$$s \geq f(\lceil b \rceil - y) \tag{8.1}$$

is valid for X^{MI}.
ii. The polyhedron

$$s + y \geq b$$
$$s + fy \geq f\lceil b \rceil$$
$$s \geq 0$$

describes the convex hull of X^{MI}.

Proof. We prove here the validity result i. The convex hull result ii is proved in Section 8.10.

Let $X^1 = X^{MI} \cap \{(s,y) : y \geq \lceil b \rceil\}$ and $X^2 = X^{MI} \cap \{(s,y) : y \leq \lfloor b \rfloor\}$. For X^1, we combine the valid inequalities $0 \geq \lceil b \rceil - y$ and $s \geq 0$ with weights f and 1, respectively, to obtain $s \geq f(\lceil b \rceil - y)$. For X^2, we combine the valid inequalities $s \geq b - y$ and $0 \geq y - \lfloor b \rfloor$ with weights 1 and $1 - f$, respectively, to obtain $s \geq b - y + (1 - f)(y - \lfloor b \rfloor) = f(\lceil b \rceil - y)$.

As the SMIR inequality is valid for X^1 and X^2, it is valid for $X^{MI} = X^1 \cup X^2$. □

It is also useful to be able to recognize this set and its valid inequality when it is written differently. Setting $z = -y$ and $d = -b$, $s + y \geq b$ becomes $z \leq d + s$.

Corollary 8.1 *For the set*

$$X = \{(s, z) \in \mathbb{R}^1_+ \times \mathbb{Z}^1 : z \leq d + s\},$$

the SMIR inequality takes the form

$$z \leq \lfloor d \rfloor + \frac{s}{1 - f_d} \qquad (8.2)$$

where $f_d = d - \lfloor d \rfloor$.

Similarly, setting $x = b - s$ and eliminating s, the set $s \geq 0, s + y \geq b$ becomes the set $x \leq b, x \leq y$.

Corollary 8.2 *For the set*

$$X = \{(x, y) \in \mathbb{R}^1 \times \mathbb{Z}^1 : x \leq b, x \leq y\},$$

the SMIR inequality takes the form

$$x \leq \lfloor b \rfloor + f(y - \lfloor b \rfloor). \qquad (8.3)$$

Example 8.1 *Consider the set $X = \{(s,y) \in \mathbb{R}^1_+ \times \mathbb{Z}^1 : s + y \geq 2.25\}$ shown in Figure 8.1. From Proposition 8.1, we obtain the valid inequality*

$$s \geq (2.25 - \lfloor 2.25 \rfloor)(\lceil 2.25 \rceil - y),$$

or

$$y \geq 0.25 \, (3 - y),$$

or

$$s + 0.25y \geq 0.75,$$

which states that $s \geq 0$ when $y = 3$ and $s \geq 0.25$ when $y = 2$. In Figure 8.1, the two points $(s, y) = (0, 3)$ and $(s, y) = (0.25, 2)$ are the extreme points of $\mathrm{conv}(X^{MI})$ limiting the shaded region cut off by inequality (8.1). Observe that these two points suffice to prove that (8.1) is a facet-defining valid inequality

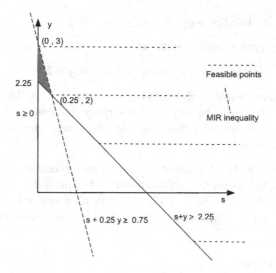

Figure 8.1. Simple mixed integer rounding (MIR) inequality.

of $\mathrm{conv}(X^{MI})$.

Now consider the set

$$Y - \{(x, y) \in \mathbb{R}^1_+ \times \mathbb{Z}^1 : x \leq 2.25, x \leq y\}.$$

Introducing $s = 2.25 - x$ with $s \geq 0$ and then eliminating x gives the set

$$\{(s, y) \in \mathbb{R}^1_+ \times \mathbb{Z}^1 : s + y \geq 2.25\}$$

considered above. Now the valid inequality $s + 0.25y \geq 0.75$ gives after substitution for s, $2.25 - x + 0.25y \geq 0.75$ or

$$x \leq 2 + 0.25(y - 2),$$

the inequality given in Corollary 8.2.
    Finally consider the set

$$Z = \{(s, z) \in \mathbb{R}^1_+ \times \mathbb{Z}^1 : z \leq -2.25 + s\}.$$

By Corollary 8.1, the SMIR inequality for Z takes the form

$$z \leq \lfloor -2.25 \rfloor + \frac{s}{1 - 0.75} \qquad \text{or} \qquad z \leq -3 + 4s.$$

8.1.2 Optimal Solutions

Consider now the associated optimization problem

$$\min\{cs + qy : s + y \geq b, s \geq 0, y \in \mathbb{Z}\},$$

and its linear programming relaxation

$$\min\{cs + qy : s + y \geq b, s \geq 0, y \in \mathbb{R}\}.$$

Both problems are clearly unbounded if $c < 0$ or $q < 0$. They are also unbounded if $c < q$ because the objective is strictly decreased when s and y are simultaneously increased and decreased by the same amount. This leads to the following observation.

Observation 8.1 *i. For the linear programming relaxation, if $c \geq q \geq 0$, the point $(s, y) = (0, b)$ is optimal, and otherwise the optimal value is unbounded. ii. For the mixed integer program, if $c \geq q \geq 0$, then either $(s, y) = (0, \lceil b \rceil)$ or $(s, y) = (f, \lfloor b \rfloor)$ is optimal, and otherwise the objective value is unbounded.*

8.2 The MIP Set

Here we consider the MIP set

$$X_K^{MI} = \{(s, y) \in \mathbb{R}_+^K \times \mathbb{Z}^1 : s_k + y \geq b_k \text{ for } k = 1, \dots, K\}.$$

Note that $X_K^{MI}(b) = \cap_{k=1}^K X_1^{MI}(b_k) \subseteq \mathbb{R}_+^K \times \mathbb{Z}^1$. Surprisingly it suffices to just add an SMIR inequality for each of the sets $X_1^{MI}(b_k)$ to get the convex hull. What is more, this is still true when several such sets X_K^{MI} intersect by having constraints linking the integer variables.

Throughout this chapter, we define $f_k = b_k - \lfloor b_k \rfloor$ for all k, and we let $f_0 = 0$.

Theorem 8.2 *i. The polyhedron*

$$
\begin{aligned}
s_k + y &\geq b_k && \text{for } 1 \leq k \leq K \\
s_k + f_k y &\geq f_k \lceil b_k \rceil && \text{for } 1 \leq k \leq K \\
s &\in \mathbb{R}_+^K, y \in \mathbb{R}^1
\end{aligned}
$$

describes the convex hull of X_K^{MI}.

ii. Let $X^i = \{(s^i, y^i) \in \mathbb{R}_+^{K_i} \times \mathbb{Z}^1 : s_k^i + y^i \geq b_k^i \text{ for } 1 \leq k \leq K_i\}$ for $1 \leq i \leq m$, let $y = (y^1, \dots, y^m) \in \mathbb{Z}^m$, and consider the set

$$W = (\cap_{i=1}^m X^i) \cap \{y : By \leq d\} \subseteq \mathbb{R}_+^{K_1} \times \cdots \times \mathbb{R}_+^{K_m} \times \mathbb{Z}^m.$$

The polyhedron

$$
\begin{aligned}
s_k^i + y^i &\geq b_k^i && \text{for } 1 \leq k \leq K_i, \ 1 \leq i \leq m \\
s_k^i + f_k^i y^i &\geq f_k^i \lceil b_k^i \rceil && \text{for } 1 \leq k \leq K_i, \ 1 \leq i \leq m \\
By &\leq d \\
y &\in \mathbb{R}^m, \ s^i \in \mathbb{R}_+^{K_i} && \text{for } 1 \leq i \leq m
\end{aligned}
$$

is integral and describes conv(W) *if B is a totally unimodular matrix and d is integer, where $f_k^i = b_k^i - \lfloor b_k^i \rfloor$ for all i, k.*

The proof of Theorem 8.2 is given in Section 8.10.

Example 8.2 *Consider the set*

$$X = \{(s,y) \in \mathbb{R}_+^3 \times \mathbb{Z}^1 : s_1 + y \geq 1.4, \ s_2 + y \geq 2.3, \ s_3 + y \geq 0.7\}.$$

By Theorem 8.2i, it suffices to add the three SMIR inequalities

$$s_1 \geq 0.4(2 - y), \quad s_2 \geq 0.3(3 - y), \quad s_3 \geq 0.7(1 - y)$$

to obtain conv(X).

Alternatively, the set X can be written as

$$s_1 + y_1 \geq 1.4, \quad s_2 + y_2 \geq 2.3, \quad s_3 + y_3 \geq 0.7 \tag{8.4}$$
$$y_1 - y_2 = 0, \quad y_2 - y_3 = 0 \tag{8.5}$$
$$(s,y) \in \mathbb{R}_+^3 \times \mathbb{Z}^3. \tag{8.6}$$

Here (8.4) and (8.6) are the constraints of three independent basic-MIP sets, and the matrix associated with the additional linking constraints (8.5) is totally unimodular. So by Theorem 8.2ii, it suffices to add the three SMIR inequalities

$$s_1 \geq 0.4(2 - y_1), \quad s_2 \geq 0.3(3 - y_2), \quad s_3 \geq 0.7(1 - y_3)$$

to obtain conv(X). *Substituting back $y = y_1 = y_2 = y_3$ by Equation (8.5), we obtain the same description of* conv(X) *as above.*

As an application of the above reformulation, the discrete lot-sizing problem with constant capacity and backlogging (*DLS-CC-B*) can be expressed, after elimination of the backlogging variables, as the intersection of n (n is the number of time periods) basic MIP sets, where the linking constraints, involving only the integer variables, correspond to the arc-node incidence matrix of a digraph. Because this matrix is totally unimodular, Theorem 8.2, part ii, can be used to describe conv($X^{DLS-CC-B}$). This is presented in Section 10.3.1.

8.3 The Mixing Set

Here we consider the mixing set

$$X_K^{MIX} = \{(s,y) \in \mathbb{R}_+^1 \times \mathbb{Z}^K : s + y_k \geq b_k \text{ for } 1 \leq k \leq K\}.$$

8.3.1 Extreme Points

Proposition 8.3 *The extreme rays of* $\mathrm{conv}(X_K^{MIX})$ *are the vectors* $(s, y) = (0, e_k)$ *for* $1 \le k \le K$ *and the vector* $(1, -e)$, *where* $e_k \in \mathbb{R}^K$ *is the kth unit vector and* $e \in \mathbb{R}^K$ *is the vector of all 1s.*
The extreme points are (s^j, y^j) *for* $0 \le j \le K$ *with* $s^j = f_j, y_k^j = \lceil b_k - f_j \rceil$ *for* $1 \le k \le K$.

Example 8.3 *Consider the set*

$$X = \{(s, y) \in \mathbb{R}_+^1 \times \mathbb{Z}^3 : s + y_1 \ge 1.4, s + y_2 \ge 2.6, s + y_3 \ge 0.7\}.$$

By Proposition 8.3, the extreme rays of $\mathrm{conv}(X)$ *are* $(s, y) = (0, 1, 0, 0)$, $(0, 0, 1, 0)$, $(0, 0, 0, 1)$, *and* $(1, -1, -1, -1)$. *The extreme points of* $\mathrm{conv}(X)$ *are the vectors* $(s^0, y^0) = (0, 2, 3, 1)$, $(s^1, y^1) = (0.4, 1, 3, 1)$, $(s^2, y^2) = (0.6, 1, 2, 1)$, *and* $(s^3, y^3) = (0.7, 1, 2, 0)$.

8.3.2 Valid Inequalities

Here the K SMIR inequalities $s + f_k y_k \ge f_k \lceil b_k \rceil$ do not suffice to give the convex hull when $K > 1$. To fit with what follows, note that when $0 \le f_k < 1$, the right-hand side of the SMIR inequality can be be rewritten as $f_k(\lfloor b_k \rfloor + 1)$.

Proposition 8.4 *Let* $T \subseteq \{1, \dots, K\}$ *with* $|T| = t$, *and suppose that* $i_1, \dots i_t$ *is an ordering of* T *such that* $0 = f_{i_0} \le f_{i_1} \le f_{i_2} \le \cdots \le f_{i_t} < 1$. *Then the mixing inequalities*

$$s \ge \sum_{\tau=1}^{t} (f_{i_\tau} - f_{i_{\tau-1}})(\lfloor b_{i_\tau} \rfloor + 1 - y_{i_\tau}) \tag{8.7}$$

and

$$s \ge \sum_{\tau=1}^{t} (f_{i_\tau} - f_{i_{\tau-1}})(\lfloor b_{i_\tau} \rfloor + 1 - y_{i_\tau}) + (1 - f_{i_t})(\lfloor b_{i_1} \rfloor - y_{i_1}) \tag{8.8}$$

are valid for X_K^{MIX}.

Proof. We prove the validity of (8.8). The proof of (8.7) is similar.
Take any $(s, y) \in X_K^{MIX}$, and define $\beta = \max_{\tau=1}^{t}\{\lfloor b_{i_\tau} \rfloor + 1 - y_{i_\tau}\}$. If $\beta \le 0$, then the inequality is satisfied at (s, y) because $s \ge 0$ and all terms in the right-hand side are nonpositive. So, it remains to prove validity for the case $\beta \ge 1$. We define $\nu = \max\{\tau \in \{1, \dots, t\} : \beta = \lfloor b_{i_\tau} \rfloor + 1 - y_{i_\tau}\}$. Using $\beta \ge \lfloor b_{i_\tau} \rfloor + 1 - y_{i_\tau}$ for $\tau \le \nu$, and $\beta \ge \lfloor b_{i_\tau} \rfloor + 2 - y_{i_\tau}$ for $\tau > \nu$, we can write

$$\sum_{\tau=1}^{t} (f_{i_\tau} - f_{i_{\tau-1}})(\lfloor b_{i_\tau} \rfloor + 1 - y_{i_\tau}) + (1 - f_{i_t})(\lfloor b_{i_1} \rfloor - y_{i_1})$$

$$\leq \sum_{\tau=1}^{\nu} (f_{i_\tau} - f_{i_{\tau-1}})(\beta) + \sum_{\tau=\nu+1}^{t} (f_{i_\tau} - f_{i_{\tau-1}})(\beta - 1) + (1 - f_{i_t})(\beta - 1)$$

$$= f_{i_\nu}\beta + (1 - f_{i_\nu})(\beta - 1)$$

$$= (\beta - 1) + f_{i_\nu} = \lfloor b_{i_\nu} \rfloor + 1 - y_{i_\nu} - 1 + f_{i_\nu} = b_{i_\nu} - y_{i_\nu}$$

$$\leq \quad s. \quad \square$$

Theorem 8.5 *i. The constraints*

$$s + y_k \geq b_k \quad \text{for } 1 \leq k \leq K, \ s \geq 0$$

and the mixing constraints (8.7),(8.8) completely describe the convex hull of X_K^{MIX}.

ii. Let $X^i = \{(s^i, y^i) \in \mathbb{R}_+^1 \times \mathbb{Z}^{K_i} : s^i + y_k^i \geq b_k^i \text{ for } 1 \leq k \leq K_i\}$ *for* $1 \leq i \leq m$, *let* $y = (y^1, \ldots, y^m) \in \mathbb{Z}^{K_1} \times \cdots \times \mathbb{Z}^{K_m}$, *and consider the set* $(\cap_{i=1}^m X^i) \cap \{y : By \leq d\} \subseteq \mathbb{R}_+^m \times \mathbb{Z}^{K_1} \times \cdots \times \mathbb{Z}^{K_m}$.

The polyhedron

$s^i + y_k^i \geq b_k^i$	for all k, i
the mixing inequalities (8.7), (8.8)	for $1 \leq i \leq m$
$By \leq d$	
$s^i \geq 0$	for $1 \leq i \leq m$

is integral if the polyhedron $\{z : Bz \leq d, l_{ij} \leq z_i - z_j \leq h_{ij} \text{ for } i, j \in \{1, \ldots, K\}, i \neq j\}$ *is integral for all integral* l_{ij}, h_{ij}. *In particular, the condition holds if* B *is the arc-node incidence matrix of a directed graph (i.e.,* B *is a* $\{0, +1, -1\}$ *matrix with at most one entry* $+1$ *and one entry* -1 *in each row), and* d *is integer.*

Example 8.4 *Consider again the set*

$$X = \{(s, y) \in \mathbb{R}_+^1 \times \mathbb{Z}^3 : s + y_1 \geq 1.4, s + y_2 \geq 2.6, s + y_3 \geq 0.7\}.$$

By Theorem 8.5, it suffices to add the following inequalities to obtain the convex hull: first three SMIR inequalities, which are the inequalities of the form (8.7) with $|T| = 1$,

$$s \geq 0.4(2 - y_1), \quad s \geq 0.6(3 - y_2), \quad s \geq 0.7(1 - y_3) \ ,$$

then the mixing inequalities (8.7) with $|T| > 1$

$$s \geq 0.4(2 - y_1) + (0.6 - 0.4)(3 - y_2)$$
$$s \geq 0.4(2 - y_1) \qquad\qquad\qquad\qquad + (0.7 - 0.4)(1 - y_3)$$
$$s \geq \qquad\qquad 0.6(3 - y_2) \quad + (0.7 - 0.6)(1 - y_3)$$
$$s \geq 0.4(2 - y_1) + (0.6 - 0.4)(3 - y_2) + (0.7 - 0.6)(1 - y_3)$$

and the mixing inequalities (8.8) with $|T| > 1$

$$s \geq 0.4(2 - y_1) + (0.6 - 0.4)(3 - y_2) \qquad\qquad\qquad\qquad +(1 - 0.6)(1 - y_1)$$
$$s \geq 0.4(2 - y_1) \qquad\qquad\qquad +(0.7 - 0.4)(1 - y_3) +(1 - 0.7)(1 - y_1)$$
$$s \geq \qquad\qquad 0.6(3 - y_2) \qquad +(0.7 - 0.6)(1 - y_3) +(1 - 0.7)(2 - y_2)$$
$$s \geq 0.4(2 - y_1) +(0.6 - 0.4)(3 - y_2) +(0.7 - 0.6)(1 - y_3) +(1 - 0.7)(1 - y_1).$$

Note that the mixing inequalities (8.8) with $|T| = 1$ need not be added because they are the original inequalities $s + y_k \geq b_k$ for all k.

As there are an exponential number of mixing inequalities, it is important to look for a separation algorithm if we want to use them computationally. The alternative would be to find an extended formulation that gives conv(X) implicitly.

8.3.3 Separation of the Mixing Inequalities

Given a point $(s^*, y^*) \in \mathbb{R}^1_+ \times \mathbb{R}^K$, it is not difficult to see that a most-violated inequality of the form (8.7) or (8.8) can be found by the following procedure.

The general idea is to construct $T = \{i_1, \ldots, i_t\}$ so as to maximize the right-hand side of (8.7) or (8.8). We build a set T such that $\max_{j=1,\ldots,K}(\lfloor b_j \rfloor + 1 - y_j^*) = \lfloor b_{i_1} \rfloor + 1 - y_{i_1}^* > \lfloor b_{i_2} \rfloor + 1 - y_{i_2}^* > \cdots > \lfloor b_{i_t} \rfloor + 1 - y_{i_t}^* > (\lfloor b_{i_1} \rfloor - y_{i_1}^*)^+$.

Separation Algorithm for the Mixing Inequalities

Reorder the variables $k = 1, \ldots, K$ so that $f_1 \leq \cdots \leq f_K$. Let $\beta = \max_{j=1,\ldots,K}(\lfloor b_j \rfloor + 1 - y_j^*)$. If $\beta \leq 0$, there is no violated inequality. Otherwise, taking $i_0 = 0$, find a subsequence $i_1, \ldots i_r$ of $\{1, \ldots, K\}$ so that:

$$i_j = \arg \max_{i:i > i_{j-1}} \{\lfloor b_i \rfloor + 1 - y_i^* : (\lfloor b_i \rfloor + 1 - y_i^*) > (\beta - 1)^+\} \qquad \text{for } j = 1, \ldots, r$$

$$\lfloor b_i \rfloor + 1 - y_i^* \leq (\beta - 1)^+ \qquad\qquad\qquad\qquad\qquad \text{for } i > i_r.$$

Note that $\beta = \lfloor b_{i_1} \rfloor + 1 - y_{i_1}^*$, and let $\gamma = \sum_{j=1}^r (f_{i_j} - f_{i_{j-1}})(\lfloor b_{i_j} \rfloor + 1 - y_{i_j}^*)$.

Case a. $\beta \leq 1$.
If $\gamma > s^*$, the mixing inequality (8.7)

$$s \geq \sum_{j=1}^r (f_{i_j} - f_{i_{j-1}})(\lfloor b_{i_j} \rfloor + 1 - y_{i_j}^*)$$

is most violated with violation $\gamma - s^*$, and otherwise no inequality is violated.

Case b. $\beta > 1$.
If $\gamma + (1 - f_{i_r})(\beta - 1) > s^*$, the mixing inequality (8.8),

$$s \geq \sum_{j=1}^{r} (f_{i_j} - f_{i_{j-1}})(\lfloor b_{i_j} \rfloor + 1 - y_{i_j}^*) + (1 - f_{i_r})(\lfloor b_{i_1} \rfloor - y_{i_1}^*),$$

is most violated with violation $\gamma + (1 - f_{i_r})(\lfloor b_{i_1} \rfloor - y_{i_1}^*) - s^*$, and otherwise no inequality is violated.

The complexity of this routine is $O(n \log n)$ as sorting the f_i requires $O(n \log n)$, and then starting from K and working backwards one can find $i_r, \ldots, i_1$ in linear time, where n is taken as the number of variables in the mixing set.

Example 8.5 *Consider again the set*

$$X = \{(s, y) \in \mathbb{R}_+^1 \times \mathbb{Z}^3 : s + y_1 \geq 1.4, s + y_2 \geq 2.6, s + y_3 \geq 0.7\},$$

and the point $(s^*, y^*) = (0.21, 1.5, 2.9, 0.8)$.
Note that $f_1 < f_2 < f_3$ *so that we do not need to reorder the variables. Here* $(\lfloor b_1 \rfloor + 1 - y_1^*, \lfloor b_2 \rfloor + 1 - y_2^*, \lfloor b_3 \rfloor + 1 - y_3^*) = (0.5, 0.1, 0.2)$. *Thus* $\beta = 0.5$, $i_1 = 1, i_2 = 3$, *and* $\gamma = f_1(\lfloor b_1 \rfloor + 1 - y_1^*) + (f_3 - f_1)(\lfloor b_3 \rfloor + 1 - y_3^*) = 0.4 \times 0.5 + (0.7 - 0.4) \times 0.2 = 0.26$.

Thus as $\beta < 1$ *and* $\gamma - s^* = 0.05 > 0$, *a most-violated inequality is*

$$s \geq 0.4(2 - y_1) + (0.7 - 0.4)(1 - y_3)$$

with violation 0.05.

8.3.4 An Extended Formulation for conv(X_K^{MIX})

Taking the convex hull of the extreme points and nonnegative multiples of the extreme rays leads to an extended formulation. More precisely, we take

- nonnegative multiples δ_j of the extreme points (s^j, y^j) for $j = 0, \ldots, K$ with $\sum_{j=0}^{K} \delta_j = 1$,
- a nonnegative multiple μ of the ray $(s, y) = (1, -e)$, and
- a nonnegative multiple of the ray $(s, y) = (0, e_k)$ defined implicitly as the slack variable of the constraint defining y_k for $k = 1, \ldots, K$.

Theorem 8.6 *An extended formulation for* conv(X_K^{MIX}) *is*

$$s = \sum_{j=0}^{K} f_j \delta_j + \mu$$

$$y_k \geq \sum_{j=0}^{K} \lceil b_k - f_j \rceil \delta_j - \mu \qquad \text{for } 1 \leq k \leq K$$

$$\sum_{j=0}^{K} \delta_j = 1$$

$$\mu \in \mathbb{R}_+^1, \delta \in \mathbb{R}_+^{K+1},$$

with $f_0 = 0$.

Note that the inequality bounding y_k from below can be rewritten, after addition of $\lfloor b_k \rfloor + 1$ times the last equality as

$$\mu + y_k + \sum_{\{j:f_j \geq f_k\}} \delta_j \geq \lfloor b_k \rfloor + 1. \tag{8.9}$$

Consider now the reformulation obtained by introducing the new variables

$\mu_k = \mu + \sum_{j:f_j \geq f_k} \delta_j$ for $k = 1, \ldots, K$, and
$\mu_{K+1} = \mu$.

The above formulation now becomes

$$s = \sum_{k=1}^{K} (f_j - f_{j-1})\mu_j + (1 - f_K)\mu_{K+1} \tag{8.10}$$

$$\mu_k + y_k \geq \lfloor b_k \rfloor + 1 \qquad \qquad \text{for } k = 1, \ldots, K \qquad (8.11)$$

$$\mu_k - \mu_{k+1} \geq 0 \qquad \qquad \text{for } k = 1, \ldots, K \qquad (8.12)$$

$$\mu_{K+1} - \mu_1 \geq -1 \tag{8.13}$$

$$\mu_{K+1} \geq 0. \tag{8.14}$$

Changing the sign of the y variables, it is easily seen that the matrix defined by the constraints (8.11)–(8.13) is the dual of a network matrix. It follows that the optimization problem over a mixing set can be solved as a minimum cost network flow problem. In addition the validity of the mixing inequalities (8.7) and (8.8) is immediate by elimination of the μ_k variables in (8.10), and the separation problem for these inequalities reduces directly to a min-cut problem, see Exercise 8.8.

Example 8.6 *Consider again the mixing set*

$$X = \{(s, y) \in \mathbb{R}_+^1 \times \mathbb{Z}^3 : s + y_1 \geq 1.4, s + y_2 \geq 2.6, s + y_3 \geq 0.7\}.$$

From Theorem 8.6, it has the tight reformulation

$$
\begin{aligned}
s &= & 0\delta_0 & +0.4\delta_1 & +0.6\delta_2 & +0.7\delta_3 & +\mu \\
y_1 &\geq & 2\delta_0 & +1\delta_1 & +1\delta_2 & +1\delta_3 & -\mu \\
y_2 &\geq & 3\delta_0 & +3\delta_1 & +2\delta_2 & +2\delta_3 & -\mu \\
y_3 &\geq & 1\delta_0 & +1\delta_1 & +1\delta_2 & +0\delta_3 & -\mu \\
1 &= & \delta_0 & +\delta_1 & +\delta_2 & +\delta_3 & \\
\end{aligned}
$$

$$\delta \in \mathbb{R}_+^4,\, \mu \in \mathbb{R}_+^1.$$

With the introduction of the μ_k variables, this becomes

$$s = 0.4\mu_1 + 0.2\mu_2 + 0.1\mu_3 + 0.3\mu_4$$

$$
\begin{aligned}
\mu_1 & & & & +y_1 & & & \geq 2 \\
& \mu_2 & & & & +y_2 & & \geq 3 \\
& & \mu_3 & & & & +y_3 & \geq 1 \\
\mu_1 & -\mu_2 & & & & & & \geq 0 \\
& \mu_2 & -\mu_3 & & & & & \geq 0 \\
& & \mu_3 & -\mu_4 & & & & \geq 0 \\
-\mu_1 & & & +\mu_4 & & & & \geq -1 \\
& & & +\mu_4 & & & & \geq 0.
\end{aligned}
$$

8.3.5 Application of the Mixing Reformulation

The solution set of $DLSI\text{-}CC$ can be expressed as a mixing set with linking constraints, involving only the integer variables and a totally unimodular matrix. Thus Theorem 8.5, part ii, can be used to describe $\mathrm{conv}(X^{DLSI\text{-}CC})$, see Section 9.4. In addition, mixing sets are crucial to the description of the constant-capacity lot-sizing problem with Wagner–Whitin costs $WW\text{-}CC$ because the solution sets are the intersection of n $DLSI\text{-}CC$ sets; see Section 9.5.

Finally we show in Section 9.7 how to use these mixing sets to improve the formulation of lot-sizing problems $WW\text{-}C$ where capacities vary over time.

8.4 Reformulation Approaches for More General Mixing Sets

For the mixing set studied in the previous section, we were able to list explicitly all the extreme points and extreme rays. For more complicated sets this is typically impossible, either because they are not easily characterized or because there are too many of them. Though more complicated than the mixing set, the sets we study in Sections 8.5 and 8.6 still have one important property, the fractional values taken by the variable s in any extreme point of the convex hull can only take a small number of values, that is, $s \bmod 1 \in \{f_0, \ldots, f_m\}$. This allows us to derive extended formulations for the set based on these values. Below we derive such extended formulations for the set X using three different arguments that we now briefly describe.

Convex Hull Approach. The idea here, just demonstrated for the mixing set, is to explicitly or implicitly represent the convex hull $\mathrm{conv}(X)$ in terms of its extreme points and extreme rays.

Enumerative Approach. We start with an extended formulation

$$Y = X \cap \{(s,\delta,\mu) \in \mathbb{R}^1 \times \mathbb{Z}_+^m \times \mathbb{Z}_+^1 : s = \sum_{i=0}^m f_i \delta_i + \mu, \sum_{i=0}^m \delta_i = 1\}$$

in which the continuous variable s is replaced by a small number of integer variables. With these, it is often easy to derive valid inequalities so as to tighten the formulation for Y. If one can derive a tight formulation for Y, this then provides an extended formulation for $\mathrm{conv}(X)$.

As a simple example of the derivation of such an inequality, consider the constraint $s + y_t \geq b_t$ from the description of the mixing set in the last section. With $f_t = b_t - \lfloor b_t \rfloor$ for $t = 1,\ldots,m$ and $f_0 = 0$, one obtains the partial reformulation

$$\sum_{i=0}^m f_i \delta_i + \mu + y_t \geq b_t$$
$$\sum_{i=0}^m \delta_i = 1$$
$$\delta \in \mathbb{Z}_+^{m+1}, \mu \in \mathbb{Z}_+^1, y_t \in \mathbb{Z}^1$$

for which one can derive the valid inequality

$$\sum_{i:f_i \geq f_t} \delta_i + \mu + y_t \geq \lfloor b_t \rfloor + 1.$$

Disjunctive Approach. Here we consider a disjunctive formulation

$$X = \cup_{t=0}^m X^t,$$

where $X^t = X \cap \{s : s - f_t \in \mathbb{Z}_+^1\}$. Suppose now that each set $\mathrm{conv}(X^t)$ has the same extreme rays as $\mathrm{conv}(X)$ and that one can find an explicit polyhedral description

$$\mathrm{conv}(X^t) = \{x : A^t x + B^t w^t \geq d^t \text{ for some } w^t\}.$$

Here x represents the variables in the original space of X^t including the variable s, and w^t are possible additional variables from an extended formulation. Then $\mathrm{conv}(X) = \mathrm{conv}(\cup_{t=0}^m \mathrm{conv}(X^t))$, and an explicit extended formulation for $\mathrm{conv}(X)$ is

$$x = \sum_{t=0}^{m} x^t \tag{8.15}$$

$$A^t x^t + B^t w^t \geq d^t \delta_t \qquad \text{for } 0 \leq t \leq m \tag{8.16}$$

$$\sum_{t=0}^{m} \delta_t = 1 \tag{8.17}$$

$$x \in \mathbb{R}^n, \delta \in \mathbb{R}_+^{m+1}, (x^t, w^t) \in \mathbb{R}^n \times \mathbb{R}^{p_t} \qquad \text{for } 0 \leq t \leq m. \tag{8.18}$$

8.5 The Continuous Mixing Set

Here we consider the continuous mixing set

$$X_K^{CMIX} = \{(s, r, y) \in \mathbb{R}_+^1 \times \mathbb{R}_+^K \times \mathbb{Z}^K : s + r_k + y_k > b_k \text{ for } 1 \leq k \leq K\}.$$

First we characterize the extreme points and extreme rays that will lead us to define finite sets of values for s and r_k.

Notation. As before $f_j = b_j - \lfloor b_j \rfloor$. We also use $F_i^j = (f_j - f_i) \bmod 1$; that is, $F_i^j = f_j - f_i$ if $f_j \geq f_i$ and $F_i^j = f_j - f_i + 1$ otherwise, and also $f_0 = 0, F_0^j = f_j$. Again, $e_k \in \mathbb{R}^K$ is the kth unit vector and $e \in \mathbb{R}^K$ is the vector of all 1s.

Proposition 8.7 *The extreme rays of* $\mathrm{conv}(X_K^{CMIX})$ *are of three types:*
$(s, r, y) = (0, 0, e_j)$ *for* $j = 1, \ldots, K$, *or*
$(1, 0, -e)$, *or*
$(0, e_j, -e_j)$ *for* $1 \leq j \leq K$.

The extreme points are of two types:

Type A: $s = 0$.
Also for all $1 \leq j \leq K$,
    either $r_j = 0$ *and* $y_j = \lceil b_j \rceil$
    or $r_j = f_j$ *and* $y_j = \lfloor b_j \rfloor$.
    When $b_j \in Z^1$, *the two cases coincide.*

Type B: For all $1 \leq i \leq K$,
$s = f_i, r_i = 0, y_i = \lfloor b_i \rfloor$.
Also for $j \neq i$,
    either $y_j = \lceil b_j - f_i \rceil$ *and* $r_j = 0$,
    or $y_j = \lfloor b_j - f_i \rfloor$ *and* $r_j = (b_j - f_i) - \lfloor b_j - f_i \rfloor = F_i^j$.
    Again when $f_j = f_i$, *the two cases coincide.*

Example 8.7 *Consider the set*

$$X = \{(s,r,y) \in \mathbb{R}_+^1 \times \mathbb{R}_+^2 \times \mathbb{Z}^2 : s + r_1 + y_1 \geq 1.4, s + r_2 + y_2 \geq 2.7\}.$$

The extreme points of type A are

$$(s,r_1,r_2,y_1,y_2) = (0,0,0,2,3), (0,0.4,0,1,3), (0,0,0.7,2,2), (0,0.4,0.7,1,2).$$

The extreme points of type B are

$$i = 1: \quad (0.4,0,0,1,3), (0.4,0,0.3,1,2)$$

$$i = 2: \quad (0.7,0,0,1,2), (0.7,0.7,0,0,2).$$

To obtain an extended formulation one possibility, following the convex hull approach, is to introduce the variables:

$\delta_i = 1$ if the fractional part of s takes the value f_i for $i = 0, \ldots, K$.
$\beta_i^j = 1$ if the fractional part of s takes the value f_i and the fractional part of r_j is F_i^j for $j = 1, \ldots, K$, $i = 0, \ldots, m$ and $i \neq j$.

Theorem 8.8 *An extended formulation for* $\mathrm{conv}(X_K^{CMIX})$ *is*

$$s = \sum_{i=0}^{K} f_i \delta_i + \mu \tag{8.19}$$

$$r_j = \sum_{i=0}^{K} F_i^j \beta_i^j + \nu_j \qquad \text{for } 1 \leq j \leq K \tag{8.20}$$

$$y_j \geq |b_j| + \sum_{i:f_i < f_j} (\delta_i - \beta_i^j) - \sum_{i:f_i > f_j} \beta_i^j - \mu - \nu_j \quad \text{for } 1 \leq j \leq K \tag{8.21}$$

$$\beta_i^j \leq \delta_i \qquad \text{for } 1 \leq j \leq K, 0 \leq i \leq K \tag{8.22}$$

$$\sum_{i=0}^{K} \delta_i = 1 \tag{8.23}$$

$$\beta \in \mathbb{R}_+^{K(m+1)}, \quad \delta \in \mathbb{R}_+^{K+1}, \quad \mu \in \mathbb{R}_+^1, \quad \nu \in \mathbb{R}_+^K. \tag{8.24}$$

Example 8.8 *Consider again the set*

$$s + r_1 + y_1 \geq 1.4$$
$$s + r_2 + y_2 \geq 2.7$$
$$s, r_1, r_2 \in \mathbb{R}_+^1, y_1, y_2 \in \mathbb{Z}^1.$$

We have that $f_0 = 0, f_1 = 0.4, f_2 = 0.7, F_0^1 = 0.4, F_2^1 = 0.7, F_0^2 = 0.7,$ *and* $F_1^2 = 0.3.$

The formulation (8.19)-(8.24) takes the form

$$s = \mu + 0.4\delta_1 + 0.7\delta_2$$

$$r_1 = \nu_1 + 0.4\beta_0^1 + 0.7\beta_2^1$$

$$r_2 = \nu_2 + 0.7\beta_0^2 + 0.3\beta_1^2$$

$$\mu + \nu_1 + y_1 - (\delta_0 - \beta_0^1) + \beta_2^1 \geq 1$$

$$\mu + \nu_2 + y_2 - (\delta_0 - \beta_0^2) - (\delta_1 - \beta_1^2) \geq 2$$

$$\delta_0 + \delta_1 + \delta_2 = 1$$

$$\beta_0^1 \leq \delta_0, \quad \beta_0^2 \leq \delta_0$$

$$\beta_1^2 \leq \delta_1$$

$$\beta_2^1 \leq \delta_2$$

$$\mu, \nu, \delta, \beta \geq 0.$$

The above formulation has $O(K^2)$ constraints and variables. The formulation that we present below without justification is of practical interest because it has only $O(K)$ variables, and also because it permits us to see the structure of the valid inequalities for X_K^{CMIX}.

Theorem 8.9 *An extended formulation for* conv(X_K^{CMIX}) *is*

$$s + r_j + F_k^j(y_j - \lfloor b_j \rfloor) \geq f_k + \alpha_j - \alpha_k \qquad \text{for } 1 \leq j \leq K \ 1 \leq j < k \leq K$$
$$(8.25)$$

$$r_j + F_k^j(y_j - \lfloor b_j \rfloor) \geq \alpha_j - \alpha_k \qquad \text{for } 1 \leq j \leq K \ 0 \leq k < j \leq K$$
$$(8.26)$$

$$s > f_k + \alpha_0 - \alpha_k \qquad \text{for } 1 \leq k \leq K \qquad (8.27)$$

$$s + r_j + (y_j - \lfloor b_j \rfloor) \geq f_j \qquad \text{for } 1 \leq j \leq K \qquad (8.28)$$

$$s \in \mathbb{R}_+^1, r \in \mathbb{R}_+^K, \alpha \in \mathbb{R}^{K+1}, \qquad (8.29)$$

where it is assumed that $f_1 \leq f_2 \leq \cdots \leq f_K.$

To obtain the projection of this formulation into the original (s, r, y) space, we need to eliminate the variables $\alpha_0, \ldots, \alpha_K$. As each occurrence of these variables is in the form $(\alpha_j - \alpha_k)$, we can associate a complete directed graph $D = (V, A)$ with nodes $V = \{0, \ldots, K\}$ and arcs $A = \{(j, k)\}$. The only way to eliminate the α variables is to find a directed cycle C in D.

Proposition 8.10 *Every valid inequality for* X_K^{CMIX}, *other than a defining inequality (8.28), is of the form*

$$\sum_{(j,k) \in C: 1 \leq j < k} [s + r_j + F_k^j(y_j - \lfloor b_j \rfloor) - f_k]$$

$$+ \sum_{(j,k) \in C: 0 \leq k < j} [r_j + F_k^j(y_j - \lfloor b_j \rfloor)] + \sum_{(0,k) \in C} [s - f_k] \geq 0,$$

where C is a directed cycle in D.

This immediately suggests a separation algorithm. For a point (s^*, r^*, y^*), calculate the corresponding value associated with each arc and take this value as arc weight. Now any negative weight directed cycle gives a violated inequality.

Example 8.9 *Consider again the set*

$$s + r_1 + y_1 \geq 1.4$$
$$s + r_2 + y_2 \geq 2.7$$
$$s, r_1, r_2 \in \mathbb{R}_+^1, y_1, y_2 \in \mathbb{Z}^1$$

The variables are ordered so that $f_1 = 0.4 \leq f_2 = 0.7$, so from Theorem 8.9 we obtain the tight extended formulation

$$s + r_1 + 0.7(y_1 - 1) \geq 0.7 + \alpha_1 - \alpha_2$$
$$r_1 + 0.4(y_1 - 1) \geq 0 + \alpha_1 - \alpha_0$$
$$r_2 + 0.7(y_2 - 2) \geq 0 + \alpha_2 - \alpha_0$$
$$r_2 + 0.3(y_2 - 2) \geq 0 + \alpha_2 - \alpha_1$$
$$s \geq 0.4 + \alpha_0 - \alpha_1$$
$$s \geq 0.7 + \alpha_0 - \alpha_2$$
$$s + r_1 + y_1 \geq 1.4$$
$$s + r_2 + y_2 \geq 2.7$$
$$s, r \geq 0.$$

Let us now modify the set by adding a third constraint

$$s + r_3 + y_3 \geq 5.9 \quad \text{with } r_3 \in \mathbb{R}_+^1, y_3 \in \mathbb{Z}^1.$$

Consider now the directed cycle $C = \{(0,2), (2,1), (1,3), (3,0)\}$. From Proposition 8.10, the inequality

$$(s - f_2) + [r_2 + F_1^2(y_2 - \lfloor b_2 \rfloor)] + [s + r_1 + F_3^1(y_1 - \lfloor b_1 \rfloor) - f_3] + [r_3 + F_0^3(y_3 - \lfloor b_3 \rfloor)] \geq 0$$

or

$$2s + r_1 + r_2 + r_3 + 0.5(y_1 - 1) + 0.3(y_2 - 2) + 0.9(y_3 - 5) \geq 1.6$$

is valid.

In Chapter 10 we show that the feasible region of *DLSI-CC-B* can be viewed as a continuous mixing set with additional constraints in the form of an arc-node incidence matrix, and the feasible region of *WW-CC-B* is the intersection of n such sets.

8.6 Divisible Capacity Mixing Sets

Here we consider two more general sets

$$X^{2DIV} = \{(s,y,z) \in \mathbb{R}^1_+ \times \mathbb{Z}^m_+ \times \mathbb{Z}^m_+ : s + y_t + Cz_t \geq b_t \text{ for } 1 \leq t \leq m\}$$

and

$$X^{DMIX} = \{(s,y) \in \mathbb{R}^1_+ \times \mathbb{Z}^{|K_1|+|K_2|} : s + y_k \geq b_k \text{ for } k \in K_1,$$
$$s + Cy_k \geq b_k \text{ for } k \in K_2\},$$

where $C \in \mathbb{Z}^1$ with $C \geq 2$.

8.6.1 The Two-Capacity Mixing Set

To derive a tight extended formulation for X^{2DIV}, we use the disjunctive approach outlined in Section 8.4. It is easily shown that in any extreme point of conv(X^{2DIV}), $s \in \{f_0, \ldots, f_m\} \pmod 1$. So we consider the set

$$X_k^{2DIV} = X^{2DIV} \cap \{(s,\sigma) \in \mathbb{R}^1 \times \mathbb{Z}^1_+ : s = \sigma + f_k\}.$$

Substituting for s, and rounding up the right-hand side, we obtain the formulation

$$\sigma + y_t + Cz_t \geq \lceil b_t - f_k \rceil \qquad \text{for } 1 \leq t \leq m$$
$$(\sigma, y, z) \in \mathbb{Z}^1_+ \times \mathbb{Z}^m_+ \times \mathbb{Z}^m_+$$

for X_k^{2DIV}.

. Now total unimodularity of the constraint matrix associated with the σ, y and slack variables indicates that the integrality constraints on σ and y can be dropped.

Proposition 8.11 $(s,y,z) \in$ conv(X_k^{2DIV}) *if and only if* $(\frac{s-f_k}{C}, \frac{y}{C}, z) \in$ conv$(X^{CMIX}(B^k))$, *where* $B_t^k = \frac{\lceil b_t - f_k \rceil}{C}$ *for* $1 \leq t \leq m$.

Taking the $O(m^2) \times O(m)$ extended formulation for conv$(X^{CMIX}(B^k))$ given by (8.25)–(8.29), and the convex hull of the union of these polyhedra using (8.15)–(8.18) gives a formulation for conv(X^{2DIV}) of size $O(m^3) \times O(m^2)$.

8.6.2 The Divisible Mixing Set

To obtain a tight extended formulation for X^{DMIX}, we use the enumerative approach outlined in Section 8.4.

We can again limit the values that s will take.

Observation 8.2 *Let* $f_k = b_k - \lfloor b_k \rfloor$ *for* $k \in K_1 \cup K_2$, $f_0 = 0$, *and* $m = |K_1 \cup K_2|$. *Then there exists an (extreme) optimal solution with*

$$s \mod 1 \in \{f_0, \dots, f_m\}.$$

Then we consider the set $Y^{DMIX} = X^{DMIX} \cap \{(s, \mu, \delta) \in \mathbb{R}^1 \times \mathbb{Z}_+^1 \times \mathbb{Z}_+^{m+1} : s = \mu + \sum_{i=0}^m f_i \delta_i, \sum_{i=0}^m \delta_i = 1\}$.

Proposition 8.12 *The inequality*

$$\mu + \sum_{i:f_i \geq f_k} \delta_i + y_k \geq \lfloor b_k \rfloor + 1 \tag{8.30}$$

is valid for Y^{DMIX} *for all* $k \in K_1$.

For $j \in K_2$, let $f_{p(j)} = \max\{f_i : f_i < f_j\}$ and let $f_{p(0)} = 0$. Set $\sigma_j = \mu + \sum_{i:f_i \geq f_j} \delta_i$ for $j \in K_2 \cup \{0\}$.

Proposition 8.13 *The inequalities*

$$\sigma_j + C y_k \geq \lfloor b_k - f_{p(j)} \rfloor + 1 \text{ for } k \in K_2 \tag{8.31}$$

are valid for Y^{DMIX} *for all* $j \in K_2 \cup \{0\}$. *In other words,* $(\frac{\sigma_j}{C}, y) \in X^{MIX}(B^j)$ *for all* $j \in K_2 \cup \{0\}$, *where* $B_k^j = \frac{\lfloor b_k - f_{p(j)} \rfloor + 1}{C}$ *for* $k \in K_2$.

Now it is possible to show that we have a tight formulation.

Theorem 8.14 *A tight extended formulation for* $\mathrm{conv}(X^{DMIX})$ *is given by*

$$s = \mu + \sum_{i=0}^m f_i \delta_i$$

$$\sum_{i=0}^m \delta_i = 1$$

$$\mu + \sum_{i:f_i \geq f_k} \delta_i + y_k \geq \lfloor b_k \rfloor + 1 \qquad \text{for } k \in K_1$$

$$\sigma_j = \mu + \sum_{i:f_i \geq f_j} \delta_i \qquad \text{for } j \in K_2 \cup \{0\}$$

$$\left(\frac{\sigma_j}{C}, y\right) \in \mathrm{conv}(X^{MIX}(B^j)) \qquad \text{for } j \in K_2 \cup \{0\}$$

$$s, y, \sigma, \delta \in \mathbb{R}_+^1 \times \mathbb{R}_+^{|K_1|+|K_2|} \times \mathbb{R}_+^{|K_2|+1} \times \mathbb{R}_+^{m+1}.$$

Example 8.10 *Consider an instance of the set* X^{DMIX} *with* $n = 4, C = 10, |K_1| = |K_2| = 2$, *namely,*

$$s + y_1 \geq 0.3$$
$$s + y_2 \geq 0.8$$
$$s + 10y_3 \geq 7.2$$
$$s + 10y_4 \geq 4.5$$
$$s \in \mathbb{R}_+^1, y \in \mathbb{Z}^4.$$

Here $m = 4$ *and* $f = (0.0, 0.3, 0.8, 0.2, 0.5)$, *so we introduce*

$$s = \mu + 0.3\delta_1 + 0.8\delta_2 + 0.2\delta_3 + 0.5\delta_4$$
$$\delta_0 + \delta_1 + \delta_2 + \delta_3 + \delta_4 = 1, \ \delta \in \mathbb{R}_+^5.$$

Proposition 8.12 gives the two inequalities

$$\mu + \delta_1 + \delta_2 + \delta_4 + y_1 \geq 1$$
$$\mu + \delta_2 + y_2 \geq 1.$$

Now we just choose one possible value of $j \in K^2 \cup \{0\}$. *With* $j = 4, f_4 = 0.5, f_{p(4)} = 0.3$, *Theorem 8.14 gives the mixing set*

$$\sigma_4 + 10y_3 \geq \lfloor 7.2 - 0.3 \rfloor + 1 = 7$$
$$\sigma_4 + 10y_4 \geq \lfloor 4.5 - 0.3 \rfloor + 1 = 5.$$
$$\sigma_4 = \mu + \delta_2 + \delta_4 \in \mathbb{R}_+^1, y_3, y_4 \in \mathbb{Z}_+^1.$$

Both of the sets studied here provide relaxations for capacitated lot-sizing problems; X^{2DIV} for the case where there are two machines of different capacity producing the same item, and X^{DMIX} for the model with production lower bounds; see Section 11.2.

8.7 The Continuous Integer Knapsack Set and the Gomory Mixed Integer Set

For the sets that we consider in the rest of this chapter, the optimization problem is NP-hard, so the best that we can hope to obtain are some families of strong valid inequalities, and separation algorithms or heuristics so that these inequalities can be used as cutting planes.

First we consider the continuous integer knapsack set

$$K^I = \{(s, y) \in \mathbb{R}_+^1 \times \mathbb{Z}_+^n : \sum_{j=1}^n a_j y_j \leq b + s\}.$$

Let $f_j = a_j - \lfloor a_j \rfloor$, and $f_b = b - \lfloor b \rfloor$.

Proposition 8.15 *The mixed integer rounding (MIR) inequality*

$$\sum_j (\lfloor a_j \rfloor + \frac{(f_j - f_b)^+}{1 - f_b})y_j \leq \lfloor b \rfloor + \frac{s}{1 - f_b} \tag{8.32}$$

is valid for K^I.

Proof. Letting $z = \sum_{j:f_j \leq f_b} \lfloor a_j \rfloor y_j + \sum_{j:f_j > f_b} \lceil a_j \rceil y_j$, we see that
$z \leq \sum_j a_j y_j + \sum_{j:f_j > f_b}(1 - f_j)y_j \leq b + s + \sum_{j:f_j > f_b}(1 - f_j)y_j = b + s'$,
where $s' = s + \sum_{j:f_j > f_b}(1 - f_j)y_j \geq 0$ and $z \in \mathbb{Z}^1$. Now by Proposition 8.1,
the SMIR inequality $z \leq \lfloor b \rfloor + \frac{s'}{1-f_b}$ is valid. Substituting for s' and z gives
the inequality. $\square$

Note that if we define the functions $F_\alpha, \bar{F}_\alpha : \mathbb{R}^1 \to \mathbb{R}^1$ for $0 < \alpha < 1$ by

$$F_\alpha(d) = \lfloor d \rfloor + \frac{(d - \lfloor d \rfloor - \alpha)^+}{1 - \alpha}, \tag{8.33}$$

and $\bar{F}_\alpha(d) = \frac{\min(d,0)}{1-\alpha}$, the MIR inequality can be written as

$$\sum_{j=1}^n F_\alpha(a_j)y_j + \bar{F}_\alpha(-1)s \leq F_\alpha(b)$$

with $\alpha = f_b = b - \lfloor b \rfloor$, the fractional part of b.

As a first example of an MIR inequality, one important class of cutting
planes for commercial MIP systems are the mixed integer Gomory cuts. These
are obtained from a row of a simplex tableau, that is, the set

$$X^{GOM} = \{(y_0, y, x) \in \mathbb{Z}^1 \times \mathbb{Z}_+^p \times \mathbb{R}_+^q : y_0 + \sum_j a_j y_j + \sum_k g_k x_k = b\}$$

with $b \notin \mathbb{Z}^1$.

Setting $s = -\sum_{g_j < 0} g_j x_j \geq 0$, we see that $y_0 + \sum_j a_j y_j \leq b + s$ with
$y_0 \in \mathbb{Z}^1, y \in \mathbb{Z}_+^p$, and $s \in \mathbb{R}_+^1$. The corresponding MIR inequality is

$$y_0 + \sum_j (\lfloor a_j \rfloor + \frac{(f_j - f_b)^+}{1 - f_b})y_j \leq \lfloor b \rfloor - \sum_{k:g_k < 0} \frac{g_k}{1 - f_b} x_k.$$

Now we eliminate the y_0 variable to obtain:

Proposition 8.16 *The mixed integer Gomory cut*

$$\sum_{j:f_j \leq f_b} f_j y_j + \sum_{j:f_j > f_b} \frac{f_b(1 - f_j)}{1 - f_b} y_j + \sum_{k:g_k > 0} g_k x_k - \sum_{k:g_k < 0} \frac{f_b g_k}{1 - f_b} x_k \geq f_b \tag{8.34}$$

is valid for X^{GOM}.

Example 8.11 *Consider the set*

$$X = \{(s,y) \in \mathbb{R}^1_+ \times \mathbb{Z}^3_+ : 7.2y_1 - 3.5y_2 + 5.4y_3 \leq 12.0 + s\},$$

and the point $(s,y) = (0, \frac{5}{3}, 0, 0)$, *which is feasible for the linear relaxation of* X.

Dividing by 7.2, the set can be rewritten as

$$X = \{(s,y) \in \mathbb{R}^1_+ \times \mathbb{Z}^3_+ : y_1 - \frac{35}{72}y_2 + \frac{3}{4}y_3 \leq \frac{5}{3} + \frac{10s}{72}\}.$$

From Proposition 8.15, the MIR inequality for a set in this form with $f_b = \frac{2}{3}$, $f_1 = 0$, $f_2 = \frac{37}{72}$, *and* $f_3 = \frac{3}{4} > f_b$, *is*

$$y_1 - y_2 + \frac{1}{4}y_3 \leq 1 + \frac{10s}{24},$$

and this cuts off the point (s^*, y^*).

Alternatively suppose that we have a row of an optimal LP tableau:

$$(y_1 - \frac{35}{72}y_2 + \frac{3}{4}y_3 - \frac{10s}{72}) + s' = \frac{5}{3}, \quad y \in \mathbb{Z}^3_+, (s, s') \in \mathbb{R}^2_+.$$

Taking the MIR inequality from above and substituting for the variable y_1, *we obtain the mixed integer Gomory cut for that row*

$$\frac{37}{72}y_2 + \frac{1}{2}y_3 + \frac{10}{36}s + s' \geq \frac{2}{3},$$

which cuts off the basic solution $y_2 = y_3 = s = s' = 0$.

8.8 The Continuous 0–1 Knapsack Set

Here we consider the *continuous 0–1 knapsack set*

$$K^B = \{(s,y) \in \mathbb{R}^1_+ \times \{0,1\}^n : \sum_{j=1}^n a_j y_j \leq b + s\}.$$

where the integer variables are further restricted to be binary. We can always assume without loss of generality that $a_j > 0$ for all $j \in N = \{1, \ldots, n\}$, because binary variables can be complemented. For instance, $-y_1 \leq b + s$ is equivalent to $\bar{y}_1 \leq (b+1) + s$, where $\bar{y}_1 = 1 - y_1 \in \{0,1\}$.

We first derive a family of strong valid inequalities, based on covers (or infeasible points).

Definition 8.1 *A set* $C \subseteq N$ *is a cover for* K^B *if*
i. $\sum_{j \in C} a_j = b + \lambda$ *with* $\lambda > 0$, *and*
ii. if $k = \arg\max\{a_j : j \in C\}$, *then* $a_k > \lambda$.

Proposition 8.17 *The MIR cover inequality*

$$s + \sum_{j \in C} \min[a_j, \lambda](1 - y_j) \geq \lambda + \lambda \sum_{j \in N \setminus C} F_\alpha(\tfrac{a_j}{a_k}) y_j \qquad (8.35)$$

is valid for K^B.

Proof. This inequality can be derived as a MIR inequality. By introducing the complementary variables $\bar{y}_j = 1 - y_j$ for $j \in C$, the knapsack constraint becomes

$$-\sum_{j \in C} a_j \bar{y}_j + \sum_{j \in N \setminus C} a_j y_j \leq -\lambda + s,$$

or after division by a_k

$$\sum_{j \in C} \frac{-a_j}{a_k} \bar{y}_j + \sum_{j \in N \setminus C} \frac{a_j}{a_k} y_j \leq \frac{-\lambda}{a_k} + \frac{s}{a_k}.$$

With $\alpha = -\frac{\lambda}{a_k} - \lceil \frac{\lambda}{a_k} \rceil = \frac{a_k - \lambda}{a_k}$, the resulting *MIR* inequality is

$$-\sum_{j \in C} \min[1, \frac{a_j}{\lambda}] \bar{y}_j + \sum_{j \in N \setminus C} F_\alpha(\frac{a_j}{a_k}) y_j \leq -1 + \frac{s}{\lambda},$$

which after multiplication by λ and substitution for $\bar{y}_j$ gives the required inequality. □

The MIR cover inequality can be strengthened by taking into account the fact that the variables y_j for $j \in N \setminus C$ are not just nonnegative integer, but are 0–1 variables.

Let $\tilde{C} = \{j \in C : a_j > \lambda\}$ and let $r = |\tilde{C}| \geq 1$. Reorder the elements so that $\tilde{C} = \{a_1, \ldots a_r\}$ with $a_1 \geq \ldots \geq a_r > \lambda$. Now let $A_j = \sum_{i=1}^{j} a_i$ for $1 \leq j \leq r$.

Let $\phi_C(u)$ be defined as follows:

$$\phi_C(u) = j - 1 \qquad\qquad \text{if } A_{j-1} \leq u \leq A_j - \lambda$$

$$\phi_C(u) = j - 1 + \frac{u - (A_j - \lambda)}{\lambda} \qquad \text{if } A_j - \lambda \leq u \leq A_j$$

$$\phi_C(u) = r - 1 + \frac{u - (A_r - \lambda)}{\lambda} \qquad \text{if } u \geq A_r - \lambda.$$

Theorem 8.18 *The inequality*

$$s + \sum_{j \in C} \min[a_j, \lambda](1 - y_j) \geq \lambda + \lambda \sum_{j \in N \setminus C} \phi_C(a_j) y_j$$

is valid and facet-defining for K^B.

In Figure 8.2, we show the functions $F_\alpha(u/a_1)$ and $\phi_C(u)$ for $a(C) = (10, 8, 5)$, $b = 19$, $\alpha = \frac{6}{10}$, and $\lambda = 4$. The functions are identical for $0 \leq u \leq 14$.

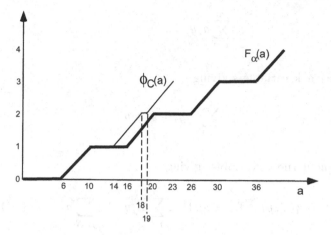

Figure 8.2. The functions F_α and ϕ_C.

Example 8.12 *Consider the continuous 0–1 knapsack set*

$$\{(s, y) \in \mathbb{R}^1_+ \times [0,1]^6 : 10y_1 + 9y_2 + 7y_3 + 16y_4 + 5y_5 + 19y_6 \le 18 + s\}.$$

Taking $C = \{1,2,3\}$, we have that $\lambda = 8$ and $a_1 = 10$. Complementing the variables in C, and dividing by a_1 leaves

$$-\bar{y}_1 - \frac{9}{10}\bar{y}_2 - \frac{7}{10}\bar{y}_3 + \frac{16}{10}y_4 + \frac{5}{10}y_5 + \frac{19}{10}y_6 \le -\frac{8}{10} + \frac{s}{10}.$$

The resulting MIR inequality is

$$-\bar{y}_1 - \bar{y}_2 - \frac{7}{8}\bar{y}_3 + \frac{3}{2}y_4 + \frac{1}{2}y_5 + \frac{15}{8}y_6 \le -1 + \frac{s}{8},$$

or

$$s + 8(1 - y_1) + 8(1 - y_2) + 7(1 - y_3) \ge 8 + 12y_4 + 3y_5 + 15y_6.$$

Now using Theorem 8.18, the inequality can be strengthened giving

$$s + 8(1 - y_1) + 8(1 - y_2) + 7(1 - y_3) \ge 8 + 13y_4 + 3y_5 + 16y_6.$$

Thus we see that in the 0–1 case the use of the lifting function ϕ_C leads to an inequality that is stronger than that obtained using just complementation and the MIR inequality.

A standard way to get further inequalities for knapsack-like sets is to convert an inequality into an equality, and then into an inequality in the other direction. This approach can be applied to sets such as K^B in the following way.

i. Introduce a nonnegative slack variable s' such that

$$\sum_{j=1}^{n} a_j y_j + s' = b + s.$$

ii. Drop the slack variable s giving

$$\sum_{j=1}^{n} a_j y_j + s' \geq b.$$

iii. Complement the y_j variables giving

$$K^{B'} = \{(s', \bar{y}) \in \mathbb{R}^1_+ \times [0,1]^n : \sum_{j=1}^{n} a_j \bar{y}_j \leq (\sum_{j \in N} a_j - b) + s'\}.$$

iv. Generate a MIR cover inequality (8.35) and convert it back into the original variables by substituting for $\bar{y}_j$ and s'.

Example 8.13 *Consider again the set K^B of Example 8.12. Complementing the y variables and introducing the slack variable s' gives*

$$10\bar{y}_1 + 9\bar{y}_2 + 7\bar{y}_3 + 16\bar{y}_4 + 5\bar{y}_5 + 19\bar{y}_6 \leq 66 - 18 + s'.$$

The cover $T = \{1,2,3,6\}$ has excess $\lambda = 1$. Applying the same procedure as in Example 8.12, we get

$$-10y_1 + 9y_2 + 7y_3 + 16\bar{y}_4 + 5\bar{y}_5 + 19y_6 \leq -1 + s'.$$

Dividing by $a_1 = 19$ with $\alpha = \frac{18}{19}$, the MIR cover inequality (8.35) gives

$$-y_1 - y_2 - y_3 + 0y_4 + 0\bar{y}_5 - y_6 \leq -1 + s'.$$

Substitution for s' and complementation gives the MIR reverse cover inequality

$$9y_1 + 8y_2 + 6y_3 + 16y_4 + 5y_5 + 18y_6 \leq 17 + s.$$

8.9 The Binary Single-Node Flow Set

Here we consider the set

$$K^F = \{(x,y) \in \mathbb{R}^n_+ \times \{0,1\}^n : \sum_{j \in N_1} x_j - \sum_{j \in N_2} x_j \leq b,$$

$$x_j \leq a_j y_j \text{ for } 1 \leq j \leq n\}.$$

which can be viewed as the flow through a single node with fixed costs on the arcs; see Figure 8.3. We now extend the definitions of cover and reverse cover.

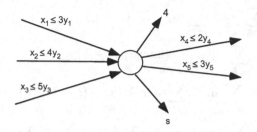

Figure 8.3. Single node flow set.

Definition 8.2 *A set* $(C_1, C_2) \subseteq (N_1, N_2)$ *is a* flow cover *for* K^F *if*
i. $\sum_{j \in C_1} a_j - \sum_{j \in C_2} a_j - b = \lambda > 0$, *and*
ii. $\bar{a} = \max_{j \in C_1} a_j > \lambda$.

Proposition 8.19 *If* (C_1, C_2) *is a flow cover for* K^F *and* (C_i, L_i, R_i) *is a partition of* N_i *for* $i = 1, 2$, *the* MIR flow cover inequality

$$\sum_{j \in C_1} \{x_j + [a_j + \lambda F(-\frac{a_j}{\bar{a}})](1 - y_j)\}$$

$$+ \sum_{j \in L_1} x_j - \sum_{j \in L_1} [a_j - \lambda F(\frac{a_j}{\bar{a}})]y_j$$

$$\leq b + \sum_{j \in C_2} a_j - \sum_{j \in C_2} \lambda F(\frac{a_j}{\bar{a}})(1 - y_j)$$

$$- \sum_{j \in L_2} \lambda F'(-\frac{a_j}{\bar{a}})y_j + \sum_{j \in R_2} x_j + s \qquad (8.36)$$

is valid for K^F, *where* $F = F_\alpha$ *(see (8.33) with* $\alpha = \frac{\bar{a} - \lambda}{\bar{a}}$).

Proof. Setting $x_j = a_j y_j - t_j$ with $t_j \geq 0$ for $j \in C_1 \cup C_2 \cup L_1 \cup L_2$ leads to the relaxation

$$\sum_{j \in C_1 \cup L_1} a_j y_j - \sum_{j \in C_2 \cup L_2} a_j y_j \leq b + s + \sum_{j \in R_2} x_j + \sum_{j \in C_1 \cup L_1} t_j.$$

Generating the MIR cover inequality (8.35) and substituting for t_j gives the inequality (8.36). $\square$

Historically various flow cover inequalities have been derived for the set K^F. The next corollary presents one of these inequalities.

Corollary 8.3 *If* $\bar{a} = \max_{j \in C_1} a_j$, *the MIR flow cover inequality (8.36) is at least as strong as the inequality*

$$\sum_{j \in C_1} x_j + \sum_{j \in C_1} [a_j - \lambda]^+ (1 - y_j)$$

$$+ \sum_{j \in L_1} x_j - \sum_{j \in L_1} (\max[a_j, \bar{a}] - \lambda) y_j$$

$$\leq b + \sum_{j \in C_2} a_j - \sum_{j \in C_2} \min[\lambda, (a_j - (\bar{a} - \lambda))^+](1 - y_j)$$

$$+ \sum_{j \in L_2} \max[a_j - (\bar{a} - \lambda), \lambda] y_j + \sum_{j \in R_2} x_j + s,$$

known as the GFC2 flow cover inequality.

It is possible to strengthen the MIR flow cover inequality in the same way that the MIR cover inequality was strengthened, as indicated in the next example.

Example 8.14 *Consider the single-node flow set*

$$x_1 + x_2 + x_3 + x_4 + x_5 + x_6 \leq 18 + s$$

$$x_1 \leq 10 y_1, x_2 \leq 9 y_2, x_3 \leq 7 y_3, x_4 \leq 16 y_4, x_5 \leq 5 y_5, x_6 \leq 19 y_6$$

$$x \in \mathbb{R}_+^6, y \in \{0, 1\}^6, s \in \mathbb{R}_+^1$$

Taking $(C_1, C_2) = (\{1, 2, 3\}, \emptyset)$ *with* $\lambda = 8$, *substitution of* $t_i = a_i y_i - x_i \geq 0$ *for* $i = 1, \ldots, 6$ *gives the continuous knapsack set*

$$10 y_1 + 9 y_2 + 7 y_3 + 16 y_4 + 5 y_5 + 19 y_6 \leq 18 + s + \sum_{j \in N_1} t_j \ ,$$

which is essentially the same set as in Example 8.12. The resulting MIR inequality is

$$-8 \bar{y}_1 - 8 \bar{y}_2 - 7 \bar{y}_3 + 12 y_4 + 3 y_5 + 15 y_6 \leq -8 + s + \sum_{j \in N_1} t_j \ ,$$

which after substitution gives the inequality (8.36) with $L_1 = \{4, 6\}$

$$x_1 + 2(1 - y_1) + x_2 + 1(1 - y_2) + x_3 + x_4 - 4 y_4 + x_6 - 4 y_6 \leq 18 + s,$$

of Proposition 8.19.

On the other hand the strengthened MIR inequality computed in Example 8.12

$$-8 \bar{y}_1 - 8 \bar{y}_2 - 7 \bar{y}_3 + 13 y_4 + 3 y_5 + 15 y_6 \leq -8 + s + \sum_{j \in N_1} t_j$$

gives the inequality

$$x_1 + 2(1 - y_1) + x_2 + 1(1 - y_2) + x_3 + x_4 - 3 y_4 + x_6 - 3 y_6 \leq 18 + s.$$

Finally we consider the reverse flow cover valid inequality. This inequality can be obtained in the same way as the reverse MIR cover inequality for the continuous 0–1 knapsack set.

Definition 8.3 (T_1, T_2) *is a reverse flow cover for* K^F *if*
i. $T_1 \subseteq N_1, T_2 \subseteq N_2$.
ii. $\sum_{j \in T_1} a_j - \sum_{j \in T_2} a_j - b = -\mu < 0$.

Proposition 8.20 *Suppose that* (T_1, T_2) *is a reverse flow cover and* $\bar{a} > \mu$. *Then the MIR reverse flow cover inequality*

$$\sum_{j \in T_1} x_j + \sum_{j \in T_1} \mu F(\frac{a_j}{\bar{a}})(1 - y_j)$$

$$+ \sum_{j \in L_1} x_j + \sum_{j \in L_1} \mu F(-\frac{a_j}{\bar{a}}) y_j$$

$$\leq \sum_{j \in T_1} a_j - \sum_{j \in T_2} [a_j + \mu F(-\frac{a_j}{\bar{a}})](1 - y_j)$$

$$+ \sum_{j \in L_2} [a_j - \mu F(\frac{a_j}{\bar{a}})] y_j + \sum_{j \in R_2} x_j + s. \qquad (8.37)$$

is valid for K^F, *where* (T_i, L_i, R_i) *is a partition of* N_i *for* $i = 1, 2$ *and* $F = F_\alpha$ *with* $\alpha = \frac{\bar{a} - \mu}{\bar{a}}$.

8.10 Some Convex Hull Proofs

Proof of Proposition 8.1
We give two proofs.

Proof 1. Let P be the formulation defined by the three inequalities. The extreme points lie at the intersection of two tight inequalities.

The intersection of $s = 0$ and $s + fy = f\lceil b\rceil$ is the point $(s, y) = (0, \lceil b\rceil) \in X_1^{MI}$.
The intersection of $s + fy = f\lceil b\rceil$ and $s + y = b$ is the point $(s, y) = (f, \lfloor b\rfloor) \in X_1^{MI}$.
The intersection of $s = 0$ and $s + y = b$ is the point $(0, b)$ that is cut off by the inequality $s + fy \geq f\lceil b\rceil$. The two (feasible) extreme points of P lie in X_1^{MI}, and thus $P = \text{conv}(X_1^{MI})$. $\square$

Proof 2. The second proof can also be used to tackle more complicated sets. We show that every facet of P is integral.

In any facet, one of the inequalities defining P must be satisfied at equality. So we must consider the three possibilities: $F_1 = \{(s, y) \in P : s = 0\}, F_2 = \{(s, y) \in P : s + fy = f\lceil b\rceil\}$, and $F_3 = \{(s, y) \in P : s + y = b\}$.

$$F_1 = \{(s,y) : \ s = 0, \ 0 \geq b - y, \ 0 \geq f(\lceil b \rceil - y)\}$$
$$= \{(s,y) : \ s = 0, \ y \geq \lceil b \rceil\}.$$
$$F_2 = \{(s,y) : \ s = f\lceil b \rceil - fy, \ f\lceil b \rceil - fy \geq 0, \ f\lceil b \rceil - fy \geq b - y\}$$
$$= \{(s,y) : \ s + fy = f\lceil b \rceil, \ \lfloor b \rfloor \leq y \leq \lceil b \rceil\}.$$
$$F_3 = \{(s,y) : \ s = b - y, \ b - y \geq f\lceil b \rceil - fy, \ b - y \geq 0\}$$
$$= \{(s,y) : \ s = b - y, \ y \leq \lfloor b \rfloor\}.$$

Viewed in the y-space, all the extreme points of these facets have y integral. These facets are of the form $s = \pi y + \mu, \alpha \leq y \leq \beta$ with α integral or $-\infty$, and β integer or $+\infty$. Thus each facet is integral, and P is integral. ☐

Proof of Theorem 8.2

We use the same argument as above. In a facet, one of the inequalities must be tight for each k, and thus each facet is defined by K equations $s_k = \pi_k y + \mu_k$ defining each of the s_k, along with the intersection of K intervals $\alpha_k \leq y \leq \beta_k$, where the α_k, β_k are integers, $-\infty$, or $+\infty$. Thus again each facet is integral and the polyhedron is integral.

Now suppose that there are the additional constraints $By \leq d$. Each face is now of the form

$$\{(s,y) : s_k = \pi_k y + \mu_k, \alpha_k \leq y \leq \beta_k \quad \text{for } 1 \leq k \leq K, By \leq d\},$$

which is integral by Theorem 6.7. ☐

Proof of Theorem 8.6

Using (8.9), the proposed formulation in the (y, μ, δ) space can be written as

$$\mu + y_k + \sum_{j:f_j \geq f_k} \delta_j \geq \lfloor b_k \rfloor + 1 \qquad \text{for } 1 \leq k \leq K$$

$$\sum_{j=0}^{K} \delta_j = 1$$

$$\mu \in \mathbb{R}^1_+, \delta \in \mathbb{R}^{K+1}_+.$$

Apart from the y variables that form an identity matrix, if the remaining variables are ordered as $\mu, \delta_{j_1}, \ldots, \delta_{j_K}$, where $f_{j_1} \geq \ldots \geq f_{j_K}$, the resulting matrix has the consecutive 1s property. Therefore the constraint matrix is totally unimodular, and the claim follows. ☐

Exercises

Exercise 8.1 Consider the set

$$X = \{(s,y) \in \mathbb{R}^1 \times \mathbb{Z}^1 : s + a_1 y \geq b_1, s + a_2 y \geq b_2\}.$$

i. Derive a valid inequality for this set, and show that it gives the convex hull.
ii. Apply to an instance with $a = (-1, 2), b = (3.2, 5.7)$.

Exercise 8.2 Consider the set

$$x_1 + y \geq 1.5$$
$$x_2 + y \geq 2.6$$
$$x \in \mathbb{R}_+^2, \ y \in \mathbb{Z}^1.$$

Find a valid inequality cutting off the point $(x_1, x_2, y) = (0.2, 1.3, 1.3)$.

Exercise 8.3 Consider the set

$$X = \{(s, y) \in \mathbb{R}_+^n \times \mathbb{Z}_+^1 : \sum_{i=1}^n s_i + y \geq b, s_i \leq u_i \text{ for } 1 \leq i \leq n\}.$$

i. Derive a family of SMIR inequalities for this set.
ii. For an instance with $n = 3$, $b = 14.3$, and $u = (2.1, 4.8, 3.4)$, find a valid inequality cutting off the point $s = (2.1, 0, 3.4), y = 8.8$.
iii.* Show that the MIR inequalities suffice to describe the convex hull.

Exercise 8.4 Consider the problem

$$\min\{g(y) + fy : y \in \mathbb{Z}^1\},$$

where $g(y) = \max\{a_1 y - b_1, a_2 y - b_2, a_3 y - b_3\}$.
i. Model as an optimization problem over a MIP set. See Exercise 8.1.
ii. Show how to solve this problem by linear programming.
iii. Consider an instance with $a = (-1, -0.5, 1), b = (-1.5, -1.25, 2.5)$. Draw the function g, and redraw with the function providing a tight formulation.
iv. Use this to describe an algorithm for the separable convex network flow problem

$$\min\{\sum_{(i,j) \in A} g_{ij}(x_{ij}) : Nx = b, x \in \mathbb{Z}_+^m\},$$

where $D = (V, A)$ is a digraph with $n \times m$ node-arc incidence matrix N, $b \in \mathbb{Z}^n$, and each function g_{ij} is piecewise linear and convex described in the same way as the function g.

Exercise 8.5 Consider the divisible knapsack set

$$X = \{(s, y) \in \mathbb{R}_+^1 \times \mathbb{Z}_+^n : s + \sum_{i=1}^m C_i y_i \geq b\},$$

where $C_1 | \cdots | C_n$, and $C_1, \ldots, C_n$ and b are positive integers.
i. Find valid inequalities for X.
ii.* Give an inequality description of conv(X).

Exercise 8.6 Consider the set

$$
\begin{aligned}
x + 5y_1 &\geq 3 \\
x + 5y_1 + 5y_2 &\geq 6 \\
x + 5y_1 + 5y_2 + 5y_3 &\geq 8 \\
x \in \mathbb{R}^1_+, \; y \in \mathbb{Z}^3_+.
\end{aligned}
$$

i. Generate a variety of valid inequalities for this set.
ii. Find a valid inequality cutting off the point $(x^*, y^*) = (1, \frac{2}{3}, \frac{1}{3}, \frac{2}{3})$.

Exercise 8.7 Consider the mixing set X_K^{MIX} with the additional constraint $s \leq d$. Find valid inequalities for this set.
   Apply your result to the instance of Example 8.4 with $s \leq 0.5$.

Exercise 8.8 Consider the mixing set of Section 8.3. Write the optimization and separation problems as network flow problems for
i. the instance of Example 8.6,
ii. the general case, and
iii. show why addition of constraints of the form $\alpha_{ij} \leq y_i - y_j \leq \beta_{ij}$ with $\alpha_{ij}, \beta_{ij} \in \mathbb{Z}^1$ does not destroy the network structure.

Exercise 8.9 Consider the set

$$
\begin{aligned}
x + 5y_1 &\quad + z_1 &&\geq 3 \\
x + 5y_1 + 5y_2 &\quad + z_2 &&\geq 6 \\
x + 5y_1 + 5y_2 + 5y_3 &\quad + z_3 &&\geq 8 \\
x \in \mathbb{R}^1_+, \; y \in \mathbb{Z}^3_+, &\quad z \in \mathbb{R}^3_+.
\end{aligned}
$$

Write down an extended formulation for this set.
   Explain how such an extended formulation can be used to find a valid inequality cutting off points (x^*, y^*, z^*) with y^* fractional.

Exercise 8.10 Consider the set

$$
\begin{aligned}
s + 5y_{11} + 10y_{12} &\geq 3 \\
s + 5y_{21} + 10y_{22} &\geq 9 \\
s \in \mathbb{R}^1_+, y \in \mathbb{Z}^4_+.
\end{aligned}
$$

Write down an extended formulation for this set.

Exercise 8.11 Consider the set X_K^{CMIX} with in addition $y \in \mathbb{Z}^K_+$ and $0 < b_1 \leq \cdots \leq b_K < 1$.
i. For $T = \{i_1, \ldots, i_t\} \subseteq \{1, \ldots, K\}$ with $i_1 < \cdots < i_t$, show that the inequalities

$$
s + \sum_{j \in T} r_j \geq \sum_{u=1}^{t} (b_{i_u} - b_{i_{u-1}})(1 - y_{i_u})
$$

are valid for X_K^{CMIX}.

ii.* Show that every nontrivial facet-defining inequality is of this form.

Exercise 8.12 Consider the following formulation for X^{CMIX}.

$$s = \sum_{i=0}^{K} f_i \delta_i + \mu \tag{8.38}$$

$$r_j = \sum_{i=0}^{K} f_i^j \beta_i^j + \nu_j \qquad \text{for } 1 \le j \le K \tag{8.39}$$

$$\mu + \nu_j + y_j + \sum_{k:f_k > f_i} \delta_k$$

$$+ \sum_{k:f_k \ge F_i^j} \beta_k^j \ge \lceil b_j - f_i \rceil \qquad \text{for } 1 \le j \le K,\ 0 \le i \le K \tag{8.40}$$

$$\sum_{i=0}^{K} \delta_i = 1 \tag{8.41}$$

$$\sum_{i=0}^{K} \beta_i^j = 1 \qquad \text{for } 1 \le j \le K \tag{8.42}$$

$$\mu \in \mathbb{R}_+^1,\ \delta \in \mathbb{R}_+^{K+1} \tag{8.43}$$

$$\nu \in \mathbb{R}_+^K,\ \beta \subset \mathbb{R}_+^{K(K+1)}, \tag{8.44}$$

where
$\delta_i = 1$ if the fractional part of s takes the value f_i;
μ is the integer part of s;
$\beta_i^j = 1$, for $i \ne j$, if the fractional part of r_j is F_i^j,
ν_j is the integer part of r_j;
plus the additional variables $\beta_j^j = 1$ if $\delta_0 = 1$ and $r_j = 0$.

i. Show that it is a valid formulation for the problem.
ii.* Show that the formulation is tight by showing that the matrix associated with the constraints (8.40)–(8.42) is totally unimodular.

Exercise 8.13 Consider a row of an optimal LP tableau

$$y_0 + \tfrac{7}{5}y_1 - \tfrac{2}{5}y_2 + \tfrac{11}{5}x_3 - \tfrac{4}{5}x_4 = \tfrac{6}{5}$$
$$y \in \mathbb{Z}_+^3,\ x \in \mathbb{R}_+^2.$$

i. Construct an MIR inequality cutting off $y_0 = \tfrac{6}{5}$, $y_1 = y_2 = x_3 = x_4 = 0$.
ii. Construct the mixed integer Gomory cut, and verify that it is the same as the MIR inequality.

Exercise 8.14 Consider the set

$$K^B = \{(x,y) \in \mathbb{R}^1_+ \times \{0,1\}^6 : 15y_1 + 11y_2 + 10y_3 + 6y_4 + 5y_5 + 2y_6 \le 19 + x\}.$$

i. Taking $C = \{2,4,5\}$, generate the corresponding MIR cover inequality.

ii. Taking the same cover, generate the strengthened inequality of Theorem 8.18.

iii. Complement the 0–1 variables, take $C = \{1,2,4\}$ and construct the MIR reverse cover inequality.

iv. By inspection, find a MIR cover inequality cutting off the point $(x^*, y^*) = (1, 0, \frac{4}{11}, \frac{1}{2}, 1, 1, 0)$.

Exercise 8.15 Consider the set

$$\begin{aligned}
K^F = \{(x,y,s) \in \mathbb{R}^6_+ \times \{0,1\}^6 \times \mathbb{R}^1_+ : \\
x_1 + x_2 + x_3 - x_4 - x_5 - x_6 \le 5 + s \\
x_1 \le 14y_1, x_2 \le 9y_2, x_3 \le 8y_3, x_4 \le 12y_4, x_5 \le 10y_5, x_6 \le 6y_6\}.
\end{aligned}$$

Taking $(C_1, C_2) = (\{2,3\}, \{5\})$, compare the MIR inequality (8.36) with the inequality in Corollary 8.3.

Exercise 8.16 Consider the set

$$\begin{aligned}
s + x_t \ge b_t && \text{for } 1 \le t \le n \\
x_t \le z_t && \text{for } 1 \le t \le n \\
s \in \mathbb{R}^1_+, \ x \in \mathbb{R}^n_+, \ z \in \mathbb{Z}^n_+,
\end{aligned}$$

denoted X^{FM} and called a *mixing set with flows*.

Suppose that $0 = b_0 < b_1 \le \cdots \le b_n$. Let $\sigma_0 = s \ge 0$ and let $\sigma_t = s + x_t - b_t \ge 0$ be the slack variable in the first set of constraints for $1 \le t \le n$.

i. Show that the mixing set X_t^{MIX}:

$$\begin{aligned}
\sigma_t + z_k \ge b_k - b_t && \text{for } k + 1 \le t \le n \\
\sigma_t \in \mathbb{R}^1_+, z_t \in \mathbb{Z}^1_+ && \text{for } k + 1 \le t \le n
\end{aligned}$$

is a valid relaxation of X^{FM}.

ii.* Show that $\text{conv}(X^{FM}) = \cap_{t=0}^n \text{conv}(X_t^{MIX}) \cap \{(\sigma, x) : \sigma_t = s + x_t - b_t \text{ for all } t\} \cap \{(x, z) : 0 \le x_t \le z_t \text{ for all } t\}$.

Exercise 8.17 Consider the set

$$X = \{(y,s) \in \mathbb{Z}^2_+ \times \mathbb{R}^1_+ : 3y_1 + 7y_2 \le 31 + s\}.$$

Derive two or more MIR inequalities for this set and check whether they are facet-defining.

---

* Starred exercises are more difficult and require more mathematical or technical developments.

Notes

Section 8.2. The MIP set is studied in Miller and Wolsey [124]. The single arc set arising in Exercise 8.3 was studied by Magnanti et al. [113], and later by Atamturk and Rajan [16]. The divisible knapsack set of Exercise 8.5 was studied by Pochet and Wolsey [142]; see also Pochet and Weismantel [136] for the 0–1 case. The modeling of a separable piecewise convex function over an integer variable is also from [124]; see also Ahuja et al. [7] for an alternative reformulation and Hochbaum and Shantikumar [90].

Section 8.3. The mixing set is studied in Günlük and Pochet [85]. The mixing inequalities first arose in the context of constant capacity lot-sizing studied by Pochet and Wolsey [139]. Part ii of Theorem 8.5 and the extended formulation are from Miller and Wolsey [124]. The dual network reformulation is from Conforti et al. [43].

Section 8.4. The enumerative approach to develop more general mixing sets was developed explicitly by Van Vyve [178]. The disjunctive approach was used by Atamtürk [13], and developed independently by Conforti and Wolsey [44]. The latter approach is based on the extended formulation for the union of polyhedra from Balas [21].

Section 8.5. The first formulation of the continuous mixing set is from Miller and Wolsey [124] and the more compact formulation from Van Vyve [179]. A third formulation was proposed by Van Vyve in [178]. The uncapacitated case is treated by Atamtürk [13]; see Exercise 8.11.

Section 8.6. The set with two divisible capacities is from Conforti and Wolsey [44]. The results on the divisible mixing set are from Van Vyve [178].

Section 8.7. Mixed integer Gomory cuts are from Gomory [77]. Disjunctive cuts were introduced by Balas in [18]. The mixed integer rounding inequality is from Nemhauser and Wolsey [126], and the alternative split cut viewpoint is presented in Cook et al. [49]. Superadditive functions such as F_α are the basis of a general duality theory for integer programming. Using superadditive functions such as ϕ_C to strengthen valid inequalities was proposed by Gu et al. [81].

Section 8.8. Cover inequalities for binary knapsack sets were studied by Balas [19], Hammer et al. [86] and Wolsey [190]. The development of the heuristic separation algorithm for cover inequalities, and the first effective computational study are due to Crowder et al. [50]. Marchand and Wolsey [117] studied the continuous 0–1 knapsack set, and in [118] they proposed a heuristic separation procedure for mixed integer programs based on aggregation of constraints, substitution, and complementation of variables followed by the generation of MIRs.

Section 8.9. Single-node flow sets were first studied in Padberg et al. [130]; see also Van Roy and Wolsey [174]. Computational results using flow cover inequalities were given in Van Roy and Wolsey [175]. The MIR flow cover inequality was obtained by Gu et al. [81]; see also a recent survey of Louveaux and Wolsey [108]. Reverse flow cover inequalities were first explicitly proposed by Stallaert [156].

Section 8.10. The proof of Proposition 8.1 based on the integrality of the facets is from Pereira and Wolsey [132]. The proof of Theorem 8.6 exemplifies the most important proof technique for most of the mixing problems which is to find an extended formulation with integer right-hand side whose associated constraint matrix is totally unimodular.

Exercises. The result of Exercise 8.16 on mixing sets with flows is from Conforti et al. [42].

Single-Item Lot-Sizing

9

Lot-Sizing with Capacities

Part III consists of three chapters dealing with many of the significant variants of the single-item lot-sizing problem, apart from the uncapacitated problems $LS\text{-}U$ and $WW\text{-}U$ already presented in detail in Chapter 7. We continue to use the classification scheme introduced in Chapter 4. Specifically we deal in this chapter with capacities, either constant or time-varying, namely, the models $PROB\text{-}\{CC,C\}$. In Chapter 10 we consider two of the most important variants, the problems with backlogging $PROB\text{-}CAP\text{-}B$ and with start-up variables $PROB\text{-}CAP\text{-}SC$, and in Chapter 11 we study several other single-item variants, including piecewise concave production costs, piecewise convex storage costs, sales, minimum length set-up sequences, production time windows, and so on.

Remember that the general problem $LS\text{-}C$ can be formulated as

$$\min \quad \sum_{t=1}^{n} p'_t x_t + \sum_{t=0}^{n} h'_t s_t + \sum_{t=1}^{n} q_t y_t$$

$$s_{t-1} + x_t = d_t + s_t \qquad \qquad \text{for } 1 \leq t \leq n$$

$$x_t \leq C_t y_t \qquad \qquad \text{for } 1 \leq t \leq n$$

$$s \in \mathbb{R}_+^{n+1}, \ x \in \mathbb{R}_+^{n}, \ y \in \{0,1\}^n.$$

For $LS - CC$, we have the same formulation but with $C_t = C$ for all t.

Note that in addition to the basic preprocessing of initial stocks and stock lower bounds presented in Section 7.2, there is an elementary preprocessing step for $LS\text{-}C$ depending on the capacities. Below we assume without loss of generality that $d_t \leq C_t$. If $d_t > C_t$, then by setting $d_t = C_t$ and adding $d_t - C_t$ to d_{t-1} for $t = n, n-1, \ldots, 1$, we obtain an equivalent problem.

We now describe briefly the contents of this chapter.

- In Section 9.1 we consider the complexity issue, showing that all the varying capacity problems $PROB\text{-}C$ are NP-hard. This is in contrast to the constant capacity problems *PROB-CC* that are all shown to have polynomial algorithms in the course of the chapter.

- In Section 9.2 we examine regeneration intervals for problems with capacities that are fundamental in describing optimal solutions of single-item problems, and form the basis for the dynamic programming algorithm for *LS-CC* and other models.
- In the following sections we examine in turn the four different constant capacity problems. First in Section 9.3 we present both a linear programming formulation and a combinatorial algorithm for *DLS-CC*.
- In Sections 9.4 and 9.5 we present valid inequalities describing the convex hull, combinatorial separation algorithms and tight extended formulations for *DLSI-CC* and *WW-CC* respectively, all based on the mixing set studied in Section 8.3.
- In Section 9.6 the main result is a dynamic programming algorithm for *LS-CC* based on regeneration intervals.
- In Section 9.7 we turn to the varying capacity problem *LS-C*, and present several classes of valid inequalities. One class for *WW-C* is based on the mixing set studied in Section 8.3, while two of the classes for *LS-C* are based on the continuous knapsack set studied in Section 8.8 and on an alternative formulation using submodularity respectively.

9.1 Complexity

The first thing to observe is that the 0–1 knapsack problem:

$$\min \sum_{j=1}^{n} q_j y_j$$
$$\sum_{j=1}^{n} C_j y_j \geq b$$
$$y \in \{0,1\}^n$$

is a special case of *DLS-C* in which the demands are $d_t = 0$ for $t = 1, \ldots, n-1$, and $d_n = b$. As *DLS-C* is in turn a special case of *DLSI-C*, *WW-C* and *LS-C*, and as the 0–1 knapsack problem is NP-hard, we obtain the complexity status of these problems.

Proposition 9.1 *Problems DLS-C, DLSI-C, WW-C and LS-C are NP-hard.*

This tells us immediately that we have little hope of finding a good characterization of $\mathrm{conv}(X^{DLS-C})$, or of $\mathrm{conv}(X^{LS-C})$.

9.2 Regeneration Intervals

As for *LS-U*, it is necessary to understand the structure of optimal solutions of *LS-C* and *LS-CC* if one wants to develop an optimization algorithm. It is again important to think of the fixed cost network flow representation. We assume for the discussion below that $s_0 = s_n = 0$, and repeat the definition of a *regeneration interval* from Chapter 7.

Definition 9.1 *The interval* $[k, l]$ *is a* regeneration interval *for a solution* $(x, y, s) \in X^{LS-C}$ *if and only if* $s_{k-1} = s_l = 0$, *but* $s_t > 0$ *for* $t = k, \dots, l - 1$ *(see Figure 9.1).*

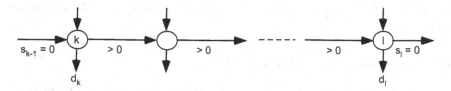

Figure 9.1. A regeneration interval for *LS-C*.

Every extreme solution of $\text{conv}(X^{LS-C})$ corresponds to a partition of $[1, n]$ into regeneration intervals. We now examine what happens within each regeneration interval. Again we use the fact that, for fixed $y \in \{0, 1\}^n$, the problem in the (x, s) variables is a network flow problem, and an optimal basic solution is acyclic, so that there exists an optimal solution in which the arcs with flow strictly between the lower and upper bound ($0 < x_t < C_t$, or $s_t > 0$) form an acyclic subgraph in the network.

Observation 9.1 *Suppose that* $[k, l]$ *is a regeneration interval forming part of an optimal solution of LS-C. Then within the interval* $[k, l]$ *there exists a solution of the following form: there exists a period* p *with* $k \leq p \leq l$ *such that* $x_j = C_j y_j$ *for all* $j \in \{k, \dots, l\} \setminus \{p\}$. *In other words there is at most one period in the regeneration interval in which* $x_p \notin \{0, C_p\}$. *Such a period is referred to as a* fractional period.

Observation 9.2 *Let* α_{kl} *be the minimum cost of a* $[k, l]$-regeneration interval *for all* $1 \leq k \leq l \leq n$, *and* $D = (V, A)$ *be an acyclic digraph with nodes* $V = \{1, \dots, n + 1\}$, *and arcs* $A = \{(k, l + 1) : 1 \leq k \leq l \leq n\}$ *with cost* α_{kl} *for* $1 \leq k \leq l \leq n$. *A minimum cost path from node 1 to* $n + 1$ *solves LS-C.*

We have already used regeneration intervals and this minimum cost path in the uncapacitated case *LS-U*. See Figure 7.6, where $\alpha_{kl} = q_k + p_k d_{kl}$.

In the general *LS-C* case, calculating the optimal cost α_{kl} of a regeneration interval is hard. However, we see below that it can be calculated efficiently in the constant capacity cases *WW-CC* and *LS-CC*. As a minimum cost path can be found in $O(n^2)$, this will obviously lead to a polynomial algorithm.

9.3 Discrete Lot-Sizing with Constant Capacities

With constant capacities, X^{DLS-CC} is of the form

$$C \sum_{u=1}^{t} y_u \geq d_{1t} \qquad \qquad \text{for } 1 \leq t \leq n \qquad (9.1)$$

$$y \in \{0,1\}^n. \qquad (9.2)$$

Clearly $X^{DLS-CC} \neq \emptyset$ if and only if $t \geq \lceil \frac{d_{1t}}{C} \rceil$ for $t = 1, \dots, n$. From now on, we assume that $X^{DLS-CC} \neq \emptyset$.

9.3.1 Valid Inequalities for *DLS-CC*

We immediately observe that Gomory fractional cuts can be used to tighten this formulation.

Proposition 9.2 *The inequalities*

$$\sum_{u=1}^{t} y_u \geq \lceil \frac{d_{1t}}{C} \rceil$$

are valid for X^{DLS-CC} for $1 \leq t \leq n$.

What is more, they can be used in place of the existing constraints and suffice to describe $\text{conv}(X^{DLS-CC})$.

Theorem 9.3 *The polyhedron*

$$\sum_{u=1}^{t} y_u \geq \lceil \frac{d_{1t}}{C} \rceil \qquad \qquad \text{for } 1 \leq t \leq n$$

$$y \in [0,1]^n$$

describes $\text{conv}(X^{DLS-CC})$.

In Chapter 12, we show that the same inequalities suffice for the multi-item variant in which at most one item is produced per period.

9.3.2 Optimization for *DLS-CC*

To solve the optimization problem without using linear programming, it suffices to use a greedy algorithm. By setting $y_j = 1$ and modifying the demands whenever $q_j < 0$, we can assume that $q_j \geq 0$ for all j. Then we complement the variables, setting $\bar{y}_j = 1 - y_j$ for $j = 1, \dots, n$, giving

$$\max \ \sum_{j=1}^{n} q_j \bar{y}_j$$

$$\sum_{u=1}^{t} \bar{y}_u \leq t - \lceil \frac{d_{1t}}{C} \rceil \qquad \qquad \text{for } 1 \leq t \leq n$$

$$\bar{y} \in \{0,1\}^n.$$

Algorithm for *DLS-CC*
1. Find an ordering $j_1, \ldots, j_n$ of $\{1, \ldots, n\}$ with $q_{j_1} \geq \ldots \geq q_{j_n} \geq 0$.
2. Initialize $\delta_t = t - \lceil \frac{d_{1t}}{C} \rceil \geq 0$ for all t.
3. For $k = 1, \ldots, n$,
   If $\min_{t:t \geq j_k} \delta_t > 0$, set $y_{j_k} = 0$ (or $\bar{y}_{j_k} = 1$) and set $\delta_t \leftarrow \max\{\delta_t - 1, 0\}$
   for all $t \geq j_k$.
   Otherwise set $y_{j_k} = 1$.
   Augment k.

Example 9.1 *Consider an instance of DLS-CC with $n = 6$, $C = 10$, $d = (0, 5, 3, 6, 8, 1)$, and $q = (34, 20, 35, 40, 33, 21)$. Initially $\delta = (1, 1, 2, 2, 2, 3)$.*

Iteration 1. $j_1 = 4, y_4 = 0, \delta = (1, 1, 2, 1, 1, 2)$,
Iteration 2. $j_2 = 3, y_3 = 0, \delta = (1, 1, 1, 0, 0, 1)$,
Iteration 3. $j_3 = 1, y_1 = 1, \delta = (1, 1, 1, 0, 0, 1)$,
Iteration 4. $j_4 = 5, y_5 = 1, \delta = (1, 1, 1, 0, 0, 1)$,
Iteration 5. $j_5 = 6, y_6 = 0, \delta = (1, 1, 1, 0, 0, 0)$,
Iteration 6. $j_6 = 2, y_2 = 1, \delta = (1, 1, 1, 0, 0, 0)$,
giving as optimal solution $y = (1, 1, 0, 0, 1, 0)$ with cost 87.

The above algorithm can be viewed as the greedy algorithm for a matroid. Its complexity as described is $O(n^2)$. It can, however, be implemented to run in $O(n \log n)$.

9.3.3 Parametric Optimization for *DLS-CC*

Here we describe an $O(n \log n)$ algorithm that solves a family of problems. For simplicity we assume that the constant capacity C has been normalized so that $C = 1$. Specifically we consider three closely related problems:

problem $Q(j)$:

$$z_j = \max \ \sum_{i=1}^{j} q_i y_i$$

$$\sum_{i=1}^{t} y_i \leq \delta_t \qquad \qquad \text{for } 1 \leq t \leq j$$

$$y \in \{0,1\}^j,$$

problem $Q'(j)$:

$$\max \sum_{i=1}^{j} q_i y_i$$

$$\sum_{i=1}^{t} y_i \leq \delta_t \qquad \text{for } 1 \leq t \leq j-1$$

$$y \in \{0,1\}^j,$$

and problem $R(j, \beta_j)$:

$$\max \sum_{i=1}^{j} q_i y_i$$

$$\sum_{i=1}^{t} y_i \leq \delta_t \qquad \text{for } 1 \leq t \leq j-1$$

$$\sum_{i=1}^{j} y_i = \beta_j$$

$$y \in \{0,1\}^j.$$

Our goal is to solve the family of problems $Q(j)$ and $R(j, \beta_j)$ for $j = 1, \ldots, n$ efficiently. *DLS-CC* is equivalent to $Q(n)$, with $\delta_t = t - \lceil \frac{d_{1t}}{C} \rceil$ for all t.

Observation 9.3 *The feasible region $X(j)$ of problem $Q(j)$ is the set of independent sets of a matroid. It follows that every maximal feasible solution of $X(j)$ is of the same cardinality,*

$$\rho_j = \max\{\sum_{i=1}^{j} y_i : y \in X(j)\} = \min_{t=0,\ldots,j} [\delta_t + (j-t)].$$

Observation 9.4 *$R(j, \beta_j)$ is feasible if and only if $\beta_j \leq \rho_j$.*

Suppose now that y' is an optimal solution of $Q'(j)$, or in other words $(y'_1, \ldots, y'_{j-1})$ is an optimal solution of $Q(j-1)$ and $y'_j = 1$. Each iteration of the algorithm computing $Q(j)$ and $R(j, \beta_j)$ for increasing values $j = 1, \ldots, n$ consists of two simple steps.

i. Convert an optimal solution y' to $Q'(j)$ into an optimal solution $\tilde{y}$ of $R(j, \beta_j)$; necessarily $\tilde{y} \leq y'$ (if $\beta_j \leq \rho_j$).
ii. Convert $\tilde{y}$ into an optimal solution y^* of $Q(j)$; necessarily $y^* \leq \tilde{y}$.

Because of the matroid structure, the construction of $\tilde{y}$ from y' and of y^* from $\tilde{y}$ just involves removal of the least profitable elements.

Example 9.2 *Consider an instance with $n = 6$, $q = (34, 20, 35, 40, 33, 21)$, $\delta = (1, 1, 2, 2, 2, 3)$, and $\beta = (1, 2, 2, 3, 3, 3)$. Here problem $Q(6)$ is precisely the problem solved in Example 9.1.*

$j = 1.\ Q'(1):\ \ y_1' = 1$
$\qquad\quad R(1, 1): \tilde{y}_1 = 1$
$\qquad\quad Q(1):\ \ y_1^* = 1,\ z_1 = 34$

$j = 2.\ Q'(2):\ \ y_1' = 1,\ y_2' = 1$
$\qquad\quad R(2, 2): \tilde{y}_1 = 1,\ \tilde{y}_2 = 1$
$\qquad\qquad\quad$ *As $\delta_2 = 1$, we must remove the least profitable variable.*
$\qquad\qquad\quad$ *As $q_1 = 34 > q_2 = 20$, we set $y_2 = 0$.*
$\qquad\quad Q(2): y_1^* = 1,\ z_2 = 34$

$j = 3.\ Q'(3):\ \ y_1' = 1,\ y_3' = 1$
$\qquad\quad R(3, 2): \tilde{y}_1 = 1,\ \tilde{y}_3 = 1$
$\qquad\quad Q(3):\ \ y_1^* = 1,\ y_3^* = 1,\ z_3 = 69$

$j = 4.\ Q'(4):\ \ y_1' = 1,\ y_3' = 1,\ y_4' = 1$
$\qquad\quad R(4, 3): \tilde{y}_1 = 1,\ \tilde{y}_3 = 1,\ \tilde{y}_4 = 1$
$\qquad\qquad\quad$ *As $\delta_4 = 2$, we remove the least profitable variable. We set $y_1 = 0$.*
$\qquad\quad Q(4): y_3^* = 1,\ y_4^* = 1,\ z_4 = 75$

$j = 5.\ Q'(5):\ \ y_3' = 1,\ y_4' = 1,\ y_5' = 1$
$\qquad\quad R(5, 3): \tilde{y}_3 = 1,\ \tilde{y}_4 = 1,\ \tilde{y}_5 = 1$
$\qquad\qquad\quad$ *As $\delta_5 = 2$, we remove the least profitable variable. We set $y_5 = 0$.*
$\qquad\quad Q(5): y_3^* = 1,\ y_4^* = 1,\ z_5 = 75$

$j = 6.\ Q'(6):\ \ y_3' = 1,\ y_4' = 1,\ y_6' = 1$
$\qquad\quad R(6, 3): \tilde{y}_3 = 1,\ \tilde{y}_4 = 1,\ \tilde{y}_6 = 1$
$\qquad\quad Q(6):\ \ y_3^* = 1,\ y_4^* = 1,\ y_6^* = 1,\ z_6 = 96.$

Observe that this agrees with the solution of Example 9.1.

9.4 Discrete Lot-Sizing with Initial Stock and Constant Capacities

Here $X^{DLSI-CC}$ takes the form

$$s_0 + C \sum_{u=1}^{t} y_u \geq d_{1t} \qquad\qquad \text{for } 1 \leq t \leq n \qquad (9.3)$$

$$s_0 \in \mathbb{R}_+^1, y \in \{0, 1\}^n. \qquad (9.4)$$

Perhaps surprisingly, it is not known if problem *DLSI-CC* with an initial stock variable can also be solved faster than $O(n^2 \log n)$.

9.4.1 Valid Inequalities for *DLSI-CC*

Letting $z_t = \sum_{u=1}^{t} y_t$, $s = \frac{s_0}{C}$, and $b_t = \frac{d_{1t}}{C}$, the solution set $X^{DLSI-CC}$ can be rewritten as

$$
\begin{aligned}
s + z_t &\geq b_t & \text{for } 1 \leq t \leq n \\
0 \leq z_t - z_{t-1} &\leq 1 & \text{for } 2 \leq t \leq n \\
z_1 &\leq 1 \\
s \in \mathbb{R}^1_+, &\ z \in \mathbb{Z}^n_+.
\end{aligned}
$$

This is an instance of a mixing set (see Section 8.3) in which the additional constraints are of the form $Bz \leq d$ with B the arc-node incidence matrix of a digraph and d integer. From Theorem 8.5, we obtain immediately:

Theorem 9.4 conv($X^{DLSI-CC}$) *is described by* $s_0 \in \mathbb{R}^1_+, y \in [0,1]^n$, *the initial inequalities (9.3), and the inequalities*

$$
s_0 \geq C \sum_{\tau=1}^{t} (f_{i_\tau} - f_{i_{\tau-1}})(\lfloor \frac{d_{1i_\tau}}{C} \rfloor + 1 - \sum_{j=1}^{i_\tau} y_j)
$$

and

$$
s_0 \geq C \sum_{\tau=1}^{t} (f_{i_\tau} - f_{i_{\tau-1}})(\lfloor \frac{d_{1i_\tau}}{C} \rfloor + 1 - \sum_{j=1}^{i_\tau} y_j) + C(1 - f_{i_t})(\lfloor \frac{d_{1i_1}}{C} \rfloor - \sum_{j=1}^{i_1} y_j),
$$

for all $T = \{i_1, \ldots, i_t\} \subseteq \{1, \ldots, n\}$, *where* $f_\tau = \frac{d_{1\tau}}{C} - \lfloor \frac{d_{1\tau}}{C} \rfloor$ *for all* τ, *and* $0 - f_{i_0} \leq f_{i_1} \leq \ldots \leq f_{i_t} < 1$.

Example 9.3 *Given an instance of DLSI-CC with* $n = 4, C = 10$, *and* $d = (4, 3, 6, 2)$, *we have* $(d_{1t}) = (4, 7, 13, 15)$ *and* $f = (0.4, 0.7, 0.3, 0.5)$. *For* $T = \{3, 1, 2\}$, *the two inequalities of Theorem 9.4 are*

$$
s_0 \geq 3(2 - y_1 - y_2 - y_3) + 1(1 - y_1) + 3(1 - y_1 - y_2) \quad \text{and}
$$

$$
s_0 \geq 3(2 - y_1 - y_2 - y_3) + 1(1 - y_1) + 3(1 - y_1 - y_2) + 3(1 - y_1 - y_2 - y_3).
$$

To separate these inequalities in O($n \log n$), we take the separation algorithm for the mixing set of Section 8.3 using the substitutions for s, z, and b presented above.

9.4.2 Extended Formulation for *DLSI-CC*

Theorem 8.6 also leads to an extended formulation.

Theorem 9.5 *The linear program*

$$\min \; h_0 s_0 + \sum_{t=1}^{n} q_t y_t$$

$$s_0 = C \sum_{j=1}^{n} f_j \delta_j + C\mu$$

$$\sum_{u=1}^{k} y_u \geq \sum_{j=0}^{n} \lceil \frac{d_{1k}}{C} - f_j \rceil \delta_j - \mu \qquad \text{for } 1 \leq k \leq n$$

$$\sum_{j=0}^{n} \delta_j = 1$$

$$y \in [0,1]^n, \mu \in \mathbb{R}^1_+, \delta \in \mathbb{R}^{n+1}_+,$$

where $f_0 = 0$, solves DLSI-CC.

9.5 Lot-Sizing with Wagner–Whitin Costs and Constant Capacities

Here we consider the Wagner–Whitin problem with constant capacities. The objective function is written as $\sum_{t=0}^{n} h_t s_t + \sum_{t=1}^{n} q_t y_t$ with $h_t = p'_t + h'_t - p'_{t+1} \geq 0$ for all t. This is an important variant in practice because many small-bucket multi-item problems (problems with at most one item produced per period: M_1 in the classification of Chapter 12) satisfy the Wagner–Whitin cost hypothesis, and a single machine or line often has constant capacities, corresponding to a time period such as a shift, or a day. X^{WW-CC} takes the form

$$s_{k-1} + C \sum_{u=k}^{t} y_u \geq d_{kt} \qquad \text{for } 1 \leq k \leq t \leq n \qquad (9.5)$$

$$s \in \mathbb{R}^{n+1}_+, y \in \{0,1\}^n. \qquad (9.6)$$

9.5.1 Optimization for *WW-CC*

The following proposition is a direct consequence of the fact, that with Wagner–Whitin costs, it pays to produce as late as possible, once the periods with a set-up have been determined.

Proposition 9.6 *With Wagner–Whitin costs, if $[a,b]$ is a regeneration interval and $\rho_{ab} = d_{ab} - C\lfloor \frac{d_{ab}}{C} \rfloor > 0$, then period a is the fractional period, and the amount produced in period a is $x_a = \rho_{ab}$.*

We can now write the minimum cost α_{ab} of the $[a, b]$ regeneration interval as the following integer program,

$$\alpha_{ab} =$$

$$\min \ \sum_{i=a}^{b}(q_i + C\sum_{u=i}^{b-1} h_u)y_i + \sum_{i=a}^{b-1} h_i d_{i+1,b} - (\sum_{i=a}^{b-1} h_i)C\lceil\frac{d_{ab}}{C}\rceil \qquad (9.7)$$

$$\sum_{i=t}^{b} y_i \leq \lceil\frac{d_{tb}}{C}\rceil - 1 \qquad \text{for } a+1 \leq t \leq b \qquad (9.8)$$

$$\sum_{i=a}^{b} y_i = \lceil\frac{d_{ab}}{C}\rceil \qquad (9.9)$$

$$y \in \{0,1\}^{b-a+1}, \qquad (9.10)$$

where the stock available s_i for $i = a, \dots, b-1$ is eliminated by substitution of $s_i = d_{i+1,b} - C\sum_{t=i+1}^{b} y_t = d_{i+1,b} - C\sum_{t=a}^{b} y_t + C\sum_{t=a}^{i} y_t$ in the objective function, and the constraints are obtained from $s_i/C \geq 0$ tightened by Chvátal–Gomory rounding.

Observing that for varying $a = b, b-1, \dots, 1$, this is precisely the parametric problem $R(j, \beta_j)$ solved in the previous subsection in $O(b \log b)$, we have

Proposition 9.7 *There is an $O(n^2 \log n)$ algorithm for the optimization problem WW-CC.*

Example 9.4 *Consider an instance of WW-CC with $n = 6, C = 10, h_0 = 3, h = (1,1,1,1,1,1), q = (16,2,2,5,9,7)$, and $d = (4,3,5,6,5,1)$. We have that*

$$\alpha_{36} = \min 32y_3 + 25y_4 + 19y_5 + 7y_6 + 1(12 + 6 + 1) - (1 + 1 + 1)20$$
$$y_6 \leq 0$$
$$y_5 + y_6 \leq 0$$
$$y_4 + y_5 + y_6 \leq 1$$
$$y_3 + y_4 + y_5 + y_6 = 2$$
$$y \in \{0,1\}^4.$$

Applying the parametric algorithm of Subsection 9.3.3 to solve the IP (9.7)–(9.10), we obtain $\alpha_{66} = 7, \alpha_{56} = 10, \alpha_{46} = \infty, \alpha_{36} = 16, \alpha_{26} = 23, \alpha_{16} = \infty$.

In particular, for α_{36} the optimal solution is $y_3 = y_4 = 1$ giving $\alpha_{36} = 32 + 25 + 19 - 60 = 16$. Working with the initial costs, we have $y_3 = y_4 = 1, x_3 = 7, x_4 = 10, s_3 = 2, s_4 = 6, s_5 = 1$ with cost $2 + 5 + 1(2 + 6 + 1) = 16$. Two of the regeneration intervals are shown in Figure 9.2.

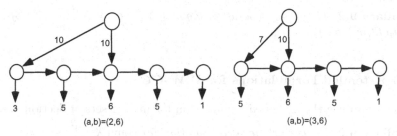

Figure 9.2. Two regeneration intervals for WW-CC.

9.5.2 Valid Inequalities for WW-CC

Letting $X_k^{DLSI-CC} = \{(s,y) \in \mathbb{R}_+^{n+1} \times \{0,1\}^n$ satisfying (9.5) for $k \leq t \leq n\}$ with k fixed, we see that

$$X_1^{DLSI-CC} = X^{DLSI-CC} \quad \text{and}$$

$$X^{WW-CC} = \bigcap_{k=1}^{n} X_k^{DLSI-CC}.$$

It follows that $\operatorname{conv}(X^{WW-CC}) \subseteq \bigcap_{k=1}^{n} \operatorname{conv}(X_k^{DLSI-CC})$, so that any valid inequality for $X_k^{DLSI-CC}$ is valid for X^{WW-CC}. Surprisingly, the two sets are equal.

Theorem 9.8

$$\operatorname{conv}(X^{WW-CC}) = \bigcap_{k=1}^{n} \operatorname{conv}(X_k^{DLSI-CC}).$$

Now we combine this with the results for mixing sets of Section 8.3.

Corollary 9.1 *All nontrivial facet-defining inequalities of* $\operatorname{conv}(X^{WW-CC})$ *are of the form*

$$s_{k-1} \geq C \sum_{\tau=1}^{t}(f_{i_\tau}^k - f_{i_{\tau-1}}^k)(\lfloor \frac{d_{ki_\tau}}{C} \rfloor + 1 - \sum_{j=k}^{i_\tau} y_j) \tag{9.11}$$

or

$$s_{k-1} \geq C \sum_{\tau=1}^{t}(f_{i_\tau}^k - f_{i_{\tau-1}}^k)(\lfloor \frac{d_{ki_\tau}}{C} \rfloor + 1 - \sum_{j=k}^{i_\tau} y_j)$$

$$+ C(1 - f_{i_t}^k)(\lfloor \frac{d_{ki_1}}{C} \rfloor - \sum_{j=k}^{i_1} y_j), \tag{9.12}$$

where $T = \{i_1, \ldots, i_t\} \subseteq \{k, \ldots, n\}$, $f_{i_\tau}^k = \frac{d_{ki_\tau}}{C} - \lfloor \frac{d_{ki_\tau}}{C} \rfloor$ *for all* τ, *and* $0 = f_{i_0}^k \leq f_{i_1}^k \leq \cdots \leq f_{i_t}^k < 1$, *and* $1 \leq k \leq n$.

Corollary 9.2 *There is a separation algorithm for* $\mathrm{conv}(X^{WW-CC})$ *of complexity* $O(n^2 \log n)$.

9.5.3 Extended Formulations for *WW-CC*

One can also use the extended formulation for mixing sets of Section 8.3.

Corollary 9.3 *A tight extended formulation for* $\mathrm{conv}(X^{WW-CC})$ *is*

$$s_{k-1} = C \sum_{t \in [k,n]} f_t^k \delta_t^k + C\mu^k \qquad \text{for all } k$$

$$\sum_{u=k}^{t} y_u \geq \sum_{\tau \in \{0\} \cup [k,n]} \lceil \frac{d_{k\tau}}{C} - f_\tau^k \rceil \delta_\tau^k - \mu^k \quad \text{for all } k, t, \ k \leq t$$

$$\sum_{t \in \{0\} \cup [k,n]} \delta_t^k = 1 \qquad \text{for all } k$$

$$\mu^k \geq 0, \ \delta_t^k \geq 0 \qquad \text{for all } t \in \{0\} \cup [k,n], \text{ and all } k$$

$$y \in [0,1]^n \ ,$$

where $f_0^k = 0$, $[k,n] = \{k, \dots, n\}$ *and* $f_\tau^k = \frac{d_{k\tau}}{C} - \lfloor \frac{d_{k\tau}}{C} \rfloor$. *The additional variables* δ_t^k *indicate that* $s_{k-1} = Cf_t^k \mod C$.

Example 9.5 *Consider an instance of WW-CC with again* $n = 4, C = 10$, *and* $d = (4, 3, 6, 2)$. *For* $k = 2$, *the constraints of the extended formulation of Corollary 9.3 take the form*

$$
\begin{array}{rcll}
s_1 & - & 3\delta_2^2 + 9\delta_3^2 + 1\delta_4^2 & + 10\mu^2 \\
y_2 & \geq & 1\delta_0^2 + \quad 0\delta_2^2 + 0\delta_3^2 + 1\delta_4^2 & - \mu^2 \\
y_2 + y_3 & \geq & 1\delta_0^2 + \quad 1\delta_2^2 + 0\delta_3^2 + 1\delta_4^2 & - \mu^2 \\
y_2 + y_3 + y_4 & \geq & 2\delta_0^2 + \quad 1\delta_2^2 + 1\delta_3^2 + 1\delta_4^2 & - \mu^2 \\
1 & = & \delta_0^2 + \quad \delta_2^2 + \delta_3^2 + \delta_4^2 &
\end{array}
$$

$$\delta^2 \in \mathbb{R}_+^4, \mu^2 \in \mathbb{R}_+^1, y \in [0,1]^4.$$

9.6 Lot-Sizing with Constant Capacities

9.6.1 Optimization: An Algorithm for *LS-CC*

For simplicity of notation, we here consider the objective function in the form:

$$\min \sum_{t=1}^{n} p_t x_t + \sum_{t=1}^{n} q_t y_t.$$

For given k and l, we present a dynamic programming algorithm to find an optimal solution to the subproblem on the regeneration interval $[k, l]$. Let $\rho_{kl} = d_{kl} - \lfloor \frac{d_{kl}}{C} \rfloor C$ with $0 \leq \rho_{kl} < C$. From Observation 9.1, there must be exactly $\lfloor \frac{d_{kl}}{C} \rfloor$ periods within the regeneration interval in which production is at full capacity, and one period in which ρ_{kl} is produced.

Let $G_k(t, \tau, \delta)$ be the value of a minimum cost solution for periods k up to t during which production occurs τ times at full capacity and $\delta \in \{0, 1\}$ times at level ρ_{kl}. If $\tau C + \delta \rho_{kl} \leq d_{kl}$, it is impossible to have $s_t > 0$ for these (τ, δ) values and so we define $G_k(t, \tau, \delta) = \infty$. Also if $\tau + \delta > t - k + 1$ for some (τ, δ) values, it is not possible to produce $\tau + \delta$ times in the interval $[k, t]$ and again $G_k(t, \tau, \delta) = \infty$.

We fix initially $G_k(k - 1, \tau, \delta) = 0$ for all (τ, δ) with $\tau \leq 0$ and $\delta \leq 0$, and $G_k(k - 1, \tau, \delta) = \infty$ otherwise. A forward recursion to compute $G_k(t, \tau, \delta)$ for the case $\rho_{kl} > 0$ is:

$$G_k(t, \tau, 0)$$
$$= \begin{cases} \infty & \text{if } \tau C \leq d_{kt} \text{ or } \tau > t - k + 1 \\ \min \left\{ \begin{array}{l} G_k(t - 1, \tau, 0), \\ G_k(t - 1, \tau - 1, 0) + q_t + p_t C \end{array} \right\} & \text{otherwise} \end{cases} \quad (9.13)$$

for $t = k, \ldots, l, \ \tau = 0, \ldots, \lfloor \frac{d_{kl}}{C} \rfloor$,

$$G_k(t, \tau, 1)$$
$$= \begin{cases} \infty & \text{if } \tau C + \rho_{kl} \leq d_{kt} \text{ or } \tau > t - k \\ \min \left\{ \begin{array}{l} G_k(t - 1, \tau, 1), \\ G_k(t - 1, \tau - 1, 1) + q_t + p_t C, \\ G_k(t - 1, \tau, 0) + q_t + p_t \rho_{kl} \end{array} \right\} & \text{otherwise} \end{cases} \quad (9.14)$$

for $t = k, \ldots, l, \ \tau = 0, \ldots, \lfloor \frac{d_{kl}}{C} \rfloor$.

Starting from $G_k(k, 1, 0) = q_k + p_k C$ if $d_k \leq C$, and $G_k(k, 0, 1) = q_k + p_k \rho_{kl}$ if $d_k \leq \rho_{kl}$, $G_k(k, 0, 0) = 0$ if $d_k = 0$ and $G_k(k, \tau, l) = \infty$ otherwise, evaluating (9.13)–(9.14) for increasing values of t and all values of τ computes $\alpha_{kl} = G_k(l, \lfloor \frac{d_{kl}}{C} \rfloor, 1)$, the value of a minimum cost solution for the regeneration interval $[k, l]$ with $\rho_{kl} > 0$.

When $\rho_{kl} = 0$ and $d_{kl} > 0$, it suffices to use the same recursion for $G_k(t, \tau, 0)$ to calculate $\alpha_{kl} = G_k(l, \lfloor \frac{d_{kl}}{C} \rfloor, 0)$.

Example 9.6 *Consider an instance of LS-CC with an interval $[k, l] = [2, 6]$, $(d_2, \ldots, d_6) = (4, 2, 7, 3, 6)$, $(p_2, \ldots, p_6) = (3, 1, 2, 1, 2)$, $(q_2, \ldots, q_6) = (24, 17, 35, 42, 11)$, and $C = 10$.*

Note that $d_{kl} = 22$ and $\rho_{kl} = 2$. Using the above recurrence, we calculate first $G_2(t, \tau, 0)$. This gives

$$G_2(2,0,0) = G_2(2,2,0) = \infty,$$
$$G_2(2,1,0) = \min[G_2(1,1,0), G_2(1,0,0) + q_2 + Cp_2]$$
$$= \min[\infty, 0 + 24 + 10 \times 3] = 54,$$
$$G_2(3,0,0) = \infty,$$
$$G_2(3,1,0) = \min[G_2(2,1,0), G_2(2,0,0) + q_3 + Cp_3]$$
$$= \min[54, \infty] = 54,$$
$$G_2(3,2,0) = \min[G_2(2,2,0), G_2(2,1,0) + q_3 + Cp_3]$$
$$= \min[\infty, 54 + 17 + 10 \times 1] = 81,$$
$$G_2(4,1,0) = \infty \ and \ G_2(4,2,0) = 81,$$
$$G_2(5,1,0) = \infty, \ and \ G_2(5,2,0) = 81.$$

Now we calculate $G_2(t, \tau, 1)$, and obtain

$$G_2(2,0,1) = G_2(2,1,1) = G_2(2,2,1) = \infty,$$
$$G_2(3,0,1) = G_2(3,2,1) = \infty,$$
$$G_2(3,1,1) = \min[G_2(2,1,1), G_2(2,0,1) + q_3 + Cp_3, G_2(2,1,0) + q_3 + \rho_{26}p_3]$$
$$= \min[\infty, \infty, 54 + 17 + 2 \times 1] = 73,$$
$$G_2(4,0,1) = G_2(4,1,1) = \infty,$$
$$G_2(4,2,1) = \min[G_2(3,2,1), G_2(3,1,1) + q_4 + Cp_4, G_2(3,2,0) + q_4 + \rho_{26}p_4]$$
$$= \min[\infty, 73 + 35 + 10 \times 2, 81 + 35 + 2 \times 2] = 120,$$
$$G_2(5,0,1) = G_2(5,1,1) = \infty \ and \ G_2(5,2,1) = 120,$$
$$G_2(6,2,1) = \min[G_2(5,2,1), G_2(5,1,1) + q_6 + Cp_6, G_2(5,2,0) + q_6 + \rho_{26}p_6]$$
$$= \min[120, \infty, 81 + 11 + 2 \times 2] = 96.$$

Working backwards, we see that an optimal solution for the interval $[2,6]$ is $x_2 = x_3 = 10, x_6 = 2$ with a cost of $q_2 + q_3 + q_6 + 10p_2 + 10p_3 + 2p_6 = 96$.

As there always exists an optimal solution consisting of a sequence of regeneration intervals, we can restate Observation 9.2.

Theorem 9.9 *Let $D = (V, A)$ be a digraph with node set $V = \{1, \ldots, n+1\}$ and arc set $A = \{(k, l+1), 1 \le k \le l \le n\}$. The shortest path problem from node 1 to node $n+1$ in the digraph D with costs α_{kl} on arcs $(k, l+1)$ with $d_{kl} > 0$, and cost 0 on arcs $(k, l+1)$ with $d_{kl} = 0$, solves LS-CC.*

Figure 9.3 illustrates the digraph D for $n = 3$.

Example 9.7 *Consider an instance of LS-CC with $n = 5, C = 10, d = (4, 2, 7, 3, 6), p = (3, 1, 2, 1, 2)$, and $q = (24, 17, 35, 42, 11)$. We have that*

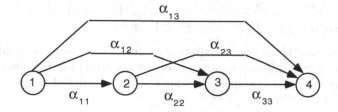

Figure 9.3. Shortest path formulation of $LS\text{-}CC$ ($n = 3$).

$$\alpha_{11} = 36, \quad \alpha_{22} = 19, \quad \alpha_{33} = 49, \quad \alpha_{44} = 45, \quad \alpha_{55} = 23,$$
$$\alpha_{12} = 42, \quad \alpha_{23} = 26, \quad \alpha_{34} = 55, \quad \alpha_{45} = 51,$$
$$\alpha_{13} = 74, \quad \alpha_{24} = 74, \quad \alpha_{35} = 78,$$
$$\alpha_{14} = 69, \quad \alpha_{25} = 77,$$
$$\alpha_{15} = 96.$$

Solving the resulting shortest path problem, an optimal solution of cost 92 consists of the regeneration intervals $[1,4],[5]$, and the corresponding solution is $x_1 = 6, x_2 = 10$, and $x_5 = 6$ with $y_1 = y_2 = y_5 = 1$.

The dynamic program for each interval is $O(n^2)$, and as there are $O(n^2)$ intervals this gives an algorithm for $LS\text{-}CC$ whose running time is $O(n^4)$. However, it is possible to reduce the running time to $O(n^3)$.

9.6.2 Valid Inequalities for $LS\text{-}CC$

As in the uncapacitated case, the inequalities (9.11) and (9.12) for the Wagner–Whitin case generalize to give valid inequalities for $LS\text{-}CC$. Given k, $T = \{i_1, \ldots, i_t\} \subseteq \{k, \ldots, n\}$ with $i_1 < \cdots < i_t$ and $V \subseteq \{k, \ldots, n\}$, we can apply the mixing procedure to the constraint system

$$s_{k-1} + \sum_{j \in V \cap [k, i_t]} x_j + C \sum_{j \in [k, i_\tau] \setminus V} y_j \geq d_{k i_\tau} \qquad \text{for } 1 \leq \tau \leq l$$

$$s_{k-1} + \sum_{j \in V \cap [k, i_t]} x_j \in R_+^1, \quad \sum_{j \in [k, i_\tau] \setminus V} y_j \in \mathbb{Z}_+^1 \qquad \text{for } 1 \leq \tau \leq t.$$

Proposition 9.10 *The inequalities*

$$s_{k-1} + \sum_{j \in V \cap [k, i_t]} x_j \geq C \sum_{\tau=1}^{t} (f_{i_\tau}^k - f_{i_{\tau-1}}^k)(\lfloor \frac{d_{k i_\tau}}{C} \rfloor + 1 - \sum_{j \in [k, i_\tau] \setminus V} y_j)$$

and

$$s_{k-1} + \sum_{j \in V \cap [k, i_t]} x_j \geq C \sum_{\tau=1}^{t} (f_{i_\tau}^k - f_{i_{\tau-1}}^k)(\lfloor \frac{d_{k i_\tau}}{C} \rfloor + 1 - \sum_{j \in [k, i_\tau] \setminus V} y_j)$$

$$+ C(1 - f_{i_t}^k)(\lfloor \frac{d_{k i_1}}{C} \rfloor - \sum_{j \in [k, i_1] \setminus V} y_j),$$

are valid for X^{LS-CC}, where $T = \{i_1, \ldots, i_t\} \subseteq \{k, \ldots, n\}$, $f_{i_\tau}^k = \frac{d_{ki_\tau}}{C} - \lfloor \frac{d_{ki_\tau}}{C} \rfloor$ for all τ, and $0 = f_0^k \leq f_{i_1}^k \leq \ldots \leq f_{i_t}^k < 1$, and $1 \leq k \leq n$.

It is clear that there are an exponential number of these inequalities, and till now no combinatorial polynomial separation algorithm is known for this family. There are also examples showing that this family of inequalities does not suffice to describe conv(X^{LS-CC}).

Example 9.8 *Consider an LS-CC instance with $n = 5$, $d_2 = 7$, $d_3 = 4$, $d_4 = 2$, $d_5 = 3$, and capacity $C = 10$.*

Taking $k = 2$, $T = \{2, 4, 5\}$, and $V = \{4\}$, we see that $\{i_1, i_2, i_3\} = \{4, 5, 2\}$ with $f_{i_1} = 0.3$, $f_{i_2} = 0.6$, $f_{i_3} = 0.7$. Using Proposition 9.10, the first inequality takes the form

$$s_1 + x_4 \geq 3(2 - y_2 - y_3) + 3(2 - y_2 - y_3 - y_5) + 1(1 - y_2).$$

9.6.3 Extended Formulation for *LS-CC*

First we consider writing down an extended formulation to calculate the optimal cost of a $[k, l]$ regeneration interval.

Let $z_t = 1$ if $x_t = C$ and $z_t = 0$ otherwise, and
$\epsilon_t = 1$ if $x_t = \rho_{kl} > 0$ and $\epsilon_t = 0$ otherwise.

The problem can now be formulated as the integer program

$$\alpha_{kl} = \min \; \sum_{t=k}^{l} q_t(z_t + \epsilon_t) \; + \; C \sum_t p_t z_t \; + \; \rho_{kl} \sum_t p_t \epsilon_t$$

$$\sum_{u=k}^{t} z_u + \sum_{u=k}^{t} \epsilon_u \geq \lceil \frac{d_{kt}}{C} \rceil \qquad \text{for } k \leq t < l \text{ with } \rho_{kt} \leq \rho_{kl}$$

$$\sum_{u=k}^{t} z_u \geq \lceil \frac{d_{kt}}{C} \rceil \qquad \text{for } k \leq t < l \text{ with } \rho_{kt} > \rho_{kl}$$

$$\sum_{u=k}^{l} z_u + \sum_{u=k}^{l} \epsilon_u = \lceil \frac{d_{kl}}{C} \rceil$$

$$\sum_{u=k}^{l} z_u = \lceil \frac{d_{kl} - \rho_{kl}}{C} \rceil$$

$$z_t + \epsilon_t \leq 1 \qquad \text{for } k \leq t \leq l$$

$$z, \epsilon \in \mathbb{Z}_+^{l-k+1}.$$

However, the corresponding matrix is totally unimodular, and therefore:

Proposition 9.11 *The linear programming relaxation has an optimal solution with z, ϵ integer, and optimal value α_{kl}.*

Now, it is possible to use such a linear program to model the cost of every arc or regeneration interval of the shortest path formulation (7.22)–(7.27). This defines the following large linear program with $O(n^3)$ constraints and variables that solves $LS\text{-}CC$.

$$\min \sum_{k=1}^{n}\sum_{l=k}^{n}\sum_{t=k}^{l} [q_t(z_t^{kl} + \epsilon_t^{kl}) \quad + p_t(Cz_t^{kl} + \rho_{kl}\epsilon_t^{kl})] \tag{9.15}$$

$$-\sum_{l=1}^{n} \phi_{1l} = -1 \tag{9.16}$$

$$\sum_{k=1}^{t-1} \phi_{k,t-1} - \sum_{l=t}^{n} \phi_{tl} = 0 \qquad \text{for } 2 \leq t \leq n \tag{9.17}$$

$$\sum_{k=1}^{n} \phi_{kn} = 1 \tag{9.18}$$

$$\sum_{u=k}^{t} z_u^{kl} + \sum_{u=k}^{t} \epsilon_u^{kl} \geq \lceil\frac{d_{kt}}{C}\rceil\phi_{kl} \qquad \text{for } 1 \leq k \leq t < l \leq n \text{ with } \rho_{kt} \leq \rho_{kl} \tag{9.19}$$

$$\sum_{u=k}^{t} z_u^{kl} \geq \lceil\frac{d_{kt}}{C}\rceil\phi_{kl} \qquad \text{for } 1 \leq k \leq t < l \leq n \text{ with } \rho_{kt} > \rho_{kl} \tag{9.20}$$

$$\sum_{u=k}^{l} z_u^{kl} + \sum_{u=k}^{l} \epsilon_u^{kl} = \lceil\frac{d_{kl}}{C}\rceil\phi_{kl} \qquad \text{for } 1 \leq k \leq l \leq n \tag{9.21}$$

$$\sum_{u=k}^{l} z_u^{kl} = \lceil\frac{d_{kl} - \rho_{kl}}{C}\rceil\phi_{kl} \qquad \text{for } 1 \leq k \leq l \leq n \tag{9.22}$$

$$z_t^{kl} + \epsilon_t^{kl} \leq \phi_{kl} \qquad \text{for } 1 \leq k \leq t \leq l \leq n \tag{9.23}$$

$$\phi \in \mathbb{R}_+^{n(n+1)/2} \tag{9.24}$$

$$z^{kl}, \epsilon^{kl} \in \mathbb{R}_+^{l-k+1} \qquad \text{for } 1 \leq k \leq l \leq n. \tag{9.25}$$

9.6.4 Résumé of Results

In Table 9.6.4 we summarize the results that are known for $PROB\text{-}CC$, where LP indicates that the separation problem can be solved as a linear program using the polynomial-size extended formulation as discussed in Section 6.1, and $\star$ indicates that an explicit description of a family of valid inequalities is known but it only gives a partial description of the convex hull of solutions.

Table 9.1. Models PROB-CC

	LS	WW	DLSI	DLS
Formulation				
CC	$O(n^3) \times O(n^3)$	$O(n^2) \times O(n^2)$	$O(n) \times O(n)$	$O(n) \times O(n)$
Separation				
CC	LP and $\star$	$O(n^2 \log n)$	$O(n \log n)$	$O(n)$
Optimization				
CC	$O(n^3)$	$O(n^2 \log n)$	$O(n^2 \log n)$	$O(n \log n)$

9.7 Lot-Sizing with Varying Capacities

Here we describe several classes of valid inequalities for the problems $WW\text{-}C$, $LS\text{-}C$ and $DLSI\text{-}C$.

9.7.1 Valid Inequalities for $WW\text{-}C$

Consider the set

$$X^{WW-C} = \{(s,y) \in \mathbb{R}_+^{n+1} \times \{0,1\}^n : s_{t-1} + \sum_{u=t}^{l} C_u y_u \geq d_{tl} \text{ for } 1 \leq t \leq l \leq n\}.$$

For $1 \leq t \leq l \leq n$, we define the following values:

$$\delta_{tl} = \min\{s_{t-1} + C_t \sum_{u=t}^{l} y_u : (s,y) \in X^{WW-C}\}. \tag{9.26}$$

Observation 9.5 *The inequality*

$$s_{t-1} + C_t \sum_{u=t}^{l} y_u \geq \delta_{tl}$$

is valid for X^{WW-C} *for* $1 \leq t \leq l \leq n$.

Calculation of δ

First note that if $\alpha = s_{t-1}$ is known, then the calculation of δ_{tl} reduces to the problem

$$\min \sum_{u=t}^{l} y_u$$

$$\sum_{u=t}^{j} C_u y_u \geq d_{tj} - \alpha \qquad \text{for } t \leq j \leq l$$

$$y \in \{0,1\}^{l-t+1}.$$

Proposition 9.12 *When the C_t are nondecreasing (i.e., $C_t \leq C_{t+1}$ for all t), the $\{\delta_{tl}\}_{1\leq t\leq l\leq n}$ can be calculated in polynomial time.*

We now indicate how the δ_{tl} can be calculated in the case where $C_t \leq C_{t+1}$ for all t.

After preprocessing so that $d_t \leq C_t$ for all t, the following simple greedy algorithm gives the optimal lexico-min 0–1 vector $y^{\alpha,t}$ for fixed t and α.

Calculation of $y^{\alpha,t}$

For $k = t, \ldots, n$, let $\phi_k(\alpha, t) = \sum_{j=t}^{k-1} C_j y_j^{\alpha,t} - (d_{tk} - \alpha)^+$.

If $\phi_k(\alpha, t) < 0$, set $y_k^{\alpha,t} = 1$, and otherwise set $y_k^{\alpha,t} = 0$.

Now we need to choose $s_{t-1} = \alpha$. One can show that there always exists an optimal solution to (9.26) with $\alpha_t < C_t$, so we have

$$\delta_{tl} = \min_{0 \leq \alpha < C_t} \{\alpha + C_t \sum_{u=t}^{l} y_u^{\alpha,t}\}.$$

The following procedure selects the values of α that one needs to consider. For fixed t, we start by computing $y_k^{\alpha,t}$ for $\alpha = 0$ and $k = t, \ldots, n$.

Iterating over α

Initialize with $\alpha = 0$.

While $\alpha < C_t$, compute $y_k^{\alpha,t}$ for $k = t, \ldots, n$.

Let $\gamma = \max_{k:\phi_k(\alpha,t)<0} \phi_k(\alpha, t)$. Note that $\gamma < 0$.

Set $\alpha \leftarrow \alpha - \gamma$ and iterate.

It can then be shown that, for each t, at most $O(n^2)$ values of s_{t-1} need be considered.

Valid Inequalities

The relaxation

$$s_{t-1} + C_t \sum_{u=t}^{l} y_u \geq \delta_{tl} \qquad \text{for } 1 \leq t \leq l \leq n$$

$$s \in \mathbb{R}_+^{n+1}, y \in \{0,1\}^n$$

is the intersection of n mixing sets, so we obtain as an improved relaxation the convex hull of this set:

$$\bigcap_{t=1}^{n} \text{conv}\left(X^{MIX}(s_{t-1}/C_t, (y_t, \ldots, y_n), (\delta_{tt}, \ldots, \delta_{tn})/C_t)\right).$$

Example 9.9 *Consider an instance with* $n = 5$, $C = (5, 7, 8, 10, 12)$, *and* $d = (3, 3, 5, 2, 4)$. *We just treat the case* $t = 1$.
For $\alpha = 0$, *we have the constraints:*

$$
\begin{aligned}
5y_1 &\geq 3 - 0 \\
5y_1 + 7y_2 &\geq 6 - 0 \\
5y_1 + 7y_2 + 8y_3 &\geq 11 - 0 \\
5y_1 + 7y_2 + 8y_3 + 10y_4 &\geq 13 - 0 \\
5y_1 + 7y_2 + 8y_3 + 10y_4 + 12y_5 &\geq 17 - 0.
\end{aligned}
$$

The algorithm for the determination of $y^{\alpha,t}$ *gives* $\phi(0,1) = (-3, -1, 1, -1, 5)$ *and* $y^{0,1} = (1, 1, 0, 1, 0)$. *Now* $\gamma = -1$, *so we pass to* $\alpha = 1$.
Altogether, iterating over α, *we obtain*

$$
\begin{aligned}
\alpha = 0, \quad & y^{0,1} = (1, 1, 0, 1, 0), \quad \gamma = 1 \\
\alpha = 1, \quad & y^{1,1} = (1, 0, 1, 0, 1), \quad \gamma = 2 \\
\alpha = 3, \quad & y^{3,1} = (0, 1, 1, 0, 0), \quad \gamma = 1, \\
\alpha = 4, \quad & y^{4,1} = (0, 1, 0, 1, 0), \quad \gamma = 2
\end{aligned}
$$

from which we can calculate
$\delta_{11} = 3, \delta_{12} = 6, \delta_{13} = 9, \delta_{14} = 11, \delta_{15} = 13$. *The corresponding valid inequalities are:*

$$
\begin{aligned}
s_0 + 5y_1 &\geq 3 \\
s_0 + 5y_1 + 5y_2 &\geq 6 \\
s_0 + 5y_1 + 5y_2 + 5y_3 &\geq 9 \\
s_0 + 5y_1 + 5y_2 + 5y_3 + 5y_4 &\geq 11 \\
s_0 + 5y_1 + 5y_2 + 5y_3 + 5y_4 + 5y_5 &\geq 13.
\end{aligned}
$$

Finally, observe that with arbitrary capacities, one can, for each t, construct a set $\{\bar{C}_l\}_{l=t}^n$ of nondecreasing capacities by setting $\bar{C}_t = C_t$ and $\bar{C}_l = \max[\bar{C}_{l-1}, C_l]$ for $l = t+1, \ldots, n$. With these modified capacities, the corresponding δ values can be calculated as above, and the resulting reformulation is then valid.

9.7.2 Simple Valid Inequalities for *LS-C*

We first show two relaxations of the set X^{LS-C} that can be used to generate valid inequalities. For both we start by aggregating the flow balance constraints for periods $k, \ldots, l$. This gives

$$
s_{k-1} + \sum_{j=k}^l x_j = d_{kl} + s_l \tag{9.27}
$$

$$
x_t \leq C_t y_t \qquad\qquad \text{for } k \leq t \leq l \tag{9.28}
$$

$$
s_{k-1}, s_l \geq 0, y_t \in \{0, 1\} \qquad \text{for } k \leq t \leq l. \tag{9.29}
$$

Relaxation 1: A Continuous 0–1 Knapsack Set

Using the nonnegativity of s_l, and replacing each x_j by its variable upper bound constraint, we obtain

$$X^1 = \{s_{k-1} \geq 0, (y_k, \ldots, y_l) \in \{0,1\}^{l-k+1} : s_{k-1} + \sum_{j=k}^{l} C_j y_j \geq d_{kl}\}.$$

Valid inequalities for such sets have been presented in Section 8.8. In fact we can work with a slightly smaller set

$$\tilde{X}^1 = \{s_{k-1} \geq 0, (y_k, \ldots, y_l) \in \{0,1\}^{l-k+1} : s_{k-1} + \sum_{j=k}^{l} \min[C_j, d_{jl}] y_j \geq d_{kl}\}.$$

Relaxation 2: A Single-Node Flow Set

Dropping the variables s_{k-1}, using the (l,S) inequalities $x_j \leq d_{jl} y_j + s_l$, and fixing $s_l = 0$ temporarily, we obtain the single-node flow set X^2 :

$$\sum_{j=k}^{l} x_j \leq d_{kl} \tag{9.30}$$

$$x_j \leq \min[C_j, d_{jl}] y_j \qquad \text{for } k \leq j \leq l \tag{9.31}$$

$$(y_k, \ldots, y_l) \in \{0,1\}^{l-k+1}. \tag{9.32}$$

Flow cover inequalities for such sets were derived in Section 8.9.

Proposition 9.13 *If $\sum_{j\in S} x_j \leq \beta_0 + \sum_{j\in S} \beta_j y_j$ is a valid inequality for X^2 with $S \subseteq \{k, \ldots, l\}$, then*

$$\sum_{j\in S} x_j \leq \beta_0 + \sum_{j\in S} \beta_j y_j + s_l$$

is a valid inequality for X^{LS-C}.

Example 9.10 *Consider an instance of LS-C with $n = 6$, $d = (4, 2, 8, 5, 7, 1)$, $C = (7, 4, 14, 5, 9, 6)$, and the point $x^* = (6, 0, 14, 0, 7, 0)$, $s^* = (2, 0, 6, 1, 1, 0)$, and $y^* = (1, 0, 1, 0, 0.875, 0)$.*
Taking $k = 3$ and $l = 6$, we obtain the continuous 0–1 knapsack set relaxation

$$s_2 + 14y_3 + 5y_4 + 8y_5 + 1y_6 \geq 21$$
$$s_2 \in \mathbb{R}^1_+, y \in \{0,1\}^4.$$

Dividing by 14 and taking the MIR inequality (8.32), we obtain the inequality

$$s_2 + 7y_3 + 5y_4 + 7y_5 + 1y_6 \geq 14$$

that is violated by 0.875.

Again with $k = 3$ and $l = 6$, the single-node flow set X^2 obtained is

$$x_3 + x_4 + x_5 + x_6 \leq 21$$
$$x_3 \leq 14y_3, x_4 \leq 5y_4, x_5 \leq 8y_5, x_6 \leq 1y_6, x \in \mathbb{R}_+^4, y \in \{0, 1\}^4.$$

Using the cover $C_1 = \{3, 5\}$ with $\lambda = 1$ in Proposition 8.19 combined with Proposition 9.13 gives the flow cover inequality

$$x_3 + x_5 \leq 21 - 13(1 - y_3) - 7(1 - y_5) + s_6$$

which is also violated by 0.875.

9.7.3 Submodular Inequalities for *LS-C*

Consider now an alternative formulation for *LS-C*, namely the set

$$\tilde{X}^{LS-C} = \{(x, s_n, y) \in \mathbb{R}_+^n \times \mathbb{R}_+^1 \times \{0, 1\}^n :$$

$$\sum_{u=t}^{n} x_u \leq d_{tn} + s_n, x_t \leq C_t y_t \quad \text{for } 1 \leq t \leq n\},$$

obtained by eliminating the entering stock variables in the equations

$$s_{t-1} + \sum_{u=t}^{n} x_u = d_{tn} + s_n.$$

Let v be the set function defined on any subset of periods

$$v(T) = \max\{\sum_{j \in T} x_j : (x, s_n, y) \in \tilde{X}^{LS-C}$$

$$\begin{aligned} y_t &= 1 && \text{for } j \in T \\ y_t &= 0 && \text{for } j \in N \setminus T \\ s_n &= 0 && \}, \end{aligned}$$

for all $T \subseteq N = \{1, \ldots, n\}$. The value $v(T)$ is easy to calculate as it is just a maximum flow (or minimum cut) in a very special network; see Figure 9.4.

Observation 9.6

$$v(T) = \min_{t \in T \cup \{n+1\}} [\sum_{u \in T : u < t} C_u + d_{tn}],$$

where $d_{n+1, n} = 0$.

What is more, v has special structure.

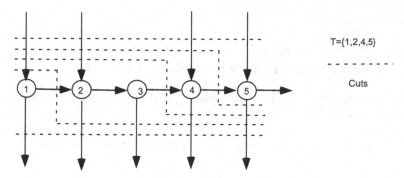

Figure 9.4. Min cuts in the calculation of $v(T)$.

Definition 9.2 *A set function* $f : P(N) \to \mathbb{R}^1$ *is* supermodular *if*

$$f(A) + f(B) \leq f(A \cup B) + f(A \cap B) \quad \text{for all } A, B \subseteq N.$$

If the set function $-f$ *is supermodular, the set function* f *is* submodular.

Observation 9.7 *Let* $\rho_j(S) = f(S \cup \{j\}) - f(S)$ *for* $\emptyset \subseteq S \subseteq N \setminus \{j\}$.
*f is supermodular (submodular) if and only if $\rho_j(S) \leq \rho_j(T)$ ($\rho_j(S) \geq \rho_j(T)$)
for all $S \subset T \subseteq N \setminus \{j\}$.*

Observation 9.8 *The set function* $v : \mathcal{P}(N) \to \mathbb{R}^1$ *is submodular and non-decreasing.*

Submodularity provides us with a valid inequality for $\tilde{X}^{LS-C}$ based on the following result.

Proposition 9.14 *The submodular inequality*

$$\sum_{j \in T} x_j + \sum_{j \in T} [v(T) - v(T \setminus \{j\})](1 - y_j) \leq v(T) + s_n$$

is valid for $\tilde{X}^{LS-C}$.

Example 9.11 *Consider an instance of LS-C with* $n = 6, d = (1, 6, 2, 4, 3, 2)$,
and $C = (12, 9, 5, 7, 6, 15)$.
    The constraints of the new formulation are $x \in \mathbb{R}_+^6, y \in \{0, 1\}^6$, $s_6 \in \mathbb{R}_+^1$,
$x_j \leq C_j y_j$ *for* $j \in \{1, \ldots, 6\}$, *and*

$$
\begin{aligned}
x_6 &\leq 2 + s_6 \\
x_5 + x_6 &\leq 5 + s_6 \\
x_4 + x_5 + x_6 &\leq 9 + s_6 \\
x_3 + x_4 + x_5 + x_6 &\leq 11 + s_6 \\
x_2 + x_3 + x_4 + x_5 + x_6 &\leq 17 + s_6 \\
x_1 + x_2 + x_3 + x_4 + x_5 + x_6 &\leq 18 + s_6.
\end{aligned}
$$

Using Observation 9.6,

$$v(\{2,4,5\}) = \min[d_{26}, C_2 + d_{46}, C_2 + C_4 + d_{56}, C_2 + C_4 + C_5]$$
$$= \min[17, 9+9, 9+7+5, 9+7+6] = 17.$$

As $v(\{2,4\}) = 16, v(\{2,5\}) = 14$ and $v(\{4,5\}) = 9$, we obtain the valid submodular inequality

$$x_2 + x_4 + x_5 + 8(1 - y_2) + 3(1 - y_4) + 1(1 - y_5) \leq 17 + s_6.$$

Lifting the Submodular Inequalities

To strengthen the submodular inequalities by a procedure known as *lifting*, we only consider inequalities for sets T for which $v(T) = \max_{j \in T} d_{jn}$. We order the values $v(T) - v(T \setminus \{j\})$ for $j \in T$ in nonincreasing order $\alpha_1 \geq \cdots \geq \alpha_{|T|}$. Similarly we order the values $C_j - (v(T) - v(T \setminus \{j\}))$ in nondecreasing order $\beta_1 \leq \cdots \leq \beta_{|T|}$.

Now consider the function ψ on $\mathbb{R}^1_+$ defined as follows.

$$\psi(u) = \sum_{i=1}^{r} \beta_i \qquad \text{if } \sum_{i=1}^{r}(\alpha_i + \beta_i) \leq u < \sum_{i=1}^{r}(\alpha_i + \beta_i) + \alpha_{r+1}$$

$$\psi(u) = \sum_{i=1}^{r} \beta_i + (u - \sum_{i=1}^{r}(\alpha_i + \beta_i) + \alpha_{r+1})$$
$$\text{if } \sum_{i=1}^{r}(\alpha_i + \beta_i) + \alpha_{r+1} \leq u < \sum_{i=1}^{r+1}(\alpha_i + \beta_i).$$

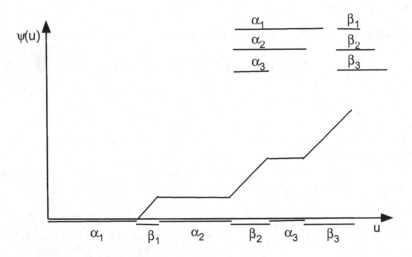

Figure 9.5. The function ψ used for lifting.

Now, for k in $N \setminus T$, consider the function $F_k(u) = \psi(u - \delta_k)$, where $\delta_k = [d_{kn} - v(T)]^+$. Based on the properties of F_k, the following proposition holds.

Proposition 9.15 *If $v(T) = \max_{j \in T} d_{jn}$, and (π_k, μ_k) satisfy $\pi_k + \mu_k u \leq F_k(u)$ for $0 \leq u \leq \min[C_k, d_{kn}]$ for all $k \in N \setminus T$, then the lifted submodular inequality*

$$\sum_{j \in T} x_j + \sum_{j \in T} [v(T) - v(T \setminus \{j\})](1 - y_j) + \sum_{k \in N \setminus T} (\pi_k x_k + \mu_k y_k) \leq v(T) + s_n$$

is valid for $\tilde{X}^{LS-C}$.

Example 9.11 continued.
Construction of the functions ψ and F_k. With $T = \{2, 4, 5\}$, $v(T) - v(T \setminus \{j\})$ takes the values $8, 3, 1$ for $j = 2, 4, 5$, so $\alpha = (8, 3, 1)$. Also as $C_j - \rho_j(T \setminus \{j\})$ takes the values $(9 - 8, 7 - 3, 6 - 1)$ for $j = 2, 4, 5$, $\beta = (1, 4, 5)$.

So $\psi(u) = 0$ for $0 \leq u \leq 8$, $\psi(u) = u - 8$ for $8 \leq u \leq 9$, $\psi(u) = 1$ for $9 \leq u \leq 12$, and so on.

Now we calculate the lifting functions. For $k = 1$, $\delta_1 = d_{1n} - v(T) = 1$, so $F_1(u) = \psi(u - 1)$. It is easily checked that the function $-3 + \frac{1}{3}u \leq F_1(u)$ on $0 \leq u \leq 12$, and in addition equality holds for $u = 9$ and $u = 12$; see Figure 9.6.

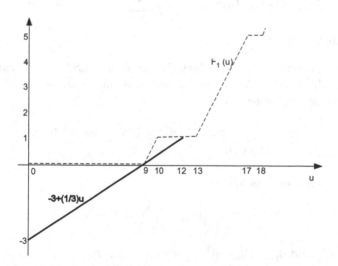

Figure 9.6. Lifting function F_1 and its support on $[0, 12]$.

For $k = 3$, $\delta_3 = 0$, and so $F_3 = \psi$. As $C_3 = 5$, the only support is $(\pi_3, \mu_3) = (0, 0)$.

For $k = 6$, $\delta_6 = 0$, and so $F_6 = \psi$. As $\min[C_6, d_6] = 2$, the only support is again $(\pi_6, \mu_6) = (0, 0)$.

Thus from Proposition 9.15 we obtain the lifted submodular inequality

$$\frac{1}{3}x_1 - 3y_1 + x_2 + x_4 + x_5 + 8(1 - y_2) + 3(1 - y_4) + 1(1 - y_5) \leq 17.$$

9.7.4 Lifted (l, S) Inequalities for $DLSI$-C

Valid Inequalities for $DLSI$-C

Here we develop a class of valid inequalities for $DLSI$-C. We then briefly indicate how they also lead to inequalities for WW-C and LS-C.

We start from the following formulation

$$s_0 + \sum_{j=1}^{t} \min[C_j, d_{jt}]y_j \geq d_{1t} \qquad \text{for all } t$$

$$s_0 \geq 0, y \in \{0, 1\}^n.$$

Now choose a subset $T \subseteq N = \{1, \ldots, n\}$ of periods, and set $y_j = 1$ for $j \in T$. The resulting set $\tilde{X}^T$ is

$$s_0 + \sum_{j \in [1,t] \setminus T} \min[C_j, d_{jt}]y_j \geq d_{1t} - \sum_{j \in [1,t] \cap T} C_j \qquad \text{for all } t \qquad (9.33)$$

$$s_0 \geq 0, y \in \{0, 1\}^{|N \setminus T|}. \qquad (9.34)$$

This can be interpreted as a new instance of $DLSI$-C involving set-up decisions in the periods $N \setminus T$, and a new demand vector $d^T \in \mathbb{R}^n_+$, so that (9.33) can be strengthened to

$$s_0 + \sum_{j \in [1,t] \setminus T} \min[C_j, d_{jt}^T]y_j \geq d_{1t}^T \quad \text{for all } t,$$

where d^T is calculated as follows.

$$d_{1t}^T = \max_{j:j \leq t}(d_{1j} - \sum_{u \in T : u \leq j} C_j)^+,$$

with $d_1^T = d_{11}^T$ and $d_t^T = d_{1t}^T - d_{1,t-1}^T$ for $t = 2, \ldots, n$.

We require that $d_{jn}^T < C_j$ for all $j \in N \setminus T$, otherwise we choose some other set T. Now we take the (l, S) inequality

$$s_0 + \sum_{j \in [1,n] \setminus T} d_{jn}^T y_j \geq d_{1n}^T$$

that is valid for $\tilde{X}^T$.

Next we reintroduce the variables y_j for $j \in T$ that have been set to 1. Specifically to lift back the variable y_k, we set it to zero, while keeping $y_j = 1$ for $j \in T \setminus \{k\}$. To see the effect of bringing in y_k, we calculate

$$\pi_k = \min s_0 + \sum_{j \in [1,n] \setminus T} d_{jn}^T y_j - d_{1n}^T \tag{9.35}$$

$$s_0 + \sum_{j \in [1,t] \setminus T} \min[C_j, d_{jn}^{T \setminus \{k\}}] y_j \geq d_{1t}^{T \setminus \{k\}} \qquad \text{for all } t \tag{9.36}$$

$$s_0 \geq 0, y \in \{0,1\}^{|N \setminus T|}. \tag{9.37}$$

Though in general the lifting coefficients are order-dependent, here it can be shown that the $\{\pi_k\}$ give a valid inequality directly.

Theorem 9.16 *The inequality*

$$s_0 + \sum_{j \in N \setminus T} d_{jn}^T y_j \geq d_{1n}^T + \sum_{k \in T} \pi_k (1 - y_k)$$

is valid and facet-defining for $\mathrm{conv}(X^{DLSI-C})$.

Example 9.12 *Consider an instance of DLSI-C with $n = 5$, $d = (4, 3, 4, 3, 6)$, and $C = (10, 5, 4, 7, 12)$. The initial formulation is*

$$
\begin{array}{llllll}
s_0 & +10y_1 & & & & \geq 4 \\
s_0 & +10y_1 & +5y_2 & & & \geq 7 \\
s_0 & +10y_1 & +5y_2 & +4y_3 & & \geq 11 \\
s_0 & +10y_1 & +5y_2 & +4y_3 & +7y_4 & \geq 14 \\
s_0 & +10y_1 & +5y_2 & +4y_3 & +7y_4 & +6y_5 \geq 20 \\
& s_0 & \geq 0, & y \in \{0,1\}^5. & &
\end{array}
$$

Setting $T = \{1, 4\}$, the new demand vector is $d^T = (0, 0, 1, 0, 2)$, and the (l, S) inequality

$$s_0 + 3y_2 + 3y_3 + 2y_5 \geq 3$$

is valid for $\tilde{X}^T$.

To calculate the lifting coefficient π_1^T, we calculate $d^{T \setminus \{1\}} = d^{\{4\}} = (4, 3, 4, 0, 2)$ and solve

$$
\begin{array}{llllll}
\min & s_0 & +3y_2 & +3y_3 & +2y_5 & -3 \\
& s_0 & & & & \geq 4 \\
& s_0 & +5y_2 & & & \geq 7 \\
& s_0 & +5y_2 & +4y_3 & & \geq 11 \\
& s_0 & +5y_2 & +4y_3 & & \geq 11 \\
& s_0 & +5y_2 & +4y_3 & +6y_5 & \geq 13 \\
& s_0 \geq 0, & y \in & \{0,1\}^3. & &
\end{array}
$$

with optimal solution $s_0 = 4, y_2 = y_3 = 1$, and $\pi_1^T = 7$.

By a similar calculation we get that $\pi_4^T = 2$ and the resulting facet-defining inequality for $\text{conv}(X^{DLSI-C})$ is

$$s_0 + 7y_1 + 3y_2 + 3y_3 + 2y_4 + 2y_5 \geq 3 + 7 + 2 = 12.$$

Note that as $X^{WW-C} = \bigcap_{k=1}^{n} X_k^{DLSI-C}$, the same approach also provides valid inequalities for $WW\text{-}C$.

A related approach of fixing set-up variables to one, setting the corresponding production variables to their capacity, and then lifting an (l, S) inequality can be applied to generate facet-defining inequalities for $LS\text{-}C$. However the lifting is more complicated as it now involves both continuous x_j and 0–1 variables y_j.

Exercises

Exercise 9.1 Consider an instance of $DLS\text{-}CC$ with $C = 7$ and $d = (4, 2, 5, 1, 4)$. Observe that the constraints $\sum_{u=1}^{t} y_u \geq \lceil \frac{d_{1t}}{C} \rceil$ imply certain lower bounds on the stocks $s_t \geq \underline{s}_t$. Use these lower bounds to change variables such that the basic unit of demand is in multiples of C units, and all the new variables are in terms of these units.

Exercise 9.2 Show how the greedy algorithm for $DLS\text{-}CC$ can be implemented to run in $O(n \log n)$.

Exercise 9.3 Consider an instance of $DLS\text{-}CC$ with $n = 6$, $C = 9$, $d = (5, 3, 1, 4, 3, 4)$, and $q = (42, 68, 31, 93, 72, 53)$.
Solve by the greedy algorithm and by linear programming.

Exercise 9.4 Consider an instance of $WW\text{-}CC$ with $n = 5, C = 10$, costs $p' = (7, 4, 2, 3, 4), h' = (1, 3, 2, 1, 3), q' = (25, 45, 26, 30, 20)$, and demands $d = (5, 3, 7, 4, 9)$ for which initial and final stock must be zero.
Find an optimal solution.

Exercise 9.5 Consider an instance of the problem $WW\text{-}CC$. Show that if $n = 5, C = 10$, and $d = (5, 3, 7, 4, 9)$, the regeneration interval $[2, 5]$ cannot be part of an optimal solution.

Exercise 9.6 Consider an instance of $LS\text{-}CC$ with $n = 5, C = 10$, costs $p' = (7, 4, 2, 3, 4), h' = (1, 3, 5, 1, 3), q' = (25, 45, 26, 30, 20)$, and demands $d = (5, 3, 7, 4, 9)$ for which initial and final stock must be zero.
Solve by dynamic programming, MIP, and using LS–LIB.

Exercise 9.7 Consider an instance of $WW\text{-}C$ with $n = 6, d = (4, 2, 8, 5, 7, 1)$, $C = (7, 4, 14, 5, 9, 6), q = (10, 20, 30, 30, 40, 10)$, and $h = (1, 2, 1, 3, 1, 0)$.
i. Calculate the minimum cost of the $[2, k]$ regeneration intervals for $k \geq 2$.
ii. Solve the instance by MIP.

Exercise 9.8 Consider an instance of *LS-C* with $n = 6$, $d = (4, 2, 8, 5, 7, 1)$, $C = (7, 4, 14, 5, 9, 6)$, and costs $h' = (1, 2, 1, 3, 1, 0)$, $q = (10, 20, 30, 30, 40, 10)$, and $p' = (4, 8, 5, 3, 6, 3)$.
i. Solve the instance by MIP.
ii. Solve with LS–LIB comparing the effectiveness of the reformulations for WW-U, LS-U, and WW-CC with $C = \max_t C_t$, and the cutting planes for LS-U, WW-CC and WW-C.
iii. Find a valid inequality for *LS-C* cutting off the point $x^* = (6, 0, 13, 0, 7, 1)$, $y^* = (\frac{6}{7}, 0, \frac{13}{14}, 0, \frac{7}{9}, \frac{1}{6})$, $s^* = (2, 0, 5, 0, 0, 0)$.
iv. Find a submodular valid inequality for *LS-C* cutting off the point $x^* = (7, 0, 12.47, 0.55, 6.98, 0)$, $y^* = (1, 0, 0.89, 0.109, 0.872, 0)$,   and $s^* = (3, 1, 5.47, 1.02, 1, 0)$.
v. Find a violated valid inequality for *LS-C* cutting off the point $x^* = (7, 0, 13.14, 0, 6.86, 0)$, $y^* = (1, 0, 1, 0, 0.857, 0)$, and $s^* = (3, 1, 6.14, 1.14, 1, 0)$.

Exercise 9.9 Prove that the matrix occurring in the integer program of Section 9.6.3 for the calculation of α_{kl} is totally unimodular using the characterization of Proposition 6.6.

Exercise 9.10 Rewrite the constraint matrix from Corollary 9.3 equivalently as

$$\sum_{u=k}^{t} y_u + \sum_{\tau \in \{0\} \cup [k,n]: f_\tau^k \geq f_l^k} \delta_\tau^k + \mu^k \geq \lfloor \frac{d_{kt}}{C} \rfloor + 1 \quad \text{for all } k, t, \ k \leq t$$

$$\sum_{t \in \{0\} \cup [k,n]} \delta_t^k = 1 \quad \text{for all } k$$

$$\mu^k \geq 0, \ \delta_t^k \geq 0 \quad \text{for all } k \text{ and } t \in \{0\} \cup [k, n]$$

$$y \in [0, 1]^n ,$$

and prove that the matrix is totally unimodular. Show that the matrix remains totally unimodular with the additional constraints

$$\sum_{u=k}^{t} y_u \geq \beta_{kt} \quad \text{for all } 1 \leq k \leq t \leq n.$$

Notes

Section 9.1 The complexity of various special cases of *LS-C* is studied in Bitran and Yanasse [28].

Section 9.2 Regeneration intervals or points of regeneration were first formally introduced to characterize the structure of optimal solutions by Manne [116]. The term itself of regeneration point has been taken from the renewal processes literature and was first applied to inventory management by Karlin [97].

Section 9.3 The reformulation of $DLS\text{-}CC$ by normalizing so that the capacity becomes one is part of the folklore; see for example, Fleischmann [68]. See Van Vyve [178] for the $O(n \log n)$ variant of the greedy algorithm, and the parametric algorithm.

Section 9.4 The treatment of $DLSI\text{-}CC$ as a mixing set with additional constraints is from Miller and Wolsey [124].

Section 9.5 The polyhedral characterization, the extended formulation, and the separation algorithm for $WW\text{-}CC$ are from Pochet and Wolsey [140]. The $O(n^2 \log n)$ optimization algorithm is from Van Vyve [178].

Section 9.6 The dynamic programming algorithm based on regeneration intervals is from Florian and Klein [71]. An improved implementation running in $O(n^3)$ was given by Van Hoesel and Wagelmans [171]. The $O(n^3) \times O(n^3)$ extended formulation for $LS\text{-}CC$ is based on Pochet and Wolsey [139].

Section 9.7 Fully polynomial approximation schemes for $LS\text{-}C$ have been given by van Hoesel and Wagelmans [172] and Chubanov et al. [39]. The valid inequalities for $WW\text{-}C$ are new, motivated by the polynomial algorithm of Bitran and Yanasse [28] in the case where the capacities are nondecreasing and the fixed costs nonincreasing. An improved $O(n^2)$ algorithm for the latter problem is given by Chung and Lin in [40]. Valid inequalities for $LS\text{-}C$ have been proposed by Pochet [134] based on flow cover inequalities. The submodular inequalities and their lifted version based on the superadditive lifting function F_k of Subsection 9.7.3 were proposed by Atamtürk and Munoz [15]. The lifted (l, S) inequalities for $DLSI\text{-}C$ and $LS\text{-}C$ are from Loparic et al. [106].

10

Backlogging and Start-Ups

Here we consider two of the most important variants of the single-item problem, namely the problems with backlogging and with start-ups. Throughout the chapter, we concentrate on the results that we believe can be used in practice. We now briefly describe its contents.

- Sections 10.1 to 10.3 treat the problem with backlogging $PROB\text{-}CAP\text{-}B$, whereas Sections 10.4 to 10.5 treat the problem with start-up variables $PROB\text{-}CAP\text{-}SC$.
- In Section 10.2 we describe a dynamic programming algorithm for $LS\text{-}U\text{-}B$, present some valid inequalities, and give a tight reformulation. For $WW\text{-}U\text{-}B$ we obtain a more compact extended formulation, a characterization of the valid inequalities, and a combinatorial separation algorithm.
- In Section 10.3 we describe the convex hull of solutions of $DLS\text{-}CC\text{-}B$ using MIR inequalities and give a combinatorial optimization algorithm. For $DLSI\text{-}CC\text{-}B$ we give a tight extended formulation based on the continuous mixing set, and for $WW\text{-}CC\text{-}B$ we generate a tight extended formulation that is unfortunately large with $O(n^3)$ rows and $O(n^2)$ variables.
- In Section 10.4 we consider the uncapacitated problems with start-up variables $LS\text{-}U\text{-}SC$ and $WW\text{-}U\text{-}SC$ giving complete descriptions of the valid inequalities and tight extended formulations.
- In Section 10.5 we consider constant capacity problems with start-up variables. We describe valid inequalities for $DLS\text{-}CC\text{-}SC$ that suffice to solve the problem by linear programming with a certain class of costs, and also an extended formulation for the general case. We also present classes of valid inequalities for $LS\text{-}CC\text{-}SC$ and $LS\text{-}C\text{-}SC$ that have combinatorial separation algorithms.
- Finally in Section 10.6, we present a tight extended formulation for the uncapacitated problem with both backlogging and start-up variables $WW\text{-}U\text{-}B, SC$.

10.1 Backlogging

We first formulate the general single-item lot-sizing problem with backlogging LS-C-B. The additional data needed is b'_t, the per unit *backlogging cost* in period t.

The new variable r_t denotes the total accumulated backlog at the end of period t, that is, $r_t = \max[0, d_{1t} - s_0 - \sum_{u=1}^{t} x_u]$. With this, we obtain the formulation

$$\min \ \sum_{t=1}^{n} p'_t x_t + \sum_{t=0}^{n} h'_t s_t + \sum_{t=1}^{n} b'_t r_t + \sum_{t=1}^{n} q_t y_t$$

$$s_{t-1} - r_{t-1} + x_t = d_t + s_t - r_t \qquad\qquad \text{for } 1 \le t \le n$$

$$x_t \le C_t y_t \qquad\qquad \text{for } 1 \le t \le n$$

$$s \in \mathbb{R}_+^{n+1}, x, r \in \mathbb{R}_+^{n}, y \in \{0,1\}^n.$$

Throughout we assume that $r_0 = 0$.

Note that the objective function can always be rewritten as

$$\min \sum_{t=0}^{n} h_t s_t + \sum_{t=1}^{n} b_t r_t + \sum_{t=1}^{n} q_t y_t \ \Big(+ \sum_{t=1}^{n} p'_t d_t\Big),$$

with $h_t = h'_t + p'_t - p'_{t+1}$ and $b_t = b'_t + p'_{t+1} - p'_t$, where we take $p'_0 = 0$. We say that a problem with backlogging has *Wagner–Whitin costs* if $p'_t + h'_t \ge p'_{t+1}$ for $t = 0, \ldots, n-1$ and $p'_t + b'_{t-1} \ge p'_{t-1}$ for $t = 2, \ldots, n$. So the problem has Wagner–Whitin costs precisely when h_t and b_t are nonnegative for all t.

10.2 Backlogging: The Uncapacitated Case

10.2.1 Extreme Points and Optimization

For LS-U-B, the structure of extreme optimal solutions is very similar to that of LS-U. Essentially whenever $x_t > 0$, then $x_t = d_{t-k,t+l}$ for some $k \ge 0$ and $l \ge 0$. The structure of such an extreme point is illustrated in Figure 10.1.

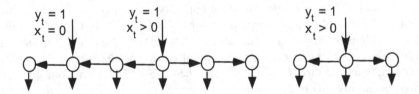

Figure 10.1. Form of an extreme point solution of LS-U-B.

This means that the dynamic programming recursion to solve LS-U-B is not much more complicated than that for LS-U.

A Dynamic Programming Algorithm for *LS-U-B*

For periods $u, v \in \{1, \ldots, n\}$, let $\phi(u, v)$ denote the minimum cost of satisfying demands for periods $v, \ldots, n$ in which the demand for period v is satisfied by production in u where the fixed cost in u is only counted if $u \geq v$, and let $G(v)$ denote the minimum cost solution of problem *LS-U-B* defined over the horizon $v, \ldots, n$.

We have

$$G(v) = \min_{u \geq v} \phi(u, v)$$

$$\phi(u, v) = \left(\sum_{t=u}^{v-1} h_t\right) d_v + \min[\phi(u, v+1), G(v+1)] \quad \text{for } u < v$$

$$\phi(u, v) = \left(\sum_{t=v}^{u-1} b_t\right) d_v + \phi(u, v+1) \qquad \text{for } u > v$$

$$\phi(u, u) = q_u + \min[\phi(u, u+1), G(u+1)] \qquad \text{for } u = v.$$

$G(1)$ and a corresponding optimal solution can be found using this recursion in $O(n^2)$. As for *LS-U*, it is possible to reduce the running time to $O(n \log n)$.

Note that when there is an initial stock variable s_0, or items can be backlogged after the end of the time horizon $r_n > 0$, minor modifications need to be made to the above recursion. Specifically u and v run from $n+1$ down to 0, $G(n+1) = \phi(u, n+1) = 0$ for all u, $\phi(u, 0) = \phi(u, 1)$ for all $u \geq 1$, and the optimal value $G(0) = \phi(0, 0) = \min[\phi(0, 1), G(1)]$.

Example 10.1 *Consider an instance of LS-U-B with $n = 4, h_0 = 6, h = (1, 1, 1, 1), b = (3, 3, 3, 10), q = (100, 81, 70, 40), \text{ and } d = (8, 4, 0, 5)$.*

In Table 10.1 we give the values of $\phi(u, v)$ and $G(v)$ as we work through calculating terms row by row. Note that $G(v)$ can be found once $\phi(v, v)$ has been calculated.

Table 10.1. $\phi(u, v)$ Values for Example 10.1

$\phi(u, v)$	$v = 5$	$v = 4$	$v = 3$	$v = 2$	$v = 1$	$v = 0$
$u = 5$	0	50	50	114	266	266
$u = 4$	0	40	40	64	136	136
$u = 3$	0	5	75	87	135	135
$u = 2$	0	10	10	91	115	115
$u = 1$	0	15	15	19	119	119
$u = 0$	0	45	40	68	112	112
$G(v)$	0	40	40	64	115	112

The individual calculations of the values $\phi(u, 2)$ and $G(2)$ in column $v = 2$ of Table 10.1 are as follows.

$$
\begin{aligned}
\phi(5,2) &= (b_2 + b_3 + b_4)d_2 + \phi(5,3) = 114. \\
\phi(4,2) &= (b_2 + b_3)d_2 + \phi(4,3) = 64. \\
\phi(3,2) &= b_2 d_2 + \phi(3,3) = 87. \\
\phi(2,2) &= q_2 + \min[\phi(2,3), G(3)] = 91. \\
G(2) &= \min[114, 64, 87, 91] = 64. \\
\phi(1,2) &= h_1 d_2 + \min[\phi(1,3), G(3)] = 19. \\
\phi(0,2) &= (h_0 + h_1)d_2 + \min[\phi(0,3), G(3)] = 68.
\end{aligned}
$$

Working backwards to find an optimal solution, $G(0) = \phi(0,1) = h_0 d_1 + G(2) = h_0 d_1 + \phi(4,2) = h_0 d_1 + (b_2 + b_3)d_2 + \phi(4,3) = h_0 d_1 + (b_2 + b_3)d_2 + b_3 d_3 + \phi(4,4) = h_0 d_1 + (b_2 + b_3)d_2 + b_3 d_3 + q_4.$
Thus an optimal solution is $s_0 = d_1, y_4 = 1, x_4 = d_{24}, r_3 = d_{23}, r_2 = d_2.$

10.2.2 Tight Formulations and Inequalities for *LS-U-B*

Valid inequalities

Some simple valid inequalities are obtained by adding backlog variables to the (l, S) inequalities for $LS - U$.

Proposition 10.1 *Given the interval* $[k, l]$ *with periods* $t_0 = k - 1 < t_1 < \cdots < t_p = l$, *the following inequality is valid for* X^{LS-U-B}:

$$
s_{k-1} + \sum_{q=1}^{p} r_{t_q} + \sum_{q=0}^{p} \sum_{u=t_q+1}^{t_{q+1}} d_{t_q+1,l} y_u \geq d_{kl}
$$

One way to show the validity of these inequalities is by applying mixing to the surrogate inequalities $s_{k-1} + r_{t_q} + M \sum_{u=k}^{t_q} y_u \geq d_{k,t_q}$ for $q = 1, \ldots, p$, obtained by summing the flow balance constraints for periods k up to t, replacing r_{k-1} and s_t by the lower bound of zero and $\sum_{u=k}^{t} x_u$ by the upper bound of $M \sum_{u=k}^{t} y_u$.

Example 10.2 *With* $[k, l] = [2, 6]$, $t_0 = 1, t_1 = 3$, $t_2 = 5$, *and* $t_3 = 6$, *we obtain the surrogates*

$$
\begin{aligned}
s_1 + r_3 + My_2 + My_3 &\geq d_{23} \\
s_1 + r_5 + My_2 + My_3 + My_4 + My_5 &\geq d_{25} \\
s_1 + r_6 + My_2 + My_3 + My_4 + My_5 + My_6 &\geq d_{26}.
\end{aligned}
$$

Proposition 10.1 gives the valid mixing inequality

$$
s_1 + r_3 + r_5 + r_6 \geq d_{23}(1 - y_2 - y_3)
$$
$$
+ d_{45}(1 - y_2 - y_3 - y_4 - y_5) + d_6(1 - y_2 - y_3 - y_4 - y_5 - y_6).
$$

To understand the coefficient of d_{45}, *note that the demands* d_4 *and* d_5 *must be satisfied either from* s_1, *or from* r_5, *or from production in the interval* $[2, 5]$.

A complete description of the valid inequalities for this problem is unknown. However, in discussing WW-U-B below, a larger class of inequalities is demonstrated.

Extended Formulations

Both the facility location and the shortest path reformulations for LS-U extend to LS-U-B. We just present the latter.

The Shortest Path Reformulation of LS-U-B.

Let $w_{tt} = 1$ if the demand for t is produced in t.
Let $\phi_{ut} = 1$ if production in u includes the future demand precisely up to period $t \geq u$.
Let $\psi_{ut} = 1$ if production in u includes backlogged demand precisely from period $t \leq u$.

The resulting reformulation Q^{LS-U-B} presented now corresponds to a shortest path formulation in the directed graph in Figure 10.2, where the regeneration interval $[k, l]$ with production in period $u \in \{k, \ldots, l\}$ is decomposed into 1 unit of flow through the sequence of arcs with flows ψ_{uk}, w_{uu}, and ϕ_{ul}.

$$\sum_{k=1}^{n} \psi_{k1} = 1 \qquad\qquad \text{Node 1}$$

$$\sum_{k=1}^{t-1} \phi_{k,t-1} - \sum_{k=t}^{n} \psi_{kt} = 0 \qquad\qquad \text{Node } t, \text{ for } 2 \leq t \leq n$$

$$-\sum_{l=1}^{t} \psi_{tl} + w_{tt} = 0 \qquad\qquad \text{Node } t', \text{ for } 1 \leq t \leq n$$

$$-w_{tt} + \sum_{l=t}^{n} \phi_{tl} = 0 \qquad\qquad \text{Node } t'', \text{ for } 1 \leq t \leq n$$

$$w_{tt} - y_t \leq 0 \qquad\qquad \text{for } 1 \leq t \leq n$$

$$x_t = \sum_{u=t+1}^{n} d_{t+1,u} \phi_{tu} + \sum_{u=1}^{t-1} d_{u,t-1} \psi_{tu} + d_t w_{tt} \qquad \text{for } 1 \leq t \leq n$$

$$s_{t-1} = \sum_{u,l:u<t,l\geq t} d_{tl} \phi_{ul} \qquad\qquad \text{for } 1 \leq t \leq n$$

$$r_t = \sum_{u,l:u>t,l\leq t} d_{lt} \psi_{ul} \qquad\qquad \text{for } 1 \leq t \leq n$$

$$\psi, \phi \in \mathbb{R}_+^{n(n+1)/2}, w \in \mathbb{R}_+^n, y \in [0,1]^n.$$

Theorem 10.2 $\operatorname{conv}(X^{LS-U-B}) = \operatorname{proj}_{x,s,r,y} Q^{LS-U-B}$.

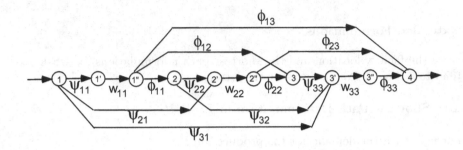

Figure 10.2. Shortest path formulation for LS-U-B.

10.2.3 Tight Formulations and Inequalities for WW-U-B

The results for WW-U-B are surprising and useful. Here it turns out that there is an exponentially large number of facet-defining inequalities, so a separation algorithm is required. However it suffices just to add $2n$ additional variables to obtain a tight extended formulation with only $O(n^2)$ constraints.

An Extended Formulation

With Wagner–Whitin costs, when some demand d_t is satisfied through inventory, it is produced in the last set-up period before t. Similarly when the demand d_t is satisfied through backlogging, it is produced in the first set-up period after t. These observations are used to construct an extended formulation. We define new variables for each period t in which $d_t > 0$.

$\alpha_t = 1$ if the demand d_t is satisfied from stock, and
$\beta_t = 1$ if the demand d_t is satisfied from backlog.

The resulting reformulation Q^{WW-U-B} is

$$\alpha_t + y_t + \beta_t = 1 \qquad \text{for all } t \text{ with } d_t > 0$$

$$s_{k-1} \geq \sum_{l=k}^{t} d_l \left(\alpha_l - \sum_{u=k}^{l-1} y_u \right) \qquad \text{for all } k,t \text{ with } k \leq t$$

$$r_k \geq \sum_{l=t}^{k} d_l \left(\beta_l - \sum_{u=l+1}^{k} y_u \right) \qquad \text{for all } k,t \text{ with } k \geq t$$

$$s \in \mathbb{R}_+^{n+1}, r, \alpha, \beta \in \mathbb{R}_+^n, y \in [0,1]^n,$$

where the first constraint says that d_t is either satisfied from stock, or from backlogging, or from production in period t. The constraints defining s_{k-1} stipulate that s_{k-1} includes the demand d_l for $l \geq k$ if $\alpha_l = 1$ and $\sum_{u=k}^{l-1} y_u = 0$. The constraint for r_k has a similar interpretation.

Theorem 10.3 $\mathrm{conv}(X^{WW-U-B}) = proj_{x,s,r,y} Q^{WW-U-B}$.

Example 10.3 *The costs in Example 10.1 are Wagner–Whitin. As $d =$ $(8, 4, 0, 5)$, the polyhedron Q^{WW-U-B} takes the form*

$$\alpha_1 + y_1 + \beta_1 = 1, \quad \alpha_2 + y_2 + \beta_2 = 1, \quad \alpha_4 + y_4 + \beta_4 = 1$$
$$s_0 \geq 8\alpha_1, \quad s_0 + 4y_1 \geq 8\alpha_1 + 4\alpha_2$$
$$s_0 + 9y_1 + 5y_2 + 5y_3 \geq 8\alpha_1 + 4\alpha_2 + 5\alpha_4$$
$$s_1 \geq 4\alpha_2, \quad s_1 + 5y_2 + 5y_3 \geq 4\alpha_2 + 5\alpha_4$$
$$s_2 + 5y_3 \geq 5\alpha_4$$
$$s_3 \geq 5\alpha_4$$
$$r_4 \geq 5\beta_4, \quad r_4 + 4y_3 + 4y_4 \geq 5\beta_4 + 4\beta_2$$
$$r_4 + 8y_2 + 12y_3 + 12y_4 \geq 5\beta_4 + 4\beta_2 + 8\beta_1$$
$$r_3 + 4y_3 \geq 4\beta_2, \quad r_3 + 8y_2 + 12y_3 \geq 4\beta_2 + 8\beta_1$$
$$r_2 \geq 4\beta_2, \quad r_2 + 8y_2 \geq 4\beta_2 + 8\beta_1$$
$$r_1 \geq 8\beta_1$$
$$s \in \mathbb{R}_+^5, r, \alpha, \beta \in \mathbb{R}_+^4, y \in [0,1]^4.$$

From this, we can also eliminate the α and β variables by projection, so as to obtain a characterization of the valid inequalities in the original space.

Valid Inequalities and the Convex Hull

Proposition 10.4 *Every facet-defining inequality of $conv(X^{WW-U-B})$ is of the form*

$$\sum_{k=1}^{n} \sum_{t=k}^{n} v_{kt}\left(s_{k-1} + \sum_{l=k+1}^{t} d_l \sum_{j=k}^{l-1} y_j\right) + \sum_{k=1}^{n} \sum_{t=1}^{k} w_{kt}\left(r_k + \sum_{l=t}^{k} d_l \sum_{j=l+1}^{k} y_j\right)$$

$$\geq \sum_{l=1}^{n} u_l d_l (1 - y_l),$$

where (v, w) is the characteristic vector of an elementary cycle in the digraph $D = (V, A)$ with $V = \{1, \ldots, n+1\}$, forward arcs $(k, t+1)$ corresponding to v_{kt}, backward arcs $(k+1, t)$ corresponding to w_{kt}, and $u_l = \sum_{k:k \leq l} \sum_{t:t \geq l} v_{kt}$.

Example 10.4 *Suppose that $n = 4$, and consider the elementary cycle obtained by taking $v_{12} = w_{22} = v_{24} = w_{44} = w_{31} = 1$ illustrated in Figure 10.3. The resulting inequality is*

$$[(s_0 + d_2 y_1) - d_1(1 - y_1) - d_2(1 - y_2)] + r_2$$
$$+ [s_1 + d_3 y_2 + d_4(y_2 + y_3) - d_2(1 - y_2) - d_3(1 - y_3) - d_4(1 - y_4)]$$
$$+ r_4 + [r_3 + d_1(y_2 + y_3) + d_2 y_3] \geq 0,$$

or

$$s_0 + s_1 + r_2 + r_3 + r_4 \geq d_1(1 - y_1 - y_2 - y_3)$$
$$+ d_2(1 - y_2)$$
$$+ d_2(1 - y_1 - y_2 - y_3)$$
$$+ d_3(1 - y_2 - y_3)$$
$$+ d_4(1 - y_2 - y_3 - y_4).$$

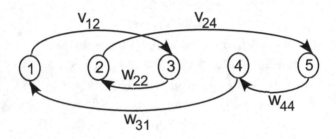

Figure 10.3. Cycle $(1, 3, 2, 5, 4, 1)$ in the separation digraph.

The Separation Problem

Separation of the point (s^*, r^*, y^*) reduces to finding a negative cost cycle in the digraph of Proposition 10.4, where the costs on the arcs are precisely

$$\left(s_{k-1}^* + \sum_{l=k}^{t} d_l \sum_{j=k}^{l-1} y_j^* \right) - \sum_{l=k}^{t} d_l(1 - y_l^*)$$

on the arcs $(k, t+1)$ corresponding to the variables v_{kt}, and

$$\left(r_k^* + \sum_{l=t}^{k} d_l \sum_{j=l+1}^{k} y_j^* \right)$$

on the arcs $(k+1, t)$ corresponding to the variables w_{kt}.

A straightforward implementation of the minimum mean cost cycle algorithm that allows one to detect a negative cost cycle (if any) has complexity $O(n^3)$.

Example 10.5 *Consider an instance of* WW-U-B *with* $n = 4$, $d = (8, 4, 0, 5)$, $h = (1, 1, 1, 1)$, $b = (3, 3, 3, 10)$, *and* $q = (100, 81, 70, 40)$. *This is the same as Example 10.1 except that* $s_0 = 0$. *After adding a priori the inequalities*

$$s_{t-1} + r_t + d_t y_t \geq d_t \qquad\qquad \text{for } t = 1, \ldots, n$$
$$s_{t-1} + r_t + r_{t+1} + d_{t,t+1} y_t + d_{t+1} y_{t+1} \geq d_{t,t+1} \quad \text{for } t = 1, \ldots, n-1,$$

we obtain a fractional solution $y^* = (0, 0.615, 0.128, 0.256)$, $s^* = (0, 0, 3.72, 0)$, *and* $r^* = (8, 1.54, 3.08, 0)$.

We obtain the cost matrix $cv = \begin{pmatrix} -8 & -9.54 & -9.54 & -9.54 \\ & -1.54 & -1.54 & -1.54 \\ & & 0 & -3.08 \\ & & & 0 \end{pmatrix}$ *for the vari-*

ables v_{kt}, *and the cost matrix* $cw = \begin{pmatrix} 8 \\ 6.46 & 1.54 \\ 9.54 & 3.59 & 3.08 \\ 9.54 & 1.54 & 0 & 0 \end{pmatrix}$ *for the variables* w_{kt}.

The minimum mean cycle is given by $v_{12} = w_{21} = 1$, *and the corresponding inequality is*

$$(s_0 + d_2 y_2) + (r_2 + d_1 y_2) \geq d_1 (1 - y_1) + d_2 (1 - y_2) \quad \text{or}$$

$$r_2 + d_{12} y_1 + d_{12} y_2 \geq d_{12} \ .$$

When this inequality is added to the linear program and one reoptimizes, one obtains an integer feasible solution $y^* = (0, 1, 0, 0)$, $s^* = (0, 5, 5, 0)$, *and* $r^* = (8, 0, 0, 0)$.

10.3 Backlogging: The Constant Capacity Case

10.3.1 Discrete Lot-Sizing with Backlogging *DLS-CC-B*

The flow conservation constraints for *DLS-C-B* can be rewritten as

$$\sum_{u=1}^{t} C_u y_u = d_{1t} + s_t - r_t,$$

or, if we eliminate the r_t variables, as

$$s_t \geq \sum_{u=1}^{t} C_u y_u - d_{1t}.$$

Similarly the objective function $\sum_{t=1}^{n}(h'_t s_t + b'_t r_t + q'_t y_t)$ can be written, after elimination of the r_t variables, as $\sum_{t=1}^{n}[(h'_t + b'_t)s_t + (q'_t - C_t \sum_{u=t}^{n} b'_u)y_t]$ plus the constant term $\sum_{t=1}^{n} b'_t d_{1t}$.

We now consider the problem with *constant* capacities and constraint set

$$s_t \geq C \sum_{u=1}^{t} y_u - d_{1t} \qquad \text{for } 1 \leq t \leq n \tag{10.1}$$

$$s \in \mathbb{R}_+^n, y \in \{0,1\}^n, \tag{10.2}$$

denoted $X^{DLS-CC-B}$, and we assume that $h'_t + b'_t \geq 0$ for all t.

The Convex Hull

Setting $\sigma_t = s_t/C \in \mathbb{R}_+, z_t = \sum_{u=1}^{t} y_u \in \mathbb{Z}_+, b_t = d_{1t}/C$, the set $X^{DLS-CC-B}$ is the intersection of n basic mixed integer sets X_1^{MI} with linking constraints $(1 \geq z_t - z_{t-1} \geq 0)$ for which the corresponding matrix is the arc-node incidence matrix of a digraph; see Section 8.2.

Theorem 10.5 conv$(X^{DLS-CC-B})$ *is obtained by adding the MIR inequalities*

$$s_t \geq C(1-f_t)(\sum_{u=1}^{t} y_u - \lfloor \tfrac{d_{1t}}{C} \rfloor) \quad \text{for } 1 \leq t \leq n$$

to the constraints (10.1), and the bound constraints $s \in \mathbb{R}_+^n, y \in [0,1]^n$, where $f_t = \tfrac{d_{1t}}{C} - \lfloor \tfrac{d_{1t}}{C} \rfloor$ for all t.

Optimization

Consider the optimization problem

$$z^{DLS-CC-B} = \min\{\sum_{i=1}^{n} q_i y_i + \sum_{i=1}^{n} h_i s_i : (x,s) \in X^{DLS-CC-B}\},$$

with $h_t \geq 0$ for all t.

To obtain an efficient algorithm for this problem, we consider the family of problems $Q(j,p)$:

$$z(j,p) = \min\{\sum_{i=1}^{j} [q_i y_i + h_i (C\sum_{u=1}^{i} y_u - d_{1i})^+] : \sum_{i=1}^{j} y_i = p, y \in \{0,1\}^j\}.$$

We now suppose that $y \in \{0,1\}^n$ is the characteristic vector of $S \subseteq N = \{1,\ldots,n\}$, and rephrase the problem.

For $S \subseteq N$, let

$$v(S) = \sum_{i \in S} q_i + \sum_{i \in N} h_i (C|S \cap N_i| - d_{1i})^+,$$

where $N_i = \{1,\ldots,i\}$ for $i = 1,\ldots,n$. Thus $z(j,p) = \min_{S \subseteq N, |S|=p} v(S)$. Moreover, v has important structure.

Observation 10.1 *The set function v is a supermodular function, as $C|S \cap N_i| - d_{1i}$ is supermodular and nondecreasing, $\max[g(S), k]$ is supermodular when g is supermodular, and the sum of supermodular functions is supermodular.*

Supermodularity was defined in Definition 9.2. The very special structure of v leads to the following important properties.

Proposition 10.6 *If $S \subseteq N_j$ is optimal for $Q(j, p)$ and $l < j$, then $S \cap N_l$ is optimal for $Q(l, |S \cap N_l|)$.*

Proposition 10.7 *If $S \subseteq N_j$ is the lex-optimal solution of $Q(j, p-1)$, then there exists $t \in N_j$ such that $S \cup \{t\}$ is the lex-optimal solution of $Q(j, p)$.*

It follows that a simple greedy algorithm can be used to solve

$$z^{DLS-CC-B} = \min_p z(n, p) = \min_{S \subseteq N} v(S)$$

by computing iteratively $v(S^p) = \min_{S \subseteq N, |S|=p} v(S)$ for $p = 1, \ldots, |N|$.

Example 10.6 *Consider an instance of DLS-CC-B with $n = 6$, $C = 10$, $d = (3, 6, 2, 3, 7, 5)$, $q' = (70, 20, 50, 30, 40, 10)$, $h' = (1, 2, 3, 2, 1, 3)$, $b' = (2, 3, 6, 4, 2, 4)$.*

After elimination of the backlog variables $\{r_t\}$ from the objective function we obtain $q = (-140, -170, -110, -70, -20, -30)$ and $h = (3, 5, 9, 6, 3, 7)$ with constant $\sum_t b'_t d_{1t} = 301$.

Using the notation $\rho_j(S) = v(S \cup \{j\}) - v(S)$ for $S \subseteq N, j \notin S$, the greedy algorithm gives:
Initialization. $S^0 = \emptyset$.

Iteration 1.
$\rho_j(S^0) = (-114, -165, -110, -70, -20, -30), S^1 = \{2\}, v(S^1) = -165$.

Iteration 2.
$\rho_j(S^1) = (48, -, 7, -34, -20, -30), S^2 = \{2, 4\}, v(S^2) = -199$.

Iteration 3.
$\rho_j(S^2) = (127, -, 86, -, 35, -2), S^3 = \{2, 4, 6\}, v(S^3) = -201$.

By the supermodularity of v, the $\rho_j(S)$ are increasing with S. Hence $\rho_j(S^3) \geq 0$ for all $j \notin S^3$, and thus $S^3 = \{2, 4, 6\}$ is optimal with original cost $301 - 201 = 100$. The corresponding solution is $y_2 = y_4 = y_6 = 1, r_1 = 3, s_2 = 1, r_3 = 1, s_4 = 6, r_5 = 1, s_6 = 4$.

To obtain a more efficient implementation, one does not need to calculate all the values $v(S)$ or $\rho_j(S)$ from scratch. Let $\gamma_i = C\lceil \frac{d_{1i}}{C} \rceil - d_{1i}$, and $T_S^\delta = \{i \in N : |S \cap N_i| - d_{1i}/C \in [-\delta - 1, -\delta)\}$ for $\delta = 0, 1$.

314   10 Backlogging and Start-Ups

Observation 10.2 $\rho_j(S \cup \{k\}) - \rho_j(S) = v(S \cup \{j,k\}) - v(S \cup \{k\}) + v(S) - v(S \cup \{j\}) = [v(S \cup \{j,k\}) - v(S)] - [v(S \cup \{j\}) - v(S)] - [v(S \cup \{k\}) - v(S)] = \sum_{i \in T_S^0 : i \geq \max[j,k]} h_i[(C + \gamma_i) - \gamma_i - \gamma_i] + \sum_{i \in T_S^1 : i \geq \max[j,k]} h_i \gamma_i.$

We can now describe the improved greedy algorithm.

Initialization.
Iteration 0. $S^0 = \emptyset$. Initialize $T_{S^0}^0$ and $T_{S^0}^1$.
Iteration 1. $\rho_j(\emptyset) = v(\{j\}) = q_j + \sum_{i \in T_{S^0}^0 : i \geq j} h_i \gamma_i$. $k_1 = \arg\min \rho_j(\emptyset)$.
$S^1 = \{k_1\}$.

Iterations $t = 2, \ldots, n$. For $j \notin S^{t-1}$, given that $S^{t-1} = S^{t-2} \cup \{k_{t-1}\}$, use the values

$$\rho_j(S^{t-1}) - \rho_j(S^{t-2}) = \sum_{i \in T_{S^{t-2}}^0 : i \geq \max[j,k_{t-1}]} h_i(C - \gamma_i)$$
$$+ \sum_{i \in T_{S^{t-2}}^1 : i \geq \max[j,k_{t-1}]} h_i \gamma_i$$

to calculate $\rho_j(S^{t-1})$. Let $k_t = \arg\min_{j \notin S^{t-1}} \rho_j(S^{t-1})$.

If $\rho_{k_t}(S^{t-1}) \geq 0$, stop. S^{t-1} is optimal.
Otherwise $S^t = S^{t-1} \cup \{k_t\}$, and update $T_{S^t}^0$ and $T_{S^t}^1$.
If $t = n$, stop. $S^n = N$ is optimal.
Otherwise increase t.

Note that the complexity is $O(n^2)$.

Example 10.7 *We return to the previous example, and calculate with the improved greedy algorithm.*

Initialization.
$d_{1t} = (3, 9, 11, 14, 21, 26), \gamma = (7, 1, 9, 6, 9, 4)$.
$S^0 = \emptyset$.
$T_{S^0}^0 = \{1, 2\}, T_{S^0}^1 = \{3, 4\}$.

Iteration 1.
$\rho_j(S^0) = (-114, -165, -110, -70, -20, -30)$, $k_1 = 2, S^1 = \{2\}, v(S^1) = -165$.
$T_{S^1}^0 = \{1, 3, 4\}, T_{S^1}^1 = \{5, 6\}$.

Iteration 2.
$\rho_j(S^1) - \rho_j(S^0) = (162, -, 117, 36, 0, 0)$,
$\rho_j(S^1) = (48, -, 7, -34, -20, -30)$, $k_2 = 4, S^2 = \{2, 4\}, v(S^2) = -199$.
$T_{S^2}^0 = \{1, 3, 5, 6\}, T_{S^2}^1 = \emptyset$.

Iteration 3.

$\rho_j(S^2) - \rho_j(S^1) = (79, -, 79, -, 55, 28),$

$\rho_j(S^2) = (127, -, 86, -, 35, -2),\ k_3 = 6, S^3 = \{2, 4, 6\}, v(S^3) = -201.$

$T^0_{S^3} = \emptyset, T^1_{S^3} = \emptyset.$

10.3.2 Discrete Lot-Sizing with Initial Stocks and Backlogging *DLSI-CC-B*

We consider the set $X^{DLSI-CC-B}$ described by

$$\{(s, r, y) \in \mathbb{R}^{n+1}_+ \times \mathbb{R}^n_+ \times \mathbb{Z}^n_+ : s_0 + C \sum_{u=1}^{t} y_u = d_{1t} + s_t - r_t \text{ for } t = 1, \ldots, n\}.$$

After elimination of the variables $\{s_t\}_{t=1}^{n}$, the set can be rewritten as

$$
\begin{aligned}
&s_0 + r_t + C z_t \geq d_{1t} &&\text{for } 1 \leq t \leq n \\
&0 \leq z_t - z_{t-1} \leq 1 &&\text{for } 2 \leq t \leq n \\
&z_1 \leq 1 \\
&s_0 \in \mathbb{R}^1_+, r \in \mathbb{R}^n_+, z \in \mathbb{Z}^n_+.
\end{aligned}
$$

Now we have a continuous mixing set X^{CMIX} with additional constraints. Based on the extended reformulation for $\text{conv}(X^{CMIX})$ given in Theorem 8.9, we can obtain the following reformulation which can be shown to be tight.

Theorem 10.8 *A tight extended formulation for* $\text{conv}(X^{DLSI-CC-B})$ *is given by*

$$
\begin{aligned}
&s + r_j + CF^j_k\left(\sum_{u=1}^{j} y_u - \lfloor b_j/C \rfloor\right) \geq Cf_k + \alpha_j - \alpha_k &&\text{for } 1 \leq j < k \leq n \\
&r_j + CF^j_k\left(\sum_{u=1}^{j} y_u - \lfloor b_j/C \rfloor\right) \geq \alpha_j - \alpha_k &&\text{for } 0 \leq k < j \leq n \\
&s \geq Cf_k + \alpha_0 - \alpha_k &&\text{for } 1 \leq k \leq n \\
&s + r_j + C\left(\sum_{u=1}^{j} y_u - \lfloor b_j/C \rfloor\right) \geq Cf_j &&\text{for } 1 \leq j \leq n \\
&s \in \mathbb{R}^1_+, r \in \mathbb{R}^n_+, y \in [0,1]^n, \alpha \in \mathbb{R}^{n+1}
\end{aligned}
$$

where $f_i = \frac{d_{1i}}{C} - \lfloor \frac{d_{1i}}{C} \rfloor, f_0 = 0, F^j_i = f_j - f_i$ *if* $f_i \leq f_j$, *and* $F^j_i = f_j - f_i + 1$ *if* $f_i > f_j$.

Observe that this formulation has $O(n^2)$ constraints and $O(n)$ variables. A complete description of $\text{conv}(X^{DLSI-CC-B})$ in the (s, r, y) space is obtained by projection; see Proposition 8.10.

10.3.3 Lot-Sizing with Wagner–Whitin Costs and Backlogging WW-CC-B

The first nontrivial result concerns a formulation.

Theorem 10.9 *With constant capacities and backlogging, a minimum cost stock and backlog minimal solution of*

$$\min \quad \sum_{t=0}^{n} h_t s_t + \sum_{t=1}^{n} b_t r_t + \sum_{t=1}^{n} q_t y_t$$

$$s_{k-1} + r_l + C \sum_{u=k}^{l} y_u \geq d_{kl} \qquad \text{for } 1 \leq k \leq l \leq n \qquad (10.3)$$

$$s \in \mathbb{R}_+^{n+1}, r \in \mathbb{R}_+^{n}, y \in \{0,1\}^n \qquad (10.4)$$

solves WW-CC-B, where a solution of (10.3)–(10.4) is called stock minimal *(resp., backlog minimal) if $s_{k-1} = \max_{l \geq k}[d_{kl} - C\sum_{u=k}^{l} y_u - r_l]^+$ (resp., if $r_l = \max_{k \leq l}[d_{kl} - C\sum_{u=k}^{l} y_u - s_{k-1}]^+$).*

Thus we let $X^{WW-CC-B}$ denote the feasible set (10.3)–(10.4).

Extended Formulation

Here we consider the structure of extreme solutions of LS-CC-B. In fact it suffices to observe that if $s_k + r_k > 0$, then k lies in some regeneration interval $[\alpha, \beta]$, and then either the fractional period (the period t in which $0 < x_t < C$) occurs after k in which case $C\sum_{u=\alpha}^{k} y_u - d_{\alpha,t} = s_k - r_k$, or the fractional period occurs in or before k in which case $s_k - r_k = d_{k+1,\beta} - C\sum_{u=k+1}^{\beta} y_u$.

Proposition 10.10 *In an extreme solution of LS-CC-B with $s_0 = s_n = r_n = 0$,*

$$s_k \mod C \in \{0, -d_{1k}, -d_{2k}, -\ldots, -d_{k-1,k}, -d_k,$$
$$d_{k+1}, d_{k+1,k+2}, \ldots, d_{k+1,n}\} \mod C,$$

and

$$r_k \mod C \in \{0, d_{1k}, d_{2k}, \ldots, d_{k-1,k}, d_k,$$
$$- d_{k+1}, -d_{k+1,k+2}, \ldots, -d_{k+1,n}\} \mod C.$$

Example 10.8 *Suppose $n = 6, d = (4,2,3,6,7,5), C = 10$, and $k = 3$. Then $s_3 \mod 10 \in \{0, -9, -5, -3, 6, 13, 18\} \mod 10 = \{0,1,5,7,6,3,8\}$ and $r_3 \mod 10 \in \{0, 9, 5, 3, -6, -13, -18\} \mod 10 = \{0,9,5,3,4,7,2\}$.*

Now observe that $X^{WW-CC-B} = \left(\bigcap_{1 \le k \le l \le n} X_{kl}^*\right) \bigcap \{(y,z) : z_{kl} = \sum_{u=k}^{l} y_u, y \in [0,1]^n\}$ where X_{kl}^* is the set

$$s_{k-1} + r_l + Cz_{kl} \ge d_{kl}$$
$$s_{k-1} \in \{\gamma_0^{s,k-1}, \ldots, \gamma_n^{s,k-1}\}$$
$$r_l \in \{\gamma_0^{r,l}, \ldots, \gamma_n^{r,l}\}$$
$$s_{k-1}, r_l \ge 0, z_{kl} \in \mathbb{Z}_+,$$

where $\gamma_i^{s,k}$ are the values $\{0, -d_{1k}, -d_{2k}, -\ldots, -d_k, d_{k+1}, \ldots, d_{k+1,n}\} \mod C$ in order, and $\gamma_i^{r,l}$ are the values $\{0, d_{1l}, d_{2l}, \ldots, d_l, -d_{l+1}, \ldots, -d_{l+1,n}\} \mod C$ in order. Using the enumerative approach described in Section 8.4, one can derive an extended formulation Q_{kl} for X_{kl}^*. Again based on total unimodularity, one obtains

Theorem 10.11 *A tight extended formulation for $X^{WW-CC-B}$ is given by*

$$s_{k-1} = C\mu_{k-1} + \sum_{i=0}^{n} \gamma_i^{s,k-1} \delta_i^{s,k-1} \qquad \text{for } 1 \le k \le n$$

$$\sum_{i=0}^{n} \delta_i^{s,k-1} = 1 \qquad \text{for } 1 \le k \le n$$

$$r_l = C\nu_l + \sum_{i=0}^{n} \gamma_i^{r,l} \delta_i^{r,l} \qquad \text{for } 1 \le l \le n$$

$$\sum_{i=0}^{n} \delta_i^{r,l} = 1 \qquad \text{for } 1 \le l \le n$$

$$\mu_k + \nu_l + \sum_{u=k+1}^{l} y_u + \sum_{\gamma_j^{s,k} > \gamma_i^{s,k}} \delta_j^{s,k} + \sum_{j \in \Gamma(k,l,i)} \delta_j^{r,l} \ge \lceil \frac{d_{kl} - \gamma_i^{s,k}}{C} \rceil$$
$$\text{for } 0 \le k < l \le n, \ 0 \le i \le n$$

$$\mu, \nu, \delta, y \ge 0, y \le 1,$$

where $\gamma^{s,k}$ and $\gamma^{r,l}$ are defined above and where $\Gamma(k,l,i) = \{j : \lceil \frac{d_{kl} - \gamma_i^{s,k}}{C} \rceil >$
$\lceil \frac{d_{kl} - \gamma_i^{s,k} - \gamma_j^{r,l}}{C} \rceil \}$.

This reformulation has $O(n^3)$ constraints and $O(n^2)$ variables.

Example 10.9 *We consider an instance of WW-CC-B with $n = 4$, $d = (2, 5, 7, 4)$, and $C = 10$. For $k = 1$ and $l = 4$, we get that*
$s_1 \mod 10 \in \{0, -2, 5, 12, 16\} \mod 10 = \{0, 8, 5, 2, 6\}$ and
$r_4 \mod 10 \in \{0, 18, 16, 11, 4\} \mod 10 = \{0, 8, 6, 1, 4\}$.

Thus we have the constraints

$$s_1 = 10\mu_1 + 8\delta_1^{s,1} + 5\delta_2^{s,1} + 2\delta_3^{s,1} + 6\delta_4^{s,1}$$

$$\delta_0^{s,1} + \delta_1^{s,1} + \delta_2^{s,1} + \delta_1^{s,1} + \delta_3^{s,1} + \delta_4^{s,1} = 1$$

$$r_4 = 10\nu_4 + 8\delta_1^{r,4} + 6\delta_2^{r,4} + 1\delta_3^{r,4} + 4\delta_4^{r,4}$$

$$\delta_0^{r,4} + \delta_1^{r,4} + \delta_2^{r,4} + \delta_1^{r,4} + \delta_3^{r,4} + \delta_4^{r,4} = 1,$$

which, using the symbol σ as shorthand for the expression $\mu_1 + \nu_4 + y_2 + y_3 + y_4$, give

$$i = 0 : \sigma + \delta_1^{s,1} + \delta_2^{s,1} + \delta_3^{s,1} + \delta_4^{s,1} + \delta_1^{r,4} + \delta_2^{r,4} \geq \lceil \frac{16 - 0}{10} \rceil = 2$$

$$i = 1 : \sigma \qquad\qquad\qquad\qquad + \delta_1^{r,4} \geq \lceil \frac{16 - 8}{10} \rceil = 1$$

$$i = 2 : \sigma + \delta_1^{s,1} \qquad\quad + \delta_4^{s,1} + \delta_1^{r,4} + \delta_2^{r,4} + \delta_3^{r,4} + \delta_4^{r,4} \geq \lceil \frac{16 - 5}{10} \rceil = 2$$

$$i = 3 : \sigma + \delta_1^{s,1} + \delta_2^{s,1} \qquad + \delta_4^{s,1} + \delta_1^{r,4} + \delta_2^{r,4} \qquad + \delta_4^{r,4} \geq \lceil \frac{16 - 2}{10} \rceil = 2$$

$$i = 4 : \sigma + \delta_1^{s,1} \qquad\qquad\qquad\qquad\qquad\qquad\qquad \geq \lceil \frac{16 - 6}{10} \rceil = 1.$$

Valid Inequalities

Although an explicit linear description of $\mathrm{conv}(X^{WW-CC-B})$ in the original space of variables is not known, $\mathrm{conv}(X^{WW-CC-B})$ can be viewed as the intersection of $2n$ sets of the form $X^{DLSI-CC-B}$ where each stock variable s_{k-1} and each backlog variables r_l plays the role of the initial stock variable, so the continuous mixing inequalities provide a very significant class of valid inequalities for $X^{WW-CC-B}$; see Section 8.5 and Subsection 10.3.2.

10.3.4 Lot-Sizing with Backlogging *LS-CC-B*

The Optimization Problem

For the optimization problem *LS-CC-B* there is a simple $O(n^4)$ algorithm generalizing the algorithm for *LS-CC* in which the optimal cost α_{kl} of a regeneration interval is calculated in $O(n^2)$. There is a much less obvious version in which all $O(n^2)$ values α_{kl} are calculated in $O(n^3)$, giving total running time for *LS-CC-B* of $O(n^3)$.

An Extended Formulation

Similarly there is an immediate $O(n^4) \times O(n^3)$ tight reformulation in which each value α_{kl} is obtained as the solution of an $O(n^2) \times O(n)$ linear program, and an improved formulation in which each linear program is $O(n) \times O(n)$ leading to an $O(n^3) \times O(n^3)$ reformulation.

10.3.5 Résumé of Results

In Table 10.3.5 we summarize the results that are known for $PROB\text{-}[U, CC]$-B, where LP indicates that the separation problem can be solved as a linear program using the polynomial-size extended formulation as discussed in Section 6.1, and $\star$ indicates that an explicit description of a family of valid inequalities is known, but it only gives a partial description of the convex hull of solutions.

Table 10.2. Models with Backlogging $PROB\text{-}[U, CC]^1$-B

	LS	WW	DLSI	DLS
Formulation	$Cons \times Vars$	$Cons \times Vars$	$Cons \times Vars$	$Cons \times Vars$
U	$O(n) \times O(n^2)$	$O(n^2) \times O(n)$	—	—
CC	$O(n^3) \times O(n^3)$	$O(n^3) \times O(n^2)$	$O(n^2) \times O(n)$	$O(n) \times O(n)$
Separation				
U	LP and $\star$	$O(n^3)$	—	—
CC	LP	LP and $\star$	$O(n^3)$	$O(n)$
Optimization				
U	$O(n \log n)$	$O(n)$	—	—
CC	$O(n^3)$	$O(n^3)$	$O(n^2 \log n)$	$O(n^2)$

10.4 Start-Up Costs: The Uncapacitated Case

Here the additional data are the start-up costs g_t for $t = 1, \ldots, n$. The cost g_t is incurred if there is a set-up in period t, but there was no set-up in $t - 1$. To model such situations, a new start-up variable z_t is added that takes the value $z_t = 1$ if there is a set-up in t and not in $t - 1$, and $z_t = 0$ otherwise. So if $n = 8$, and $y = (0, 1, 1, 1, 0, 0, 1, 0)$, then there are set-ups in periods 2,3,4,7, but there are start-ups only in periods 2 and 7.

A basic formulation is obtained by taking the formulation of LS-C, adding a term $\sum_t g_t z_t$ in the objective function, and the constraints

$$z_t \geq y_t - y_{t-1} \qquad\qquad \text{for all } t \qquad\qquad (10.5)$$
$$z_t \leq y_t \qquad\qquad \text{for all } t \qquad\qquad (10.6)$$
$$z_t \leq 1 - y_{t-1} \qquad\qquad \text{for all } t \qquad\qquad (10.7)$$
$$z \in \{0, 1\}^n. \qquad\qquad\qquad\qquad\qquad\qquad (10.8)$$

The constraints $z_t \leq 1 - y_{t-1}$ are required to ensure that $z_t = 1$ if and only if there is a start-up in t. However, these constraints can often be dropped because cost minimization and $g_t > 0$ ensure that $z_t = 0$ if $y_{t-1} = 1$.

10.4.1 A Dynamic Programming Algorithm for *LS-U-SC*

When some regeneration interval starts in period t (i.e., $x_t = d_{tk}$ for some $k \geq t$), it may be optimal to perform a start-up in period $i < t$ and to set up between i and t so as to avoid a costly start-up in period t. A simple backward dynamic programming recursion is presented here for the case where set-up and start-up costs are nonnegative. Below we assume that the unit production and storage costs have been normalized, using the flow balance constraints, so that $h_t = 0$ for all t, and the objective function to be minimized is $\sum_{t=1}^{n}(q_t y_t + g_t z_t + p_t x_t)$.

Let $G(t)$ be the value of an optimal solution for periods t up to n when production occurs in period t (i.e., $x_t = d_{tk}$ for some $k \geq t$) but with the possibility of performing a start-up earlier than period t. Thus $G(t)$ includes the start-up cost before t and the set-up costs between the start-up period and period t. On the other hand, let $G'(t)$ be the value of an optimal solution for periods t up to n when production occurs in period t , but where start-up and set-up costs in and before period t are ignored.

$G(t)$ and $G'(t)$ can be calculated recursively as follows.

$$G(t) = \min_{1 \leq \tau \leq t} \left[g_\tau + \sum_{i=\tau}^{t} q_i \right] + G'(t) \qquad \text{for } t = n, \ldots, 1 \qquad (10.9)$$

$$G'(t) = \min_{t < \tau \leq n+1} \left[p_t d_{t,\tau-1} + \min\{G(\tau), \sum_{i=t+1}^{\tau} q_i + G'(\tau)\} \right]$$
$$\text{for } t = n, \ldots, 1 \qquad (10.10)$$

with $G'(n+1) = G(n+1) = 0$ and $q_{n+1} = 0$.

To obtain $G'(t)$, we minimize the cost over the first production period τ after t. For each such τ, $x_t = d_{t,\tau-1}$ with variable cost $p_t d_{t,\tau-1}$ and the optimal cost for the remaining periods τ up to n is $G(\tau)$ if it is cheaper to start up between t and τ, or $\sum_{i=t+1}^{\tau} q_i + G'(\tau)$ if a start-up is best avoided by performing set-ups in periods $t+1$ to τ. $G(t)$ is then obtained from $G'(t)$ by finding the best choice of start-up period prior to t so as to be ready to produce in period t.

$G(1)$ provides the optimal value of *LS-U-SC* and working forward from periods 1 to n gives a corresponding optimal solution. A direct implementation of the above recursion gives an $O(n^2)$ algorithm. This can be improved to $O(n \log n)$.

Example 10.10 *Suppose that $n = 5$, $d = (4, 2, 1, 3, 6)$, $p = (4, 2, 3, 1, 1)$, $q = (2, 10, 3, 4, 7)$, and $g = (15, 10, 30, 10, 40)$.*
Below we show some of the values obtained using the recursion.

$G'(5) = p_5 d_5 = 6.$
$G(5) = \min\{15 + (2 + 10 + 3 + 4 + 7), 10 + (10 + 3 + 4 + 7), 30 + (3 + 4 + 7), 10+$

$(4 + 7), 40 + 7\} + G'(5) = 27.$
$G'(4) = \min\{p_4 d_4 + \min[G(5), q_5 + G'(5)], p_4 d_{45} + \min[0, q_5]\} = 9.$
$G(4) = 23.$
$G'(3) = 16.$
$G(3) = 39.$
. . . .
$G(1) = 65.$
Solution $y = (1, 1, 1, 1, 0), z = (1, 0, 0, 0, 0), x = (4, 3, 0, 9, 0).$

10.4.2 Tight Formulations and Inequalities for *LS-U-SC*

Valid Inequalities and Convex Hull of $X^{LS-U-SC}$

Consider the set of feasible set-up and start-up plans satisfying the constraints
(10.5)–(10.8).

Observation 10.3 *In any feasible production plan, $y_k + z_{k+1} + z_{k+2} + \cdots + z_t = 0$ for some $k < t$ if and only if $y_k = y_{k+1} = \cdots = y_t = 0$.*

This suggests the following simple valid inequality:

$$s_{k-1} \geq \sum_{u=k}^{l} d_u (1 - y_k - z_{k+1} - \ldots - z_u) \quad \text{for } k \leq l \leq n. \quad (10.11)$$

This inequality can then be generalized to give the so-called (k, l, L_1, L_2) inequalities.

Proposition 10.12 *Let $L = \{k, \ldots, l\}$ with $k \leq l \leq n$. Let S, L_1, L_2 be a partition of L with $k \in L_1$. For $j \in L_2$, define $p(j) = \max\{i \in L_1 \cup L_2 : i < j\}$. Then the (k, l, L_1, L_2) inequality*

$$s_{k-1} + \sum_{i \in S} x_i \geq \sum_{j=k}^{l} d_j \left[1 - \sum_{i \in L_1, i \leq j} y_i - \sum_{i \in L_2, i \leq j} (z_{p(i)+1} + \cdots + z_i)\right] (10.12)$$

is valid for $X^{LS-U-SC}$.

Proof. To show validity, it suffices to show that if

$$\sum_{i \in L_1, i \leq j} y_i + \sum_{i \in L_2, i \leq j} (z_{p(i)+1} + \cdots + z_i) = 0,$$

then $y_i = 0$ for all $i \in [k, j] \cap (L_1 \cup L_2)$. Clearly $y_i = 0$ for all $i \in [k, j] \cap L_1$. Now consider $i \in [k, j] \cap L_2$. Let $p^1(i) = p(i)$ and $p^t(i) = p(p^{t-1}(i))$. Find the smallest i such that $p^t(i) \in L_1$. Such a t exists as $p^\tau(i) \in L_1 \cup L_2, p^\tau(i) < p^{\tau-1}(i)$, and $k \in L_1$. Now

$$(y_{p^t(i)} + z_{p^t(i)+1} + \cdots + z_{p^{t-1}(i)}) + (z_{p^{t-1}(i)+1} + \cdots + z_{p^{t-2}(i)})$$
$$+ \cdots + (z_{p^1(i)+1} + \cdots + z_i) = 0,$$

implying, by Observation 10.3, that $y_i = 0$. $\square$

Example 10.11 *Consider LS-U-SC and take $k = 2$, $l = 8$, $S = \{4, 6\}$, $L_1 = \{2, 7\}$, and $L_2 = \{3, 5, 8\}$. The $(2, 8, \{2, 7\}, \{3, 5, 8\})$ inequality is*

$$
\begin{aligned}
s_1 + x_4 + x_6 \geq\ & d_2(1 - y_2) \\
& + d_3(1 - y_2 - z_3) \\
& + d_4(1 - y_2 - z_3) \\
& + d_5(1 - y_2 - z_3 - z_4 - z_5) \\
& + d_6(1 - y_2 - z_3 - z_4 - z_5) \\
& + d_7(1 - y_2 - z_3 - z_4 - z_5 - y_7) \\
& + d_8(1 - y_2 - z_3 - z_4 - z_5 - y_7 - z_8),
\end{aligned}
$$

where each term j $(j = 2, \ldots, 8)$ in the right-hand side determines whether d_j has to be produced in periods of S or before period k. For example, the last term implies that if $y_2 + z_3 + z_4 + z_5 + y_7 + z_8 = 0$, then $y_2 = y_3 = y_5 = y_7 = y_8 = 0$, so demand d_8 must be produced in S or before period 2, and thus $s_1 + x_4 + x_6 \geq d_8$.

Theorem 10.13 *The convex hull of $X^{LS-U-SC}$ is completely described by the flow balance and bound constraints, the start-up constraints (10.5)–(10.7) and the (k, l, L_1, L_2) inequalities (10.12) for all $1 \leq l \leq n$ and all disjoint subsets L_1, L_2 of $\{1, \ldots, l\}$ with $k = \min\{i : i \in L_1\} < \min\{i : i \in L_2\}$.*

The separation problem for the (k, l, L_1, L_2) inequalities can be solved for fixed l as a shortest path problem over an acyclic network with $O(l^2)$ arcs. Solving for each l leads to an $O(n^3)$ algorithm.

Extended Formulations for *LS-U-SC*

The facility location reformulation seen earlier can be applied to *LS-U-SC* by defining new variables w_{it} which represent the fraction of the demand of period t produced in period i. Now, again based on Observation 10.3, the following constraints

$$
\sum_{i=i_1}^{i_2} w_{it} \leq y_{i_1} + z_{i_1+1} + \cdots + z_{i_2}, \tag{10.13}
$$

where $1 \leq i_1 \leq i_2 \leq t \leq n$ can easily be seen to be valid.
This leads to the extended formulation $Q^{LS-U-SC}$

$$
\sum_{s=1}^{t} w_{st} = 1 \qquad\qquad \text{for } 1 \leq t \leq n
$$

$$
w_{st} \leq y_s \qquad\qquad \text{for } 1 \leq s \leq t \leq n
$$

$$z_t \geq y_t - y_{t-1} \qquad\qquad \text{for } 1 \leq t \leq n$$

$$z_t \leq y_t \qquad\qquad \text{for } 1 \leq t \leq n$$

$$\sum_{i=k}^{t} w_{it} \leq y_k + z_{k+1} + \ldots + z_t \qquad\qquad \text{for } 1 \leq k < t \leq n$$

$$x_u = \sum_{t=u}^{n} d_t w_{ut} \qquad\qquad \text{for } 1 \leq t \leq n$$

$$w \in \mathbb{R}^{n(n+1)/2}, y, z \in [0,1]^n,$$

which can be shown to be tight.

Theorem 10.14 $conv(X^{LS-U-SC}) = proj_{x,y,z} Q^{LS-U-SC}$.

10.4.3 Wagner–Whitin Costs WW-U-SC

Here the formulations are remarkably compact, and thus very useful in practice. Let $X^{WW-U-SC}$ denote the set of solutions of

$$s_{k-1} + \sum_{u=k}^{l} d_{ul} y_u \geq d_{kl} \qquad\qquad \text{for } 1 \leq k \leq l \leq n \qquad (10.14)$$

$$z_t \geq y_t - y_{t-1} \qquad\qquad \text{for } 1 \leq t \leq n \qquad (10.15)$$

$$z_t \leq y_t \qquad\qquad \text{for } 1 \leq t \leq n \qquad (10.16)$$

$$s \in R_+^{n+1}, y, z \in \{0,1\}^n. \qquad\qquad\qquad\qquad\qquad (10.17)$$

We consider again the inequalities (10.11). Clearly they dominate the (l, S) inequalities (10.14) as $z_t \leq y_t$ for all t.

Theorem 10.15 *The inequalities (10.11),(10.15),(10.16), and the bound constraints $s \in \mathbb{R}_+^{n+1}, y \in [0,1]^n$ describe $conv(X^{WW-U-SC})$ in the (s, y, z) space.*

10.5 Start-Up Costs: The Capacitated Case

10.5.1 The Discrete Lot-Sizing Problem DLS-CC-SC

This problem is also known in the literature as the *discrete lot-sizing and scheduling problem*. Using the tight formulation for DLS-CC derived in Subsection 9.3.1, we can assume that $C = 1$ and $d_t \in \{0,1\}$ for all t. Now we obtain the formulation

$$\min \sum_{t=1}^{n} q_t y_t + \sum_{t=1}^{n} g_t z_t \qquad\qquad (10.18)$$

$$\sum_{u=1}^{t} y_u \geq d_{1t} \qquad\qquad \text{for all } t \qquad (10.19)$$

$$z_t \geq y_t - y_{t-1} \qquad\qquad \text{for all } t \qquad (10.20)$$

$$z_t \leq y_t \qquad\qquad \text{for all } t \qquad (10.21)$$

$$y, z \in \{0,1\}^n. \qquad\qquad (10.22)$$

We let $X^{DLS-CC-SC}$ denote the feasible region (10.19)–(10.22), and note that $s_t = \sum_{u=1}^{t} y_u - d_{1t}$. Note also that in contrast to LS-U-SC, in this model we are not allowed to start up in a period without producing.

Valid Inequalities for DLS-CC-SC

Proposition 10.16 *Consider an interval $[t,l] \subseteq [1,n]$ with $d_l = 1$. Let $d_{tl} = p > 0$, and let $t_1 < t_2 < \cdots < t_p = l$ be the periods in the interval $[t,l]$ with nonzero (unit) demand. The inequality*

$$s_{t-1} + \sum_{u=t}^{t+p-1} y_u + \sum_{u=t+1}^{t+p-1} [d_{ul} - (t+p-u)]z_u + \sum_{u=t+p}^{l} d_{ul}z_u \geq d_{tl} \quad (10.23)$$

is valid for $X^{DLS-CC-SC}$.

Proof. First we rewrite the inequality as

$$s_{t-1} + \sum_{j=1}^{p}(y_{t+j-1} + z_{t+j} + \cdots + z_{t_j}) \geq p.$$

Given a feasible solution, let τ_j for $j = 1, \ldots, p$ be the period in which the demand of period t_j is produced with $\tau_1 < \tau_2 < \cdots < \tau_p$ and $\tau_j \leq t_j$. Let $k = \max\{j : \tau_j < t+j-1\}$. It follows that

$$y_{t+j-1} + z_{t+j} + \cdots + z_{t_j} \geq 1 \qquad\qquad (10.24)$$

for $j > k$, and, as $\tau_k < t+k-1$, that the demands for periods $t_1, \ldots, t_k$ are all produced by period $t+k-2$ at latest. But this means that $s_{t-1} + \sum_{j=1}^{k}(y_{t+j-1} + z_{t+j} + \cdots + z_{t_j}) \geq s_{t-1} + \sum_{j=1}^{k-1} y_{t+j-1} \geq k$, where the last inequality follows from flow conservation. Summing (10.24) for $j = k+1, \ldots, p$ and adding to this last inequality establishes validity. $\square$

An Extended Formulation for DLS-CC-SC

Suppose again that $d_t \in \{0,1\}$ for all t, $d_{1n} = p$ and $1 \leq t_1 < t_2 < \cdots < t_p \leq n$ are the demand periods.

Let $y_{t,i} = 1$ if there is production in period t and this production is used to satisfy the unit demand in period t_i, and $y_{t,i} = 0$ otherwise with $i \leq t \leq t_i$ and $1 \leq i \leq p$.

Let $z_{t,i} = 1$ if there is production in period t, this production is used to satisfy the unit demand in period t_i, and there is a start-up in period t with $i \leq t \leq t_i$ and $1 \leq i \leq p$.

This leads to the following formulation $Q^{DLS-CC-SC}$ that can be obtained by projecting into the (y, z) space a shortest path formulation derived from a dynamic program solving problem $DLS\text{-}CC\text{-}SC$.

$$\sum_{t=i}^{t_i} y_{t,i} = 1 \qquad \text{for } 1 \leq i \leq d_{1n} \qquad (10.25)$$

$$y_{t,1} = z_{t,1} \qquad \text{for } 1 \leq t \leq t_1 \qquad (10.26)$$

$$y_{t,i} \geq z_{t,i} \qquad \text{for } 2 \leq i \leq d_{1n}, \ i < t \leq t_{i-1} + 1 \qquad (10.27)$$

$$y_{t,i} = z_{t,i} \qquad \text{for } 2 \leq i \leq d_{1n}, \ t_{i-1} + 1 < t \leq t_i \qquad (10.28)$$

$$y_{i-1,i-1} \geq y_{i,i} \qquad \text{for } 2 \leq i \leq d_{1n} \qquad (10.29)$$

$$z_{t,i} \geq y_{t,i} - y_{t-1,i-1} \qquad \text{for } 2 \leq i \leq d_{1n}, \ i < t \leq \min[t_{i-1} + 1, t_i - 1] \qquad (10.30)$$

$$z_{t,i} \geq y_{t,i} - y_{t-1,i-1} \qquad \text{for } 2 \leq i \leq d_{1n}, t = t_{i-1} + 1 = t_i \qquad (10.31)$$

$$z_{t,i} + \sum_{u=t-1}^{t_{i-1}} y_{u,i-1} \leq \sum_{u=t}^{t_i} y_{u,i} \qquad \text{for } 2 \leq i \leq d_{1n}, i < t \leq t_{i-1} + 1 \qquad (10.32)$$

$$y_{t,i}, z_{t,i} \geq 0 \qquad \text{for } i \leq t \leq t_i, \ 1 \leq i \leq p. \qquad (10.33)$$

Here each constraint has a natural interpretation. (10.25) states that each demand in period t_i must be produced in the interval $[i, t_i]$. (10.26) states that the first production period is the first start-up period. (10.27) states that if one starts up in a period, one must produce in that period. (10.28) states that if $i > 1$ and if the demand for t_i is produced in $t \in [t_{i-1} + 2, t_i]$, then t must also be a start-up period. (10.29) states that if $i > 1$ and the demand for t_i is produced in i, the demand for t_{i-1} must be produced in $i - 1$. (10.30) and (10.31) are standard start-up constraints. Finally (10.32) states that for $i > 1$, (a) producing demand $i - 1$ in $[t - 1, t_{i-1}]$ and starting up and producing demand i in t are mutually exclusive because together they imply that $y_{t-1,i-1} = y_{t,i} = 1$ which implies that $z_{t,i} = 0$, and (b) if either occurs, then demand i cannot be produced before period t.

Theorem 10.17 *The linear program*

$$\min \ \sum_{t=1}^{n} q_t y_t + \sum_{t=1}^{n} g_t z_t$$

$$y_t = \sum_i y_{ti} \qquad \text{for all } t$$

$$z_t = \sum_i z_{ti} \qquad \text{for all } t$$

$$\{(y_{ti}, z_{ti})_{i \leq t \leq t_i, 1 \leq i \leq p}\} \in Q^{DLS-CC-SC}$$

solves DLS-CC-SC.

When the total batch production costs are nondecreasing $q_t \geq q_{t+1}$ for all t, a compact formulation is known just involving the original variables. This special case is denoted WW in Table 10.5.3.

Theorem 10.18 If $q_t \geq q_{t+1}$ for all t, the the linear program

$$\min \{qy + gz \; : \; (s, y, z) \text{ satisfy } (10.23) \qquad \text{for all intervals } [t, t_i]$$

$$s_t = \sum_{u=1}^{t} y_u - d_{1t} \qquad \text{for all } t$$

$$y_t \geq z_t \geq y_t - y_{t-1} \qquad \text{for all } t$$

$$s \in \mathbb{R}_+^n, y, z \in [0, 1]^n \qquad \}$$

solves DLS-CC-SC.

Example 10.12 Consider an instance of DLS-CC-SC after normalization of the demands with $n = 10$ and $d = (0, 0, 1, 0, 1, 1, 0, 0, 1, 0)$. Taking $t = 2$ and $i = 3$, the $[t, t_i] = [2, 6]$ inequality (10.23) is of the form

$$s_1 \geq 1(1 - y_2 - z_3) + 1(1 - y_3 - z_4 - z_5) + 1(1 - y_4 - z_5 - z_6), \text{ or}$$
$$s_1 + y_2 + y_3 + y_4 + z_3 + 2z_4 + 2z_5 + z_6 \geq 3.$$

10.5.2 Capacitated Lot-Sizing with Start-Up Costs LS-C-SC

Valid Inequalities

We take as starting set $X^{LS-C-SC}$:

$$s_{t-1} + x_t = d_t + s_t \qquad \text{for all } t$$
$$x_t \leq C_t y_t \qquad \text{for all } t$$
$$z_t \geq y_t - y_{t-1} \qquad \text{for all } t$$
$$z_t \leq y_t \qquad \text{for all } t$$
$$z_t \leq 1 - y_{t-1} \qquad \text{for all } t$$
$$s \in \mathbb{R}_+^{n+1}, x \in \mathbb{R}_+^n, y, z \in \{0, 1\}^n.$$

Note that for any $k \in N = [1, n]$ and $S \subseteq [k, n]$, the function

$$g_k(S) = \max_{j \in [k,n]} (d_{kj} - \sum_{i \in [k,j] \setminus S} C_i)^+ \tag{10.34}$$

gives a lower bound on s_{k-1} provided that there is no set-up in any period in S.

Observation 10.4 *If $S \subseteq [k,n]$, $u = \min\{i \in S\}$, and $v = \max\{i \in S\}$, the inequality*

$$s_{k-1} \geq g_k(S)(1 - y_u - z_{u+1} - \cdots - z_v)$$

is valid for $X^{LS-C-SC}$.

The function g_k has special structure.

Proposition 10.19 *For g_k defined as in (10.34),*
i. g_k is nonnegative and nondecreasing.
ii. g_k is supermodular on $\{k, \ldots, n\}$; see Definition 9.2.
iii. $\beta_{uv} = g_k([u,v]) + g_k([u+1, v-1]) - g_k([u+1, v]) - g_k([u, v-1]) \geq 0$ for $k \leq u \leq v \leq n$, where $g_k([a,b]) = 0$ for $a > b, a < k$ or $b > n$.
iv. $g_k([u,v]) = \sum_{u < \bar{u} \leq \bar{v} \leq v} \beta_{\bar{u}\bar{v}}$ for $k \leq u \leq v \leq n$.

This structure leads to the *left supermodular* inequalities generalizing the inequality of Observation 10.4.

Proposition 10.20 *Let $F \subseteq \{(u,v) : k \leq u \leq v \leq n\}$. The inequality*

$$s_{k-1} \geq \sum_{(u,v) \in F} \beta_{uv}(1 - y_u - \sum_{j=u+1}^{v} z_j) \tag{10.35}$$

is valid for $X^{LS-C-SC}$.

Proof. Consider a point $(x, s, y, z) \in X^{LS-C-SC}$ and let $T = \{j \in [k,n] : y_j = 0\}$ be the periods from k onwards in which there is no set-up. If $T = \emptyset$, then $y_j = 1$ for all $j \in [k,n]$, so $1 - y_u - \sum_{j=u+1}^{v} z_j \leq 0$ for all $(u,v) \in F$, and the inequality is implied by $s_{k-1} \geq 0$.

Otherwise suppose that $T = \cup_{i=1}^{p}[u_i, v_i]$ is the union of disjoint intervals with $v_i + 1 < u_{i+1}$ for all i. Consider a pair $(u,v) \in F$. If $[u,v]$ does not lie in $[u_i, v_i]$ for some i, a set-up occurs in $[u,v]$ and thus $1 - y_u - \sum_{j=u+1}^{v} z_j \leq 0$. Thus we have

$$\sum_{(u,v) \in F} \beta_{uv}(1 - y_u - \sum_{j=u+1}^{v} z_j)$$
$$\leq \sum_{i=1}^{p} \sum_{(u,v) \in F, [u,v] \subseteq [u_i, v_i]} \beta_{uv}(1 - y_u - \sum_{j=u+1}^{v} z_j)$$
$$\leq \sum_{i=1}^{p} \sum_{(u,v) \in F, [u,v] \subseteq [u_i, v_i]} \beta_{uv}$$
$$\leq \sum_{i=1}^{p} \sum_{(u,v) : [u,v] \subseteq [u_i, v_i]} \beta_{uv} = \sum_{i=1}^{p} g_k([u_i, v_i])$$
$$\leq g_k(\sum_{i=1}^{p}[u_i, v_i]) \text{ by supermodularity}$$
$$= g_k(T) \leq s_{k-1} \text{ by definition (10.34) of } g_k . \qquad \square$$

Example 10.13 *Suppose that* $n = 5, C = (4, 3, 6, 7, 6),$ *and* $d = (2, 1, 5, 3, 2).$
Taking $k = 2,$ *we obtain*

$$g_2([k, l]) = \begin{array}{c|cccc} & l = 2\ 3\ 4\ 5 \\ \hline k = 2 & 1\ 6\ 9\ 11 \\ k = 3 & 3\ 6\ 8 \\ k = 4 & 0\ 2 \\ k = 5 & 0 \end{array} \quad and \quad (\beta_{kl}) = \begin{array}{c|cccc} & l = 2\ 3\ 4\ 5 \\ \hline k = 2 & 1\ 2\ 0\ 0 \\ k = 3 & 3\ 3\ 0 \\ k = 4 & 0\ 2 \\ k = 5 & 0 \end{array}$$

For example, $g_2([3, 4]) = \max(d_2 - C_2, d_{23} - C_2, d_{24} - C_2, d_{25} - C_2 - C_5)^+ = \max(1 - 3, 6 - 3, 9 - 3, 11 - 9)^+ = 6,$ and $\beta_{23} = g_2([2, 3]) - g_2([2, 2]) - g_2([3, 3]) = 6 - 1 - 3 = 2.$

Taking $k = 2$ and $F = \{(2, 2), (2, 3), (3, 3), (3, 4), (4, 5)\},$ the valid inequality (10.35) is of the form

$$s_1 \geq 1(1 - y_2) + 2(1 - y_2 - z_3) + 3(1 - y_3) + 3(1 - y_3 - z_4) + 2(1 - y_4 - z_5).$$

Separation for Left Supermodular Inequalities

Observe that $\beta_{uv} > 0$ only if $g_k([u, v]) > 0$ and $g_k([u+1, v-1]) = 0$. It follows that the $\beta_{uv} > 0$ lie on a frontier involving at most $2(n - k + 1)$ pairs (u, v). This frontier is illustrated in Example 10.13. By following this frontier, it is easy to find a most violated inequality for fixed k in $O(n)$.

Similar inequalities can be obtained by working backwards, and looking for lower bounds on the stock at the end of an interval. Specifically

$$h_l(T) = \max_{j=1,\ldots,l} \left(\sum_{i \in [j,l] \cap T} C_i - d_{jl} \right)^+$$

is a lower bound on the end-stock s_l if production takes place at full capacity in periods $T \subseteq [1, l]$.

Observing that $y_v - \sum_{j=u+1}^{v} z_j = 1$ if and only if $y_j = 1$ for all $j \in [u, v]$, we obtain the basic inequality

$$s_l + \sum_{j \in T} (C_j y_j - x_j) \geq h_l(T)(y_v - \sum_{j=u+1}^{v} z_j),$$

where $u = \min\{i : i \in T\}, v = \max\{i : i \in T\}$. The set function $h_l(T)$ is again supermodular, and the above inequality can be generalized in the same way as in Proposition 10.20 leading to a family of *right supermodular* inequalities.

Valid Inequalities for *LS-CC-SC*

Using the disjunction $y_j + \sum_{u=j+1}^{l} z_u = 0$ (implying no set-ups in the interval $[j, l]$), or $y_j + \sum_{u=j+1}^{l} z_u \geq 1$, one obtains:

Proposition 10.21 *The (k,j,l) inequality*

$$s_{k-1} + C(\sum_{u=k}^{j} y_u + \sum_{u=j+1}^{l} z_u) \geq d_{k,j-1} + \min[d_{jl}, C] \qquad (10.36)$$

is valid for $X^{LS-CC-SC}$.

The family of *left extended klSI* inequalities is obtained by applying the mixing procedure to these inequalities.

Example 10.14 *Let $n = 4, d = (7,4,5,6)$, and $C = 10$. With $(k,j,l) = (1,3,4)$, the inequality (10.36) gives*

$$s_0 + 10(y_1 + y_2 + y_3 + z_4) \geq d_{12} + C = 21,$$

and with $(k,j,l) = (1,2,3)$ gives

$$s_0 + 10(y_1 + y_2 + z_3) \geq d_1 + d_{23} = 16.$$

We also have the standard surrogates

$$s_0 + 10(y_1 + y_2 + y_3 + y_4) \geq d_{14} = 22, \quad \text{and}$$

$$s_0 + 10y_1 \geq 7.$$

Mixing these four inequalities gives a left extended klSI inequality:

$$s_0 \geq 1(3 - y_1 - y_2 \quad y_3 - z_4) + (2-1)(3 - y_1 - y_2 - y_3 - y_4)$$
$$+(6-2)(2 - y_1 - y_2 - z_3) + (7-6)(1 - y_1).$$

Optimization for *LS-CC-SC*

It is easily seen that the dynamic programming recursion for *LS-CC* of Section 9.6.1 extends to include start-up costs.

10.5.3 Résumé of Results

In Table 10.5.3 we summarize the results that we know for $PROB\text{-}[U,CC]\text{-}SC$. $\star\star\star$ indicates that nothing is known about the specific question. *LP* indicates that separation by linear programming is possible because there is a tight polynomial-size extended formulation, and $\star$ indicates that an explicit description of a family of valid inequalities is known but it only gives a partial description of the convex hull of solutions. The reader is asked to work out the complexity of the separation problem for *WW-U-SC* in Exercise 10.13.

Table 10.3. Models with Start-Up Costs $PROB\text{-}[U,CC]^1\text{-}SC$

	LS	WW	DLS
Formulation	$Cons \times Vars$	$Cons \times Vars$	$Cons \times Vars$
U	$O(n^2) \times O(n^2)$	$O(n^2) \times O(n)$	–
CC	$\star\star\star$	$\star\star\star$	$O(n^2) \times O(n^2)$ $(WW)\ O(n^2) \times O(n)$
Separation			
U	$O(n^3)$	$Exercise 10.13$	–
CC	$O(n^2)$ and $\star$	$\star\star\star$	LP and $\star$
Optimization			
U	$O(n \log n)$	$O(n)$	–
CC	$O(n^4)$	$\star\star\star$	$O(n^2)$ $(WW)\ O(n \log n)$

10.6 Backlogging and Start-Ups $WW\text{-}U\text{-}B, SC$

Consider the problem

$$\min \ \sum_{t=0}^{n} h_t s_t + \sum_{t=1}^{n} b_t r_t + \sum_{t=1}^{n} q_t y_t + \sum_{t=1}^{n} g_t z_t$$

$$s_{t-1} - r_{t-1} + x_t = d_t + s_t - r_t \qquad \text{for all } t$$
$$x_t \leq My_t \qquad \text{for all } t$$
$$z_t - w_{t-1} = y_t - y_{t-1} \qquad \text{for all } t$$
$$z_t \leq y_t \qquad \text{for all } t$$
$$x \in \mathbb{R}_+^n, \ s,r \in \mathbb{R}_+^{n+1}, \ y,z,w \in \{0,1\}^n, \ r_0 = 0.$$

The extended formulation $Q^{WW-U-B,SC}$ below, generalizing that for *WW-U-B* in Subsection 10.2.3, is based on two observations. If d_u is satisfied from stock, and there is no switch-off in the interval $[t, u-1]$, then there is no set-up in $[t, u]$, and s_{t-1} contains the demand d_u. Similarly if d_u is satisfied by backlogging, and there is no start-up in $[u+1, t]$, then there is no set-up in $[u, t]$ and r_l contains the demand d_u.

$$\alpha_t + y_t + \beta_t = 1 \qquad \text{for all } t \text{ with } d_t > 0$$
$$\beta_{t+1} + z_{t+1} \geq \beta_t \qquad \text{for all } t$$
$$s_{t-1} \geq \sum_{u=t}^{k} d_u \left(\alpha_u - \sum_{j=t}^{u-1} w_j \right) \qquad \text{for all } t,k \text{ with } t \leq k$$
$$r_t \geq \sum_{u=k}^{t} d_u \left(\beta_u - \sum_{j=u+1}^{t} z_j \right) \qquad \text{for all } k,t \text{ with } k \leq t$$

$$z_t - w_{t-1} = y_t - y_{t-1} \qquad\qquad \text{for all } t$$
$$z_t \leq y_t \qquad\qquad \text{for all } t$$
$$x \in \mathbb{R}_+^n, \ s, r \in \mathbb{R}_+^{n+1}, \ y, z, w \in [0,1]^n, \ r_0 = 0.$$

Theorem 10.22 *The linear program:*

$$\min\{hs + br + qy + gz : (x, s, r, y, z, w, \alpha, \beta) \in Q^{WW-U-B-SC}\}$$

solves WW-U-$\{B, SC\}$.

Using the same approach as in Subsection 10.2.3, a complete linear description of conv($X^{WW-U-B,SC}$) is obtained by separating out the α, β variables, and the separation problem can be solved in $O(n^3)$ by finding a minimum mean cost cycle.

Exercises

Exercise 10.1 Consider an instance of *LS-U-B* with $s_0 = 0$, $n = 5$, $h' = (0,5,1,1,2,1)$, $p' = (2,4,3,5,3)$, $b' = (3,3,3,5,4)$, $q = (70, 81, 70, 80, 63)$, and $d = (8, 0, 7, 5, 11)$.
i. Find an optimal solution by dynamic programming.
ii. Solve using LS–LIB and an extended formulation.

Exercise 10.2 Derive a shortest path algorithm for *LS-U-B* resembling that for *LS-U*. In particular, indicate how to compute the cost α_{kl} of any regeneration interval $[k, l]$.

Exercise 10.3 i.* Derive an $O(n \log n)$ dynamic programming algorithm for *LS-U-B*.
ii.* Derive an $O(n)$ dynamic programming algorithm for *WW-U-B*.

Exercise 10.4 Derive a facility location reformulation for *LS-U-B* by starting from the known facility location formulation for *LS-U*.

Exercise 10.5 Consider an instance of *WW-U-B* with $n = 5$, $h = (9, 1, 3, 2, 1, 4)$, $b = (3, 3, 5, 5, 8)$, $q = (80, 21, 39, 50, 73)$, and $d = (8, 0, 7, 5, 11)$.
i. Find an optimal solution by dynamic programming.
ii. Solve using LS–LIB and an extended formulation.
iii. Consider the fractional solution given by $x^* = (0, 18.5, 12.5, 0, 0)$, $y^* = (0, 1, 0.125, 0, 0)$, $s^* = (0, 0, 10.5, 16, 11, 0)$, $r^* = (8, 0, 0, 0, 0)$. Find a valid inequality cutting off this point.

Exercise 10.6 Consider an instance of *DLS-CC-B* with $n = 5$, $h = (5, 1, 1, 2, 1)$, $b = (3, 3, 3, 5, 9)$, $q = (20, 31, 15, 19, 23)$, $d = (6, 0, 7, 5, 4)$, and $C = 10$.
i. Find an optimal solution.
ii. Find a valid inequality cutting off the point $y^* = (1, 0, 0.3, 0.5, 0.4)$, $s^* = (4, 4, 0, 0, 0)$, $r^* = 0$.

Exercise 10.7 Consider an instance of $WW\text{-}CC\text{-}B$ with $n = 5, h = (0, 5, 1, 1, 2, 1), b = (3, 3, 3, 5, 9), q = (40, 51, 65, 59, 43), d = (6, 0, 7, 5, 4)$, and $C = 10$. Find a valid inequality cutting off the point $y^* = (1, 0.3, 0, 0.9, 0), s^* = (0, 4, 7, 0.9, 4, 0), r^* = (0, 0, 0.9, 0, 0), x = (10, 3, 0, 9, 0)$.

Exercise 10.8 Consider the extended formulation for $WW\text{-}CC\text{-}B$ defined in Theorem 10.11.
i. Show that it is a valid formulation for $WW\text{-}CC\text{-}B$.
ii. Prove the theorem by showing that the underlying matrix (without the equality constraints defining the s_{k-1} and r_l variables) is totally unimodular.

Exercise 10.9 Consider an instance of $LS\text{-}CC\text{-}B$ with $n = 6$, and data defined by $h' = (0,1,2,1,1,3,1)$, $b' = (1,1,1,1,1,0)$, $p' = (1,3,5,5,3,5)$, $q = (15, 12, 32, 31, 24, 17)$, $d = (8, 3, 5, 6, 1, 4)$, $s_0 = s_6 = r_6 = 0$, and $C = 10$.
i. Solve with a MIP solver.
ii. Solve with LS–LIB using extended formulations of appropriate relaxations.

Exercise 10.10 Consider an instance of $LS\text{-}U\text{-}SC$ with $n = 5$, and data $h = (5,1,1,1,1,1), p = (5, 3, 8, 4, 6), g = (30, 30, 40, 40, 30), q = (40, 51, 65, 59, 43)$, and $d = (6, 0, 7, 5, 4)$.
i. Find an optimal solution by dynamic programming.
ii. Solve using LS–LIB and an extended formulation.

Exercise 10.11 Prove Theorem 10.14 by showing that every (k, l, L_1, L_2) inequality is valid for $Q^{LS-U-SC}$.

Exercise 10.12 Consider an instance of $WW\text{-}U\text{-}SC$ with $n = 5$, and data defined by $h = (5, 4, 4, 3, 1, 2), g = (30, 30, 40, 40, 30), q = (40, 51, 65, 59, 43)$, and $d = (6, 0, 7, 5, 4)$.
i. Find an optimal solution by dynamic programming with $s_0 = 0$.
ii. Find an optimal solution by dynamic programming with $s_0 \geq 0$.
iii. Solve using LS–LIB and an extended formulation.

Exercise 10.13 i. Describe a separation algorithm for $WW\text{-}U\text{-}SC$ based on the inequalities (10.11), and analyze its complexity.
ii. For the instance of Exercise 10.12, find a valid inequality cutting off the point $x^* = (6, 4.364, 7.447, 4.189)$, $y^* = (0.273, 0.273, 0.465, 0.465, 0)$, $z^* = (0.273, 0, 0.192, 0, 0)$, $s^* = (0, 0, 4.364, 4.811, 4, 0)$.

Exercise 10.14 Derive a polynomial time dynamic programming algorithm for $DLS\text{-}CC\text{-}SC$.

Exercise 10.15 Consider an instance of $DLS\text{-}CC\text{-}SC$ with $n = 10, g = (30, 43, 48, 40, 30, 20, 15, 34, 12, 33)$, $q = (80, 65, 61, 59, 43, 34, 33, 32, 29, 27)$, and $d = (0, 0, 1, 0, 0, 1, 0, 1, 0, 1)$.
i. Find an optimal solution.
ii. Solve as a linear program using LS–LIB.
iii. Find a valid inequality cutting off the point $y^* = (\frac{1}{3}, \frac{1}{3}, \frac{1}{3}, \frac{1}{4}, \frac{1}{4}, \frac{1}{2}, \frac{1}{2}, \frac{1}{2}, \frac{1}{2}, \frac{1}{2})$, $z^* = (\frac{1}{3}, 0, 0, 0, 0, \frac{1}{4}, 0, 0, 0, 0)$.

Exercise 10.16 Consider an instance of $LS\text{-}CC\text{-}SC$ with $n = 5$, $C = 10$, $p = 0$, $h = (5, -4, 4, -3, -1, 2)$, $g = (30, 30, 20, 40, 30)$, $q = (40, 71, 25, 59, 13)$, and $d = (6, 0, 7, 5, 4)$.
i. Find a left supermodular inequality cutting off the point $x^* = (10, 0, 10, 1, 1)$, $y^* = (1, 0, 1, 0.1, 0.1)$, $z^* = (1, 0, 1, 0, 0)$, $s^* = (0, 4, 4, 7, 3, 0)$.
ii. Find a left extended $klSI$ inequality cutting off the above point.

Exercise 10.17 Consider an instance of $WW\text{-}U\text{-}B, SC$ with $n = 5$, $h = (5, 4, 4, 3, 1, 2)$, $b = (5, 6, 4, 3, 10)$, $g = (30, 30, 40, 40, 30)$, $q = (40, 51, 65, 59, 43)$, and $d = (6, 0, 7, 5, 4)$.
i) Solve with a MIP solver.
ii) Solve with a MIP solver, using a tight extended formulation for the $WW\text{-}U\text{-}B$ relaxation from LS–LIB.
iii) Solve as a linear program by using an appropriate reformulation.

* Starred exercises are more difficult and require more mathematical or technical developments.

Notes

Section 10.2. Zangwill [196, 198] studied the uncapacitated lot-sizing problem with backlogging $LS\text{-}U\text{-}B$ and developed dynamic programming recurrences for the problem. As for $LS\text{-}U$, a faster implementation is possible as shown in Aggarwal and Park [3], Federgrün and Tzur [63], and van Hoesel [164]. The observation that the facility location and shortest path reformulations are tight follows from Barany et al. [22]. The extended formulation and convex hull description of $WW\text{-}U\text{-}B$ are from Pochet and Wolsey [140]. A description of the minimum mean cost cycle algorithm can be found in Ahuja et al. [7].

Section 10.3. The convex hull description of $DLS\text{-}CC\text{-}B$ is from Miller and Wolsey [124], and the optimization algorithm is due to Van Vyve [176]. The reformulation for $DLSI\text{-}CC\text{-}B$ was proposed by Miller and Wolsey [125], and shown to be tight by Van Vyve [178]. A proof of the validity of the formulation (10.3)–(10.4) for $WW\text{-}CC\text{-}B$ and the tight formulations for $WW\text{-}CC\text{-}B$ and $LS - CC - B$ are from Van Vyve [178, 180].

Section 10.4. The basic dynamic programming algorithm for $LS\text{-}U\text{-}SC$, as well as a faster $O(n \log n)$ version, appears in van Hoesel [164]. More generally he shows that there is an $O(n \log n)$ algorithm for $LS\text{-}U\text{-}B, SC$ and an $O(n)$ algorithm for $WW\text{-}U\text{-}B, SC$. The convex hull description of $LS\text{-}U\text{-}SC$ by (k, l, S_1, S_2) inequalities is from van Hoesel et al. [170]. The extended formulation was proposed in Wolsey [192] and shown to be tight in [170]. That for $WW\text{-}U\text{-}SC$ is from Pochet and Wolsey [140].

Section 10.5. The valid inequalities for $DLS\text{-}CC\text{-}SC$ and Theorem 10.18 are from van Eijl and van Hoesel [163], and the extended formulation is from van Hoesel and Kolen [165]. Other classes of valid inequalities such as hole-bucket inequalities are described in van Eijl [162]. For optimization an $O(n^2)$ dynamic programming algorithm is given in Fleischmann [67] and an algorithm of the same complexity is obtained as a special case of a multi-item recursion of Salomon [147]. With nondecreasing set-up costs (WW), faster algorithms are presented in van Hoesel et al. [166].

The left supermodular and left extended $klSI$ inequalities for $LS\text{-}C\text{-}SC$ and $LS\text{-}CC\text{-}SC$, respectively, are from Constantino [46]. The paper also contains two related families of inequalities, the right supermodular and the right extended $klSI$ inequalities as well as combinatorial separation algorithms.

Section 10.6. The results for $WW\text{-}U\text{-}B, SC$ are from Agra and Constantino [6].

11

Single-Item Variants

Here we consider several important variants of the single-item problem. After listing the variants, we examine each of them in turn and give results on valid inequalities, tight formulations, and so on. Throughout the chapter, we concentrate on variants for which reformulation results are available.

The variants treated concern either changes in demand, production constraints and costs, or stock constraints and costs.

First we consider variants in the demands.

- In Section 11.1 we consider a problem with potential sales limited by a fixed upper bound (in place of fixed demands) SL, and derive both a valid inequality description, and a tight extended formulation in a slightly restricted case. This is also known as the problem with *lost sales*.

Next we consider different production options.

- In Section 11.2, we suppose that if there is production in a period, at least a certain amount must be produced. We classify this as (Constant) Lower Bounds LB.
- In Section 11.3, we suppose that there are restrictions related to the production sequence rather than just the production in one period. We study the case where production is at full capacity, except in the first and last periods of a production sequence, which we call Almost Full Capacity Production or AFC, and also the case with lower bounds on the total amount produced during a production sequence, which we call Minimum Runs or MR.
- In Section 11.4 we consider bounds on the length of a production sequence, or Restricted Length Set-Up Sequences RLS.
- In Section 11.5 we treat the modeling of Piecewise Concave Production Costs CP.
- In Section 11.6, we consider the problem with production time windows TWP, in which each customer demand has to be satisfied from products

manufactured within a given time window. This variant is usually considered to model product perishability constraints, and implicitly or explicitly leads to bounds on the stock levels.

Next we examine two different options for the stocks.

- In Section 11.7, we consider the effect of limited storage capacities, or Upper Bounds on Stocks SUB.
- In Section 11.8, we generalize by allowing for Safety Stocks or Piecewise Convex Stock Costs SS.

Finally we consider a model incorporating several of these features simultaneously, and a model that allows us to treat uncertainty.

- In Section 11.9 we provide a tight extended formulation for a model with backlogging, piecewise concave production costs, and sales.
- In Section 11.10 we formulate a lot-sizing model on a tree which allows us to model a stochastic lot-sizing problem with a tree of scenarios.

Other variants including fixed costs on stocks, limits on the number of set-ups, and constant set-up times are treated in the Exercises and Notes.

11.1 Sales or Variable Demand (SL)

Here the additional data are upper bounds u_t on the potential sales in period t, and a unit selling price γ_t.

If v_t is a variable representing the quantity sold (distinct from the fixed demand d_t) in period t, a formulation for the profit maximization problem $LS\text{-}C\text{-}SL$ is

$$\max \ \sum_{t=1}^{n} \gamma_t v_t - \sum_{t=1}^{n} p_t x_t - \sum_{t=0}^{n} h_t s_t - \sum_{t=1}^{n} f_t y_t$$

$$s_{t-1} + x_t = d_t + v_t + s_t \qquad \text{for all } t$$

$$x_t \leq C_t y_t \qquad \text{for all } t$$

$$v_t \leq u_t \qquad \text{for all } t$$

$$s \in \mathbb{R}_+^{n+1}, x, v \in \mathbb{R}_+^n, y \in \{0,1\}^n.$$

In this model d_t may represent a fixed dependent demand or an independent demand. The constant objective term $\sum_{t=1}^{n} \gamma_t d_t$ can be added without changing the problem. Note that this model also allows one to treat the case of *lost sales* with a linear penalty cost. Specifically the slack variable in the constraint $v_t \leq u_t$ is the amount lost.

In discussing this model, we allow the demands d_t to be negative in certain cases. This can arise, for example, as the result of a purchase order for the item from an outside vendor.

11.1.1 The Uncapacitated Case: Sales and Arbitrary Demands

Here we allow demands to be negative and take $s_0 = 0$. We consider the set $X^{LS-U-SL}$ in the form

$$\sum_{j=1}^{t} x_j \geq \sum_{j=1}^{t} v_j + d_{1t} \qquad \text{for all } t \qquad (11.1)$$

$$x_t \leq M y_t \qquad \text{for all } t \qquad (11.2)$$

$$v_t \leq u_t \qquad \text{for all } t \qquad (11.3)$$

$$x, v \in \mathbb{R}_+^n, y \in \{0,1\}^n. \qquad (11.4)$$

By analyzing the structure of optimal extreme point solutions, and extending the definition of a regeneration interval, a polynomial dynamic program that finds an optimal sequence of regeneration intervals can be constructed to solve *LS-U-SL*. Here we describe valid inequalities for $X^{LS-U-SL}$.

Consider a subset $R \subseteq N$. We then calculate new nonnegative demands $d^R \in \mathbb{R}_+^n$, based on the idea that $v_t = u_t$ for $t \in R$, and $v_t = 0$ otherwise. Specifically, we set the cumulative demand

$$d_{1t}^R = \max_{j:j\leq t}[d_{1t} + \sum_{j\in[1,t]\cap R} u_j]^+ \quad \text{for } t = 1,\ldots,n,$$

and from this we calculate the demand for individual time periods $d_1^R = d_{11}^R$ and $d_t^R = d_{1t}^R - d_{1,t-1}^R$ for $t = 2,\ldots,n$. We can now describe a generalization of the (l, S) inequality.

Proposition 11.1 *For* $1 \leq l \leq n$, *with* $R, S \subseteq L = \{1,\ldots,l\}$ *and* $d_{1,l-1}^R < d_{1l}^R$, *the* (l, S, R) *inequality*

$$\sum_{j\in L\backslash S} x_j + \sum_{j\in S} d_{jl}^R y_j \geq \sum_{j\in R} v_j + d_{1l} \qquad (11.5)$$

is valid for $X^{LS-U-SL}$.

The proof of validity is very similar to the proof of validity for the (l, S) inequality, and is based on the fact that $d_{jl}^R = d_{1l}^R - d_{1,j-1}^R$ is the maximum of the demands in $\{1,\ldots,l\}$ and sales in $\{1,\ldots,l\} \backslash R$ that can be satisfied from production in j when $y_j = 1$. In fact these inequalities suffice to give $\text{conv}(X^{LS-U-SL})$.

Theorem 11.2 $\text{conv}(X^{LS-U-SL})$ *is completely described by the initial constraints (11.1),(11.3), the (l, S, R) inequalities (11.5), and the bounds* $x, v \in \mathbb{R}_+^n, y \in [0,1]^n$.

Example 11.1 *Consider an instance with* $n = 5, d = (3, -4, 2, -1, 2)$, *and* $u = (2, 3, 4, 3, 5)$. *Taking* $R = \{2, 4\}$, *we obtain that* $(d_{11}^R, d_{12}^R, d_{13}^R, d_{14}^R, d_{15}^R) =$

$(3, 3, 4, 6, 8)$ *and thus* $d^R = (3, 0, 1, 2, 2)$. *Now a sample of the* (l, S, R) *inequalities with* $R = \{2, 4\}$ *is:*

$$l = 1, S = \{1\} \qquad 3y_1 \geq 3$$
$$l = 3, S = \{3\} \qquad x_1 + x_2 + y_3 \geq 1 + v_2$$
$$l = 4, S = \{2, 4\} \qquad x_1 + 3y_2 + x_3 + 2y_4 \geq 0 + v_2 + v_4$$
$$l = 5, S = \{2, 4, 5\} \qquad x_1 + 5y_2 + x_3 + 4y_4 + 2y_5 \geq 2 + v_2 + v_4.$$

Unfortunately a fast combinatorial separation algorithm for the (l, S, R) inequalities is not known in this general case.

11.1.2 The Uncapacitated Case: Sales and Nonnegative Demands

Here we suppose that $d_t \geq 0$ for all t, and that there is no initial and no ending stock (i.e., $\sum_{j=1}^n x_j = \sum_{j=1}^n v_j + d_{1n}$), and derive an extended formulation for $\text{conv}(X^{LS-U-SL})$. In this case the extreme points of $\text{conv}(X^{LS-U-SL})$ are readily characterized.

Proposition 11.3 *Every extreme point is characterized by three sets* $I, J, K \subseteq \{1, \ldots, n\}$, *where* $I \subseteq J$ *and* $I = \{t_1, \ldots, t_q\}$ *with* $t_1 < t_2 < \cdots < t_q$. *The corresponding extreme point is:*
$y_t = 1$ *for* $t \in J$, *and* $y_t = 0$ *otherwise;* $v_t = u_t$ *for* $t \in K$, *and* $v_t = 0$ *otherwise;* $x_t = \sum_{i=t_j}^{t_{j+1}-1}(v_i + d_i)$ *if* $t = t_j \in I$, *and* $x_t = 0$ *for* $j \notin I$.

It is now easy to derive an extended formulation generalizing the facility location formulation (7.14)–(7.18) for *LS-U*.

Let $\alpha_{it} = 1$ if the demand d_t is produced in period i, and $\alpha_{it} = 0$ otherwise.

Let $\beta_{it} = 1$ if the demand $d_t + u_t$ is produced in period i, and $\beta_{it} = 0$ otherwise.

The corresponding formulation $Q^{LS-U-SL}$ is

$$\max \quad \sum_{t=1}^n \gamma_t v_t - \sum_{t=1}^n c_t x_t - \sum_{t=1}^n f_t y_t$$

$$\sum_{i:i\leq t}(\alpha_{it} + \beta_{it}) = 1 \qquad \text{for all } t$$

$$\alpha_{it} + \beta_{it} \leq y_i \qquad \text{for all } i, t \text{ with } i \leq t \text{ if } d_t > 0$$

$$\beta_{it} \leq y_i \qquad \text{for all } i, t \text{ with } i \leq t \text{ if } d_t = 0, u_t > 0$$

$$x_i = \sum_{t:t\geq i} d_t \alpha_{it} + \sum_{t:t\geq i}(d_t + u_t)\beta_{it} \qquad \text{for all } i$$

$$v_t = u_t \sum_{i:i\leq t} \beta_{it} \qquad \text{for all } t$$

$$x, v \in \mathbb{R}_+^n, \ \alpha, \beta \in \mathbb{R}_+^{n(n+1)/2}, \qquad y \in [0, 1]^n.$$

Theorem 11.4 $\text{conv}(X^{LS-U-SL}) = proj_{x,y}(Q^{LS-U-SL})$.

11.2 Lower Bounds on Production (*LB*)

We suppose that the constraints

$$x_t \geq L y_t \quad \text{for } t = 1, \dots, n$$

have been added to a given single-item lot-sizing model, and that L divides C exactly. If not, we can obtain a valid relaxation with this property by taking $C \leftarrow L\lceil \frac{C}{L} \rceil$ or taking C very large. It also means that we can rescale so that $L = 1$ and C is an integer.

11.2.1 A Wagner–Whitin Relaxation *WW-CC-LB*

For a fixed period t, we have the two sets of valid balance constraints

$$s_{l-1} + \sum_{u=l}^{t} x_u = d_{lt} + s_t \qquad \text{for } l = 1, \dots, t$$

$$s_t + \sum_{u=t+1}^{l} x_u = d_{t+1,l} + s_l \qquad \text{for } l = t+1, \dots, n.$$

Using the variable lower bounds $x_t \geq L y_t$ on the first set of inequalities, and the variable upper bounds $x_t \leq C y_t$ on the second, we obtain the relaxation

$$s_t - L \sum_{u=l}^{t} y_u \geq -d_{lt} \qquad \text{for } l = 1, \dots, t$$

$$s_t + C \sum_{u=t+1}^{l} y_u \geq d_{t+1,l} \qquad \text{for } l = t+1, \dots, n.$$

Still with t fixed, setting $z_l = -\sum_{u=l}^{t} y_u$ for $1 \leq l \leq t$ and $z_l = \sum_{u=t+1}^{l}$ for $t+1 \leq l \leq n$, the result is a divisible mixing set

$$
\begin{aligned}
s_t + L z_l &\geq b_l & \text{for } l = 1, \dots, t \\
s_t + C z_l &\geq b_l & \text{for } l = t+1, \dots, n \\
s_t \in \mathbb{R}_+^1, z &\in \mathbb{Z}^n
\end{aligned}
$$

with some additional constraints. Thus one can use the results of Section 8.6 to generate a tightened extended formulation, or valid inequalities.

Example 11.2 *Consider an instance with* $n = 5, C = 8, L = 4$, *and* $d = (7, 3, 2, 2, 6)$. *For* $t = 2$, *we obtain the surrogates*

$$
\begin{aligned}
s_2 - 4y_1 - 4y_2 &\geq -10 \\
s_2 \qquad - 4y_2 &\geq -3 \\
s_2 \qquad + 8y_3 &\geq 2 \\
s_2 \qquad + 8y_3 + 8y_4 &\geq 4 \\
s_2 \qquad + 8y_3 + 8y_4 + 8y_5 &\geq 10.
\end{aligned}
$$

11.2.2 A Wagner–Whitin Relaxation with Backlogging WW-CC-B, LB

With backlogging, constructing similar surrogates to those above we obtain for fixed t

$$s_t + r_{l-1} - L\sum_{u=l}^{t} y_u \geq -d_{lt} \qquad \text{for } 1 \leq l \leq t,$$

$$s_t + r_l + C\sum_{u=t+1}^{l} y_u \geq d_{t+1,l} \qquad \text{for } t+1 \leq l \leq n.$$

This can be viewed as a continuous mixing set with divisible capacities and additional constraints, generalizing the sets X^{CMIX} and X^{DMIX} studied in Sections 8.5 and 8.6, respectively.

11.3 Lower Bounds on Production in a Set-Up Sequence

11.3.1 Almost Full Capacity Production (AFC)

Here production is at full capacity in all but the first and last periods of a set-up sequence. This may arise in multi-item models (classified as $PM = M_1, M_2$ in Chapter 12) due to start-up or cleaning times, or the possibility of producing a second item in the period.

A simple formulation is

$$x_t \geq C_t(y_{t-1} + y_t + y_{t+1} - 2),$$

but a more effective formulation using the start-up and switch-off variables is

$$x_t \geq C_t(y_t - z_t - w_t).$$

11.3.2 Minimum Production Level per Set-Up Sequence (MR)

Suppose now that once a start-up occurs, a minimum amount P, called a minimum run, must be produced, and it must be produced in the next α periods (because of almost full capacity production, or for other reasons).

A basic formulation is given by

$$\sum_{u\in[t,t+\alpha-1]} x_u \geq Pz_t.$$

By considering intervals $[k,l]$ whose length exceeds α, but whose demand is less than P, we obtain the valid inequality

$$r_{k-1} + s_l \geq (P - d_{kl})^+ \sum_{u=k}^{l-\alpha+1} z_u$$

if we can assume that there is at most one start-up in the interval $[k, l-\alpha+1]$.

11.4 Restricted Length Set-Up Sequences (*RLS*)

Here we consider the set-up and start-up sequences for a single-item independently of the production quantity. Thus the base model is

$$y_t - z_t = y_{t-1} - w_{t-1} \qquad \text{for all } t \qquad (11.6)$$
$$z_t \le y_t \qquad \text{for all } t \qquad (11.7)$$
$$y, z, w \in [0, 1]^n \qquad (11.8)$$
$$y, z, w \in \mathbb{Z}^n. \qquad (11.9)$$

11.4.1 Varying Length Sequences

Suppose that a sequence of set-ups starting in period t must last for between a minimum of $\alpha_t \ge 1$ and a maximum of β_t periods, and also that if the item is switched off in t, then it remains off for between a minimum of $\gamma_t \ge 1$ and a maximum of δ_t periods. In the (y, z, w) space, each of the four cases has a simple formulation.

If there is a start-up in t, there must be a set-up for every period in the interval $[t, t + \alpha_t - 1]$. This gives

$$z_t \le y_l \quad \text{for } t \le l \le t + \alpha_t - 1.$$

If there is a switch off in t, there must be a start-up in the interval $[t - \beta_t + 1, t]$, and hence

$$w_t \le \sum_{l=t-\beta_t+1}^{t} z_l.$$

If there is a switch-off in t, there is no set-up in any period in the interval $[t + 1, t + \gamma_t]$, which gives

$$w_t \le 1 - y_l \quad \text{for } t + 1 \le l \le t + \gamma_t.$$

If there is a switch-off in t, there must be a start-up in the interval in the interval $[t + 2, t + \delta_t + 1]$, and we have the inequality

$$w_t \le \sum_{l=t+2}^{t+\delta_t+1} z_l.$$

To obtain a tight formulation with such restricted length sequences, it is once again possible to model the problem as a unit flow in a network.

Let $\xi_{tl} = 1$ if there is a start-up in t and the following switch-off is in l with necessarily $t + \alpha_t - 1 \le l \le t + \beta_t - 1$, and $\omega_{tl} = 1$ if there is a switch-off in t and the following start-up is in l with necessarily $t + \gamma_t + 1 \le l \le t + \delta_t + 1$. Then we have the flow conservation constraints

$$\sum_{t:t\leq l} \xi_{tl} = w_l \qquad\qquad \text{for all } l$$

$$\sum_{t:t\leq l} \omega_{tl} = z_l \qquad\qquad \text{for all } l$$

$$\sum_{l:l\geq t} \xi_{tl} = z_t \qquad\qquad \text{for all } t$$

$$\sum_{l:l\geq t} \omega_{tl} = w_t \qquad\qquad \text{for all } t$$

$$\sum_{\tau:\tau\leq t}\sum_{l:l\geq t} \xi_{\tau l} = y_t \qquad\qquad \text{for all } t$$

$$\xi,\omega,y,z,w \in \{0,1\}^n \ ,$$

as well as constraints defining the initial state at time $t = 0$ or $t = 1$ giving an entering flow of one unit into the network.

Example 11.3 *See Figure 11.1 for an example with $n = 8$ and time independent sequence length values $\alpha = 2, \beta = 3, \gamma = 1, \delta = 4$. A flow through a node t at the upper level implies that $z_t = 1$ and a flow through a node τ at the lower level means that $w_\tau = 1$. The flow, indicated by the arrows, that is, $\xi_{12} = \omega_{25} = \xi_{57} = 1$, corresponds to a feasible set-up sequence with $y = (1,1,0,0,1,1,1,0)$.*

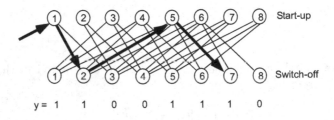

Figure 11.1. Restricted set-up sequences.

11.4.2 Constant Length Sequences

Here we assume that all four parameters $\alpha_t, \beta_t, \gamma_t, \delta_t$ take constant values $\alpha, \beta, \gamma, \delta$. Now each of the four inequalities proposed in Section 11.4.1 can be strengthened. For example, if there is a start-up in the interval $[t - \alpha + 1, t]$, the item must still be set up in period t giving

$$\sum_{u=1}^{\alpha} z_{t-u+1} \leq y_t.$$

The next result gives a complete linear description of the convex hull of the set of feasible sequences. It can be proved by verifying that (11.10)–(11.13) plus the bound constraints (11.8) provides a valid formulation and by showing that the corresponding matrix is totally unimodular.

Proposition 11.5 *Adding the constraints*

$$\sum_{u=1}^{\alpha} z_{t-u+1} \leq y_t \tag{11.10}$$

$$\sum_{u=1}^{\beta} z_{t-u+1} \geq y_t \tag{11.11}$$

$$\sum_{u=1}^{\gamma} z_{t+u} \leq 1 - y_t \tag{11.12}$$

$$\sum_{u=1}^{\delta} z_{t+u} \geq 1 - y_t \tag{11.13}$$

to the basic formulation (11.6)–(11.8) gives the convex hull of the constant restricted length sequence model.

In certain circumstances, one may want to have restricted sequence lengths when only modeling in the set-up space. The valid inequalities providing lower bounds on z_t are

$$z_t \geq y_t - y_{t-1}, \quad z_t \geq 0,$$

so the projection of the minimum time inequalities (11.10) and (11.12) gives

$$\sum_{j\in S}(y_j - y_{j-1}) \leq y_t \qquad \text{for } \emptyset \subset S \subseteq [t - \alpha + 1, t] \quad \text{and}$$

$$\sum_{j\in T}(y_j - y_{j-1}) \leq 1 - y_t \qquad \text{for } \emptyset \subset T \subseteq [t + 1, t + \gamma],$$

respectively.

Similarly for the maximum time constraints, the inequalities providing upper bounds on z_t are

$$z_t \leq y_t, z_t \leq 1 - y_{t-1}.$$

Applied to the inequalities (11.11) and (11.13), the inequalities obtained by projection are

$$\sum_{j\in S} y_j + \sum_{j\in[t-\beta+1,t]\setminus S}(1 - y_{j-1}) \geq y_t \qquad \text{for } \emptyset \subset S \subseteq [t - \beta + 1, t] \quad \text{and}$$

$$\sum_{j\in T} y_j + \sum_{j\in[t+1,t+\delta]\setminus T}(1 - y_{j-1}) \geq 1 - y_t \quad \text{for } \emptyset \subset T \subseteq [t + 1, t + \delta].$$

Separation of these inequalities is very simple.

11.5 Piecewise Concave Production Costs (CP)

Suppose that the production cost function is $p(x_t) = \min_{k=1}^{K}[q_t^k \delta(x_t) + p_t^k x_t]$ where $\delta(x) = 1$ if $x > 0$ and $\delta(x) = 0$ otherwise, and $0 \leq q_t^1 \leq q_t^2 \leq \ldots q_t^K$ and $p_t^1 \geq \ldots \geq p_t^K \geq 0$. See Figure 11.2.

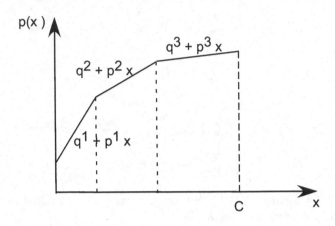

Figure 11.2. Piecewise linear concave production costs.

Introducing variables x_t^k and y_t^k to model the production level and set-ups corresponding to each cost segment k, we obtain the following formulation for *LS-C-CP*.

$$\min \sum_{k=1}^{K}\sum_{t=1}^{n} p_t^k x_t^k + \sum_{k=1}^{K}\sum_{t=1}^{n} q_t^k y_t^k + \sum_{t=0}^{n} h_t s_t$$

$$s_{t-1} + \sum_{k} x_t^k = d_t + s_t \qquad \text{for all } t \qquad (11.14)$$

$$x_t^k \leq C_t y_t^k \qquad \text{for all } k, t \qquad (11.15)$$

$$\sum_{k} y_t^k \leq 1 \qquad \text{for all } t \qquad (11.16)$$

$$s \in \mathbb{R}_+^{n+1}, \ x \in \mathbb{R}_+^{Kn}, \ y \in \{0,1\}^{Kn}. \qquad (11.17)$$

Note that if items can be bought from the outside at cost β_t per unit, this can be modeled by setting $p_t^1 = \beta_t, q_t^1 = 0$. Observe also that in the uncapacitated case, constraints (11.16) are automatically satisfied when $q_t^k > 0$ for all k, t.

For the uncapacitated problem *LS-U-CP* with $s_0 = 0$, one obtains the constraint set:

$$s_{t-1} + \sum_k x_t^k = d_t + s_t \qquad\qquad \text{for all } t$$

$$x_t^k \le My_t^k \qquad\qquad \text{for all } k,t$$

$$s \in \mathbb{R}_+^{n+1},\ x \in \mathbb{R}_+^{Kn},\ y \in \{0,1\}^{Kn},\ s_0 = 0,$$

denoted $X^{LS-U-CP}$.

Proposition 11.6 *The generalized* (l, S) *inequalities*

$$\sum_{k=1}^K \sum_{j \in S_k} x_j^k + \sum_{k=1}^K \sum_{j \in L \setminus S_k} d_{jl} y_j^k \ge d_{1l},$$

with $S_k \subseteq L = \{1, \dots, l\}$ *for* $k = 1, \dots, K$, *are valid for* $X^{LS-U-CP}$. *Adding these inequalities to the initial formulation gives* conv($X^{LS-U-CP}$).

It is also possible to generalize the facility location and shortest path extended formulations for *LS-U* to this more general case; see Section 11.9 below.

11.6 Production Time Windows (*TWP*)

Here we consider the problem in which, instead of the standard demands, one has a set of orders. Order k for $1 \le k \le K$ consists of a quantity $D^k > 0$ and a time window $[b^k, e^k]$ with $1 \le b^k \le e^k \le n$ in which the D^k units must be produced. Delivery is in period e^k. By regrouping orders with the same time window, one can assume that $K \le n(n+1)/2$.

Two variants of the problem arise: the case of *distinguishable* or *client-specific* orders in which the time windows must be individually respected, and the case of *indistinguishable* or *nonclient-specific* orders.

We now attempt to clarify the distinction. Below the "bottles" play the role of "raw materials" provided by the clients. Suppose that production involves filling bottles of liquid gas. The bottles are provided empty by the client in period b^k and he picks up the full bottles in period e^k. In the distinguishable case, the bottles carry the name of the client, so they cannot be mixed up and the client will leave in e^k with the same bottles that he provided in b^k. On the other hand, in the indistinguishable case, all that matters is that the client retrieves D^k units of full (but possibly different) bottles in period e^k.

We first introduce some notation and then we present two MIP formulations of the distinguishable order version. Note that with distinguishable orders, it is natural to take $s_0 = 0$.

$D_{tl} = \sum_{k:t \le b^k, e^k \le l} D^k$ is the amount that must be produced in the interval $[t, l]$.

$\Delta_t = D_{1t} - D_{1,t-1} = \sum_{k:e^k=t} D^k$ is the amount that must be delivered in

period t, and $\Delta_{tl} = \sum_{u=t}^{l} \Delta_u$.
$\Gamma_t = D_{tn} - D_{t+1,n} = \sum_{k:b^k=t} D^k$ is the amount that becomes available for production in period t, and $\Gamma_{tl} = \sum_{u=t}^{l} \Gamma_u$.

In addition to the standard variables x_t, s_t, y_t, we introduce

z_t^k is the amount of order k produced in period $t \in [b^k, e^k]$.

The following formulation of $LS\text{-}CC\text{-}TWP$ is now self-explanatory.

$$\min \quad \sum_{t=1}^{n} p_t' x_t + \sum_{t=0}^{n} h_t' s_t + \sum_{t=1}^{n} q_t y_t \tag{11.18}$$

$$s_{t-1} + x_t = \Delta_t + s_t \qquad\qquad \text{for } 1 \le t \le n \quad (11.19)$$

$$\sum_{u=b^k}^{e^k} z_u^k = D^k \qquad\qquad \text{for } 1 \le k \le K \quad (11.20)$$

$$\sum_{\{k:u\in[b^k,e^k]\}} z_u^k = x_u \qquad\qquad \text{for } 1 \le u \le n \quad (11.21)$$

$$x_u \le C y_u \qquad\qquad \text{for } 1 \le u \le n \quad (11.22)$$

$$s \in \mathbb{R}_+^{n+1}, x \in \mathbb{R}_+^n, y \in \{0,1\}^n, z \ge 0, s_0 = 0. \tag{11.23}$$

Eliminating the z_t^k variables using the max flow–min cut theorem on the transportation problem defined by the constraints (11.20)-(11.21), and eliminating the s_t variables, we obtain the equivalent formulation:

$$\min \quad \sum_{t=1}^{n} p_t x_t + \sum_{t=1}^{n} q_t y_t \tag{11.24}$$

$$\sum_{u=t}^{l} x_u \ge D_{tl} \qquad\qquad \text{for } 1 \le t \le l \le n \quad (11.25)$$

$$x_u \le C y_u \qquad\qquad \text{for } 1 \le u \le n \quad (11.26)$$

$$x \in \mathbb{R}_+^n, \ y \in \{0,1\}^n. \tag{11.27}$$

11.6.1 An Algorithm for $WW\text{-}U\text{-}TWP$ and Extended Formulation for $WW\text{-}CC\text{-}TWP$

Here we suppose that the costs are Wagner–Whitin. First we describe a dynamic programming algorithm for the uncapacitated problem $WW\text{-}U\text{-}TWP$, and then we give a tight extended formulation for the constant capacity case $WW\text{-}CC\text{-}TWP$.

Let $H(t)$ be the cost of an optimal solution of $WW\text{-}U\text{-}TWP$ in which the horizon is the interval $[1,t]$ and only the orders with $e^k \le t$ are considered. One obtains the following simple recurrence,

$$H(l) = \min_{\{t:t\leq l, D_{t+1,l}=0\}} [H(t-1) + f_t + p_t \Delta_{tl}],$$

using the fact that if t is the last production period in the interval $[1,l]$, all that must be delivered in the interval $[t,l]$ will be produced in t because it pays to produce as late as possible, and on the other hand if $D_{t+1,l} > 0$, then there must be production within the interval $[t+1,l]$.

To describe a tight extended formulation, we work in the (s,y) space. By combining (a) the constraints (11.25), and (b) the aggregate balance constraints $s_{t-1} + \sum_{u=t}^{l} x_u \geq \Delta_{tl}$ derived from (11.19) with the variable upper bound constraints (11.26), we obtain the relaxation:

$$\min \quad \sum_{t=0}^{n} h_t s_t + \sum_{t=1}^{n} q_t y_t \tag{11.28}$$

$$s_{t-1} + C\sum_{u=t}^{l} y_t \geq \Delta_{tl} \qquad \text{for } 1 \leq t \leq l \leq n \tag{11.29}$$

$$C\sum_{u=t}^{l} y_t \geq D_{tl} \qquad \text{for } 1 \leq t \leq l \leq n \tag{11.30}$$

$$s \in \mathbb{R}_{+}^{n+1}, y \in \{0,1\}^n, \ s_0 = 0. \tag{11.31}$$

From the analysis of stock-minimal solutions for WW-CC, we can observe that this is a correct formulation for WW-CC-TWP. Specifically note that, without the constraints (11.30), the feasible region resembles the basic formulation of WW-CC with Δ_{tl} in place of d_{tl}, and thus the convex hull is known; see Section 9.5. On the other hand Chvátal–Gomory rounding applied to the constraints (11.30) gives the valid inequalities

$$\sum_{u=t}^{l} y_u \geq \lceil \frac{D_{tl}}{C} \rceil \quad \text{for } 1 \leq t \leq l \leq n.$$

This leads to a tight formulation.

Theorem 11.7 *The linear program*

$$\min \quad \sum_{t=0}^{n} h_t s_t + \sum_{t=1}^{n} q_t y_t$$

$$(s,y) \in \text{conv}(X^{WW-CC}(\Delta))$$

$$\sum_{u=t}^{l} y_u \geq \lceil \frac{D_{tl}}{C} \rceil \qquad \text{for } 1 \leq t \leq l \leq n$$

$$s \in \mathbb{R}_{+}^{n+1}, y \in [0,1]^n, \ s_0 = 0,$$

solves WW-CC-TWP, where $X^{WW-CC}(\Delta)$ is the set X^{WW-CC} with demand vector Δ.

In the uncapacitated case, the formulation is simpler.

Corollary 11.1 *The linear program*

$$\min \quad \sum_{t=0}^{n} h_t s_t + \sum_{t=1}^{n} q_t y_t$$

$$s_{t-1} \geq \sum_{\{k:b^k < t \leq e^k \leq l\}} D^k(1 - y_t - \ldots - y_{e^k}) \qquad \text{for } 1 \leq t \leq l \leq n$$

$$\sum_{u=b^k}^{e^k} y_u \geq 1 \qquad \qquad \text{for } k = 1, \ldots, K$$

$$s \in \mathbb{R}_+^{n+1}, y \in [0,1]^n, \ s_0 = 0 ,$$

solves WW-U-TWP.

11.6.2 Indistinguishable Time Windows *LS-C-TWP(I)* and an Equivalent Problem

Here we consider the problem with indistinguishable orders $LS\text{-}C\text{-}TWP(I)$. The crucial order information is all contained in the arrival quantities Γ_t and the delivery quantities Δ_t that satisfy:

$$\Gamma_{1t} \geq \Delta_{1t} \qquad \qquad \text{for } 1 \leq t \leq n-1 \qquad (11.32)$$
$$\Gamma_{1n} = \Delta_{1n} \qquad \qquad (11.33)$$
$$\Gamma, \Delta \in \mathbb{R}_+^n. \qquad \qquad (11.34)$$

For this case the basic MIP formulation simplifies to

$$\min \quad \sum_{t=1}^{n} p_t x_t + \sum_{t=1}^{n} f_t y_t \qquad (11.35)$$

$$\sum_{u=1}^{l} x_u \geq \Delta_{1l} \qquad \qquad \text{for } 1 \leq l \leq n \qquad (11.36)$$

$$\sum_{u=1}^{l} x_u \leq \Gamma_{1l} \qquad \qquad \text{for } 1 \leq l \leq n \qquad (11.37)$$

$$x_u \leq C_u y_u \qquad \qquad \text{for } 1 \leq u \leq n \qquad (11.38)$$
$$x \in \mathbb{R}_+^n, y \in \{0,1\}^n. \qquad (11.39)$$

Now we show that this problem is equivalent to the distinguishable order problem with special time windows.

Definition 11.1 *A set of distinguishable orders has* noninclusive *time windows $[b^k, e^k]_{k=1}^K$ if there is no pair k, κ with $b^k < b^\kappa \leq e^\kappa < e^k$.*

This property turns out to be very useful.

Observation 11.1 *Given an availability vector Γ and a delivery vector Δ satisfying (11.32)–(11.34), there is a unique set of orders D^k with non-inclusive time windows associated with the (Γ, Δ) pair.*

Algorithm to Compute the Orders
Initialization Set $L_t = \Gamma_t, R_t = \Delta_t$ for all t. $k = 1$
While $L, R \neq 0$
    Set $\sigma = \min\{t : L_t > 0\}, \tau = \min\{t : R_t > 0\}$.
    Set $D^k = \min\{L_\sigma, R_\tau\}, b^k = \sigma, e^k = \tau$.
    $L_\sigma \leftarrow L_\sigma - D^k, R_\tau \leftarrow R_\tau - D^k$
    $k \leftarrow k + 1$
end-While.

Clearly there are at most $2n - 1$ orders and they are uniquely defined.

It follows that an instance of *LS-U-TWP* with noninclusive time windows can be treated as if the orders were indistinguishable; that is, it suffices to solve the relaxation *LS-U-TWP(I)*, and conversely given a problem with indistinguishable orders but arbitrary time windows, the time windows can be modified so as to be noninclusive using the above algorithm.

11.6.3 A Dynamic Programming Algorithm for *LS-U-TWP(I)*

Here we see some of the consequences of the noninclusive time window property.

Observation 11.2 *i. A set of noninclusive time windows can be ordered so that for all k either $b^k < b^{k+1}$ and $e^k \leq e^{k+1}$, or $b^k = b^{k+1}$ and $e^k < e^{k+1}$.*
ii. With noninclusive time windows ordered as in i, there exists an optimal solution in which order k is produced before (or at the same time) as order $k + 1$ for all k.
iii. In the uncapacitated case there exists an optimal solution in which each order k is produced in a single period.

Now we can describe the dynamic programming algorithm. We take the objective function in the form

$$\min \sum_t p_t x_t + \sum_t q_t y_t,$$

and we assume that the orders $k = 1, \ldots, K$ are numbered from earliest to latest as in i of Observation 11.2.

Using iii of the same observation, we define the following quantities.

$H(t, k)$ is the value of an optimal solution for periods $1, \ldots, t$ in which the demands $D^1, \ldots, D^k$ are produced in or before period t. Note that $H(t, k) = \infty$ if $b^k > t$.

$G(t, k)$ is the value of an optimal solution for periods $1, \ldots, t$ in which the demands $D^1, \ldots, D^{k-1}$ are produced in or before period t and D^k is produced in t. Also $G(t, k) = \infty$ if $e^k < t$ or $b^k > t$.

The recursion is

$$H(t, k) = \min[H(t-1, k),\ G(t, k)] \quad \text{for } t, k \text{ with } b^k \le t \tag{11.40}$$
$$G(t, k) = \min[H(t-1, k-1) + q_t + p_t D^k,\ G(t, k-1) + p_t D^k]$$
$$\text{for } t, k \text{ with } t \in [b^k, e^k], \tag{11.41}$$

where the first equation just uses the observation that order k is produced either before, or in period t, and the second the fact that order $k - 1$ is produced either before, or in t. Obviously this provides an $O(n^2)$ algorithm for the problem.

11.6.4 A Tight Extended Formulation for $LS\text{-}U\text{-}TWP(I)$

We now use the above dynamic program to get an extended formulation. Specifically the recursion suggests the linear program

$$\max\ H(n, K)$$
$$H(t, k) - H(t-1, k) \le 0 \qquad \text{for all } k, t \text{ with } b^k \le t$$
$$H(t, k) - G(t, k) \le 0 \qquad \text{for all } k, t \text{ with } t \in [b^k, e^k]$$
$$G(t, k) - G(t, k-1) \le p_t D^k \qquad \text{for all } k, t \text{ with } t \in [b^k, e^k]$$
$$G(t, k) - H(t-1, k-1) \le q_t + p_t D^k \quad \text{for all } k, t \text{ with } t \in [b^k, e^k].$$

Let the dual variables be $v_{tk}, w_{tk}, x_{tk}, z_{tk}$, respectively. The dual of this linear program is then

$$\min\ \sum_{t,k} [(q_t + p_t D^k) z_{tk} + p_t D^k x_{tk}] \tag{11.42}$$

$$z_{tk} + x_{tk} - x_{t,k+1} - w_{t,k} = 0 \qquad \text{for all } k, t \text{ with } b^k \le t \tag{11.43}$$

$$v_{tk} - v_{t+1,k} + w_{tk} - z_{t+1,k+1} = 0 \quad \text{for all } k, t \text{ with } t \in [b^k, e^k] \tag{11.44}$$

$$v_{n,K} + w_{n,K} = 1 \tag{11.45}$$

$$v, w, x, z \ge 0. \tag{11.46}$$

This can be seen as a shortest path problem. An interpretation of the variables is:

$z_{tk} = 1$ if there is production in t and order k is the first order produced in t (i.e., order $k - 1$ is produced earlier).

$x_{tk} = 1$ if orders k and $k - 1$ are produced in t.

$w_{tk} = 1$ if there is production in t and the last order produced is order k.

$v_{t,k} = 1$ if last order produced in or before t is order k.

To obtain a complete formulation, we just need to add:

$$1 \geq y_t \geq \sum_k z_{tk} \qquad\qquad \text{for all } t \qquad (11.47)$$

$$x_t = \sum_k D^k(z_{tk} + x_{tk}) \qquad\qquad \text{for all } t. \qquad (11.48)$$

Theorem 11.8 *Let $X^{LS-U-TWP(I)}$ be the set of feasible solutions of (11.36)–(11.39) of the uncapacitated problem LS-U-TWP(I). A tight extended formulation for $\mathrm{conv}(X^{LS-U-TWP(I)})$ is given by the polyhedron (11.43)–(11.48).*

11.7 Upper Bounds on Stocks (SUB)

Here we consider the case with upper bounds $s_t \leq \overline{S}_t$ for all periods t, and we assume that $\overline{S}_n = 0$.

11.7.1 Equivalence to LS-CAP-$TWP(I)$

As $s_{t-1} \leq d_t + \overline{S}_t$, we can assume without loss of generality that $\overline{S}_{t-1} \leq d_t + \overline{S}_t$ for all t. Now $x \in \mathbb{R}_+^n$ is a feasible production vector if and only if $s_t = \sum_{u=1}^t x_u - d_{1t} \geq 0$ for all t, and

$$s_t = \sum_{u=1}^t x_u - d_{1t} \leq S_t \quad \text{for all } t.$$

Setting $\Delta_t = d_t \geq 0$ and $\Gamma_t = d_t + \overline{S}_t - \overline{S}_{t-1} \geq 0$, we have that Γ and Δ satisfy (11.32)–(11.34), and thus we obtain precisely the formulation (11.35)–(11.39) of LS-CAP-$TWP(I)$.

Because of the equivalence with LS-U-$TWP(I)$, LS-U-SUB can be solved by the dynamic programming recurrence (11.40)–(11.41).

For the constant capacity problem LS-CC-SUB, it is not difficult to generalize the concept of regeneration interval appropriately, and derive a polynomial shortest path algorithm; see Exercise 11.8.

11.7.2 Valid Inequalities for LS-U-SUB

Let $X^{LS-U-SUB} = X^{LS-U} \cap \{(s, y) : s_t \leq \overline{S}_t \text{ for all } t\}$. Obviously we have the surrogate constraints

$$M \sum_{u=k}^t y_u \geq d_{kt} - \overline{S}_{k-1}.$$

This gives immediately the inequality

$$\sum_{u=k}^{t} y_u \geq 1$$

which is valid whenever $\overline{S}_{k-1} < d_{kt}$.

Another obvious valid inequality is the variable upper bound constraint

$$x_t \leq (d_t + \overline{S}_t)y_t. \tag{11.49}$$

Other stronger inequalities are variants of the (l, S) inequalities.

Proposition 11.9 *Given an interval $[k, l]$ and $S \subseteq [k, l]$, the inequalities*

$$s_{k-1} + \sum_{u \in S} x_u \leq \overline{S}_{k-1} + \sum_{u \in S} [d_{ku} + \overline{S}_u - \overline{S}_{k-1}]y_u \tag{11.50}$$

and

$$s_{k-1} + \sum_{u \in S} x_u$$
$$\leq \overline{S}_{k-1} + \sum_{u \in S} \min[d_{ku} + \overline{S}_u - \overline{S}_{k-1}, d_{kl} - \overline{S}_{k-1}, d_{ul}]y_u + s_l \tag{11.51}$$

are valid for $X^{LS-U-SUB}$.

Example 11.4 *Consider an instance of LS-U-SUB with $n = 5$, $d = (3, 4, 2, 1, 2)$, $\overline{S} = (3, 3, 4, 3, 5)$, and the fractional solution $y^* = (1, 0.25, 1, 0, 0)$, $x^* = (6, 1, 5, 0, 0)$, $s^* = (3, 0, 3, 2, 0)$.*

It is easily checked that the inequality (11.51) with $k = 2$ and $S = \{2\}$, namely

$$s_1 + x_2 \leq 3 + (4 - 3)y_2 + s_2,$$

cuts off this point. Note that with $k = t = 2$, the inequality (11.49) gives

$$y_2 \geq 1,$$

which is also violated.

11.7.3 Valid Inequalities for WW-CC-SUB and WW-CC-B, SUB

Let $X^{WW-CC-SUB} = X^{WW-CC} \cap \{(s, y) : s_t \leq \overline{S}_t \text{ for all } t\}$. Again we use the surrogate constraints

$$C \sum_{u=k}^{t} y_u \geq d_{kt} - \overline{S}_{k-1}.$$

These can be strengthened by the corresponding Gomory fractional cuts.

Proposition 11.10 *The convex hull* $\mathrm{conv}(X^{WW-CC-SUB})$ *is obtained by adding the constraints*

$$\sum_{u=k}^{t} y_u \geq \lceil \frac{d_{kt} - \overline{S}_{k-1}}{C} \rceil \quad \text{for } 1 \leq k \leq t \leq n$$

to $\mathrm{conv}(X^{WW-CC})$.

With backlogging, the corresponding surrogate constraints are

$$r_t + C \sum_{u=k}^{t} y_u \geq d_{kt} - \overline{S}_{k-1} \quad \text{for } 1 \leq k \leq t.$$

Fixing t, we can generate mixing inequalities; see Section 8.3.

Finally we note that stock upper bounds can be viewed as a limiting case of the convex storage cost functions examined in the next section.

11.8 Safety Stocks or Piecewise Convex Storage Costs (SS)

Suppose that the inventory costs $H(s)$ in each period are as shown in Figure 11.3, with safety stock levels $SS^1, SS^2, \ldots, SS^L$, nondecreasing slopes $h, h + h^1, \ldots, h + \sum_{l=1}^{L} h^l$, where $h^l \geq 0$ for $l = 1, \ldots, L$, and intercept $H(0) = H^0$. Such a model allows more flexibility than one in which a fixed minimum (safety) stock level is imposed at the end of each period, and linear holding costs are incurred for stocks above these minima.

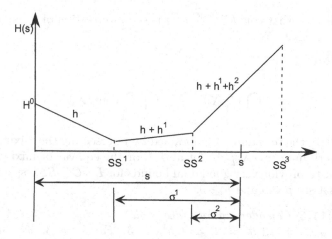

Figure 11.3. Piecewise convex inventory holding costs.

Introducing the variables σ_t^l to be the amount, if any, by which s_t exceeds SS^l, we obtain the following formulation for $LS\text{-}C\text{-}SS$

$$\min \ \sum_{t=1}^{n} p_t x_t + \sum_{t=0}^{n} h s_t + \sum_{t=1}^{n} \sum_{l=1}^{L} h^l \sigma_t^l + \sum_{t=1}^{n} q_t y_t$$

$$s_{t-1} + x_t = d_t + s_t \qquad\qquad \text{for } 1 \le t \le n$$
$$x_t \le C_t y_t \qquad\qquad\qquad \text{for } 1 \le t \le n$$
$$\sigma_t^l \ge s_t - SS^l \qquad\qquad\quad \text{for } 1 \le t \le n, 1 \le l \le L$$
$$s \in \mathbb{R}_+^{n+1}, \ x \in \mathbb{R}_+^n, \ \sigma \in \mathbb{R}_+^{Ln}, y \in \{0,1\}^n,$$

with feasible region $X^{LS\text{-}C\text{-}SS}$.

11.8.1 Mixing Set Relaxations for $LS\text{-}CC\text{-}SS$

Combining the constraint $\sigma_t^l \ge s_t - SS^l$ with the surrogate constraint $s_{k-1} + C \sum_{u=k}^{t} y_u \ge d_{kt}$, obtained by summing the flow balance constraints for periods k up to t, gives, for fixed k and l, a mixing set X_{kl}^{MIX} of the form

$$\sigma_{k-1}^l + C z_{kt} \ge d_{kt}^l \qquad\qquad \text{for } k \le t \le n$$
$$\sigma_{k-1}^l \in \mathbb{R}_+^1, z_{kt} \in \mathbb{Z}_+^1 \qquad\qquad \text{for } k \le t \le n,$$

where $z_{kt} = \sum_{u=k}^{t} y_u$ and $d_{kt}^l = (d_{kt} - SS^l)^+$ for $k \le t \le n$.

If in addition we write $SS^0 = 0$, $\sigma_t^0 = s_t$ for all t and $d_{kt}^0 = d_{kt}$ for all k, t, it follows that

$$X^{LS\text{-}CC\text{-}SS} \subseteq \bigcap_{l=0}^{L} \bigcap_{k=1}^{n} X_{kl}^{MIX}.$$

In the same way that $\mathrm{conv}(X^{WW\text{-}CC})$ is the intersection of n mixing sets, it can be shown that

Proposition 11.11

$$\mathrm{conv}\Big(\bigcap_{l=0}^{L} \bigcap_{k=1}^{n} X_{kl}^{MIX}\Big) = \bigcap_{l=0}^{L} \bigcap_{k=1}^{n} \mathrm{conv}(X_{kl}^{MIX}).$$

In practice the reformulation obtained by intersecting the mixing reformulations describing $\mathrm{conv}(X_{kl}^{MIX})$ for all k, l, or the equivalent mixing inequalities, appear to provide very good dual bounds for $LS\text{-}CC\text{-}SS$, especially when the costs satisfy $p = 0$ and $h \ge 0$.

Example 11.5 *Consider an instance with* $n = 5, d = (5, 8, 4, 7, 2), L = 1, SS^1 = 5, C = 12, h = 2,$ *and* $h^1 = 3$. *For* $k = 3$, *we have* $d_{33}^1 = (d_3 - SS^1)^+ = 0, d_{34}^1 = (d_{34} - SS^1)^+ = 6$ *and* $d_{35}^1 = (d_{35} - SS^1)^+ = 8,$ *so we obtain the surrogate inequalities*

$$\begin{aligned}
\sigma_2^1 + 12y_3 &\geq 0 \\
\sigma_2^1 + 12y_3 + 12y_4 &\geq 6 \\
\sigma_2^1 + 12y_3 + 12y_4 + 12y_5 &\geq 8,
\end{aligned}$$

from which we obtain mixing inequalities such as

$$\sigma_2^1 \geq 6(1 - y_3 - y_4),$$
$$\sigma_2^1 \geq 6(1 - y_3 - y_4) + 2(1 - y_3 - y_4 - y_5).$$

This way of modeling piecewise convex storage costs extends to the case with $h < 0$ and also to the case of piecewise linear backlog costs. Now we allow the holding costs to vary with time. In such cases, the storage and backlog cost function $H_t(s)$ is defined for $s \in \mathbb{R}$ and all periods t. One approach is to move the origin to the point $\underline{SS}_t = \arg\min_l H_t(SS^l)$ at which the function $H_t(s)$ is minimized. Now with new variables $s_t' = (s_t - r_t - \underline{SS}_t)^+$ and $r_t' = (r_t - s_t + \underline{SS}_t)^+$, we have

$$s_t' - r_t' = s_t - r_t - \underline{SS}_t \qquad \text{for all } t$$

which in turn gives the modified demand vector $\tilde{d}$, with $\tilde{d}_t = d_t + \underline{SS}_t - \underline{SS}_{t-1}$, for the new balance equations from which we obtain standard surrogates

$$s_{k-1}' + r_t' + C \sum_{u=k}^{t} y_u \geq \tilde{d}_{kt}.$$

Now for breakpoints with $SS^l > \underline{SS}_t$, we can introduce the new variables $\sigma_t^l \geq s_t' - (SS^l - \underline{SS}_t)$, and for breakpoints with $SS^l < \underline{SS}_t$, we introduce in similar fashion the variables $\rho_t^l \geq r_t' - (\underline{SS}_t - SS^l)$.

Now the storage and backlog costs in the objective function can be written as a nonnegative combination of the variables $s_t', r_t', \sigma_t^l, \rho_t^l$. The surrogates now give rise to continuous mixing sets based on constraints such as

$$s_{k-1}' + r_t' + C \sum_{u=k}^{t} y_u \geq \tilde{d}_{kt} \quad \text{for } 1 \leq k \leq t \leq n$$

and

$$\sigma_{k-1}^l + \rho_t^{l'} + C \sum_{u=k}^{t} y_u \geq \tilde{d}_{kt} - (SS^l - \underline{SS}_{k-1}) - (\underline{SS}_t - SS^{l'})$$
$$\text{for } 1 \leq k \leq t \leq n \text{ and all } l, l'.$$

Example 11.6 *Consider the same instance as above but with* $h = -2$*. Introducing* $s_t' - r_t' = s_t - SS^1$ *as* $r_t = 0$*, we obtain standard flow balance constraints with the modified demand vector* $\tilde{d} = (10, 8, 4, 7, 2)$*. We also have* $r_t' \leq 5$ *as* $s_t \geq 0$*. The backlog cost for* r_t' *is now* $-h = 2$*, and the storage cost for* s_t' *is* $h + h^1 = 1$*, so the objective function can be written (modulo a constant) as*

$$\sum_t (2r_t' + 1s_t').$$

11.9 A Model with Backlogging, Sales Markets, and Concave Production Costs

Problem $LS\text{-}U\text{-}B, SL, CP$ can be formulated as

$$\min \quad \sum_{k=1}^{K}\sum_{t=1}^{n}(p_t^k x_t^k + q_t^k y_t^k)$$

$$+ \sum_{t=0}^{n} h_t s_t + \sum_{t=1}^{n} b_t r_t - \sum_{l=1}^{L}\sum_{t=1}^{n} e_t^l v_t^l \qquad (11.52)$$

$$s_{t-1} - r_{t-1} + \sum_{k} x_t^k = d_t + \sum_{l} v_t^l + s_t - r_t \qquad \text{for all } t \qquad (11.53)$$

$$v_t^l \leq V_t^l \qquad \text{for all } l, t \qquad (11.54)$$

$$x_t^k \leq M y_t^k \qquad \text{for all } k, t \qquad (11.55)$$

$$s \in \mathbb{R}_+^{n+1}, \quad r \in \mathbb{R}_+^{n}, \quad x \in \mathbb{R}_+^{Kn}, \quad v \in \mathbb{R}_+^{Ln}, \quad y \in \{0,1\}^{Kn}, \qquad (11.56)$$

where we assume that $0 \leq q_t^1 \leq q_t^2 \leq \cdots q_t^K$, $p_t^1 \geq p_t^2 \geq \cdots p_t^K \geq 0$ and $e_t^1 \geq e_t^2 \cdots \geq e_t^L$ for all t. Note that if $q_t^k = 0$, x_t^k can be viewed as the amount bought in from outside in period t. Variables v_t^l can be viewed as potential sales to different markets $l = 1, \dots, L$ at unit price e_t^l in period t.

An Extended Formulation

Observation 11.3 *Following standard network-flow-based arguments for the uncapacitated lot-sizing problem, there exists an optimal solution in which if $x_t^k > 0$, then $x_t^k = \sum_{u=t-a}^{t+b}(d_u + \sum_{l \in S(u)} V_u^l)$ for some $a, b \geq 0$, and $S(u) \subseteq \{1, \dots, L\}$ for all u.*

Define $\bar{V}_t^l = \sum_{\lambda=1}^{l} V_t^\lambda$ for $0 \leq l \leq L$ and $1 \leq t \leq n$. In particular, $\bar{V}_t^0 = 0$ for all t. Observe that, because $e_t^1 \geq e_t^2 \cdots$ for all t, there always exists an optimal solution with $\sum_{\lambda=1}^{L} v_t^\lambda = \bar{V}_t^l$ for some $l \geq 0$ and all t.

We use the following variables in our extended formulation.

$\alpha_{ut}^{kl} = 1$ if production takes place in period u using production type k to satisfy a demand of $d_t + \bar{V}_t^l$ in period t.
$z_{ut}^{k} = 1$ if production takes place in period u using production type k to satisfy some demand in period t.

We now consider the following formulation:

$$\min \sum_{k=1}^{K}\sum_{t=1}^{n}(p_t^k x_t^k + q_t^k y_t^k)$$

$$+ \sum_{t=0}^{n} h_t s_t + \sum_{t=1}^{n} b_t r_t - \sum_{l=1}^{L}\sum_{t=1}^{n} e_t^l v_t^l \qquad (11.57)$$

$$z_{ut}^k = \sum_{l=0}^{L} \alpha_{ut}^{kl} \qquad \text{for } 1 \le u, t \le n, \text{ for all } k$$
$$(11.58)$$

$$\sum_{k=1}^{K}\sum_{u=1}^{n} z_{ut}^k = 1 \qquad \text{for } 1 \le t \le n \qquad (11.59)$$

$$y_t^k \ge z_{tt}^k \qquad \text{for } 1 \le t \le n, \text{ for all } k$$
$$(11.60)$$

$$z_{ut}^k \ge z_{u,t+1}^k \qquad \text{for } 1 \le u \le t \le n, \text{ for all } k$$
$$(11.61)$$

$$z_{ut}^k \ge z_{u,t-1}^k \qquad \text{for } 1 \le t \le u \le n, \text{ for all } k$$
$$(11.62)$$

$$v_t^l = \sum_{k=1}^{K}\sum_{u=1}^{n}\sum_{\lambda=l}^{L} V_t^\lambda \alpha_{ut}^{kl} \qquad \text{for } 1 \le t \le n, \text{ for } 1 \le l \le L$$
$$(11.63)$$

$$x_u^k = \sum_{l=0}^{L}\sum_{t=1}^{n}(d_t + \bar{V}_t^l)\alpha_{ut}^{kl} \qquad \text{for } 1 \le u \le n, \text{ for all } k$$
$$(11.64)$$

$$s_{t-1} = \sum_{k=1}^{K}\sum_{l=0}^{L}\sum_{u,\tau:u<t\le\tau}(d_\tau + \bar{V}_\tau^l)\alpha_{u\tau}^{kl} \qquad \text{for } 1 \le t \le n \qquad (11.65)$$

$$r_t = \sum_{k=1}^{K}\sum_{l=0}^{L}\sum_{u,\tau:\tau\le t<u}(d_\tau + \bar{V}_\tau^l)\alpha_{u\tau}^{kl} \qquad \text{for } 1 \le t \le n \qquad (11.66)$$

$$\alpha \in \mathbb{R}_+^{K(L+1)n^2}, \ z \in \mathbb{R}_+^{Kn^2}, \qquad y \in \{0,1\}^{Kn}. \qquad (11.67)$$

Note that the constraints $z_{ut}^k \ge z_{u,t+1}^k$ and $z_{ut}^k \ge z_{u,t-1}^k$ are justified by Observation 11.3.

This formulation can be shown to be equivalent to the minimum cost path in a network representing the sequences of regeneration intervals $[a, b]$ as defined in Observation 11.3. From this we obtain the following result.

Theorem 11.12 *The linear program (11.57)–(11.67) solves problem LS-U-B,SL,CP.*

11.10 Stochastic Lot-Sizing on a Tree

Here we present a problem of lot-sizing on a rooted tree, and then show by example how this model enables us to tackle a single-item stochastic lot-sizing problem.

Given a rooted directed out-tree $T = (N, A)$, let $D(v)$ be the direct successors of v, $S(v)$ the set of all successors of v, and $P(j, k)$ with $k \in S(j)$ the set of nodes on the path from j to k. Node $r = 1 \in N$ is the root. $L = \{v \in N : S(v) = \emptyset\}$ are the leaves. We add a dummy node 0 and an arc $(0, 1)$, and let $p(v)$ be the unique predecessor of v, for all $v \in N$.

The lot-sizing problem on a tree $LS\text{-}C\text{-}TREE$ is defined as the following mixed integer program,

$$\min \sum_{v \in N} (P'_v x_v + Q_v y_v) + \sum_{v \in N \cup \{0\}} H'_v s_v \qquad (11.68)$$

$$s_{p(v)} + x_v = d_v + s_v \quad \text{for all } v \in N \qquad (11.69)$$

$$x_v \leq C_v y_v \quad \text{for all } v \in N \qquad (11.70)$$

$$s \in \mathbb{R}_+^{|N|+1}, x \in \mathbb{R}_+^{|N|}, y \in [0, 1]^{|N|}, \qquad (11.71)$$

with production costs P'_v, fixed costs Q_v and demands d_v for all $v \in N$, and storage costs H'_v for all $v \in N \cup \{0\}$. Note the special form of the balance constraint in which the flow s_v out of node $v \in N \setminus L$ is the inflow to each direct successor node $w \in D(v)$.

Eliminating the x_v variables by substitution using (11.69), the objective function can be rewritten as:

$$\min \sum_{v \in N \cup \{0\}} H_v s_v + \sum_{v \in N} Q_v y_v + K_1,$$

where $H_v = H'_v + P'_v - \sum_{w \in D(v)} P'_w$ and $K_1 = \sum_{v \in N} P'_v d_v$.

We now consider an example of a stochastic lot-sizing problem with a tree of scenarios, and show how it can be modeled as a problem of lot-sizing on a tree.

Example 11.7 *Consider an instance with three periods and four scenarios, and an initial stock variable s_0.*

In period 1 the demand $d_1 = 3$ and the set-up cost $q_1 = 18$ are known with certainty.

In period 2 there are two possible events. Either the demand and set-up cost will be $d_2 = 4, q_2 = 18$ with probability 0.5, or they will be $d_3 = 3, q_3 = 24$ with probability 0.5. Which of these two events occurs is known before the production decision in period 2.

In period 3, if the outcome in period 2 was $d_2 = 4, q_2 = 18$, then with probability 1/3 it will be $d_4 = 7, q_4 = 30$, and with probability 2/3 it will be $d_5 = 1, q_5 = 30$. On the other hand, if the outcome in period 2 was

$d_3 = 3, q_3 = 24$, *then with probability 1/5 it will be* $d_6 = 2, q_6 = 20$, *and with probability 4/5 it will be* $d_7 = 5, q_7 = 25$. *Again the period 3 outcome is known before the production decision in period 3 is taken.*

The storage costs are $h_0 = 4$ *and* $h_t = 1$ *in periods* $t = 1, 2, 3$ *whatever the outcomes. The production capacity is* $C = 10$ *throughout.*

The corresponding tree and optimal solution are shown in Figure 11.4.

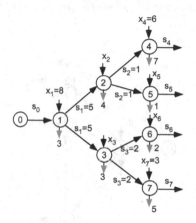

Figure 11.4. Scenario and lot-sizing tree

Taking into account the probabilities, the expected cost to be minimized is

$$\begin{aligned}
\min \quad & 4s_0 + (1s_1 + 18y_1) + 0.5(1s_2 + 18y_2) + 0.5(1s_3 + 24y_3) \\
& + 0.5(1/3)(1s_4 + 30y_4) + 0.5(2/3)(1s_5 + 30y_5) \\
& + 0.5(1/5)(1s_6 + 20y_6) + 0.5(4/5)(1s_7 + 25y_7) \\
= \; & 4s_0 + s_1 + 0.5s_2 + 0.5s_3 + (1/6)s_4 + 1/3s_5 + 0.1s_6 + 0.4s_7 \\
& + 18y_1 + 9y_2 + 12y_3 + 5y_4 + 10y_5 + 2y_6 + 10y_7.
\end{aligned}$$

Note that the coefficients H_v *for* $v \geq 1$ *are precisely the probabilities of outcome* d_v *occurring (because we have taken* $h = 1$ *throughout).*

11.10.1   Mixing Set Relaxations with Constant Capacities

First we rewrite the constraints (11.69) defining *LS-C-TREE* in the form

$$s_0 + \sum_{u \in P(1,v)} x_u \geq d_{1v} \quad \text{for all } v \in V,$$

where $d_{uv} = \sum_{w \in P(u,v)} d_w$ for all $u \in V$ and $v \in S(u) \cup \{u\}$.

Now for each $U \subset V$, we use surrogate inequalities to construct obtain a mixing set for the problem $LS - CC - TREE$. Specifically, using the capacity constraints $x_v \leq Cy_v$, we obtain the mixing set $X^{MIX}(U)$:

$$s_U + Cz_{U,v} \geq d_{1v} \quad \text{for all } v \in V$$
$$s_U \geq 0, z_{U,v} \in \mathbb{Z}^+ \quad \text{for all } v \in V,$$

where $s_U = s_0 + \sum_{u \in U} x_u$, and $z_{U,v} = \sum_{u \in P(1,v) \setminus U} y_u$ for all $v \in V$.

Example 11.8 *Suppose that the scenario tree has just two levels as shown in Figure 11.5,*

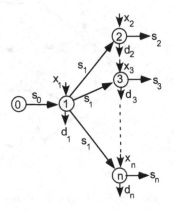

Figure 11.5. Scenario tree with two levels.

and suppose in addition that $d_1 > 0$ and $0 < d_2 \leq d_3 \leq \cdots \leq d_n$.
  The mixing set $X^{MIX}(\emptyset)$ is

$$s_0 + Cy_1 \geq d_1$$
$$s_0 + C(y_1 + y_t) \geq d_1 + d_t \quad \text{for } t = 2, \ldots, n$$
$$s_0 \in \mathbb{R}^1_+, y_1 \in \mathbb{Z}^1_+, y_1 + y_t \in \mathbb{Z}^1_+ \quad \text{for } t = 2, \ldots, n,$$

the mixing set $X^{MIX}(\{1\})$ is

$$s_0 + x_1 \geq d_1$$
$$s_0 + x_1 + Cy_t \geq d_{1t} \quad \text{for } t = 2, \ldots, n$$
$$s_0 + x_1 \in \mathbb{R}^1_+, y_t \in \mathbb{Z}^1_+ \quad \text{for } t = 2, \ldots, n,$$

or alternatively, using $s_1 = s_0 + x_1 - d_1$,

$$\{(s_1, y) \in \mathbb{R}^1_+ \times \{0,1\}^{n-1} : s_1 + Cy_t \geq d_t \quad \text{for } t = 2, \ldots, n\},$$

and similarly the mixing set $X^{MIX}(\{1, v\})$ for nodes $v \in \{2, \ldots, n\}$ can easily be rewritten as the mixing set

$$s_v + Cy_t \geq d_t - d_v \quad \text{for } t = v+1, \ldots, n$$
$$s_v \in \mathbb{R}^1_+, y_t \in \mathbb{Z}^1_+ \quad \text{for } t = v+1, \ldots, n.$$

11.10.2 Valid Inequalities for *LS-CC-TREE*

As the feasible region $X^{LS-CC-TREE}$ lies in the intersection of the mixing sets $X^{MIX}(U)$ for $U \subseteq V$, the corresponding mixing inequalities are valid inequalities for problem *LS-CC-TREE*, and for fixed U one can separate each of these families of inequalities in time polynomial in $n = |V|$.

Example 11.7 cont. *We give an example of three types of valid inequalities arising just from consideration of the first two periods. The inequality*

$$s_0 + 7y_1 + 1y_2 + 3y_3 \geq 7$$

is a a mixing inequality for the set $X^{MIX}(\emptyset)$. The inequality

$$s_0 + 7y_1 + 1y_2 + x_3 \geq 7$$

is a mixing inequality for the set $X^{MIX}(\{3\})$, and the inequality

$$s_0 + x_1 + 1y_2 + x_3 \geq 7 \quad \text{or} \quad s_3 + 1y_2 \geq 1$$

is a mixing inequality for the set $X^{MIX}(\{1,3\})$.

Finally note that the convex hulls of the mixing set relaxations defined in Section 11.10.1 suffice to define $\text{conv}(X^{LS-CC-TREE})$ when $s_0 = 0$ and the scenario tree has just two levels (see Exercise 11.11).

Exercises

Exercise 11.1 Prove the validity of the (l, S, R) inequalities for the lot-sizing set with sales $X^{LS-U-SL}$.

Exercise 11.2 We say that a *hot start-up* occurs if it is at most α periods since the item was switched off (last produced), and it is a *cold start-up* otherwise.
i. Model such start-ups by adding to the basic (y_t, z_t, w_t) model.
ii. Generate tight valid inequalities.
iii. Give an extended reformulation as a unit flow in a network.

Exercise 11.3 Consider an instance of *LS-U-SUB* with data $n = 5$, $d = (3,4,2,1,3)$, $\overline{S} = (6,3,4,3,5)$, and the fractional solution $y^* = (1,1,0,0.75,0)$, $x^* = (3,7,0,3,0)$, $s^* = (0,0,3,1,3,0)$. Find a valid inequality cutting off this point.

Exercise 11.4 Solve the instances of $WW\text{-}CC\text{-}SS$ in Examples 11.5 and 11.6 where $q = (20, 25, 20, 44, 22)$ just by reformulations and linear programming.

Exercise 11.5 (Lot-Sizing with Warm-Starts)
Consider the following variant of $WW\text{-}CC$. If a set-up takes place in period t, it is a *cold* set-up if $x_{t-1} < Q$ in which case one has to pay a set-up cost q_t, or it is a *warm* set-up if $x_{t-1} \geq Q$ in which case there is no set-up cost, but there is a *keep warm* cost of $\alpha(C - x_{t-1})$. Here $Q \leq C$.
i. Formulate as a MIP.
ii. Show that an optimal solution consists of sequences of set-ups of the form: "Cold, Warm, ..., Warm, Warm" or "Cold, Off, ...,Off". Use this to derive a dynamic programming algorithm for the problem.

Exercise 11.6 (Lot-Sizing with a Perishable Product)
Consider the following variant of $LS\text{-}U$. The product becomes unusable after τ periods for some positive integer τ.
i. Derive a polynomial algorithm for this problem.
ii. Propose an extended formulation for this problem.

Exercise 11.7 (Lot-Sizing with Delivery Time Windows)
Suppose that production, storage, and set-up costs are as usual, but that demand consists of a set of client orders of size $\{D^v\}_{v \in V}$ each with an associated time window $[b_v, e_v]$ during which the client must receive the order. Suppose that there is no backlogging. If the order v is produced during the $[b_v, e_v]$ interval, no storage costs are paid. On the other hand, if some of the order is produced before period b_v, storage costs will be paid until period b_v and that part of the order is delivered in period b_v. Note that there are at most $n(n+1)/2$ distinct time windows, so we can assume that $|V| \leq n(n+1)/2$.
i. Formulate as an MIP.
ii. Show that if the problem is uncapacitated, each order D^v is produced completely in one period, so there is no order splitting. Use this to derive a polynomial algorithm for the case of Wagner–Whitin costs.
iii. With Wagner–Whitin costs, show that the following inequalities are valid and provide a tight formulation. For each interval $[k, l]$,

$$s_{k-1} \geq \sum_{v \in V : k \leq b_v, e_v \leq l} D^v (1 - y_k - \ldots - y_{e_v}).$$

Exercise 11.8 (Lot-Sizing with Constant Capacities and Upper Bounds on Stocks)
Define a regeneration interval $[t, l]$ for $LS\text{-}CC\text{-}SUB$ as a partial solution with $s_{t-1} \in \{0, \bar{S}_{t-1}\}$, $s_l \in \{0, \bar{S}_l\}$ and $0 < s_u < \bar{S}_u$ for $t \leq u < l$. Show
i. how, for each of these four possibilities, the cost of a minimum cost partial solution can be calculated in polynomial time by dynamic programming,
ii. how these costs can be used to solve $LS\text{-}CC\text{-}SUB$, and
iii. how to derive a tight extended formulation.

Exercise 11.9 (Lot-Sizing with Constant Start-Up Times)
Consider WW-CC-SC with in addition a loss of capacity, or start-up time $(x_t \leq Cy_t - Lz_t$ for all $t)$. Derive a polynomial dynamic programming algorithm to solve the problem.
Hint: Build a recurrence based on $G(t, k, p, q, \delta)$, the minimum cost of a production plan for periods $t, \ldots, n$, where k is the first regeneration point $(s_{k-1} = 0)$, p is the number of set-ups, q is the number of start-ups, and $y_{t-1} = \delta \in \{0, 1\}$ indicates the start-up state.

Exercise 11.10 (Lot-Sizing with a Restricted Number of Set-Ups)
Consider an uncapacitated problem with the additional constraint that there can be at most K periods of production.
i. Show that $\mathrm{conv}(X^{WW-U}) \cap \{(s, y) : \sum_{u=1}^{n} y_u \leq K\}$ is an integral polyhedron.
ii. Show by example that $\mathrm{conv}(X^{LS-U}) \cap \{(s, x, y) : \sum_{u=1}^{n} y_u \leq K\}$ is not integral.

Exercise 11.11 (Lot-Sizing on a Tree)
Consider the special case of the two-level lot-sizing problem on a tree examined in Example 11.8 with $s_0 = 0$, $Q_1 = 0$ and no limit on the production in period 1, so that the first period decision is just to choose the end stock level.
i. With constant capacities in period 2, formulate the set of feasible solutions as a mixing set with flows as described in Exercise 8.16.
ii. Use this to show that under the above conditions the mixing set relaxations presented in Section 11.10.1 suffice to describe $\mathrm{conv}(X^{LS-CC-TREE})$.

Notes

Section 11.1 The uncapacitated lot-sizing model with sales LS-U-SL is treated in Loparic et al. [107]. It is also shown there that, when the demands are nonnegative, the separation problem for the (l, S, R) inequalities can be reduced to the problem of maximizing a quadratic Boolean function in which all the quadratic terms have nonnegative coefficients, which is in turn reducible to a maximum flow problem. In Aksen et al. [8], an $O(n^2)$ dynamic programming algorithm is presented for uncapacitated lot-sizing with lost sales which is precisely the problem LS-U-SL with nonnegative demands.

Section 11.2 The lot-sizing model with constant lower bounds on production is treated in Van Vyve [178]. See also Constantino [47] for results on a multi-item single-period submodel.

Section 11.3 The results for a model with almost full capacity production are from Belvaux and Wolsey [26].

Section 11.4 Minimum and maximum length set-up sequence, or up/down times arise in many production planning problems as well as in job scheduling and unit commitment problems. The inequalities of Proposition 11.5 have been rediscovered many times, see, for instance, Lee et al. [103], Belvaux and Wolsey [26], and Kondili et al. [99]. The proof of Proposition 11.5 is due to Malkin [114].

Section 11.5 The formulation of piecewise linear concave production costs is standard. Aghezzaf and Wolsey [5] show that one obtains the convex hull for LS-U-CP with the generalized (l, S) inequalities. With varying capacities, piecewise linear production costs and general holding costs, Shaw and Wagelmans [150] present a computationally effective pseudopolynomial dynamic programming algorithm.

Section 11.6 The study of the two different lot-sizing problems with production time window appears in the thesis of Brahimi [29]; see also Brahimi [30], and Dauzère-Pérès et al. [54], including MIP formulations for LS-U-TWP, the DP algorithm for WW-U-TWP, the equivalence of LS-U-$TWP(I)$ with the noninclusive time window problem, and an $O(n^4)$ DP algorithm for $LS - U - TWP(I)$. The other results and reformulations are from Wolsey [195].

Section 11.7 Love [111] demonstrated an $O(n^3)$ DP algorithm for LS-U-SUB. The equivalence with the production time window problem with indistinguishable orders implying the existence of an $O(n^2)$ algorithm is from Wolsey [195]. The valid inequalities for the LS-U-SUB model are due to Atamtürk and Küçükyavuz [14]. The tight formulation for the Wagner–Whitin constant capacity model is from Pochet and Wolsey [140]. See Exercise 11.8 for an algorithm for LS-CC-SUB.

Section 11.8 Piecewise convex storage costs are treated implicitly in Miller and Wolsey [125].

Section 11.9 The formulation for LS-U-B, SL, PC is from Verweij and Wolsey [184].

Section 11.10 A family of valid inequalities for stochastic uncapacitated lot-sizing on a scenario tree, called (Q, S_Q)-inequalities, have been studied recently by Guan et al. [83]. The (Q, S_Q)-inequalities are a subset of the mixing inequalities presented here.

For the special two-level tree discussed in Example 11.8 with no initial stock, it follows from Exercise 11.11 that the convex hull is obtained by adding the mixing set relaxations $\mathrm{conv}(X(1)) \cap_{v=2}^{n} \mathrm{conv}(X(1,v))$ in the constant capacity case. A similar result for the uncapacitated case using only (Q, S_Q) inequalities appears in Guan et al. [82]. However, for an uncapacitated three-level problem, the mixing inequalities do not suffice to give the convex hull of solutions.

Other Extensions and Exercises. Formulations for several other extensions have been addressed in the literature. Fixed costs on stocks have been treated by Ortega and van Vyve [129] and the general case with bounds and fixed costs is treated in Atamtürk and Küçükyavuz [14]. Dynamic programming algorithms for lot-sizing of a perishable good are presented in Hsu [92], and formulations and valid inequalities are examined in Ortega [128]; see Exercise 11.6. In the simplest case, this is a special case of the problem with production time windows. Bounds on the number of set-ups are discussed in Aghezzaf and Wolsey [4]; see Exercise 11.10. A problem with demand time windows is discussed in Lee et al. [102] (see Exercise 11.7), and a problem with constant start-up times is treated in Vanderbeck [182] (see Exercise 11.9).

Multi-Item Lot-Sizing

12

Multi-Item Single-Level Problems

So far, we have only studied the formulations of single-item models. Part IV consists of two chapters dealing with multi-item models. We derive formulations for the most important classes of multi-item models and extend the classification scheme introduced in Chapter 4.

Specifically in this chapter we present multi-item models with single-level product structure. In such models, the different items interact because their manufacture involves the use of common equipment or resources. In Chapter 13 we deal with multi-item multi-level models, where the different items may also interact because of the product structure. In these models, some items are inputs for the production of other items, so their production is inevitably linked.

We now consider the formulation of single-level multi-item problems. Typically we consider that all production takes place on one machine, or in one department or location. Such problems were discussed in the chapter on decomposition and algorithms, as they are the simplest problems that are encountered in practice, and often decompose very naturally.

These problems typically have a feasible solution set of the form

$$Z^{MI} = (\prod_{i=1}^{NI} X_i^{LS-C}) \bigcap Y^L,$$

where X_i^{LS-C} represents the solution set of the item i lot-sizing subproblem and Y^L is the set resulting from constraints dealing with the interaction between items, either because the items are competing for scarce resources such as machines, manpower, or finance, or because there are switch-over times or costs that depend on the order of production (set-up) of the items. Thus sometimes Y^L decomposes into

$$Y^L = Y^{PM} \bigcap Y^{PQ},$$

where Y^{PM} is the production mode set modeling the restrictions on set-ups and start-ups, and Y^{PQ} the production quantity set indicating the restrictions on the production levels. As both production mode and production quantity constraints are typically by time period, Y^{PM} and Y^{PQ} often decompose further into

$$Y^{PM} = \prod_{t=1}^{NT} Y_t^{PM} \quad \text{and} \quad Y^{PQ} = \prod_{t=1}^{NT} Y_t^{PQ}.$$

In this chapter we study ways to model the linking set Y^L, or more specifically the sets Y^{PM} and Y^{PQ}. This leads us to extend the single-item classification introduced in Chapter 4 to include a classification of the production constraints linking items, leading to

$$\{PROB-CAP-VAR\}-PM-PQ.$$

One important aspect of the linking constraints is the restriction that they impose on the set of production, or set-up sequences. Thus in Figure 12.1 we show graphically three different production plans covering four periods. The length of each rectangle indicates the time spent in producing the item during the given period.

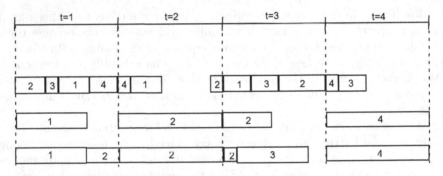

Figure 12.1. Three different production plans.

We observe that in the first sequence many items are produced in each period, while in the last two sequences there are at most one, respectively two, set-ups in each period. This turns out to be an important distinction in our study of the linking sets Y^L.

We now describe the contents of the chapter.

- We start in Section 12.1 by extending our single-item classification scheme to multi-item problems by describing the sort of constraints encountered in each time period. Specifically we distinguish between a *production mode* classification that essentially characterizes the constraints on set-ups and start-ups linking the items in each period, and a *production quantity* classification that describes the sort of resource or budget type constraints that limit the total amount produced in each period.

- Next we present in Section 12.2 formulations for the case where only one set-up is allowed per period, problems often referred to as *small bucket* problems. We present tight formulations when each individual item is of the form *DLS-CC*, *DLS-CC-B*, and for submodels consisting just of set-ups and start-ups or sequence-dependent changeovers. Problems with discrete lot-sizing and start-ups are often called *discrete lot-sizing and scheduling* problems.
- Problems with two or more set-ups are presented in Section 12.3. The case of at most two set-ups, sometimes referred to as the *proportional lot-sizing problem*, forms the basis for the general case.
- Production quantity constraints are presented in Section 12.4. Here the single-period problem is already NP-hard, so all that is known are various families of valid inequalities. These constraints often arise in applications and are difficult to treat effectively in practice.
- Finally in Section 12.5 we consider a multi-item constant capacity problem with a family set-up variable in addition to, or in place of, the individual item set-up variables. The uncapacitated version with both family and item set-up costs is known as the *joint replenishment* problem.

12.1 Joint Resource Classification

Many items may require the same resources. Such a resource may be a machine, a specialized set of workers, a facility, and so on. Each of these may lead to a *production mode PM* constraint restricting set-ups and/or a *production quantity PQ* constraint limiting the amount that can be produced.

As earlier, we use the compact notation n and m to represent, respectively, the number of time periods and number of items in canonical multi-item models, whereas we use NT and NI for specific production planning instances.

12.1.1 Production Mode Classification

First we consider the structure of the single-period sets Y_t^{PM}. In the first field PM, we limit the choice to one out of six versions

$$PM = [M_1, M_1-SC, M_1-SQ, M_k, M_k-SC, M_k-SQ]^1,$$

where M_k indicates at most k set-ups, and M_∞ indicates no restriction on the number of set-ups.

M_1: This is the simplest case in which there is at most one set-up per period. Here there is just the additional constraint

$$\sum_{i=1}^{m} y_t^i \leq 1 \quad \text{for all } t.$$

M_1-SC: This is the same as M_1, except for the additional presence of start-up and/or switch-off variables defined exactly as in Chapter 10:

$$z_t^i = 1 \quad \text{if } y_t^i = 1 \quad \text{and} \quad y_{t-1}^i = 0$$

for the start-up variables and

$$w_t^i = 1 \text{ if } y_t^i = 1 \text{ and } \quad y_{t+1}^i = 0$$

for the switch-off variables.

M_1-SQ: This is the same as M_1-SC, except for the additional presence of sequence-dependent changeover variables indicating that one switches from a set-up of item i in period $t - 1$ to a set-up of item j in period t. For this one introduces the variables

$$\chi_t^{ij} = 1 \quad \text{if} \quad y_{t-1}^i = 1 \quad \text{and} \quad y_t^j = 1.$$

When more than one item can be produced per period, we need to make precise what we mean by a start-up. First we assume that only one item is set up at any moment during a period. Also, as before, an item can only be produced if it is set up. Now we say that an item i *starts up* in period t if it is set up in period t, and item i has not been set up continuously since before the end of period $t - 1$. Thus in the first production plan shown in Figure 12.1, item 1 has a start-up in period 2, but item 4 does not.

We now continue with the production mode options.

M_k-SC: Here at most $k \geq 2$ items can be set up in a period, and start-up variables are present.

M_k-SQ: Here at most $k \geq 2$ items can be set up in a period, and sequence-dependent changeover variables are present.

M_∞: Here any number of items can be set up in a period.

The production mode classification is thus

$$PM = [M_1, M_1-SC, M_1-SQ, M_k, M_k-SC, M_k-SQ]^1.$$

12.1.2 Production Quantity Classification

In the field PQ, there is a choice of at most one out of six versions

$$PQ = [PC, PC-SU, PC-ST, PC-SQ, PC-U, PC-FAM]^1 .$$

PC: This is the simplest case in which there is just limited production capacity. Here there is just the additional resource capacity constraint:

$$\sum_{i=1}^{m} a^i x_t^i \le L_t \quad \text{for all } t,$$

where a^i is the amount of capacity, often production time, consumed per unit of item i produced.

PC-SU: This is similar to the previous case PC, except for the addition of set-up times that reduce the available production capacity of the joint resource. Here one adds

$$\sum_{i=1}^{m} a^i x_t^i + \sum_{i=1}^{m} b^i y_t^i \le L_t \quad \text{for all } t,$$

where b^i is the amount of capacity consumed per set-up of item i.

PC-ST: This is similar to the previous case PC-SU, except for the replacement of set-up times by start-up times. Here one adds

$$\sum_{i=1}^{m} a^i x_t^i + \sum_{i=1}^{m} c^i z_t^i \le L_t \quad \text{for all } t,$$

where c^i is the amount of capacity consumed per start-up of item i. Note that in this case the production mode M_k-SC is necessarily required to complete the model.

PC-SQ: This is similar to the previous case PC-ST, except for the replacement of set-up times by sequence-dependent changeover times. Here one adds

$$\sum_{i=1}^{m} a^i x_t^i + \sum_{i,j=1}^{m} c^{ij} \chi_t^{ij} \le L_t \quad \text{for all } t,$$

where c^{ij} is the amount of capacity consumed per changeover from item i to item j. Note that in this case the production mode $M_k - SQ$ is also required.

PC-U: This indicates that there is no joint limit on the production level.

PC-FAM: This indicates that there are family set-up variables associated to the multi-item production constraint of the form:

$$\sum_{i=1}^{m} a^i x_t^i \le L_t Y_t \quad \text{for all } t,$$

where Y_t is a 0-1/integer variable indicating a family set-up or number of set-ups in period t.

The production mode classification is thus

$$PQ = [PC, PC-SU, PC-ST, PC-SQ, PC-U, PC-FAM]^1.$$

We have already seen that several problems involve modeling of both the production mode and the production quantity constraints.

The combined *PM-PQ* classification is now

$$[M_1, M_1-SC, M_1-SQ, M_k, M_k-SC, M_k-SQ]^1 \; / $$

$$[PC, PC-SU, PC-ST, PC-SQ, PC-U, PC-FAM]^1.$$

We now consider how to model, or remodel these different cases. In the next two sections, we consider different production modes. First the case with one set-up per period, and then with more than one set-up. After that we consider the production quantity constraints.

12.2 Production Mode Models: One Set-Up

12.2.1 Single Set-Up Constraint: M_1

The basic single level multi-item problem with a single set-up per period and backlogging $LS\text{-}C\text{-}B/M_1$ can be formulated as

$$\min \; \sum_{i,t} p_t'^i x_t^i + \sum_{i,t} h_t'^i s_t^i + \sum_{i,t} b_t'^i r_t^i + \sum_{i,t} q_t^i y_t^i$$

$$s_{t-1}^i - r_{t-1}^i + x_t^i = d_t^i + s_t^i - r_t^i \qquad \text{for all } i, t$$

$$x_t^i \le C_t^i y_t^i \qquad \text{for all } i, t$$

$$\sum_i y_t^i \le 1 \qquad \text{for all } t$$

$$s \in \mathbb{R}_+^{m(n+1)}, \; r, x \in \mathbb{R}_+^{mn}, \; y \in \{0,1\}^{mn},$$

where, as before, the variables r_0^i do not exist. This problem is traditionally called *small bucket* because M_1 is often used to model problems with small time windows/periods/buckets during which the machine set-up status remains constant. For certain special cases, the convex hull of solutions is known.

Discrete Lot-Sizing with Constant Capacities

The reformulation that is tight for a single item carries over to this multi-item case, and again leads to a network flow problem.

Theorem 12.1 *The polyhedron*

$$\sum_{u=1}^{t} y_u^i \geq \lceil \frac{d_{1t}^i}{C^i} \rceil \qquad\qquad \text{for all } i, t$$

$$\sum_{i} y_t^i \leq 1 \qquad\qquad \text{for all } t$$

$$y \in [0,1]^{mn}$$

gives $conv(X^{DLS-CC/M_1})$. *Optimization over this polyhedron is a network flow problem.*

As in Chapter 9, it suffices to introduce the variables $z_t^i = \sum_{u=1}^{t} y_u^i$. Figure 12.2 shows that the resulting problem is a network flow problem (with lower bounds on the z_t^i flows).

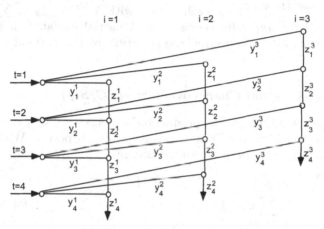

Figure 12.2. Small bucket discrete lot-sizing as a network flow.

Discrete Lot-Sizing with Constant Capacities and Backlogging

Theorem 12.2 *Optimizing over the polyhedron*

$$r_t^i + C^i \sum_{u=1}^{t} y_u^i \geq d_{1t}^i \qquad\qquad \text{for all } i, t$$

$$r_t^i + C^i f_t^i \sum_{u=1}^{t} y_u^i \geq C^i f_t^i \lceil \frac{d_{1t}^i}{C^i} \rceil \qquad\qquad \text{for all } i, t$$

$$\sum_{i} y_t^i \leq 1 \qquad\qquad \text{for all } t$$

$$r \in \mathbb{R}_+^{mn}, \ y \in [0,1]^{mn}$$

solves $DLS\text{-}CC\text{-}B/M_1$, where $f_t^i = \frac{d_{1t}^i}{C^i} - \lfloor \frac{d_{1t}^i}{C^i} \rfloor$ for all i,t.

This result also follows directly by applying Theorem 8.2 to the system

$$\tilde{r}_t^i + z_t^i \geq b_t^i, \tilde{r}_t^i \geq 0, z_t^i \in \mathbb{Z} \quad \text{for all } t,$$

where $\tilde{r}_t^i = r_t^i/C^i, b_t^i = d_{1t}^i/C^i$ and $z_t^i = \sum_{u=1}^t y_u^i$ as above. Again using Figure 12.2, the matrix arising from the additional constraints

$$\sum_i y_t^i \leq 1 \qquad\qquad \text{for all } t,$$

$$z_t^i - z_{t-1}^i = y_t^i \qquad\qquad \text{for all } i,t$$

is a network matrix, and so is totally unimodular.

Note that in practice the reformulation of $PROB\text{-}CAP\text{-}VAR/M_1$ obtained by intersecting an exact or approximate formulation for $\text{conv}(X_i^{PROB-CAP-VAR})$ for each item i with $Y_t^{M_1} = \{y_t : \sum_i y_t^i \leq 1\}$ for all periods t is typically very strong. A partial explanation is that the latter constraints do not destroy the structure of the optimal solutions of $X_i^{PROB-CAP-VAR}$.

12.2.2 Start-Ups and Changeovers M_1-$\{SC, SQ\}$

In this single set-up model, the production sequences are very simple. It suffices to indicate which item, if any, is set up in each period. With start-ups and changeovers, the sets to be studied are, respectively,

$$Y^{M_1-SC} = \{ (y,z) \in \{0,1\}^{mn} \times \{0,1\}^{mn} :$$

$$\sum_i y_t^i \leq 1 \qquad\qquad \text{for all } t$$

$$z_t^i \geq y_t^i - y_{t-1}^i, \ z_t^i \leq y_t^i, \ z_t^i \leq 1 - y_{t-1}^i, \qquad \text{for all } i,t\}$$

and

$$Y^{M_1-SQ} = \{ (y,\chi) \in \{0,1\}^{mn} \times \{0,1\}^{m^2n} :$$

$$\sum_i y_t^i \leq 1 \qquad\qquad \text{for all } t$$

$$\chi_t^{ij} \geq y_{t-1}^i + y_t^j - 1, \ \chi_t^{ij} \leq y_{t-1}^i, \ \chi_t^{ij} \leq y_t^j \quad \text{for all } i,j,t\}.$$

For simplicity below, we assume that a dummy job 0 (machine idle) has been added, so that the machine restriction will now be written as an equality

$$\sum_{i=0}^m y_t^i = 1 \quad \text{for all } t.$$

To understand the tight formulations for these sets, we consider a simple example.

Example 12.1 *Consider the five-period production sequence 1-2-2-4-3 (i.e., $y_1^1 = y_2^2 = y_3^2 = y_4^4 = y_5^3 = 1$). Graphically this can be represented as a path through a directed graph as shown in Figure 12.3.*

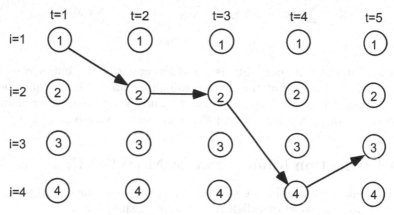

Figure 12.3. A five-period production sequence.

This immediately suggests an alternative formulation for Y^{M_1-SQ}, namely

$$1 = \sum_i y_1^i \tag{12.1}$$

$$y_{t-1}^i = \sum_j \chi_t^{ij} \qquad \text{for all } i, \text{and } t \geq 2 \tag{12.2}$$

$$\sum_i \chi_t^{ij} = y_t^j \qquad \text{for all } j, \text{and } t \geq 2 \tag{12.3}$$

$$y \in [0,1]^{mn}, \ \chi \in [0,1]^{m^2 n}. \tag{12.4}$$

As this is the standard formulation for the flow of one unit through a network, the resulting coefficient matrix is known to be totally unimodular, and we have

Proposition 12.3 *The polyhedron (12.1)–(12.4) gives conv(Y^{M_1-SQ}).*

To link the start-up and changeover variables, we have

$$z_t^j = \sum_{i:i\neq j} \chi_t^{ij} = y_t^j - \chi_t^{jj}.$$

Therefore to find conv(Y^{M_1-SC}), it suffices to find the projection of conv(Y^{M_1-SQ}) onto the (y_t^i, χ_t^{jj}) space, and then substitute for χ_t^{jj}.

Proposition 12.4 conv(Y^{M_1-SC}) *is given by*

$$\sum_i y_t^i = 1 \qquad\qquad\qquad \text{for all } t$$

$$z_t^i \geq y_t^i - y_{t-1}^i \qquad\qquad\qquad \text{for all } i, t$$

$$z_t^i \leq y_t^i \qquad\qquad\qquad\qquad \text{for all } i, t$$

$$z_t^i + \sum_{j:j \neq i} (y_t^j - z_t^j) \leq 1 - y_{t-1}^i \qquad \text{for all } i, t$$

$$y, z \in [0, 1]^{mn}.$$

Note that the last inequality is a strengthening of the initial constraint $z_t^i \leq 1 - y_{t-1}^i$. It says that the three possibilities "item i is set up in period $t - 1$," "item i is started up in period t," and "some item other than i is produced in both periods $t - 1$ and t" are mutually exclusive.

12.3 Production Modes: Two or More Set-Ups

When dealing with problems involving both more than one set-up per period and start-ups, we make two slightly restrictive assumptions.

i. The last item set up in one period is always the first item set up in the following period.
ii. Each item is set up at most once in each period.

Thus the five-period production set-up sequences

$$\{12\}\{2\}\{24\}\{43\}\{3\}$$

and

$$\{142\}\{2\}\{2534\}\{4513\}\{3\}$$

are allowed. On the other hand the sequence

$$\{12\}\{2\}\{24\}\{13\}\{3\}$$

is not allowed because item 4 is set up last in period 3, and a different item 1 is set up first in period 4. However, it can be replaced by the feasible sequence

$$\{12\}\{2\}\{24\}\{413\}\{3\}$$

with zero production of item 4 in period 4. Also the sequence

$$\{142\}\{2\}\{2532\}\{2513\}\{3\}$$

is not allowed because item 2 is set up twice in period 3. However, it is equivalent in certain circumstances to the feasible sequence

$$\{142\}\{20\}\{0532\}\{2513\}\{3\},$$

where item 0 is a dummy or idle item.

The general case M_k-SC, with any number $k > 2$ of set-ups per period, is closely related to that with at most two set-ups $M_2 - SC$, so we start with the latter case.

12.3.1 Two Set-Ups: M_2

Here we allow up to two items to be set up in each period. Our assumptions imply that there is at most one start-up per period.

Proposition 12.5 *With the introduction of an "idle" item " 0", the set of feasible set-up sequences Y^{M_2-SC} is described by the following constraint set,*

$$z_t^i \geq y_t^i - y_{t-1}^i \qquad\qquad \text{for all } i,t$$

$$\sum_i (y_t^i - z_t^i) = 1 \qquad\qquad \text{for all } t$$

$$\sum_i z_t^i \leq 1 \qquad\qquad \text{for all } t$$

$$z_t^i + z_{t-1}^i \leq y_t^i \qquad\qquad \text{for all } i,t$$

$$y, z \in \{0,1\}^{mn}.$$

In this formulation, the basic constraint $\sum_i y_t^i \leq 2$ is decomposed into $\sum_i (y_t^i - z_t^i) = 1$ stating that there is exactly one item set up but not started up in each period, and $\sum_i z_t^i \leq 1$ stating that at most one item can be started up. The constraint that $z_t^i + z_{t-1}^i \leq y_t^i$ says that i can start up at most once in periods $t - 1$ and t, and if it starts up in either period, then it is set up in period t. This particular tightening of the defining constraint $z_t^i \leq y_t^i$ is only valid in the two set-up case.

To understand whether this formulation can be improved, we again look at the problem with changeovers, namely the set Y^{M_2-SQ} which can be formulated as Y^{M_2-SC} plus additional constraints on the (y, z, χ) variables that ensure the following.

$$\chi_t^{ij} = 1 \text{ if } y_t^i = y_t^j = z_t^j = 1 \text{ for all } i,j,t, i \neq j,$$
$$\chi_t^{ii} = 1 \text{ if } y_t^i = 1 \text{ and } \sum_j z_t^j = 0 \text{ for all } i,t.$$

To obtain a tight formulation of Y^{M_2-SQ}, we consider again a unit flow model, but with a different interpretation. Here the polyhedron (Q) represents the flow of a single unit from item to item over time

$$\sum_i \chi_t^{ij} = \delta_t^j \qquad\qquad \text{for } 1 \leq j \leq m, \ 1 \leq t \leq n$$

$$\sum_j \chi_t^{ij} = \delta_{t-1}^i \qquad\qquad \text{for } 1 \leq i \leq m, \ 1 \leq t \leq n$$

$$\sum_i \delta_0^i = 1$$

$$\chi \in [0,1]^{m^2 n}$$

shown in Figure 12.4, where δ_t^i is the flow through node (i, t) and represents the set-up status at the end of period t, that is, $\delta_t^i = 1$ if the machine is set up for item i at the end of period t, and χ_t^{ij} is the flow from node $(i, t-1)$ to node (j, t), and the path shown corresponds to the four period set-up sequence $(12)(2)(24)(43)$.

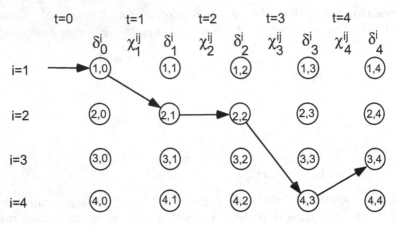

Figure 12.4. M_2: Four-period production sequence.

In this formulation, the set-up variable y_t^i is obtained from the equation $y_t^i = \delta_{t-1}^i + \delta_t^i - \chi_t^{ii}$. As for M_1-SQ, to convert this into a tight formulation for M_2-SC, we can use the equation $z_t^j = \sum_{i:i\neq j} \chi_t^{ij}$ to link the start-up and changeover variables, and then we need to take the projection of Q.

Proposition 12.6 conv(Y^{M_2-SC}) *is given by*

$$\sum_i (y_t^i - z_t^i) = 1 \qquad \text{for } 1 \le t \le n \qquad (12.5)$$

$$z_t^i + z_{t-1}^i \le y_t^i \qquad \text{for } 1 \le i \le m,\ 1 \le t \le n \qquad (12.6)$$

$$y_t^i - z_t^i \le y_{t-1}^i \qquad \text{for } 1 \le i \le m,\ 1 \le t \le n \qquad (12.7)$$

$$y_t^i + \sum_{j:j\neq i} (y_{t+1}^j - z_{t+1}^j - z_t^j) \le 1 \qquad \text{for } 1 \le i \le m,\ 1 \le t \le n \qquad (12.8)$$

$$y, z \in [0,1]^{mn}. \qquad (12.9)$$

Note that here (12.8) is a constraint strengthening the initial formulation. It says that "item i is set up during period t" and "some item other than i is set up throughout period t" are mutually exclusive.

12.3.2 Any Number of Set-Ups

Surprisingly, passing from $k = 2$ set-ups to an arbitrary number of set-ups turns out to be very simple. The crucial observations are:

i. Any item that is not set up first or last in a period t satisfies $z_t^i = y_t^i$ whether it is set up or not.

ii. The previous unit flow model for $k = 2$ models precisely those items that are first and last.

Sequence-Independent Variables M_k-SC

From the above observations, we obtain:

Proposition 12.7 $\mathrm{conv}(Y^{M_\infty-SC})$ *is the* (y, z, w)-*projection of the polyhedron*

$$\sum_{j=1}^{m} (y_t^j - z_t^j) = 1 \qquad \text{for all } t$$

$$y_t^i - z_t^i = y_{t-1}^i - w_{t-1}^i \qquad \text{for all } i, t \geq 2$$

$$y_t^i - z_t^i \geq \phi_t^{ii} \qquad \text{for all } i, t$$

$$y_t^i - w_t^i \geq \phi_t^{ii} \qquad \text{for all } i, t$$

$$\phi_t^{ii} \geq y_t^i - z_t^i - w_t^i \qquad \text{for all } i, t$$

$$y_t^i + \sum_{j:j\neq i} \phi_t^{jj} \leq 1 \qquad \text{for all } i, t$$

$$y, z, w, \phi \in [0,1]^{mn}.$$

In this formulation, the additional 0–1 variable ϕ_t^{ii} takes the value 1 when item i is the last item set up in period $t - 1$, is the only item set up in period t, and is the first item set up in period $t + 1$. This implies that, when $\phi_t^{ii} = 1$, no item is started up or switched off in period t.

Observation 12.1 *To obtain* $\mathrm{conv}(Y^{M_k-SC})$ *for* $k > 2$, *it suffices to add the constraints*

$$\sum_{j=1}^{m} z_t^j \leq k - 1 \quad \text{for all } t$$

to the formulation of $\mathrm{conv}(Y^{M_\infty-SC})$.

Sequence-Dependent Variables M_k-SQ

Here we allow more generality in that each item can be produced more than once per period. This occurs for instance when there are production batches of bounded or fixed size. For simplicity, consider just one time period, and suppose that dummy products, denoted by $i = 0$ and $i = m+1$, are used so as to determine the first and last items produced in the period. The formulations described here can be easily extended to multiple time periods.

A first formulation for Y^{M_k-SQ} can be obtained by adapting the known unit flow formulation (12.1)–(12.4) for Y^{M_1-SQ}. It suffices essentially to re-define or interpret the variable χ_t^{ij} as the batch changeover from product i produced as batch number $t-1$ to product j produced as batch number t. So, the index t now represents the batch instead of the time period. The only modification in the formulation is that Equation (12.2) becomes a $\geq$-type constraint to reflect the fact that the number of batches processed is not fixed.

Unfortunately, this unit flow formulation involves a large number of variables, especially in multi-period models with many batches per period. Therefore we suggest another formulation with fewer variables and constraints. We now temporarily modify our standard interpretation for the variables y.

Let y^i be the number of set-ups of item i in the period, χ^{ij} be the number of times production is changed from i to j, and q be an a priori upper bound on the number of times any item is produced in the period.

Note that in uncapacitated models we can take $q = 1$, but if items are produced in batches of a fixed maximum size, one may wish to produce the same item several times within a period.

Copying well-known formulations for the prize-collecting traveling salesman and vehicle routing problems, we obtain the formulation

$$\sum_{j\neq0} \chi^{0j} = 1 \tag{12.10}$$

$$\sum_{j} \chi^{ij} = y^i \qquad\qquad \text{for } i \neq 0, m+1 \tag{12.11}$$

$$\sum_{i} \chi^{ij} = y^j \qquad\qquad \text{for } j \neq 0, m+1 \tag{12.12}$$

$$\sum_{i\neq m+1} \chi^{i,m+1} = 1 \tag{12.13}$$

$$y^i, \chi^{ij} \in \{0, 1, \dots, q\} \qquad\qquad \text{for all } (i,j) \tag{12.14}$$

Subtour elimination constraints. $\tag{12.15}$

Here the constraint (12.10) says that there is a first job, and the second set of constraints (12.11) that the number of times item i is produced is equal to the number of changeovers from i to some other item (itself included). The other constraints are similar.

As we must deal with integer rather than 0–1 variables, the subtour elimination constraints are somewhat non-standard.

Proposition 12.8 *i. A vector (y, χ) satisfying the system (12.10)–(12.14) provides a feasible production sequence if and only if the induced directed multigraph does not contain a disconnected directed Eulerian (i.e., all nodes have even degree) subgraph.*
ii. The valid inequality

$$\sum_{(i,j)\in E(S)} \chi^{ij} \le \sum_{i\in S} y^i - \frac{1}{q}y^k \quad \text{for } k \in S \tag{12.16}$$

eliminates such a disconnected multigraph on the node set $S \subseteq \{1,\ldots,m\}$, where $E(S)$ is the set of pairs (i,j) with $i,j \in S$.

Example 12.2 *i. Suppose that $m = 4, q = 2$ and consider the set-up sequence $1,2,2,4,3,3$. This sequence is represented by $y^1 = y^4 = 1, y^2 = y^3 = 2$, and $\chi^{01} = \chi^{12} = \chi^{22} = \chi^{24} = \chi^{43} = \chi^{33} = \chi^{35} = 1$.*
ii. Suppose now that $m = 6, q = 2$, and consider the vector $y^1 = y^2 = 2, y^3 = y^4 = y^6 = 1, \chi^{02} = \chi^{13} = \chi^{14} = \chi^{22} = \chi^{26} = \chi^{31} = \chi^{41} = \chi^{67} = 1$ that satisfies the system (12.10)–(12.14). However, as shown in Figure 12.5, it is not a feasible production sequence.

Taking $S = \{1,3,4\}$ and $k = 1$ in Proposition 12.8, we obtain the inequality

$$\chi^{13} + \chi^{31} + \chi^{14} + \chi^{41} + \chi^{34} + \chi^{43} \le y^1 + y^3 + y^4 - \frac{1}{2}y^1,$$

which is violated by one unit, and thus cuts off the above vector.

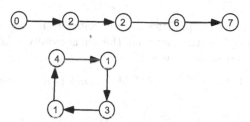

Figure 12.5. A "non-sequence".

The exponential number of valid inequalities (12.16) can be added as cutting planes at nodes of the branch-and-cut tree having either fractional or integer (y,χ) values. In the latter case it is trivial to identify violated inequalities as shown in Example 12.2.

12.4 Production Quantity Constraints PQ

12.4.1 Product Resource Constraints PC and PC-SU

The solution set of the typical single-level problem LS-C/PC-SU can be formulated as

$$s^i_{t-1} + x^i_t = d^i_t + s^i_t \qquad\qquad\qquad \text{for all } i, t \qquad (12.17)$$

$$x^i_t \le C^i_t y^i_t \qquad\qquad\qquad\qquad \text{for all } i, t \qquad (12.18)$$

$$\sum_i a^i x^i_t + \sum_i b^i y^i_t \le L_t \qquad\qquad \text{for all } t \qquad (12.19)$$

$$s \in \mathbb{R}^{m(n+1)}_+, \; x \in \mathbb{R}^{mn}_+, \; y \in \{0,1\}^{mn}. \qquad\qquad (12.20)$$

Viewed as the set $\left(\prod^m_{i=1} X^{LS-C}_i\right) \cap \left(\prod^n_{t=1} Y^{PC-SU}_t\right)$, a natural candidate choice for the capacity constraint set Y^{PC-SU} is

$$\sum^m_{i=1} a^i x^i + \sum^m_{i=1} b^i y^i \le L$$

$$x^i \le C^i y^i \qquad\qquad\qquad\qquad \text{for all } i$$

$$x \in \mathbb{R}^m_+, y \in \{0,1\}^m,$$

where the index t has been dropped for simplicity of notation to represent the single-period model Y^{PC-SU}. When $b_i = 0$ for all i, we write Y^{PC}.

Valid Inequalities for Y^{PC} and Y^{PC-SU}

We now present valid inequalities for variants of the set Y^{PC-SU}. All the inequalities are based on the fact that the set is closely related to the single-node flow set studied in Section 8.9.

Proposition 12.9 *For Y^{PC}, if $S \subseteq \{1, \ldots, m\}$ is a cover with $\sum_{i \in S} a^i C^i - L = \lambda > 0$, then*

$$\sum_{i \in S} a^i x^i + \sum_{i \in S} (a^i C^i - \lambda)^+ (1 - y^i) < L \qquad (12.21)$$

is valid for Y^{PC}.

Example 12.3 *Suppose that $m = 3$, production of each item is at constant capacity with $C^1 = 5, C^2 = 4, C^3 = 3$, the resource consumption rates are $a^1 = 1, a^2 = 2, a^3 = 3$, respectively, set-up times are $b^i = 0$ for $i = 1, 2, 3$, and the total machine availability is $L = 20$. The machine constraint in each period is then:*

$$x^1 + 2x^2 + 3x^3 \le 20$$

$$x^1 \le 5y^1, x^2 \le 4y^2, x^3 \le 3y^3$$

$$x \in \mathbb{R}^3_+, y \in \{0,1\}^3,$$

or

$$x^1 + 2x^2 + 3x^3 \le 20$$

$$x^1 \le 5y^1, 2x^2 \le (2 \times 4)y^2, 3x^3 \le (3 \times 3)y^3$$

$$x \in \mathbb{R}^3_+, y \in \{0,1\}^3.$$

For the flow cover $C = \{1, 2, 3\}$ with $\lambda = 2$, the flow cover inequality (12.21)

$$x^1 + 2x^2 + 3x^3 \leq 20 - (5 - 2)(1 - y^1) - (8 - 2)(1 - y^2) - (9 - 2)(1 - y^3)$$

is obtained.

A more restricted single-period production quantity set is obtained by also incorporating some knowledge from the lot-sizing model. Now consider the set

$$\tilde{Y}^{PC-SU} = Y^{PC-SU} \cap \{(\sigma, x, y) \in \mathbb{R}_+^m \times \mathbb{R}_+^m \times \mathbb{R}_+^m : x^i \leq \rho^i y^i + \sigma^i \text{ for all } i\}.$$

Proposition 12.10 *If $T \subseteq \{1, \ldots, m\}$ is a cover with $\sum_{i \in T}(a^i \rho^i + b^i) - L = \lambda > 0$, the inequality*

$$\sum_{i \in T} a^i x^i + \sum_{i \in T} b^i y^i + \sum_{i \in T} [a^i \rho^i + b^i - \lambda]^+ (1 - y^i) \leq L + \sum_{i \in T} a^i \sigma^i \quad (12.22)$$

is valid for $\tilde{Y}^{PC-SU}$.

In particular suppose that the single-item problems are uncapacitated (C^i is large for all i). One obvious choice for ρ^i and σ^i is given by the most basic (l, S) inequality

$$x_t^i \leq d_t^i y_t^i + s_t^i.$$

However, the (l, S) inequalities allow a much larger choice. Specifically consider the two inequalities

$$x_t^i + \sum_{u \in S_i} x_u^i \leq d_{tl_i}^i y_t^i + \sum_{u \in S_i} d_{ul_i}^i y_u^i + s_{l_i}^i \quad \text{and}$$

$$\sum_{u \in S_i} x_u^i \leq \sum_{u \in S_i} d_{ul_i}^i y_u^i + s_{l_i}^i,$$

where $S_i \subseteq \{t + 1, \ldots, l_i\}$. If we take σ_t^i to be the slack variable of the second inequality, the first can be written as

$$x_t^i \leq d_{tl_i}^i y_t^i + \sigma_t^i, \quad \sigma_t^i \geq 0,$$

and thus this inequality can be used in Proposition 12.10.

Suppose now that we wish to find a separation algorithm for a point (x^*, y^*, s^*) with respect to the set $\tilde{Y}^{PC-SU}$. Note that once l_i is chosen for each $i \in T$, and assuming that there is no violated (l, S) inequality for item i, the smallest value for σ_t^i is obtained by setting $S_i = \{u \in \{t + 1, \ldots, l_i\} : x_u^{*i} > d_{ul_i}^i y_u^{*i}\}$.

Another family of valid inequalities for $\tilde{Y}^{PC-SU}$ can be obtained from the reverse flow cover inequalities; see Section 8.9.

Proposition 12.11 *If $T \subseteq \{1, \ldots, m\}$ is a reverse cover with $L - \sum_{i \in T}(a^i \rho^i + b^i) = \mu > 0$, and $T' \subseteq N \setminus T$, the inequality*

$$\sum_{i \in T} a^i x^i + \sum_{i \in T} b^i y^i + \sum_{i \in T'} [a^i x^i + (b^i - \mu)y^i] \le \sum_{i \in T}(a^i \rho^i + b^i) + \sum_{i \in T} a^i \sigma^i$$

$$(12.23)$$

is valid for $\tilde{Y}^{PC-SU}$.

Example 12.4 *Suppose that $m = 3$, and for some period t, $a = (1, 2, 1)$, $b = (4, 10, 7)$, $L = 21$, $d = (6, 2, 5)$ and the capacities are large. The original resource constraint is*

$$x^1 + 2x^2 + x^3 + 4y^1 + 10y^2 + 7y^3 \le 21,$$

and we use the (l, S) inequalities $x^i \le d^i y^i + s^i$ to define the set $\tilde{Y}^{PC-SU}$. Taking $T = \{1, 2\}$ as flow cover, $\sum_{i \in T}(a^i d^i + b^i) = 1 \times 6 + 4 + 2 \times 2 + 10 = 24$, so $\lambda = 3$, and we obtain the flow cover inequality (12.22),

$$x^1 + 2x^2 + 4y^1 + 10y^2 + 7(1 - y^1) + 11(1 - y^2) \le 21 + s^1 + 2s^2.$$

Now consider a second instance with $m = 4$, $a = (1, 1, 1, 1)$, $b = (2, 2, 1, 3)$, $d = (5, 4, 5, 6)$, and $L = 16$. Taking $T = \{2, 3\}$ as a reverse cover with $\mu = 16 - (4 + 2) - (5 + 1) = 4$, and $T' = \{1\}$, we obtain the reverse flow cover inequality (12.23)

$$x^1 - 2y^1 + x^2 + 2y^2 + x^3 + y^3 \le 12 + s^2 + s^3.$$

12.5 Family Set-Ups: *PC-FAM*

Consider the problem $LS\text{-}C, PC\text{-}FAM$:

$$\min \quad \sum_i \sum_t h_t^i s_t^i + \sum_i \sum_t q_t^i y_t^i + \sum_t Q_t Y_t$$

$$s_{t-1}^i + x_t^i = d_t^i + s_t^i \qquad\qquad \text{for all } i, t$$

$$\sum_i x_t^i \le L_t Y_t \qquad\qquad \text{for all } t$$

$$x_t^i \le C_t^i y_t^i \qquad\qquad \text{for all } i, t$$

$$s \in \mathbb{R}_+^{m(n+1)}, \ x \in \mathbb{R}_+^{mn}, \ y \in \{0, 1\}^{mn}, Y \in \{0, 1\}^n,$$

where Y_t here represents the family set-up if any of the items $\{1, \ldots, m\}$ are produced in period t and L_t is the maximum amount that can be produced in period t. Note that we assume that the costs have been rewritten so that $p_t^i = 0$ for all i, t.

Here we consider specifically a *no individual set-up* variant, denoted LS^*-$U, PC\text{-}FAM$ with uncapacitated production of each item ($C_t^i = L_t$) and no individual set-up costs ($q_t^i = 0$ for all i, t). This has the simpler formulation

$$\min \quad \sum_i \sum_t h_t^i s_t^i + \sum_t Q_t Y_t$$

$$s_{t-1}^i + x_t^i = d_t^i + s_t^i \qquad\qquad \text{for all } i, t$$

$$\sum_i x_t^i \le L_t Y_t \qquad\qquad \text{for all } t$$

$$s \in \mathbb{R}_+^{m(n+1)}, \; x \in \mathbb{R}_+^{mn}, \; Y \in \{0,1\}^n.$$

We now derive a relaxation for LS^*-$U, PC\text{-}FAM$ taking into account the family set-up variables. We restrict our attention to the problem in which the budget is constant $L_t = L$ in each period.

We construct aggregated products consisting of the items $\{i, i+1, \ldots, m\}$. Let $S_t^i = \sum_{j \ge i} s_t^j$, $X_t^i = \sum_{j \ge i} x_t^j$, $D_t^i = \sum_{j \ge i} d_t^j$, and $H_t^i = h_t^i - h_t^{i-1}$ for all i, t. Now the storage costs can be rewritten as

$$\sum_i \sum_t h_t^i s_t^i = \sum_i \sum_t H_t^i S_t^i.$$

As $S_{t-1}^i + X_t^i = D_t^i + S_t^i$ and $X_t^i \le L Y_t$, a relaxation for the aggregate products is obtained by taking any formulation of $\mathrm{conv}(X^{WW-CC})$. Thus the formulation

$$(S^i, Y) \in \mathrm{conv}(X^{WW-CC}) \qquad\qquad \text{for all } i \qquad (12.24)$$

$$s_t^i = S_t^i - S_t^{i+1} \qquad\qquad \text{for all } i, t \qquad (12.25)$$

is a relaxation of the problems LS^*-$U, PC\text{-}FAM$ and $LS\text{-}C, PC\text{-}FAM$.

As it is a relaxation, it does not follow automatically that the s obtained from (12.25) and the production vector x obtained from the flow balance constraints will be feasible. However, for certain objective functions this relaxation can be shown to solve the original problem.

Theorem 12.12 *The polyhedron*

$$\{(S^1, \ldots, S^m, Y) \in \mathbb{R}_+^n \times \cdots \times \mathbb{R}_+^n \times \mathbb{Z}_+^n : (S^i, Y) \in \mathrm{conv}(X^{WW-CC}) \text{ for all } i\}$$

is integral.

With Wagner–Whitin costs ($h_t^i \ge 0$ for all i, t) and ordered storage costs ($H_t^i = h_t^i - h_t^{i-1} \ge 0$ for all i, t), the linear program

$$\min\{\sum_i \sum_t H_t^i S_t^i + \sum_t Q_t Y_t : (s, S, Y) \text{ satisfy } (12.24) \text{ and } (12.25)\}$$

solves WW^*-$U, PC\text{-}FAM$.

The same result holds when the family set-up variables Y_t represent batches, and are restricted to unbounded or bounded integer values.

Exercises

Exercise 12.1 Consider an instance of $DLC\text{-}CC/M_1$ with $NI = 3, NT = 4$,

$$(d_t^i) = \begin{pmatrix} 0\ 5\ 3\ 8 \\ 7\ 0\ 4\ 2 \\ 0\ 0\ 6\ 7 \end{pmatrix}, (h^i) = (1,2,3), (q^i) = (30,10,50), \text{ and } (C^i) = (10,10,12).$$

i. Convert to a network flow problem.

ii. Solve by a network flow or linear programming algorithm.

Exercise 12.2 Consider an instance of $DLC\text{-}CC\text{-}B/M_1$ with the same data as in Exercise 12.1 and backlogging costs $b^i = (2,4,4)$. Solve using LS–LIB.

Exercise 12.3 Consider an instance of a multi-item discrete lot-sizing problem with sequence-dependent changeover costs in which at most one item can be produced per period. The instance has 10 items and 35 periods. The demands have been normalized so $d_t^i \in \{0,1\}$. Below we list the periods in which there is demand for the 10 items and the changeover costs. The unit storage costs are 10 per unit per period independent of the item and the time period. Solve using tight formulations for $DLS\text{-}CC\text{-}SC$ and $M_1\text{-}SQ$, and appropriate equations linking the start-up and changeover variables.

$$
\begin{array}{l}
i = 1\ |16\ 17\ 21\ 25\ 30 \\
i = 3\ |9 \\
i = 4\ |14\ 33 \\
i = 5\ |18\ 27\ 31\ 35 \\
i = 6\ |9\ 17\ 34 \\
i = 7\ |7\ 9\ 20\ 21\ 24\ 30 \\
i = 8\ |21\ 25\ 28 \\
i = 9\ |24 \\
i = 10|10\ 11\ 12
\end{array}
\quad (c^{ij}) =
\begin{pmatrix}
78 & 86 & 172\ 211\ 134\ 146\ 232\ 229\ 191 \\
165 & & 193\ 166\ 214\ 139\ 182\ 159\ 153\ 204 \\
214\ 170 & & 149\ 166\ 223\ 124\ 212\ 198\ 148 \\
157\ 115\ 178 & & 72\ 125\ 197\ 234\ 204\ 224 \\
164\ 104\ 132\ 169 & & 90\ 77\ 147\ 108\ 87 \\
163\ 210\ 133\ 188\ 92 & & 112\ 220\ 213\ 217 \\
112\ 89\ 167\ 145\ 83\ 124 & & 122\ 211\ 141 \\
133\ 132\ 235\ 123\ 185\ 181\ 232 & & 122\ 76 \\
214\ 150\ 217\ 100\ 213\ 104\ 230\ 170 & & 133 \\
231\ 160\ 96\ 147\ 215\ 210\ 210\ 164\ 215 &
\end{pmatrix}.
$$

Exercise 12.4 Consider a problem with mode constraints $WW\text{-}U/M_1$ with $NT = 15, NI = 6$,

$$(d_t^i) = \begin{pmatrix}
0\ 0\ 0\ 0\ 0\ 0\ 83\ 108\ 114\ 121\ 110\ 124\ 104\ 86\ 87 \\
0\ 0\ 0\ 0\ 0\ 0\ 122\ 101\ 89\ 108\ 101\ 109\ 106\ 108\ 76 \\
0\ 0\ 0\ 0\ 0\ 0\ 83\ 82\ 112\ 109\ 119\ 85\ 99\ 80\ 123 \\
0\ 0\ 0\ 0\ 0\ 0\ 101\ 81\ 117\ 76\ 103\ 81\ 95\ 105\ 102 \\
0\ 0\ 0\ 0\ 0\ 0\ 111\ 98\ 97\ 80\ 98\ 124\ 78\ 108\ 109 \\
0\ 0\ 0\ 0\ 0\ 0\ 107\ 105\ 75\ 93\ 115\ 113\ 111\ 105\ 85
\end{pmatrix}.$$

$(q^i) = (600,1000,800,400,800,400)$ and $(h^i) = (3,1,1,3,2,2)$.

Solve this problem using a tight extended formulation for $\text{conv}(X^{WW-U})$ or using LS–LIB.

Exercise 12.5 Consider a problem with mode constraints $WW\text{-}U/M_k$ with the same data as in the previous exercise. How much does the increased flexibility of $k = 2,3$ bring?

What happens if there are in addition start-up costs $g = (30,20,30,40,40,10)$?

Exercise 12.6 Consider an instance with production quantity constraints $LS\text{-}U/PC$ and the same data as above, and a production limit of $L = 500$ for the joint production constraint. Solve this instance.
Solve also with the addition of set-up times $ST = (30, 20, 30, 40, 40, 10)$.

Exercise 12.7 Consider the following planning and scheduling problem involving the allocation and sequencing of $NI = 10$ products with sequence-dependent costs on $NK = 10$ parallel facilities. There is a single time period $(NT = 1)$. Let $I = \{1, \ldots, NI\}$ be the set of products, and $K = \{1, \ldots, NK\}$ be the set of facilities.

For each product i, there is some minimum demand d^i to satisfy, there is no initial inventory, and any amount larger than or equal to the demand d^i (i.e., backlogging is not allowed) can be sold at a unit selling price p^i:

$$(d^i) = (221, 118, 81, 25, 12, 56, 16, 78, 76, 24),$$

$$(p^i) = (0.01, 0.01, 0.01, 0.01, 0.01, 0.09, 0.08, 0.09, 0, 10, 0, 11).$$

For each machine or facility k, the set $PA_k \subseteq I$ is the subset of products that can be processed on facility k. Products are processed in batches, and each batch outputs a single product. When one batch of product i is processed on facility k, there is a minimum batch size MIN_k^i and a maximum batch size MAX_k^i (because of required maintenance or cleaning operations, or tool lifetime) representing the minimum and maximum quantity of product i produced. Several batches of the same product can be produced on a facility.
$PA_1 = \{1, 2, 8\}$, $PA_2 = \{1, 2, 4, 5, 10\}$, $PA_3 = \{1, 3, 4\}$,
$PA_4 = \{1, 2\}$, $PA_5 = \{1, 2\}$, $PA_6 = \{1, 6, 7, 8, 10\}$,
$PA_7 = \{1, 3, 9\}$, $PA_8 = \{1, 2\}$, $PA_9 = \{1, 2, 3\}$,
$PA_{10} = \{1, 3, 8, 10\}$,

$$MIN_k^i = 5, \quad MAX_k^i = 120 \quad \text{for all } i, k \text{ with } i \in PA_k.$$

For each facility k, the set $TR_k \subseteq PA_k \times PA_k$ defines the feasible transitions or product changeovers on facility k. In this particular problem instance we assume that $TR_k = PA_k \times PA_k$.
For each feasible transition $(i, j) \in TR_k$, the changeover cost is TC_k^{ij}. Note that when the maximum batch size MAX_k^i for a batch of product i on facility k is reached, then a changeover from product i to itself is required to continue to produce the same product i.

$$(TC_1^{ij}) = \begin{pmatrix} & 1 & 2 & 8 \\ \hline 1 & 0 & 0.25 & 0.5 \\ 2 & 0.5 & 0 & 0.5 \\ 3 & 0.5 & 0.5 & 0 \end{pmatrix}, \quad (TC_2^{ij}) = \begin{pmatrix} & 1 & 2 & 4 & 5 & 10 \\ \hline 1 & 0 & 0.5 & 48 & 48 & 48 \\ 2 & 0.5 & 0 & 48 & 48 & 48 \\ 4 & 48 & 48 & 0 & 0.5 & 0.5 \\ 5 & 48 & 48 & 0.5 & 0 & 0.5 \\ 10 & 48 & 48 & 0.5 & 0.5 & 0 \end{pmatrix},$$

$$(TC_3^{ij}) = \begin{pmatrix} & 1 & 3 & 4 \\ \hline 1 & 0 & 8 & 48 \\ 3 & 8 & 0 & 48 \\ 4 & 48 & 48 & 0 \end{pmatrix}, \quad (TC_4^{ij}) = \begin{pmatrix} & 1 & 2 \\ \hline 1 & 0 & 1 \\ 2 & 1 & 0 \end{pmatrix}, \quad (TC_5^{ij}) = \begin{pmatrix} & 1 & 2 \\ \hline 1 & 0 & 8 \\ 2 & 8 & 0 \end{pmatrix},$$

$$(TC_6^{ij}) = \begin{pmatrix} & 1 & 6 & 7 & 8 & 10 \\ \hline 1 & 0 & 16 & 16 & 16 & 16 \\ 6 & 16 & 0 & 1 & 16 & 1 \\ 7 & 16 & 1 & 0 & 16 & 1 \\ 8 & 8 & 16 & 16 & 0 & 16 \\ 10 & 16 & 1 & 1 & 16 & 0 \end{pmatrix}, \quad (TC_7^{ij}) = \begin{pmatrix} & 1 & 3 & 9 \\ \hline 1 & 0 & 72 & 72 \\ 3 & 72 & 0 & 8 \\ 9 & 72 & 8 & 0 \end{pmatrix},$$

$$(TC_8^{ij}) = \begin{pmatrix} & 1 & 2 \\ \hline 1 & 0 & 2 \\ 2 & 2 & 0 \end{pmatrix}, \quad (TC_9^{ij}) = \begin{pmatrix} & 1 & 2 & 3 \\ \hline 1 & 0 & 2 & 12 \\ 2 & 2 & 0 & 12 \\ 3 & 12 & 12 & 0 \end{pmatrix}, \quad (TC_{10}^{ij}) = \begin{pmatrix} & 1 & 3 & 8 & 10 \\ \hline 1 & 0 & 1 & 8 & 48 \\ 3 & 1 & 0 & 8 & 48 \\ 8 & 8 & 8 & 0 & 48 \\ 10 & 48 & 48 & 48 & 0 \end{pmatrix}.$$

Initially each machine k is ready to produce the item $ST_k \in PA_k$ without any preparation or changeover cost,

$$(ST_k) = (8, 2, 1, 1, 2, 6, 3, 1, 3, 10).$$

To set a boundary condition and assign a cost to the machine status at the end of the schedule, it costs EC_k^i when item i is the last item produced on machine k, where

$$EC_k^i = \min_{j:(i,j) \in TR_k} TC_k^{ij} \quad \text{for all } i, k \text{ with } i \in PA_k.$$

Finally, the production time of a batch is proportional to the batch size. In other words, for $i \in PA_k$, PT_k^i is the production time per unit of product i on facility k. The capacity or production time available on facility k is CAP_k.

$$(PT_k^i) = \begin{pmatrix}
i,k & 1 & 2 & 3 & 4 & 5 & 6 & 7 & 8 & 9 & 10 \\
\hline
1 & 5.348 & 6.024 & 11.905 & 11.236 & 4.0 & 6.061 & 6.897 & 9.524 & 9.524 & 11.111 \\
2 & 5.319 & 6.024 & & 10.989 & 4.0 & & & 9.091 & 9.091 & \\
3 & & & 12.346 & & & & 6.897 & & 8.696 & 11.111 \\
4 & & 5.650 & 12.658 & & & & & & & \\
5 & & 5.650 & & & & & & & & \\
6 & & & & & & 6.061 & & & & \\
7 & & & & & & 6.061 & & & & \\
8 & 5.465 & & & & & 6.494 & & & & 15.385 \\
9 & & & & & & & 7.407 & & & \\
10 & & 6.024 & & & & 6.061 & & & & 11.765
\end{pmatrix}$$

and

$$(CAP_k) = (480, 480, 480, 480, 672, 672, 672, 480, 480, 480).$$

i. Use the classification scheme to propose a mathematical formulation for this problem, and solve the problem.

ii. Solve the variant with changeover times TT_k^{ij} instead of changeover costs, and with

$$TT_k^{ij} = \gamma \, TC_k^{ij} \quad \text{for all } i, j, k \text{ and for some } \gamma \in \mathbb{R}_+.$$

iii. Propose a formulation for the multi-period extension.

Exercise 12.8 For an instance of $PC\text{-}SU$ with $L = 500$, $ST = (30, 40)$, and $(d_9^1, d_{10}^1, d_{11}^1) = (114, 121, 110)$, $(d_9^2, d_{10}^2, d_{11}^2) = (117, 76, 103)$, the following partial fractional solution is obtained for two of the items $(x_9^1, x_9^2) = (245.6, 193.0)$, $(y_9^1, y_9^2) = (0.71, 1)$, $(s_9^1, s_{10}^1, s_{11}^1) = (164.7, 117.9, 7.9)$, and $(s_9^2, s_{10}^2, s_{11}^2) = (76.0, 0, 81.0)$.
Find a valid inequality cutting off this solution.

Exercise 12.9 Consider the single-period submodel (12.18)–(12.20) with $a^i = 1, b^i = b, C_t^i = B$ for all i with subscript t dropped. Show that the following inequalities are valid

i. $\quad\quad\quad x^i \leq (B - b)y^i.$

ii. $\quad \sum_{i \in S} x^i + b \sum_{i=1}^{NI} y^i \leq B - (B - b\eta)(1 - \sum_{i \in \{1,\dots,NI\} \setminus S} y^i),$
where $\eta = \lfloor \frac{B}{b} \rfloor$ and $S \subseteq \{1, \dots, NI\}$.

iii. $\quad\quad b \sum_{i=1}^{NI} y^i \leq bq - \sum_{i \in \{1,\dots,NI\} \setminus S} [x^i - (B - bq)y^i]$
with $q = \min\{\eta, |S|\}$.

Exercise 12.10 Suppose that there are two items that a retailer requires over the next four periods. The required amounts are $d^1 = (0, 0, 6, 6)$ for item 1 and $d^2 = (4, 4, 0, 0)$ for item 2. To be sure that the items arrive on time, the retailer can rent one or more standard trucks to deliver the items in each period. The rental cost per period is 100 and the capacity of a truck is 10. Once delivered, items can be stored. The per unit storage costs for the items are 1 and 100, respectively. Decide on an optimal rental and storage plan.

Notes

Section 12.1 A first very partial classification of multi-item problems was proposed in Wolsey [194]. The classification used here is a little more general, and some names and notation have changed. In particular M_1 and M_2 were earlier referred to as small-bucket models, and M_k for $k > 2$ was a big-bucket model. Drexl and Kimms [58] classify certain single-level multi-item problems; specifically their classification consists of the capacitated lot-sizing problem $CLSP$ (called $WW\text{-}U/PC$ or $WW\text{-}C/PC$ in our classification), the discrete lot-sizing and scheduling problem $DLSP$ (called $DLS\text{-}C/M_1\text{-}SC/PC$), the continuous set-up lot-sizing problem $CSLP$ (called $WW\text{-}C/M_1\text{-}SC/PC$), the proportional lot-sizing and scheduling problem $PLSP$ (called WW-C/M_2-SC/PC), and the general lot-sizing and scheduling problem $GLSP$ (called WW-C/M_k-SC/PC). A taxonomy of multi-item lot-sizing models, along with a classification of the literature, is also provided in Kuik et al. [101].

Section 12.2 The results for DLS-CC/M_1 are part of the folklore. The tight formulation for DLS-CC-B/M_1 is from Miller and Wolsey [125]. With sequence-dependent changeover variables, the formulation for M_1-SQ is from Constantino [45].

Section 12.3 The formulation of M_2-SQ is from Belvaux and Wolsey [26]. That for M_k-SQ was proposed by Surie and Stadtler [157] and shown to be tight by Waterer [189]. The model in which the number of set-ups per period can exceed one appeared in the Chesapeake models [17]. Different ways used to tackle these problems include column generation in Kang et al. [95] and cutting planes in Belvaux and Wolsey [26]. The prize-collecting traveling salesman problem and subtour elimination constraints are discussed in Balas [20].

Section 12.4 The flow and reverse flow cover inequalities for PC and PC-SU were proposed by Miller et al. [121, 123]. In [122] they gave a tight formulation for the special case with constant demands and set-up times. A related model with lower bounds was studied by Constantino [47].

   Though not discussed in this book, numerous specialized heuristics have been proposed for the multi-item big bucket problem LS-U, PC-M_∞ and its variants. One of the earliest is from Dixon and Silver [57]. Very simple heuristics were also proposed by Maes and van Wassenhove [112]. A Lagrangian relaxation approach was developed by Diaby et al. [56] and column generation based heuristics can be found in Cattrysse et al. [34, 35]. An extensive literature survey including heuristics is provided in Kuik et al. [101]. See also Salomon [147].

Section 12.5 The problem LS-C/PC-FAM was considered by Kao [96]. The special case with unrestricted capacities ($C_t^i = M$) and unrestricted family budget ($L_t = M$), known as the joint replenishment problem, was shown to be NP-hard in Arkin et al. [11]. Federgrün and Tzur [64] proposed a partitioning heuristic, a variant of relax-and-fix, showed that it can produce ϵ-optimal solutions, and devised a branch-and-bound algorithm for the problem. Recently a 2-approximation primal-dual algorithm for it has been given by Levi et al. [105].

   For the no-individual set-up problem LS^*-U/PC-FAM, Federgrün et al. [62] discuss a family of progressive interval heuristics, variants of relax-and-fix, that provide polynomial approximation schemes, and show that they can be effectively used to solve large problems. The special polynomial case treated in Theorem 12.12 was shown to be polynomial for a fixed number of items by Anily and Tzur [9] based on a dynamic programming recursion. The theorem itself is from Anily et al. [10].

Exercises Exercise 12.3 is an instance based on an article of Fleischmann [68] that has been tackled in Wolsey [194]. Exercise 12.7 is the first test problem in the Chesapeake problem set [17]. Exercise 12.9 is based on Constantino [47] in which he showed that adding the families of valid inequalities (i) and (ii)/[respectively, (i) and (iii)] suffices to give the convex hull of solutions when y^i are integer (resp. 0–1) variables. Exercise 12.10 is an instance taken from Anily and Tzur [9].

13

Multi-Level Lot-Sizing Problems

Here we consider production systems in which two or more items are produced, and at least one item is required as an input (component, subcomponent, part) of another. To represent the component or *product structure*, also known as the *bill of materials*, we use a directed acyclic graph $G = (V, A)$. Here the nodes V represent the items, and an arc $(i, j) \in A$ with associated value $r^{ij} > 0$ indicates that r^{ij} units of item i are needed in the production of each unit of item j. See Figure 13.1. Throughout $D(i) = \{j : (i, j) \in A\}$ denotes the set of direct successors of $i \in V$, $S(i)$ the set of all successors, and $P(i)$ the set of predecessors.

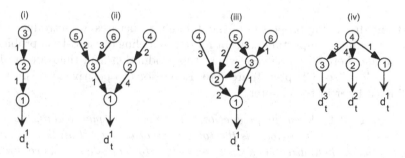

Figure 13.1. Different production structures.

It is also usual to distinguish different forms of product structure. In Figure 13.1, structure (i) is called *production in series*, (ii) is an *assembly structure*, (iii) is a *general product structure*, and (iv) is a *distribution or arborescence structure*. Note that it is possible to redefine the items so that $r^{ij} \equiv 1$ on all arcs in the series, assembly and distribution structures, but not for the general structure.

Example 13.1 *For the assembly structure (ii) in Figure 13.1, if we define a new item 5′ to be 2 units of the old item 5, 6′ to be 3 units of item 6, 4′ to be 8 units of item 4, and 2′ to be 4 units of item 2, the new product structure is the same as before, but $r^{ij} = 1$ for all arcs $(i, j) \in A$.*

Similarly for the distribution structure (iv), it suffices to take the new item 3′ to be $\frac{1}{3}$ of the old, 2′ to be $\frac{1}{4}$ of the old, and adjust the demands $d_t^{3'} = 3d_t^3, d_t^{2'} = 4d_t^2, d_t^{1'} = d_t^1$. The new digraphs are shown in Figure 13.2.

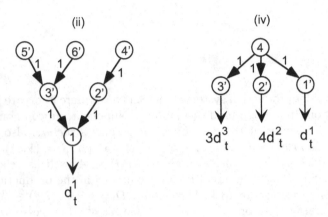

Figure 13.2. Modified items and structures.

Before discussing properties of the different structures, we introduce what turns out to be an important concept for our modeling of multi-level problems. We consider here the base case where the production lead-times or production time lags (see Chapter 2) are zero. Extensions to positive lead-times are considered later in the chapter.

Definition 13.1 *When the production lead-times are equal to zero, the* echelon stock *of item i in period t is the total stock of item i within the system at time t, whether held directly as stock, or as the stock of other items containing one or more units of item i.*

Thus in Figure 13.2, the echelon stock of item 4′ in the assembly structure is the sum of the stocks of items 4′, 2′ and 1, and the echelon stock of item 4 in the distribution structure is the sum of the stocks of items 4, 3′, 2′ and 1′.

As a classification scheme for multi-level problems, we just add ML-$[S, A, G, D]^1$ where S, A, G and D denote series, assembly, general, and distribution structure respectively.

We now outline the contents of this chapter.

- In Section 13.1 we consider the lot-sizing in series problem. For the uncapacitated problem, we demonstrate a polynomial dynamic programming
  algorithm. We then discuss two reformulations: one based on *echelon stocks*
  permits a partial decomposition into single-item subproblems, and the second *multi-commodity reformulation* generalizes that for the single-item
  problem. Also some valid inequalities are derived, and we present the
  nested property of solutions for certain classes of objective functions.
- In Sections 13.2 and 13.3 we show how the echelon stock reformulation
  can be extended and can also take into account lead-times in production
  for the assembly and general product structures, respectively.
- In Section 13.4 we extend the multi-level production model to include
  transportation to one or more demand areas. We indicate by example how
  the concept of echelon stock generalizes so that the various inequalities and
  reformulations derived earlier can also be used to tighten the formulations
  of these problems.

13.1 Production in Series *ML-S*

If there are m items, the product structure digraph is $G = (V, A)$ with $V = \{1, \ldots, m\}$ and $A = \{(2,1), (3,2), \ldots, (m, m-1)\}$. Thus $D(i) = i - 1$ and $S(i) = \{1, \ldots, i-1\}$ for $i \geq 2$, $D(1) = S(1) = \emptyset$, and $P(i) = \{i+1, \ldots, m\}$ for $i < m$. As indicated above we can take $r^{ij} = 1$ on all the arcs.

The basic formulation of ML-S/LS-C, without time lags or produc-
tion/operations lead-times (see Chapter 2) and with n time periods, is:

$$\min \quad \sum_{i=1}^{m}\sum_{t=1}^{n} p_t^i x_t^i \;+\; \sum_{i=1}^{m}\sum_{t=0}^{n} h_t^i s_t^i \;+\; \sum_{i=1}^{m}\sum_{t=1}^{n} q_t^i y_t^i \tag{13.1}$$

$$s_{t-1}^i + x_t^i = x_t^{i-1} + s_t^i \qquad \text{for } 2 \leq i \leq m,\ 1 \leq t \leq n \tag{13.2}$$

$$s_{t-1}^1 + x_t^1 = d_t^1 + s_t^1 \qquad \text{for } 1 \leq t \leq n \tag{13.3}$$

$$x_t^i \leq C_t^i y_t^i \qquad \text{for } 1 \leq i \leq m,\ 1 \leq t \leq n \tag{13.4}$$

$$s \in \mathbb{R}_+^{m(n+1)}, \quad x \in \mathbb{R}_+^{mn}, \qquad y \in \{0,1\}^{mn}. \tag{13.5}$$

In this formulation, the term x_t^{i-1} in the flow conservation constraint (13.2) is
the dependent demand of item i in period t. The flow conservation constraint
(13.3) takes the classical form with the external (or independent) demand d_t^1 of
item 1 in period t. Note that in some applications there is also an independent
demand term d_t^i for intermediate items in the right-hand side of (13.2).

We see in Figure 13.3 that ML-S/LS-C can be viewed as a fixed charge
network flow problem.

As before we can always eliminate the production variables x_t^i, or the stock
variables s_t^i, from the expression of the objective function. Thus often in this
chapter we assume that $p_t^i = 0$ for all i and t.

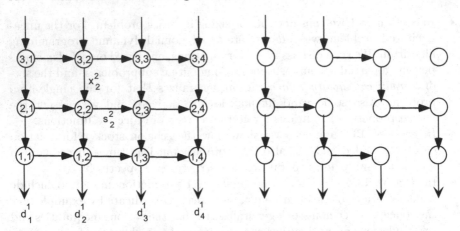

Figure 13.3. $ML\text{-}S$ as fixed charge network flow.

13.1.1 Optimization for $ML\text{-}S/LS\text{-}U$

From the fixed charge network flow representation of $ML\text{-}S$, extreme solutions correspond to acyclic support subgraphs as illustrated in the right part of Figure 13.3, and one obtains immediately the following.

Proposition 13.1 *For the uncapacitated lot-sizing in series problem $ML\text{-}S/LS\text{-}U$, there exists an optimal solution with $s_{t-1}^i x_t^i = 0$ for all i,t, and if $x_t^i > 0$, then $x_t^i = d_{\alpha\beta}^i$ for some $t \leq \alpha \leq \beta$.*

From this it is possible to develop a dynamic programming algorithm to solve $ML\text{-}S/LS - U$. The recursion is based on finding the optimal cost of converting a quantity $d_{\alpha\beta}^1$ of item i available in period t into final product item 1 so as to satisfy the demands $d_\alpha^1, \dots, d_\beta^1$ over the interval $[\alpha, \beta]$.

Let $H(i, t, \alpha, \beta)$ with $1 \leq t \leq \alpha \leq \beta \leq n$ be the minimum cost of such a transformation, and assume that $H(i, t, \alpha, \beta) = +\infty$ in all other cases.

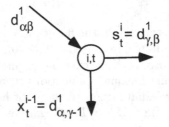

Figure 13.4. Production/stock division of the entering flow.

What can happen to the flow of $d^1_{\alpha\beta}$ units flowing through the node (i,t)? Typically it splits into two parts; see Figure 13.4. An amount $d^1_{\alpha,\gamma-1}$ of item i is used up to produce the same quantity of item $i-1$ in period t, and an amount $d^1_{\gamma\beta}$ of item i is stocked till period $t+1$.

The resulting recursion is

$$H(i,t,\alpha,\beta) = \min\Big(h^i_t d^1_{\alpha\beta} + H(i,t+1,\alpha,\beta),$$

$$\min_{\gamma=\alpha+1,\ldots,\beta} \big[q^{i-1}_t + H(i-1,t,\alpha,\gamma-1)$$

$$+ h^i_t d^1_{\gamma\beta} + H(i,t+1,\gamma,\beta)\big],$$

$$q^{i-1}_t + H(i-1,t,\alpha,\beta) \Big) \qquad \text{if } d_\alpha > 0$$

$$H(i,t,\alpha,\beta) = H(i,t,\alpha+1,\beta) \qquad\qquad \text{if } d_\alpha = 0, \alpha < \beta$$

$$H(i,t,\alpha,\alpha) = 0 \qquad\qquad\qquad\qquad \text{if } d_\alpha = 0.$$

Note that if $d_\alpha = 0$ and $\alpha < \beta$, $H(i,t,\alpha,\beta) = H(i,t,\alpha+1,\beta)$ for all i,t.

Now, to obtain a complete recursion solving ML-S/LS-U, we initialize the recursion with

$$H(1,t,\alpha,\beta) = h^1_t d^1_{\max[t+1,\alpha],\beta} + H(1,t+1,\max[t+1,\alpha],\beta) \quad \text{and}$$

$$H(1,t,t,t) = 0 \quad \text{for all } 1 < t < \alpha < \beta \le n.$$

Then we add a preproduct $m+1$ representing the raw material used in the production at level m. The total requirement d^1_{1n} of this preproduct is available at time $t=1$, and it can be stored at no cost; that is, $h^{m+1}_t = 0$ for all t. We compute $H(i,t,\alpha,\beta)$ for $1 \le t \le \alpha \le \beta \le n$ and $2 \le i \le m+1$ using the above recursion. The optimal value of problem ML-S/LS-U is given by $H(m+1,1,1,n)$.

Example 13.2 *We consider an instance of ML-S/LS-U with $m = n = 3$,*

$$(h^i_t) = \begin{pmatrix} 8 & 8 & 8 \\ 4 & 2 & 2 \\ 1 & 1 & 1 \end{pmatrix}, \quad (q^i_t) = \begin{pmatrix} 145 & 74 & 52 \\ 78 & 34 & 83 \\ 26 & 79 & 98 \end{pmatrix} \quad \text{and} \quad d = (0, 19, 24).$$

The values of $H(i,t,\alpha,\beta)$ and $\gamma(i,t,\alpha,\beta)$ from the dynamic programming recursion are shown in Table 13.1, where $\gamma(i,t,\alpha,\beta) \in \{\alpha,\ldots,\beta+1\}$ is the optimal way of splitting the demand $d^1_{\alpha\beta}$ in the computation of $H(i,t,\alpha,\beta)$. The entries are calculated column by column in the order shown. As $d_1 = 0$, we only need to consider values of $\alpha \ge 2$. We now present a sample of the calculations arising in generating the values in Table 13.1.

Table 13.1. Values of $H(i, t, \alpha, \beta)$

t, α, β	$i = 1$		$i = 2$		$i = 3$		$i = 4$	
	H	γ	H	γ	H	γ	H	γ
3,3,3	0	3	52	4	135	4	233	4
2,3,3	184		100	3	134	4	135	3
2,2,3	184		174	3	208	4	306	4
2,2,2	0		79	3	113	3	211	3
1,3,3	368		270	3	158	3	135	3
1,2,3	528		346	2	251	2	277	4
1,2,2	152		155	2	132	2		

$$H(1,1,2,3) = h_1^1 d_{23} + H(1,2,2,3) = 8 \times 43 + 184 = 528$$

$$H(2,3,3,3) = q_3^1 + H(1,3,3,3) = 52, \ \gamma = 4$$

$$H(2,2,2,3) = \min\{q_2^1 + H(1,2,2,2) + h_2^2 d_3 + H(2,2,3,3), q_2^1 + H(1,2,2,3)\}$$
$$= \min\{74 + 0 + 2 \times 24 + 52, \ 74 + 184\} = 174, \ \gamma = 3$$

$$H(3,1,2,3) = \min\{h_1^3 d_{23} + H(3,2,2,3),$$
$$q_1^2 + H(2,1,2,2) + h_1^3 d_3 + H(3,2,3,3),$$
$$q_1^2 + H(2,1,2,3)\}$$
$$= \min\{1 \times 43 + 208, 78 + 155 + 1 \times 24 + 100, 78 + 346\} = 251,$$
$$\gamma = 2.$$

The optimal value is given by $H(4,1,1,3) = H(4,1,2,3) = 277$. Working backwards to find an optimal solution, we have that $\gamma(4,1,2,3) = 4$ and thus $x_1^3 = d_{23}$, $y_1^3 = 1$, and we move to state (3,1,2,3).

As $\gamma(3,1,2,3) = 2$, we see that $s_1^3 = d_{23}$, and we move to state (3,2,2,3).
As $\gamma(3,2,2,3) = 4$, we see that $x_2^2 = d_{23}$, $y_2^2 = 1$, and we move to state (2,2,2,3).
As $\gamma(2,2,2,3) = 3$, we have that $x_2^1 = d_2$, $y_2^1 = 1$ moving to state (1,2,2,2) and $s_2^2 = d_3$ moving to state (2,3,3,3).
In state (1,2,2,2), $H(1,2,2,2) = 0$ and the amount d_2 has been delivered.
In state (2,3,3,3), $\gamma(2,3,3,3) = 4$, so we have that $x_3^1 = d_3$, $y_3^1 = 1$ moving to the end state (1,3,3,3).

Thus the complete production schedule is is $x_1^3 = d_{23}$, $x_2^2 = d_{23}$, $x_2^1 = d_2$, $x_3^1 = d_3$.

13.1.2 The Echelon Stock Reformulation for ML-S

For the production in series model, the echelon stock of i in t is just the sum of the stocks of all items containing i, namely the items j with $j \leq i$.

Let the variable e_t^i represent the echelon stock of i in t. As $e_t^i = \sum_{j=1}^{i} s_t^j$, we can rewrite the formulation (13.1)–(13.5) with the variables e_t^i in place of the variables s_t^i. Summing the constraints (13.2) for $j = 2, \ldots, i$ plus the constraint (13.3), we obtain the equation

$$\sum_{j=1}^{i} s_{t-1}^j + x_t^i = d_t^1 + \sum_{j=1}^{i} s_t^j$$

which gives

$$e_{t-1}^i + x_t^i = d_t^1 + e_t^i,$$

and the constraint $s_t^i \geq 0$ gives

$$e_t^i \geq e_t^{i-1}.$$

Also $\sum_{i,t} h_t^i s_t^i = \sum_{i,t} h_t^i (e_t^i - e_t^{i-1}) = \sum_{i,t} (h_t^i - h_t^{i+1}) e_t^i$. Thus we obtain the reformulation

$$\min\{\sum_{i,t}(h_t^i - h_t^{i+1})e_t^i + \sum_{i,t} q_t^i y_t^i : (x^i, y^i, e^i) \in \tilde{X}_i^{LS-C}, e_t^i \geq e_t^{i-1} \text{ for all } i, t\},$$

where $\tilde{X}_i^{LS-C}$ is the single item set

$$e_{t-1}^i + x_t^i = d_t^1 + e_t^i \qquad\qquad\qquad \text{for all } t$$
$$x_t^i \leq C_t^i y_t^i \qquad\qquad\qquad\qquad\qquad \text{for all } t$$
$$e^i \in \mathbb{R}_+^{(n+1)}, \; x^i \in \mathbb{R}_+^n, \; y^i \in \{0,1\}^n,$$

studied in Parts II and III. Note that whereas the original formulation contains a lot-sizing subproblem only for the final product item 1, this new formulation includes a single-item lot-sizing subproblem for each item i. $H_t^i = h_t^i - h_t^{i+1}$ is called the *echelon holding cost* of item i in period t.

Taking Z^{ML} to be the feasible region, we can write

$$Z^{ML} = \prod_{i=1}^{m} \tilde{X}_i^{LS-C} \cap \{e \in R^{m(n+1)} : e_t^i \geq e_t^{i-1} \text{ for all } i, t\},$$

where we take $e_t^0 = 0$ for all t. Using the reformulations of the earlier chapters, we can immediately obtained the improved formulation

$$Q^{ML} = \prod_{i=1}^{m} \text{conv}(\tilde{X}_i^{LS-C}) \cap \{e \in R^{m(n+1)} : e_t^i \geq e_t^{i-1} \text{ for all } i, t\}$$

in the uncapacitated *ML-S/LS-U* and constant capacity *ML-S/LS-CC* cases, respectively.

13.1.3 Multi-Commodity Reformulations: Uncapacitated Case

Another way to obtain a tighter formulation of $ML\text{-}S/LS\text{-}U$ is by using a multi-commodity reformulation with the following variables:

z_{ut}^j is the amount of component j, measured as a fraction of the demand d_t, produced in period u destined to end up in the finished product delivered in period t with $u \leq t$, and
w_{ut}^j is the stock of component j, measured as a fraction of the demand d_t, in stock at the end of period u destined to end up in the finished product delivered in period t with $u \leq t$.

This leads to the formulation

$$\min \ \sum_{j=1}^{m}\sum_{t=0}^{n} h_t^j s_t^j + \sum_{j=1}^{m}\sum_{t=1}^{n} q_t^j y_t^j$$

$$w_{u-1,t}^j + z_{ut}^j = z_{ut}^{j-1} + w_{ut}^j \qquad \text{for } 2 \leq j \leq m, \ 1 \leq u \leq t \leq n$$

$$w_{u-1,t}^1 + z_{ut}^1 = \delta_{ut} + w_{ut}^1 \qquad \text{for } 1 \leq u \leq t \leq n$$

$$x_u^j = \sum_{t=u}^{n} d_t z_{ut}^j \qquad \text{for } 1 \leq j \leq m, \ 1 \leq u \leq n$$

$$s_u^j = \sum_{t=u+1}^{n} d_t w_{ut}^j \qquad \text{for } 1 \leq j \leq m, \ 0 \leq u \leq n-1$$

$$z_{ut}^j \leq y_u^j \qquad \text{for } 1 \leq j \leq m, \ 1 \leq u \leq t \leq n$$

$$z \in \mathbb{R}_+^{mn(n+1)/2}, \quad w \in \mathbb{R}_+^{mn(n+1)/2}, \quad y \in \{0,1\}^{mn},$$

where $\delta_{ut} = 1$ if $u = t$ and $\delta_{ut} = 0$ otherwise.

Eliminating the multi-commodity stock variables and using the equivalent form of the objective function in the (x, y) variables, one obtains a facility location variant

$$\min \ \sum_{j=1}^{m}\sum_{t=1}^{n} \tilde{p}_t^j x_t^j + \sum_{j=1}^{m}\sum_{t=1}^{n} q_t^j y_t^j \tag{13.6}$$

$$\sum_{u=1}^{t} z_{ut}^j = d_t^1 \qquad \text{for } 1 \leq j \leq m, \ 1 \leq t \leq n \tag{13.7}$$

$$z_{ut}^j \leq y_u^j \qquad \text{for } 1 \leq j \leq m, \ 1 \leq u \leq t \leq n \tag{13.8}$$

$$\sum_{u=1}^{k} z_{ut}^j \geq \sum_{u=1}^{k} z_{ut}^{j-1} \qquad \text{for } 2 \leq j \leq m, \ 1 \leq k \leq t \leq n \tag{13.9}$$

$$x_u^j = \sum_{t=u}^{n} d_t z_{ut}^j \qquad \text{for } 1 \leq j \leq m, \ 1 \leq u \leq n \tag{13.10}$$

$$z \in \mathbb{R}_+^{mn(n+1)/2}, \ y \in \{0,1\}^{mn}. \tag{13.11}$$

Here the constraint (13.9) is a strengthened version of the constraint $e_t^i \geq e_t^{i-1}$. Therefore the linear relaxation of (13.6)–(13.11) is at least as strong as the echelon stock reformulation Q^{ML} derived in the last subsection. However it can be shown that solving the linear relaxation of (13.6)–(13.11) does not always lead to an integer solution.

13.1.4 Valid Inequalities for ML-S/LS-U

Here we use the echelon stock reformulation to suggest a first class of valid inequalities for ML-S/LS-U. We then interpret these inequalities as a comparison of inflows and outflows over a subgraph of the product structure digraph. This allows us to obtain several more general classes of inequalities.

As we have shown that $(x^j, e^j, y^j) \in X^{LS-U} \subset X^{WW-U}$ for each component j, the valid inequality (7.33) immediately provides us with a first class of valid inequalities

$$e_{k-1}^j = \sum_{i=1}^{j} s_{k-1}^i \geq \sum_{u=k}^{l} d_u(1 - y_k^j - \cdots - y_u^j). \tag{13.12}$$

We now give a direct interpretation of these inequalities. If $y_k^j = \cdots = y_u^j = 0$, there is no production of item j in the interval $[k, u]$, and thus the d_u units of item j that form part of the final delivery of d_u units of item 1 in period u must already have been produced prior to period k and so must form part of the echelon stock e_{k-1}^j.

Now observe that if $y_k^{j_k} = y_{k+1}^{j_{k+1}} = \cdots = y_u^{j_u} = 0$ for some sequence of items $j_k, j_{k+1}, \cdots, j_u$ with $j \geq j_k \geq j_{k+1} \geq \ldots \geq j_u \geq 1$, it is still impossible for any units of item j produced in the interval $[k, u]$ to end up in the final product d_u delivered in period u. Therefore the following holds.

Proposition 13.2 *If $j \geq j_{k,u} \geq j_{k+1,u} \geq \cdots \geq j_{u,u} \geq 1$ for all $u = k, \ldots, l$, the inequality*

$$e_{k-1}^j = \sum_{i=1}^{j} s_{k-1}^i \geq \sum_{u=k}^{l} d_u(1 - y_k^{j_{k,u}} - \cdots - y_u^{j_{u,u}}) \tag{13.13}$$

is valid for $X^{ML-S/LS-U}$.

Example 13.3 *Taking $j = 3, k = 2$, and $l = 5$, the first inequality (13.12) takes the form:*

$$e_1^3 \geq d_2(1 - y_2^3) + d_3(1 - y_2^3 - y_3^3) + d_4(1 - y_2^3 - y_3^3 - y_4^3)$$
$$+ d_5(1 - y_2^3 - y_3^3 - y_4^3 - y_5^3);$$

see Figure 13.5a.

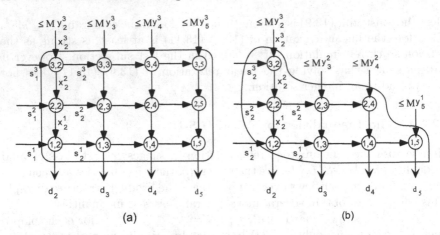

Figure 13.5. Valid inequalities: inflow versus outflow.

If we use the same nonincreasing item sequence $\{j_2, \ldots, j_5\} = \{3, 2, 2, 1\}$ for each u, we obtain

$$e_1^3 \geq d_2(1 - y_2^3) + d_3(1 - y_2^3 - y_3^2) + d_4(1 - y_2^3 - y_3^2 - y_4^2)$$
$$+ d_5(1 - y_2^3 - y_3^2 - y_4^2 - y_5^1);$$

see Figure 13.5b.

Finally, if we use a different sequence for each u, that is, the sequence $\{2\}$ for $u = 2$, $\{3, 1\}$ for $u = 3$, $\{3, 2, 1\}$ for $u = 4$, and $\{2, 2, 1, 1\}$ for $u = 5$, we obtain

$$e_1^3 \geq d_2(1 - y_2^2) + d_3(1 - y_2^3 - y_3^1) + d_4(1 - y_2^3 - y_3^2 - y_4^1)$$
$$+ d_5(1 - y_2^2 - y_3^2 - y_4^1 - y_5^1).$$

Comparing the inequalities and their representations in Figure 13.5, we see that in each inequality we have the constant term $d_2 + d_3 + d_4 + d_5 = d_{25}$ which is the outflow from the subgraph, whereas the inflows are either the horizontal flows in $e_1^3 = s_1^3 + s_1^2 + s_1^1$, or the vertical inflows with the set-up variable y_t^j times the amount of outflow that needs to be covered by inflow on that vertical arc. For the first two inequalities in the example, it is the maximum amount of flow entering on that arc that can be used to satisfy the demands $d_2, \ldots, d_5$. The third example is less obvious because each demand d_u is treated individually in the inequality (13.13).

This inflow–outflow viewpoint allows us to generalize even further. We demonstrate just by example.

Example 13.4 *One possibility is to use more general subgraphs. For instance for the subgraph shown in Figure 13.6, we have modified the horizontal inflows, and we obtain*

$$s_1^3 + s_2^2 + s_2^1 \geq d_3(1 - y_2^3 - y_3^3) + d_4(1 - y_2^3 - y_3^3 - y_4^2)$$
$$+ d_5(1 - y_2^3 - y_3^3 - y_4^2 - y_5^2).$$

A second possibility is to replace the set-up variables y_u^j due to the vertical arcs by the corresponding production levels x_u^j (flows). Replacing the period 3 variables in the second inequality of Example 13.3 gives

$$e_1^3 + x_3^2 \geq d_2(1 - y_2^3) + d_3(1 - y_3^3) + d_4(1 - y_3^3 - y_4^2) + d_5(1 - y_3^3 - y_4^2 - y_5^1).$$

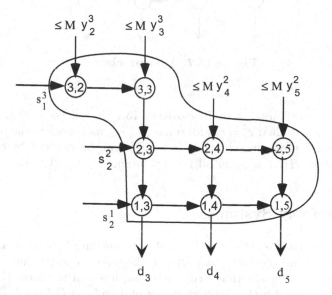

Figure 13.6. Valid inequalities: modifying stock and set-up inflows.

13.1.5 Nested Solutions

Under certain circumstances, optimal solutions have even more structure.

Definition 13.2 *A solution to a multi-level lot-sizing problem is said to be nested when, for all i, t, if $x_t^i > 0$, then $x_t^j > 0$ for all $j \in D(i)$.*

A nested solution for the series product structure is shown in Figure 13.7.

Proposition 13.3 *If either*
i. the echelon holding costs are Wagner–Whitin for each item, that is, $p_t^i + H_t^i - p_{t+1}^i \geq 0$ for all i, t, and the set-up costs q_t^j are nonincreasing in t for each j, or
ii. all objective coefficients are independent of time, that is, $p_t^i = p^i, h_t^i = h^i, q_t^i = q^i$ for all i, t,
then there exists an optimal solution that is nested for $ML\text{-}S/LS\text{-}U$.

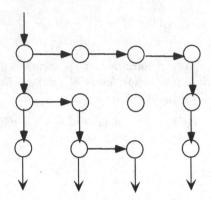

Figure 13.7. A nested solution.

Note that combined with Proposition 13.1, it follows that in a nested solution, if $x_t^i > 0$, then $x_t^i = d_{t\beta}$ for some $t \le \beta \le n$. Therefore the complexity of the dynamic program solving $ML\text{-}S/LS\text{-}U$ can be reduced by computing only the values $H(i, t, t, \beta)$ for all $1 \le i \le m$ and $1 \le t \le \beta \le n$.

13.2 Assembly Systems

For assembly systems, $|D(i)| = 1$ for all intermediate products, and $S(i)$ is the set of items on the path from $D(i)$ to the unique end-product containing i. The arcs of the production structure digraph are of the form $(i, D(i))$. In Figure 13.1ii, item 3 is the direct successor of item 5, so $D(5) = 3$. Both items 1 and 3 contain item 5, so $S(5) = \{1, 3\}$. Here, assuming that the utilization factor $r^{ij} = 1$ for all $j \in D(i)$, we obtain the mixed integer programming formulation

$$\min \ \sum_{i=1}^{m} \sum_{t=1}^{n} p_t^i x_t^i + \sum_{i=1}^{m} \sum_{t=0}^{n} h_t^i s_t^i + \sum_{i=1}^{m} \sum_{t=1}^{n} q_t^i y_t^i$$

$$s_{t-1}^i + x_t^i = x_t^{D(i)} + s_t^i \qquad \text{for } 2 \le i \le m, \text{ and all } t$$

$$s_{t-1}^1 + x_t^1 = d_t^1 + s_t^1 \qquad \text{for all } t$$

$$x_t^i \le C_t^i y_t^i \qquad \text{for } 1 \le i \le m, \text{ and all } t$$

$$s \in \mathbb{R}_+^{m(n+1)}, \ x \in \mathbb{R}_+^{mn}, \ y \in \{0, 1\}^{mn}$$

This problem is no longer a fixed charge network flow problem. However, in the uncapacitated case the extreme points still have a very simple structure.

Proposition 13.4 *For ML-A/LS-U, there exists an optimal solution with* $s_{t-1}^i x_t^i = 0$, *and if* $x_t^i > 0$, *then* $x_t^i = d_{\alpha\beta}^1$ *for* $t \le \alpha \le \beta$.

In practice it is again of interest to use the echelon stock reformulation. Now clearly $e_t^i = s_t^i + e_t^{D(i)} = s_t^i + \sum_{j \in S(i)} s_t^j$ is the *echelon stock*, and $H_t^i = h_t^i - \sum_{j:D(j)=i} h_t^j$ is the *echelon holding cost*. The reformulation obtained is thus

$$\min \quad \sum_{i=1}^{m} \sum_{t=1}^{n} p_t^i x_t^i + \sum_{i=1}^{m} \sum_{t=0}^{n} H_t^i e_t^i + \sum_{i=1}^{m} \sum_{t=1}^{n} q_t^i y_t^i$$

$$e_{t-1}^i + x_t^i = d_t^1 + e_t^i \qquad \text{for all } i, t$$

$$x_t^i \leq C_t^i y_t^i \qquad \text{for all } i, t$$

$$e_t^i \geq e_t^{D(i)} \qquad \text{for all } i, t$$

$$e \in \mathbb{R}_+^{m(n+1)}, \ x \in \mathbb{R}_+^{mn}, \ y \in \{0,1\}^{mn}.$$

This assembly problem can be treated in much the same way as the series model. One can generalize the multi-commodity reformulation, and the valid inequalities (13.13), and the echelon stock reformulation allows us to apply our knowledge about single item formulations to each component.

13.2.1 Nested Solutions

Nested solutions also arise naturally in assembly problems.

Proposition 13.5 *For ML-A, if the echelon holding costs are Wagner–Whitin for each item, that is, $p_t^i + H_t^i - p_{t+1}^i \geq 0$ for all i, t and the fixed costs are nonincreasing over time $q_t^i \geq q_{t+1}^i$ for all i, t, then there exists an optimal solution that is nested.*

When, in addition, the set-up costs are nonnegative, this suggests a dynamic programming recursion for *ML-A/WW-U* that is polynomial in m, but exponential in n.

Specifically, let $G^i(y)$ be the minimum cost for component i and all its predecessors of satisfying the external demands d_t^1 for all t at the level of component i, when $y^i = y \in \{0,1\}^n$ are the set-up periods for item i. We obtain

$$G^i(y) = \sum_t H_t^i e_t^i + \sum_t q_t^i y_t + \sum_{j:D(j)=i} \min_z \{G^j(z) : z \leq y, z \in \{0,1\}^n\},$$

where e_t^i is uniquely defined by producing as late as possible, that is, $e_{t-1}^i = \max_{k \geq t} [d_{tk}(1 - y_t - \cdots - y_k)]$.

13.2.2 Lead-Times and Echelon Stocks

Suppose now that the production of component i requires a lead-time γ^i, consisting of a nonnegative integer number of periods, and that x_t^i is the quantity whose production is started in period t and which becomes available in period $t + \gamma^i$. Thus we have the model:

$$\min \sum_{i=1}^{m}\sum_{t=1}^{n} p_t^i x_t^i + \sum_{i=1}^{m}\sum_{t=0}^{n} h_t^i s_t^i + \sum_{i=1}^{m}\sum_{t=1}^{n} q_t^i y_t^i$$

$$s_{t-1}^i + x_{t-\gamma^i}^i = x_t^{D(i)} + s_t^i \qquad \text{for } 2 \le i \le m, \text{ and all } t$$

$$s_{t-1}^1 + x_{t-\gamma^1}^1 = d_t^1 + s_t^1 \qquad \text{for all } t$$

$$x_t^i \le C_t^i y_t^i \qquad \text{for all } i, t$$

$$s \in \mathbb{R}_+^{m(n+1)}, \ x \in \mathbb{R}_+^{mn}, \ y \in \{0,1\}^{mn}.$$

How should we now define the echelon stock of i?

To preserve the decomposition into single-item lot-sizing problems, it is easily checked that it suffices to define

$$e_t^j = s_t^j + e_{t+\gamma^{D(j)}}^{D(j)}.$$

This echelon stock of component j in period t can be interpreted as the total amount of component j held in inventory, as component j or as part of its successors, that can be used to satisfy the final demand of item 1 from period $t + \tilde{\gamma}^j$ on, where $\tilde{\gamma}^j = \sum_{i \in S(j)} \gamma^i$ is defined as the cumulative production lead-time from component j till component 1. This concept leads to the new echelon stock reformulation:

$$e_{t-1}^i + x_{t-\gamma^i}^i = d_{t+\tilde{\gamma}^i}^1 + e_t^i \qquad \text{for all } i, t \qquad (13.14)$$

$$x_t^i \le C_t^i y_t^i \qquad \text{for all } i, t \qquad (13.15)$$

$$e_t^i \ge e_{t+\gamma^{D(i)}}^{D(i)} \qquad \text{for all } i, t \qquad (13.16)$$

$$e \in \mathbb{R}_+^{m(n+1)}, \ x \in \mathbb{R}_+^{mn}, \ y \in \{0,1\}^{mn}. \qquad (13.17)$$

Example 13.5 *Consider the assembly system of Figure 13.8 with five items, and lead-times $\gamma = (1, 3, 3, 2, 1)$.*

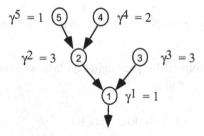

Figure 13.8. Assembly structure with lead times.

Just taking the flow conservation equation for item 4 and its successors we have $\gamma^4 = 2, \gamma^2 = 3$, and $\gamma^1 = 1$ and $\tilde{\gamma}^4 = 4$ giving

$$s^4_{t-1} + x^4_{t-2} = x^2_t + s^4_t$$
$$s^2_{t+2} + x^2_t = x^1_{t+3} + s^2_{t+3}$$
$$s^1_{t+3} + x^1_{t+3} = d^1_{t+4} + s^1_{t+4}.$$

With $e^1_t = s^1_t, e^2_t = s^2_t + e^1_{t+1}, e^4_t = s^4_t + e^2_{t+3} = s^4_t + s^2_{t+3} + s^1_{t+4}$, the sum of the three balance constraints gives

$$e^4_{t-1} + x^4_{t-2} = d^1_{t+4} + e^4_t$$

which is precisely of the form (13.14).

13.3 General Systems

For general production systems the values r^{ij} cannot all be set to 1, so these values need to be taken into account in the formulations.

The initial formulation takes the form

$$\min \sum_{i=1}^{m}\sum_{t=1}^{n} p^i_t x^i_t + \sum_{i=1}^{m}\sum_{t=0}^{n} h^i_t s^i_t + \sum_{i=1}^{m}\sum_{t=1}^{n} q^i_t y^i_t$$

$$s^i_{t-1} + x^i_t = \sum_{j\in D(i)} r^{ij} x^j_t + s^i_t \qquad\qquad \text{for } 2 \leq i \leq m, \text{ and all } t$$

$$s^1_{t-1} + x^1_t = d^1_t + s^1_t \qquad\qquad\qquad \text{for all } t$$

$$x^i_t \leq C^i_t y^i_t \qquad\qquad\qquad\qquad \text{for all } i, t$$

$$s \in \mathbb{R}^{m(n+1)}_+, \ x \in \mathbb{R}^{mn}_+, \ y \in \{0,1\}^{mn},$$

and defining the echelon stock of item i in period t as

$$e^i_t = s^i_t + \sum_{j\in D(i)} r^{ij} e^j_t,$$

the echelon stock reformulation becomes

$$\min \sum_{i=1}^{m}\sum_{t=1}^{n} p^i_t x^i_t$$

$$+ \sum_{i=1}^{m}\sum_{t=0}^{n}\left(h^i_t - \sum_{j:D(j)=i} r^{ji} h^j_t\right) e^i_t + \sum_{i=1}^{m}\sum_{t=1}^{n} q^i_t y^i_t \qquad (13.18)$$

subject to

$$e_{t-1}^i + x_t^i = R(i)d_t^1 + e_t^i \qquad\qquad \text{for all } i, t \qquad (13.19)$$

$$x_t^i \le C_t^i y_t^i \qquad\qquad \text{for all } i, t \qquad (13.20)$$

$$e_t^i \ge \sum_{j \in D(i)} r^{ij} e_t^j \qquad\qquad \text{for all } i, t \qquad (13.21)$$

$$e \in \mathbb{R}_+^{m(n+1)}, \; x \in \mathbb{R}_+^{mn}, \; y \in \{0,1\}^{mn}, \qquad\qquad (13.22)$$

where $R(i)$ is the total number of item i in one unit of item 1, the final product. More generally let $R(i, j)$ be the total amount of item i in one unit of item j, so that $R(i) = R(i, 1)$. The values $R(i, j)$ can be calculated recursively as $R(i, j) = \sum_{l \in D(i)} r_{il} R(l, j)$. It then follows that

$$e_t^i = s_t^i + \sum_{j \in D(i)} r^{ij} e_t^j = \sum_{j \in V} R(i, j) s_t^j.$$

Example 13.6 *For the general structure iii) in Figure 13.1, $R(1, 1) = 1, R(2, 1) = 2, R(3, 2) = 2, R(3, 1) = 5, R(4, 1) = 6, R(5, 1) = 19, R(6, 1) = 5$. Similarly $R(2, 2) = 1, R(3, 2) = 2, R(5, 2) = 3R(3, 2) + 2R(2, 2) = 8$, and $R(5, 3) = 3$.*

Taking the flow conservation equations for item 5 and its successors, we have

$$s_{t-1}^5 + x_t^5 = 3x_t^3 + 2x_t^2 + s_t^5$$
$$3s_{t-1}^3 + 3x_t^3 = 6x_t^2 + 3x_t^1 + 3s_t^3$$
$$8s_{t-1}^2 + 8x_t^2 = 16x_t^1 + 8s_t^2$$
$$19s_{t-1}^1 + 19x_t^1 = 19d_t^1 + 19s_t^1.$$

With $e_t^5 = s_t^5 + 3e_t^3 + 2e_t^2 = s_t^5 + 3(s_t^3 + 2e_t^2 + 1e_t^1) + 2e_t^2 = \cdots = s_t^5 + 3s_t^3 + 8s_t^2 + 19s_t^1$, we obtain

$$e_{t-1}^5 + x_t^5 = 19d_t^1 + e_t^5,$$

which is precisely the balance equation (13.19).

Finally, for general product structures with nonzero lead-times, the echelon stock of component i in period t is defined by

$$e_t^i = s_t^i + \sum_{j \in D(i)} r^{ij} e_{t+\gamma^j}^j \,,$$

and the echelon stock reformulation can be derived in much the same way as before.

Example 13.7 *For the general structure iii) in Figure 13.1, and assuming that $\gamma^i = 1$ for all i, we obtain for item 3 and its successors*

$$\begin{aligned} e_t^3 &= s_t^3 + 2e_{t+1}^2 + 1e_{t+1}^1 \\ &= s_t^3 + 2(s_{t+1}^2 + 2e_{t+2}^1) + 1e_{t+1}^1 \\ &= s_t^3 + 2s_{t+1}^2 + 4s_{t+2}^1 + 1s_{t+1}^1 \end{aligned}$$

and the flow balance constraints for item 3 in the echelon stock formulation take the form

$$e_{t-1}^3 + x_{t-\gamma^3}^3 = [d_{t+1}^1 + 4d_{t+2}^1] + e_t^3 \ .$$

In all these cases, the echelon stock reformulation allows us to apply our knowledge about single-item formulations to each item in the product structure.

13.4 Production and Distribution

Our goal in this section is to indicate how the reformulations and valid inequalities derived for both single-level and multi-level problems can be extended to include both production and distribution. For simplicity we consider a supply chain for a single item consisting of two production centers each consisting of a single production facility and of two sales areas with storage facilities; see Figure 13.9. The ideas extend directly to multi-item and multi-level production and distribution systems.

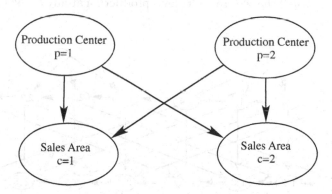

Figure 13.9. Schematic production-location network.

To describe the sample problem, we introduce the following notation:

d_t^c is the demand at sales area $c \in \{1', 2'\}$ in period t;
h_t^p is the storage cost at production site $p \in \{1, 2\}$ at the end of period t;
h_t^c is the storage cost at sales area $c \in \{1', 2'\}$ at the end of period t;
$k_t^{p,c}$ is the per unit transportation cost between production center p and sales area c in period t;

and variables:

s_t^p is the stock at production site $p \in \{1, 2\}$ at the end of period t;
σ_t^c is the stock at sales area $c \in \{1', 2'\}$ at the end of period t;
$v_t^{p,c}$ is the amount sent from production center p to sales area c in period t.

Again for simplicity we assume that there are no shipments between pairs of production centers or pairs of sales areas and no time lags. The resulting formulation is:

$$\min \quad \sum_{p,t} [h_t^p s_t^p + q_t^p y_t^p] + \sum_{c,t} h_t^c \sigma_t^c + \sum_{p,c,t} k_t^{p,c} v_t^{p,c} \tag{13.23}$$

$$s_{t-1}^p + x_t^p = \sum_c v_t^{p,c} + s_t^p \qquad \text{for all } p, t \tag{13.24}$$

$$x_t^p \le C^p y_t^p \qquad \text{for all } p, t \tag{13.25}$$

$$\sigma_{t-1}^c + \sum_p v_t^{p,c} = d_t^c + \sigma_t^c \qquad \text{for all } c, t \tag{13.26}$$

$$x_t^p, s_t^p, \sigma_t^c, v_t^{p,c} \in \mathbb{R}_+, \ y_t^p \in \{0, 1\} \qquad \text{for all } p, c, t. \tag{13.27}$$

A four-period instance is shown in Figure 13.10. Note that for our simple case the problem is a fixed charge network flow problem. Clearly this would no longer be the case if there were multi-level production at any of the production sites.

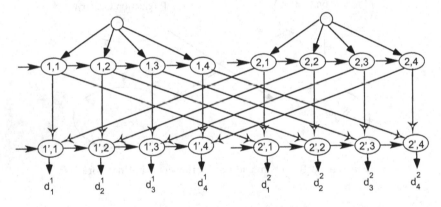

Figure 13.10. Schematic production-location network.

We now consider a few of the ways to obtain tightened reformulations and/or valid inequalities.

13.4.1 Production Center and Sales Area Aggregation

We describe the main steps of a reformulation procedure.

i. *Aggregation/Relaxation.* Aggregating the two production centers, and also aggregating the two sales areas, we obtain the relaxation:

$$s_{t-1} + \sum_p x_t^p = v_t + s_t \qquad \text{for all } t$$

$$x_t^p \leq C^p y_t^p \qquad \text{for all } p, t$$

$$\sigma_{t-1} + v_t = d_t + \sigma_{t-1} \qquad \text{for all } t$$

$$x_t^p, s_t, \sigma_t, v_t \in \mathbb{R}_+, \; y_t^p \in \{0,1\} \qquad \text{for all } p, t,$$

where $s_t = \sum_p s_t^p$, $\sigma_t = \sum_c \sigma_t^c$, $d_t = \sum_c d_t^c$, and $v_t = \sum_{p,c} v_t^{p,c}$. This can be viewed as a two-level production in series system (see Figure 13.11a) in which there are two production modes at the upper level, and no fixed costs at the lower level.

ii. *Defining Echelon Stocks.* Letting $e_t^2 = s_t + \sigma_t$, the relaxation can now be rewritten as

$$e_{t-1}^2 + \sum_p x_t^p = d_t + e_t^2 \qquad \text{for all } t \qquad (13.28)$$

$$x_t^p \leq C^p y_t^p \qquad \text{for all } p, t \qquad (13.29)$$

$$\sigma_{t-1} + v_t = d_t + \sigma_t \qquad \text{for all } t \qquad (13.30)$$

$$e_t^2 \geq \sigma_t \qquad \text{for all } t \qquad (13.31)$$

$$x_t^p, v_t, e_t^2, \sigma_t \in \mathbb{R}_+, \; y_t^p \in \{0,1\} \qquad \text{for all } p, t. \qquad (13.32)$$

Observe that dropping the constraints (13.30) and (13.31), we now have a single-item lot-sizing problem with two different production possibilities.

iii. *Deriving Mixing Sets.* Combining (13.28) and (13.29), we obtain

$$e_{t-1}^2 + \sum_p C^p y_t^p \geq d_t,$$

which after introduction of $\bar{C} = \max\{C^1, C^2\}$, $Y_t = \sum_p y_t^p$, $Z_k = \sum_{u=t}^k Y_u$, $b_k = (\sum_{u=t}^k d_u^{1,2})/\bar{C}$, and $\tau = e_{t-1}^2/\bar{C}$ gives the mixing set

$$\tau + Z_k \geq b_k \qquad \text{for } t \leq k \leq n$$

$$\tau \in \mathbb{R}_+, \; Z_k \in \mathbb{Z}_+^1 \qquad \text{for } t \leq k \leq n.$$

for which valid inequalities or extended reformulations are described in Section 8.3.

This three step procedure indicates one of the many possibilities for generating valid inequalities or extended formulations for such a problem.

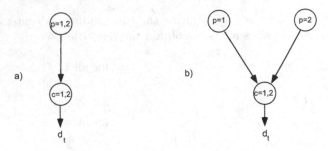

Figure 13.11. Different aggregations of the production-location network.

13.4.2 Sales Area Aggregation

Here we keep the production centers separate, but again aggregate together the sales areas.

i. *Aggregation/Relaxation.* Using similar notation to that above, we obtain

$$s_{t-1}^p + x_t^p = v_t^p + s_t^p \qquad\qquad \text{for all } p, t \qquad (13.33)$$
$$x_t^p \le C^p y_t^p \qquad\qquad \text{for all } p, t \qquad (13.34)$$
$$\sigma_{t-1} + \sum_p v_t^p = d_t + \sigma_t \qquad\qquad \text{for all } t \qquad (13.35)$$
$$x_t^p, v_t^p, s_t^p, \sigma_t \in \mathbb{R}_+, \ \ y_t^p \in \{0, 1\} \qquad\qquad \text{for all } p, t \ , \qquad (13.36)$$

where $v_t^p = \sum_c v_t^{p,c}$ for all p, t. The resulting two-level feasible region is shown in Figure 13.11b. Note that, although it resembles the assembly product structure digraph, the region (13.33)–(13.36) is a pure fixed charge network flow problem, and thus it is *not* the same as the assembly model.

ii. *Defining Echelon Stocks.* To link the production at sites $p = 1, 2$ to the final demand, one possibility is the introduction of the echelon stock variables

$$e_t^p = s_t^p + \sigma_t \quad \text{for } p = 1, 2, \text{ and all } t.$$

The set can now be rewritten as

$$e_{t-1}^p + x_t^p + v_t^{\bar p} = d_t + s_t^p \qquad\qquad \text{for all } p, t$$
$$x_t^p \le C^p y_t^p \qquad\qquad \text{for all } p, t$$
$$\sigma_{t-1} + \sum_p v_t^p = d_t + \sigma_t \qquad\qquad \text{for all } t$$
$$x_t^p, v_t^p, e_t^p, \sigma_t \in \mathbb{R}_+, \ \ y_t^p \in \{0, 1\} \qquad\qquad \text{for all } p, t,$$

where $\bar p = \{1, 2\} \setminus \{p\}$.

iii. *Deriving Mixing Sets.* For each p, we now have a single-item lot-sizing problem, except for the presence of the transportation inflow $v_t^{\bar{p}}$ from the other center. Fixing an interval $[t, l]$, it suffices to define $\tau = (e_{t-1}^p + \sum_{u=t}^l v_t^{\bar{p}})/C^1$, $Z^k = \sum_{u=t}^k y_u^p$, and $b_k = \sum_{u=t}^k d_u$ to obtain the mixing set

$$\tau + Z_k \geq b_k \qquad \text{for } t \leq k \leq l$$
$$\tau \in \mathbb{R}_+, \; Z_k \in \mathbb{Z}_+^1 \qquad \text{for } t \leq k \leq l,$$

which again indicates one possible way to obtain valid inequalities.

Many other aggregations allow us to generate valid inequalities for production and distribution problems, and as a result the choice is far from easy. Given an LP solution for the initial formulation, the goal in deciding how to reformulate or separate is to isolate a part of the network for which the flows on the entering stock and transportation variables are small, and the values of the entering set-up variables with positive flows are fractional. As in deriving valid inequalities in Section 13.1, it may be appropriate to take entering arcs corresponding to different time periods and different production/distribution levels.

Other single-item subproblems may also be of interest for more complicated problems, such as the very general uncapacitated model with sales, backlogging, and buying studied in Section 11.9. The model with sales is particularly useful when there are upper bounds on the transportation flows in each period, and the solution values are close to this upper bound. With time lags in production or transportation, it is again appropriate to define the echelon stock variables taking these lags into account in the same way as in Section 13.2.

Exercises

Exercise 13.1 Consider the following instance of production in series with $n = NT = 5, m = NI = 3$. Storage and set-up costs vary by item, but are constant over time. Specifically $(h^i) = (4, 12, 16)$, $(q^i) = (16, 45, 94)$ and the demands for item 1, the final product, are $(d_t^1) = (11, 15, 19, 0, 24)$.
i. Solve by dynamic programming.
ii. Solve by mixed integer programming with and without a single item reformulation.

Exercise 13.2 Consider the following instance of assembly production with $n = NT = 10$ and $m = NI = 7$. The assembly structure is indicated by the successor products as follows: $D(6) = D(7) = 3, D(4) = D(5) = 2, D(2) = D(3) = 1$. The storage costs are $(h^i) = (4, 12, 16, 17, 19, 24, 21)$, the set-up costs $(q^i) = (16, 45, 94, 23, 24, 35, 47)$, and the final demands for item 1 are $(d_t^1) = (11, 15, 19, 0, 24, 56, 12, 0, 9, 22)$.
Solve as a mixed integer program.

Exercise 13.3 Consider the same instance of assembly production as above, but with a budget constraint so that the total production in each period cannot exceed 100 units.

Exercise 13.4 Write out a multi-commodity reformulation for the single-item multi-level problem with assembly product structure.

Exercise 13.5 Consider an instance of $ML\text{-}S/WW\text{-}U$ with $n = NT = 4, m = NI = 2$. Storage and set-up costs vary by item, and over time, but the Wagner–Whitin cost condition is fulfilled. Specifically $(h^1_t) = (0, 2, 1, 1, 0)$, $(h^2_t) = (0, 4, 3, 2, 0)$, $(q^1_t) = (0, 4, 6, 2)$, $(q^2_t) = (0, 4, 4, 2)$, $p^i_t = 0$ for all i, t, and the demands for item 1, the final product, are $d^1_t = 1$ for all t.
i. Solve the linear relaxation of the basic formulation (13.1)–(13.5).
ii. Cut the fractional solution obtained by a valid inequality of type (13.12).
iii. Solve the linear relaxation of the echelon stock reformulation, improved by a tight formulation for each single-item lot-sizing subproblem $WW\text{-}U$.
iv. Cut the fractional solution obtained by a valid inequality (13.13).
v. Solve by using the multi-commodity reformulation, and compare the quality of the formulations.

Notes

Section 13.1. The dynamic program for the uncapacitated lot-sizing in series problem is due to Zangwill [198]. Echelon stocks were introduced by Clark and Scarf [41] in the study of (s, S) policies. Valid inequalities for multi-level lot-sizing including the series model were studied in detail in Pochet [133]. For the production in series model, the different inequalities proposed can all be seen as generalizations of the inflow–outflow inequalities proposed by Van Roy and Wolsey [173].

By generalizing further the two inequalities derived in Example 13.4, one obtains a family of *dicut* inequalities that are essentially equivalent to the projection of the multi-commodity reformulation of Subsection 13.1.3 into the original (x, y, s) space; see Rardin and Wolsey [144]. Nested schedules for series lot-sizing were studied by Love [110]; see also Pochet [133].

Section 13.2. Veinott [183] established Proposition 13.4 based on the study of Leontief substitution systems, thereby significantly generalizing the results of Zangwill [197] on concave cost flows in networks. Veinott also showed that for distribution structures this leads to a dynamic programming recursion polynomial in the number of periods, but exponential in the number of end-products. In fact the complexity of the uncapacitated assembly problem $ML\text{-}A, LS\text{-}U$ is still not known in spite of the development of several formulations that have been conjectured to be tight. See Bussieck et al. [33] for a recent counterexample. However a 2-approximation primal–dual algorithm is given in Levi et al. [105].

The idea of using echelon stocks for the assembly lot-sizing problem and the dynamic program for the nested case can be found in Crowston and Wagner [51]. Computational results using the echelon stock reformulation and Lagrangian relaxation are reported in Afentakis et al. [2]. Lead-times are cited regularly, but there does not appear to be any literature on the reformulation of problems combining echelon stocks and lead-times.

Numerous heuristics have been proposed for multi-level problems, and in particular problems with assembly structure, including Billington et al. [27], Tempelmeier and Helber [159], Tempelmeier and Derstoff [158], Dellaert and Jeunet [55], and Stadtler [154], among others.

Section 13.3. Afentakis and Gavish [1] proposed solving a problem with general product structure by reformulating the problem as an assembly problem with additional equality constraints. Pochet and Wolsey [138] applied the echelon stock approach directly to the general product structure.

Section 13.4. The idea of extending the use of echelon stocks to include both production and distribution was initially explored in the Liscos project [48]. Recently polynomial dynamic programming recursions have been developed for a constant capacity model with a single production level and a multi-level distribution system, and also for serial supply chains by Van Hoesel et al. [167, 168]. The special distribution system consisting of one warehouse and multiple retailers has been tackled using time partitioning heuristics in Federgrün and Tzur [65].

Exercises. Exercise 13.5 showing that the multi-commodity formulation for $ML\text{-}S/WW\text{-}U$ is not tight is from Pochet and Wolsey [140].

Part V

Problem Solving

14

Test Problems

In this final chapter, we present six more cases covering a wide range of problems. The first three are presented in considerable detail. For these three cases, the description of each problem involves six parts:

i. A verbal description of the context and the problem;
ii. A classification, complete or partial, of the problem based on the description;
iii. An initial problem formulation;
iv. A discussion of possible reformulation and solution strategies;
v. A report on computational results with one or more formulations or algorithms;
vi. A discussion of some algorithmic or modeling questions.

In addition for each of these three cases, we suggest in the problem description some study questions to be used to continue the analysis of reformulations and algorithms, or to put the case study into a managerial context. The exercises at the end of the chapter are designed to help in tackling some questions left open or not addressed in the text.

The last three cases are presented more briefly, starting directly from an initial MIP formulation and concentrating more on the technical details.

All models and data for these cases are available on the book Web site.

In Table 14.1 we indicate for all eight cases (including the two from Chapter 5) what structures appear in each instance according to our classification scheme, and in Table 14.2 whether and what formulations, reformulations, and heuristics are used in our treatment of each case.

Table 14.1. Classification of Test Cases

Name	PROB-CAP-VAR	PM	PQ	ML
Consumer Goods Production	DLS-CC-B	M_1	–	–
Cleaning Liquids Bottling	WW-CC-SC,LB	M_1	–	–
Making and Packing	LS-CC-B,ST(C)	M_1-SC	PC-ST	D
Storage Rack Production	WW-U	M_∞	PC-ST, FAM	A
Insulating Board Extrusion	LS-C	M_∞-SQ	PC-SQ	–
Pigment Sequencing	DLS-CC-SC	M_1-SQ	PC-U	–
Process Manufacturing	LS-C-B,SC,CLT,AFC, MR,RLS,SS,SUB	M_2-SC	PC-CLT	D
Powder Production	LS-U-B,LB,PER	M_∞	PC	D

Table 14.2. Reformulation of Test Cases

Name	Reformulation	Xform LIB	Xcut LIB	XHeur LIB
Consumer Goods Prod.	–	DLSCCB	–	–
Cleaning Liquids Bottl.	–	WWUSC,WWCC	–	RF, RINS, EXCH
Making and Packing	Echelon	WWUB,WWCCB	–	RF
Storage Rack Prod.	Echelon	WWU	–	RF,RINS
Insulating Board Extr.	VRP, M_∞-SQ	WWU, WWCC	WWU, WWCC	–
Pigment Sequencing	M_1-SQ	DLSCCSC	–	–
Process Manufacturing	M_2-SC, AFC, MR, RLS, SS	WWUSCB	–	–
Powder Production	Echelon, PER	WWUB, WWUCLB	WWUB	RF,EXCH

14.1 Making and Packing

14.1.1 Problem Description

General Context

A large company from the consumer goods industry is considering investing in new automated technology for one of its high-volume product families. This family contains several product variants that must be first produced and then packed into several packing formats and sizes.

The current production process is not flexible enough, and a large portion of the production capacity is lost in changeover times at the production and packing stages. In the past, to increase the productivity or reduce the impact of these changeovers, the company has used large production lots and regular cyclic schedules consisting of producing each product variant at regular time intervals. This has resulted in increased working capital tied up in stocks, and reduced flexibility to react to market demand.

The new production line contains several automated and non-dedicated production and packing machines, and limited storage capacity between production and packing. The increased flexibility of the line would come from several factors: the ability to produce and pack several products at the same time (because each production or packing machine can work independently of the other machines), the reduced changeover times between products or packagings due to the new automated technology, and the increased intermediate storage capacity. Nevertheless, the remaining changeover times and the diversity of products to make and pack still restrict productivity of the line. Therefore, there is again a balance to be reached in the operation of the new line between high-capacity utilization (large production lots) and reduced inventory, depending on the product mix and demand levels.

The objective of this case is to build a model in order to optimize the planning and operation of the new line, and analyze the global line capacity for the current and foreseeable product mix and demand levels. The output of this case (global line capacity, inventory levels, customer service, etc.,) will serve as an input to a more detailed operations simulation model, and ultimately to the financial investment decision model.

Problem Description

This case is inspired by a real case study, but we only use generic terms such as products, machines, and so on in our description.

- We consider a two-stage (two-level) problem. The first stage is the production stage and the second is the packing stage. There are three different types of products to be packed into five different packagings, making 15 end products in total. The three bulk products (bulks) can be stored before packing in three dedicated storage tanks (i.e., one tank for each bulk), with a large storage capacity. We do not consider here the storage capacity because sufficiently large tanks can be built relatively cheaply, so this is a secondary design question. Also, the raw materials for the production stage are not considered in this study because they do not influence the global line capacity, and their procurement is reliable.
- There are five non-dedicated making machines producing the items to be packed, and each of them can only produce one type of product per planning period (i.e., per day). This last restriction is imposed to ease the planning and operation of the line, but could be relaxed to increase the line flexibility and global line capacity. In a planning period, the capacity of a machine is reduced if there is a product changeover with respect to the previous period. This changeover time is a constant for each machine; that is, it is independent of the items produced before and after the changeover, and independent of the time period in the planning horizon. The five making machines are partly specialized in the sense that each machine cannot produce the same subset of bulk products.

424   14 Test Problems

- There are three partly specialized packing lines that can produce at most one of the 15 end items per planning period. This last restriction could also be relaxed. The capacity of a packing line is reduced in the current period if there is a changeover. Again, the changeover time is constant for each packing line. The packing rates are machine- and product-dependent.

- One or more of the six identical robots feed each packing line. The number of robots assigned to a packing line (one robot cannot be assigned simultaneously to several lines) is kept fixed during one planning period, and determines the total production capacity on that packing line. In other words, the packing capacity of a line is limited by the number of robots assigned to the line times the capacity of a single robot.

- These robots are fed themselves by three identical feeders, whose function is to get the products out of their intermediate storage tank and to supply the robots. Each feeder can only be connected to one tank (i.e., process one of the three product types) per planning period, but can feed several robots. Several feeders can be connected to a single tank. The feeder capacity may depend on the product type. Therefore, the global packing capacity in a planning period for a product (global means aggregated over all packing lines and all packaging types) is limited by the number of feeders assigned to the product times the capacity of a single feeder for that product.

- The company is using a make-to-stock policy for this product family, with a planning horizon of 15 days, not only to cover the very short production and procurement cycles, but mainly to allow some grouping of demands in order to reduce the number of changeovers. The plant is operated 24 hours per day. Hence there are 15 periods in this planning problem, each representing one day of operation. Demand forecasts for the next two weeks are available and are usually quite reliable.

- The objective is to meet the forecasted demand at minimum cost, while respecting the capacity restrictions of the line. Given the fact that the demand has to be satisfied, most of the costs are constant over such a short planning horizon. Moreover, the end products are not produced long in advance, and their inventory is very limited. Therefore, the cost will be modeled as the sum of end-of-day inventory levels over all bulk products and all planning periods (assuming that all bulks have similar production costs and added values).

- For the end products, backlogging of forecast demand is allowed, because it allows further grouping of demands and improves capacity utilization, but it is penalized because it leads to a deterioration of the customer service level. In the company, it is usually assumed that backlogging one unit of demand for one day costs as much as stocking one unit of intermediate bulk for eight days.

The product flow through the making machines, storage tanks, feeders, robots, and packing lines is illustrated in Figure 14.1.

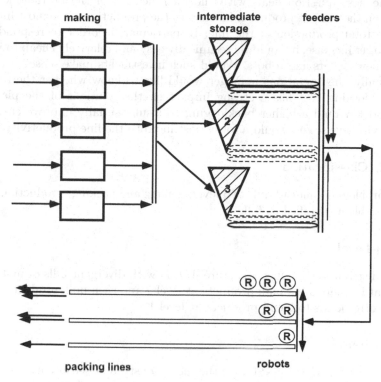

Figure 14.1. Making and packing product flow.

Study Questions

The first objective of the case study is to construct a model and an algorithm in order to solve this production planning problem, and furthermore:

- Analyze whether the new production line has enough capacity to meet the current demand, with or without backlogging;
- Analyze the impact on the customer service level of a uniform 10% increase in the end product demand.

Additional questions have been raised in the problem description.

- Current practice in the company is to fix machine assignments, as well as feeder and robot assignments, for a whole shift or planning period. This simplifies the teamwork organization, but is also a consequence of the lack of flexibility of the current technology. Because the new technology allows more flexibility in running the lines, the question of whether this assignment restriction has strong implications on the global line productivity needs to be addressed.

- The next question deals with the identification of the bottleneck stage given the current product mix, that is, the production stage most limiting the total production of the line. In particular, in order to respond to a market increase, it would be technically easy and relatively cheap to invest in new feeders and robots. Would such investments make sense?
- Finally, the operations manager would like to know whether the relative backlogging cost has a strong impact on the solutions of the planning model. Would a higher backlogging cost substantially improve customer service levels, and would a lower cost improve the line productivity?

14.1.2 Classification

This problem is a multi-level (two-level: making and packing) production planning problem classified as follows.

Multi-Level

- A distribution product structure ML-D with divergent bills of materials, that is, there are 3 bulk products at level 2 and each bulk product at level 2 corresponds to 5 end products at level 1.

Multi-Item

- Structure $PM = [M_1\text{-}SC]$ for the mode constraints because each machine is set up to produce only one item per planning period, and start-up variables need to be introduced to model the constant start-up time. This holds at both levels, for the making and packing machines.
- Structure $PQ = [PC\text{-}ST]$ for the resource constraints because the capacity of each making machine or packing line is reduced by a constant start-up time in the case of a changeover.

    Additional resource constraints exist on the assignment of feeders to bulk products (or to the dedicated storage tanks) and on the feeding capacity per bulk product from the tanks to the packing lines. Similarly, robots must be assigned to packing lines, and there exists a robot capacity constraint for each packing line. These constraints are specific to this problem, and cannot be classified.
- Both at the making and the packing levels, there are multiple machines in parallel.

Single-Item

- Structure WW-CC-$B, ST(C)$ for each end product (level 1), for the following reasons:
    - The objective function consists of the sum of the backlogging variables and satisfies the nonspeculative WW cost condition;

- The packing capacity per period is constant over time;
- There exists a constant start-up time on each packing line;
- The end products can be back-ordered.

Because of the resource constraints PQ linking the different items, it is likely that the optimal solutions of the problem will not be stock minimal solutions. Therefore, even if the objective satisfies the WW condition, the single item subproblems could also be classified as $LS\text{-}CC\text{-}B, ST(C)$ (see the discussion and remarks in Section 4.3 about the choice between classification LS or WW in multi-item problems).

- In each period, observe that the packing capacity of each end product is limited by the number of robots assigned, and the global packing capacity of the five end products obtained from each intermediate bulk product is limited by the number of feeders assigned to the intermediate product.
- Structure $WW\text{-}CC\text{-}B, ST(C)$ can also be built for each intermediate bulk product (level 2) by using the echelon stock reformulation. Again, because of the linking resource constraints destroying the stock minimal structure of optimal solutions, the classification $LS\text{-}CC\text{-}B, ST(C)$ can also be considered.

14.1.3 Initial Formulation

Using the modeling approach outlined in Chapter 1, and the structures identified during the classification process, we construct a first formulation of the problem.

Objects and Indices	Mathematical Notation
Bulk products	Object: set of bulks I_{bulk} with $|I_{bulk}| = 3$
	Index: $i \in I_{bulk}$
End products	Object: set of end products I_{end}
	with $|I_{end}| = 15$
	Index: i (or j) $\in I_{end}$
Daily time periods	Object: time periods
	Index: $t = 1, \ldots, NT$ and $NT = 15$
Making machines	Object: set of making machines K_{bulk}
	with $|K_{bulk}| = 5$
	Index: $k \in K_{bulk}$
Packing lines	Object: set of packing lines K_{end}
	with $|K_{end}| = 3$
	Index: $k \in K_{end}$

Remarks:

- Note that we do not explicitly define objects for the intermediate storage tanks, the feeders and the robots so as to keep the notation as simple as possible.

Data	Mathematical Notation
Set of end products obtained or packed from each bulk	$\forall i \in I_{bulk}$ [-]: $succ(i) \subset I_{end}$
Demand forecasts (end products)	$\forall\, i \in I_{end},\, t = 1,\ldots,NT$ [units of i]: D_t^i
Production rate	$\forall\, (i,k) \in (I_{bulk} \times K_{bulk}) \cup (I_{end} \times K_{end})$, [units of i/hour of k]: PR_k^i
Machine changeover time	$\forall\, k \in K_{bulk} \cup K_{end}$ [hours]: ST_k
Feeder rate	$\forall\, i \in I_{bulk}$ [units of $succ(i)$/hour]: FR^i
Robot rate	$\forall\, i \in I_{end}$ [units of i/hour, rob.]: RR^i
Working hours per period	[hours]: NH
Number of feeders	[-]: NF
Number of robots	[-]: NR
Relative backlogging cost (wrt storage cost)	[-]: $RBACK = 8$

Remarks:

- Note that the production rate PR_k^i represents the maximum rate of production of item i on machine k ($k \in K_{bulk}$ is a making machine if $i \in I_{bulk}$ and $k \in K_{end}$ is a packing line if $i \in I_{end}$), assuming that the machine is not limited by feeder capacity or robot capacity, and is not performing a changeover. This rate is product- and machine-dependent, with $PR_k^i = 0$ whenever product i cannot be produced or packed on machine k.
- The feeder rate FR^i of bulk i represents the maximum packing rate cumulated over all end products associated with bulk i, per feeder connected to the storage tank of bulk i. The robot rate RR^i of end product i is the maximum packing rate of end product i on a single packing line, per robot assigned to the packing line.

Variables	Mathematical Notation
Production lot size	$\forall\, (i,k) \in (I_{bulk} \times K_{bulk}) \cup (I_{end} \times K_{end})$, $t = 1,\ldots,NT$ [units of i]: $x_{kt}^i \geq 0$
Production set-up (machine or line assignment)	$\forall\, (i,k) \in (I_{bulk} \times K_{bulk}) \cup (I_{end} \times K_{end})$, $t = 1,\ldots,NT$ [-]: $y_{kt}^i \in \{0,1\}$
Machine start-up	$\forall\, k \in K_{bulk} \cup K_{end}$, $t = 1,\ldots,NT$ [-]: $z_{kt} \in \{0,1\}$
Inventory level	$\forall\, i \in I_{bulk} \cup I_{end}$, $t = 1,\ldots,NT$ [units of i]: $s_t^i \geq 0$
Backlogging level	$\forall\, i \in I_{end}$, $t = 1,\ldots,NT$ [units of i]: $r_t^i \geq 0$
Number of feeders assigned (to bulk and period)	$\forall\, i \in I_{bulk}$, $t = 1,\ldots,NT$ [-]: $\eta_t^i \in \mathbb{Z}_+$
Number of robots assigned (to packing line and period)	$\forall\, k \in K_{end}$, $t = 1,\ldots,NT$ [-]: $\pi_{kt} \in \mathbb{Z}_+$

Remarks:

- Note that the backlogging variables are only defined for end products. Inventory and backlogging variables represent the inventory and backlogging level at the end of each time period.
- Note also that the start-up variables z_{kt} need not be defined for each product i. Because the start-up time is constant over the products, we only need to identify whether there is a start-up on each machine in each time period.

Constraints	Mathematical Notation
ML-D/WW-CC-B Demand satisfaction (flow conservation) Set-up enforcement (single item capacity)	$\forall\, i \in I_{bulk} \cup I_{end}$, $\quad t = 1, \ldots, NT$ [units of i]: $dem\_sat_t^i$ $\forall\, (i,k) \in (I_{bulk} \times K_{bulk}) \cup (I_{end} \times K_{end})$, $\quad t = 1, \ldots, NT$ [units of i]: vub_{kt}^i
PM = [M_1-SC] Machine modes (one product at a time) Start-up enforcement (per product and mach.)	$\forall\, k \in K_{bulk} \cup K_{end}$, $\quad t = 1, \ldots, NT$ [-]: $mode_{kt}$ $\forall\, (i,k) \in (I_{bulk} \times K_{bulk}) \cup (I_{end} \times K_{end})$, $\quad t = 1, \ldots, NT$ [-]: $start - up_{kt}^i$
PQ = [PC-ST] Machine capacity (with start-up times)	$\forall\, k \in K_{bulk} \cup K_{end}$, $\quad t = 1, \ldots, NT$ [mach. hours]: $capa_{kt}$
Feeder assignment (to bulk) Feeder capacity (for each bulk and period) Robot assignment (to line) Robot capacity (for each line and period)	for $t = 1, \ldots, NT$ [-]: $feed\_ass_t$ $\forall\, i \in I_{bulk}$, $\quad t = 1, \ldots, NT$ [feed. hours]: $feed\_cap_t^i$ for $t = 1, \ldots, NT$ [-]: rob_ass_t $\forall\, k \in K_{end}$, $\quad t = 1, \ldots, NT$ [rob. hours]: rob_cap_{kt}

Remarks:

- We have listed the set of constraints that can be identified from the problem description. They are directly derived from the structure ML-$D/PM = [M_1\text{-}SC]/PQ = [PC\text{-}ST]/WW\text{-}CC\text{-}B, ST(C)$. Next we will formulate each constraint separately.

Objective function	Mathematical Notation
Minimize sum of inventory and backlogging	[-]: inv_back

The Complete Formulation

The complete mathematical formulation of the model is the following.

$$inv\_back \quad := \min \sum_{i \in I_{end}} \sum_{t=1}^{NT} r_t^i + \sum_{i \in I_{bulk}} \sum_{t=1}^{NT} \frac{1}{RBACK} s_t^i \qquad (14.1)$$

$$dem\_sat_t^i \quad := s_{t-1}^i + r_t^i + \sum_{k \in K_{end}} x_{kt}^i = D_t^i + r_{t-1}^i + s_t^i$$

$$\text{for } i \in I_{end}, 1 \le t \le NT \quad (s_0^i = r_0^i = 0) \qquad (14.2)$$

$$dem\_sat_t^i \quad := s_{t-1}^i + \sum_{k \in K_{bulk}} x_{kt}^i = \sum_{k \in K_{end}} \sum_{j \in succ(i)} x_{kt}^j + s_t^i$$

$$\text{for } i \in I_{bulk}, 1 \le t \le NT \quad (s_0^i \text{ fixed }) \qquad (14.3)$$

$$vub_{kt}^i \quad := x_{kt}^i \le (NH * PR_k^i) y_{kt}^i$$

$$\text{for } (i,k) \in I_{end} \times K_{end} \cup I_{bulk} \times K_{bulk}, 1 \le t \le NT$$

$$\qquad (14.4)$$

$$mode_{kt} \quad := \sum_i y_{kt}^i \le 1$$

$$\text{for } k \in K_{end} \cup K_{bulk}, 1 \le t \le NT \qquad (14.5)$$

$$start\_up_{kt}^i \quad := z_{kt} \ge y_{kt}^i - y_{k,t-1}^i$$

$$\text{for } (i,k) \in I_{end} \times K_{end} \cup I_{bulk} \times K_{bulk}, 1 \le t \le NT$$

$$\qquad (14.6)$$

$$capa_{kt} \quad := \sum_{i \in I_{bulk}} \frac{1}{PR_k^i} x_{kt}^i \le NH - ST_k z_{kt}$$

$$\text{for } k \in K_{bulk}, 1 \le t \le NT \qquad (14.7)$$

$$capa_{kt} \quad := \sum_{i \in I_{end}} \frac{1}{PR_k^i} x_{kt}^i \le NH - ST_k z_{kt}$$

$$\text{for } k \in K_{end}, 1 \le t \le NT \qquad (14.8)$$

$$feed\_ass_t \quad := \sum_{i \in I_{bulk}} \eta_t^i = NF$$

$$\text{for } 1 \le t \le NT \qquad (14.9)$$

$$feed\_cap_t^i \quad := \sum_{j \in succ(i)} \sum_{k \in K_{end}} \frac{1}{FR^i} x_{kt}^j \le NH \eta_t^i$$

$$\text{for } i \in I_{bulk}, 1 \le t \le NT \qquad (14.10)$$

$$rob\_ass_t \quad := \sum_{k \in K_{end}} \pi_{kt} = NR$$

$$\text{for } 1 \le t \le NT \qquad (14.11)$$

$$rob\_cap_{kt} \quad := \sum_{i \in I_{end}} \frac{1}{RR^i} x_{kt}^i \le NH \pi_{kt}$$

$$\text{for } k \in K_{end}, 1 \le t \le NT \qquad (14.12)$$

where the variables must satisfy the following additional restrictions

$$r_t^i, s_t^i \in \mathbb{R}_+^1, \ x_{kt}^i \in \mathbb{R}_+^1, \ y_{kt}^i \in \{0, 1\}$$
$$\text{for } (i, k) \in I_{end} \times K_{end} \cup I_{bulk} \times K_{bulk}, \ 1 \leq t \leq NT \qquad (14.13)$$

$$z_{kt} \in \{0, 1\} \qquad \text{for } k \in K_{end} \cup K_{bulk}, \ 1 \leq t \leq NT \qquad (14.14)$$

$$\pi_t^k, \eta_t^i \in \mathbb{Z}_+ \qquad \text{for } k \in K_{end}, \ i \in I_{bulk}, \ 1 \leq t \leq NT \ . \qquad (14.15)$$

In the above formulation, constraints (14.2) and (14.3) are the classical flow balance constraints in multi-level lot-sizing, and where we assume that there is some initial inventory of the bulk products. Constraint (14.4) forces the set-up, constraint (14.5) defines the modes, and constraint (14.6) defines the start-ups. Constraints (14.7) and (14.8) are the machine capacity constraints with start-up times. Constraint (14.9) defines the feeder assignments, and constraint (14.10) is the feeder capacity constraint. Similarly, constraint (14.11) defines the robot assignments, and constraint (14.12) is the robot capacity constraint. Note that the feeder capacity and robot capacity constraints (14.10) and (14.12) are usual resource capacity constraints, except that the capacity appearing at the right-hand side depends on the number of feeders or robots assigned. Finally, the objective function (14.1) is the weighted sum of inventory and backlogging levels.

14.1.4 Reformulations and Algorithms

Using this initial formulation, the first line of results in Table 14.3 shows the performance of the default branch-and-cut algorithm of Xpress-MP on this model instance with a run time limit of 7200 seconds, where the field "(Bin/Int)" indicates the number of binary and integer variables. The LP relaxation is very weak: the initial lower bound on bulk inventory and end product backlogging is zero, see column "LP" in Table 14.3. However, Xpress-MP generates strong cuts improving the initial lower bound at the root node (see column "XLP" in Table 14.3) and, after 7200 seconds, there is a remaining gap of 19%. Column "Best Lower Bound" indicates the smallest LP relaxation value among the active or remaining nodes after 7200 seconds, and column "Best Upper Bound" indicates the value of the best feasible solution found, after 7200 seconds. The gap is defined as usual as a percentage given by the formula $100 \times (best\ UB - best\ LB)/best\ UB$.

Using an echelon stock reformulation for the bulk level (i.e., defining variables $es_t^i = s_t^i + \sum_{j \in succ(i)} s_t^j$, $er_t^i = \sum_{j \in succ(i)} r_t^j$, for $i \in I_{bulk}$ and $t = 1, \ldots, NT$), and aggregating the making across all machines (i.e., defining variables $ax_t^i = \sum_{k \in K_{bulk}} x_{kt}^i$ and $ay_t^i = \sum_{k \in K_{bulk}} y_{kt}^i$ for $i \in I_{bulk}$) one gets the following valid formulation (relaxation) from constraints (14.2)–(14.4).

$$edem\_sat_t^i \quad := es_{t-1}^i + er_t^i + ax_t^i = (\sum_{j \in succ(i)} D_t^j) + er_{t-1}^i + es_t^i$$

$$\text{for } i \in I_{bulk}, \ 1 \le t \le NT \quad (es_0^i \text{ fixed} \ ; \ er_0^i = 0) \quad (14.16)$$

$$evub_t^i \quad := ax_t^i \le (\max_{k \in K_{bulk}} NH * PR_k^i) ay_t^i$$

$$\text{for } i \in I_{bulk}, \ 1 \le t \le NT \quad (14.17)$$

$$er_t^i, es_t^i \in \mathbb{R}_+^1, \ ax_t^i \in \mathbb{R}_+^1, \ ay_t^i \in \mathbb{Z}_+$$

$$\text{for } i \in I_{bulk}, \ 1 \le t \le NT \ . \quad (14.18)$$

The same aggregation process at the end product level gives the following similar formulation:

$$edem\_sat_t^i \quad := s_{t-1}^i + r_t^i + ax_t^i = D_t^i + r_{t-1}^i + s_t^i$$

$$\text{for } i \in I_{end}, \ 1 \le t \le NT \quad (s_0^i = r_0^i = 0) \quad (14.19)$$

$$evub_t^i \quad := ax_t^i \le (\max_{k \in K_{end}} NH * PR_k^i) ay_t^i$$

$$\text{for } i \in I_{end}, \ 1 \le t \le NT \quad (14.20)$$

$$r_t^i, s_t^i \in \mathbb{R}_+^1, \ ax_t^i \in \mathbb{R}_+^1, \ ay_t^i \in \mathbb{Z}_+$$

$$\text{for } i \in I_{end}, \ 1 \le t \le NT \ . \quad (14.21)$$

In these reformulations (14.16)–(14.18) and (14.19)–(14.21), all items are seen to involve constant capacity and backlogging. We first ignore the capacities and add the extended reformulation for WW-U-B for all items, using LS–LIB. We set the approximation parameter to 4 at the bulk level and to 8 at the end product level. The results can be found in Table 14.3. Due to the size of the resulting formulation, far fewer nodes can be explored within the time limit. However, significantly better bounds and better solutions are obtained.

Table 14.3. Reformulation Results for Making and Packing

K_{U-B}^{bulk}	K_{U-B}^{end}	K_{CC-B}^{end}	Cons (Model Cuts)	Vars (Bin/ Int)	LP	XLP	Best Lower Bound	Best Upper Bound	Number of Nodes	Total Time (secs)
0	0	0	1,692 (705)	1,635	0	17,760	18,443	22,843	108,500	7,200
4	8	0	4,678 (705)	2,175	18,039	18,444	19,375	22,548	12,200	7,200
4	8	6	24,028 (3,867)	9,375 (705)	20,115	20,377	20,701	21,463	4,000	7,200

Next, we tighten the formulation even more by adding the reformulation for WW-CC-B at the end product level ((14.19)–(14.21)), using 6 as value of the approximation parameter. In order to reduce and control the size of the resulting formulation, we add some of the constraints as model cuts, which means that only the violated ones are generated at the end of the root node linear relaxation. The number of model cuts added is indicated by "(Model Cuts)" in the column "Cons" of Table 14.3. The results in Table 14.3 show that even fewer nodes are explored in 7200 seconds, but better bounds and solutions are found. The resulting gap is 3–4%, which is acceptable for the purposes of the company. It is worthwhile pointing out that, in this instance, good solutions are typically available after a few hundred seconds.

Although we obtain good enough solutions to answer to our design questions with reasonable accuracy, the time needed to obtain these solutions or to guarantee their quality is long, and does not allow us to run many scenarios to perform sensitivity analysis. To reduce this run-time without losing too much in solution quality, our final test consists of the implementation of a time decomposition relax-and-fix heuristic (see Section 3.6). The results are given in Table 14.4, where the column "R&F Bin" indicates the number of non relaxed time periods at each iteration, and the column "R&F Fix" indicates the number of time periods whose solution is fixed at the end of each iteration (see the description of the relax-and-fix LS–LIB procedure in Section 5.3).

For instance, with $bin = 4$ and $fix = 2$:

- We first solve the relaxed program in which the integer variables for periods 1 to 4 are not relaxed (i.e., they keep their integrality status), whereas all variables for later periods are relaxed into continuous variables;
- We then keep the solution obtained for the periods 1 and 2 by fixing the corresponding integer variables at their current values;
- We resolve the program where now the integer variables for periods 3 to 6 are not relaxed;
- Then we fix the solution for periods 3 and 4;
- And so on, up to the last subproblem to be solved in which all binary and integer variables for periods 1 to 12 are fixed, and variables for periods 13 to 15 are not relaxed.

The relaxed subproblems involve fewer binary and integer variables (see column "(Bin/Int)" in Table 14.4, and thus can be solved faster. We have thus solved each relaxed MIP subproblem to optimality. The time indicated in the table is the total time needed to solve all the relaxed subproblems (i.e., up to the end of the planning horizon). The number of nodes is the sum of the number of branch-and-bound nodes explored over all subproblems. Note that we relax some variables in each subproblem, but in fact only the first subproblem – before any fixing – is a true relaxation of the initial problem. Therefore, the best global lower bound one can obtain with this method is the optimal solution value of the first relaxation. The best upper bound is the optimal solution value of the last subproblem solved.

First we have used the relax-and-fix heuristic on the initial formulation; see the first two lines in Table 14.4. We obtain good solutions quickly if the relaxed subproblems do not contain too many binary and integer variables (parameter bin). We do not report the results obtained with $bin = 8$ and $fix = 2$ because we were not able to solve the first subproblem to optimality in less than 1800 seconds. Observe also that we get poor lower bounds (11,920 and 16,974) with this approach based on the initial formulation. So, we get good solutions quickly without being able to guarantee their quality (the final gap is about 27%).

Table 14.4. Reformulation and Relax-and-Fix Results for Making and Packing

K_{U-B}^{bulk}	K_{U-B}^{end}	K_{CC-B}^{end}	R& F Fix	R& F Bin	Cons (Model Cuts)	Vars (Bin/ Int)	Best Lower Bound	Best Upper Bound	Number Nodes	Total Time (secs)
0	0	0	2	4	1,692 (188)	1,635	11,920	21,553	2,590	51
0	0	0	2	6	1,692 (282)	1,635	16,974	21,515	28,557	467
4	8	0	2	4	4,678 (188)	2,175	19,539	21,261	5,113	1,655
4	8	6	2	4	24,028 (3,784)	9,375 (188)	20,774	21,188	1,148	1,201

In the last two rows of Table 14.4, we analyze the combined impact of the relax-and-fix heuristic and the reformulations. We have tested the heuristic with the various reformulations WW-U-B and WW-CC-B described above. Note that we have included these reformulations for the whole planning horizon in all subproblems, and not just for the non relaxed part of the horizon. So our initial linear relaxations are as good with the time decomposition relax-and-fix approach as with the direct (single problem) approach. We observe the complementary role of relax-and-fix and reformulations, and we obtain the overall best lower and upper bounds with a final gap of about 2% in 1200 seconds.

14.1.5 Analysis of Capacity and Customer Service Level

We discuss here the most important study questions mentioned in the case description. The other questions are suggested as exercises in Section 14.6.5. Recall that the first objective of the case study is to analyze whether the new production line has enough capacity to meet the current demand, with or without backlogging.

In the best solution obtained, the production facility has enough capacity to meet current demand, without unsatisfied demand at the end of the planning horizon. However there is some backlogging in the first periods of the horizon because there is no initial end product inventory. With the current demand, and the current flexibility (i.e., one product per making line or packing line, per day), it seems necessary to hold end product inventory. However, very little intermediate product inventory is necessary to be able to meet demand. Table 14.5 summarizes some information about this best solution.

Table 14.5. Making and Packing Solution

Objective function value	21,188
Backlogging [units*days]	20,919
Intermediate Inventory [units*days]	2,154
End product Inventory [units*days]	117,691
Average making capacity utilization [%]	57
Average packing capacity utilization [%]	78
Average feeder capacity utilization [%]	47
Average robot capacity utilization [%]	56
Number of making changeovers [-]	17
Number of packing changeovers [-]	37

The next question raised is to analyze whether there is enough capacity to be able to satisfy a uniform 10% increase in the end product demand, and what would be the impact on the customer service level of such a demand increase.

With this increase in demand, we solved the problem with the extended reformulations WW-U-B and WW-CC-B, and with the initial formulation using a relax-and-fix time decomposition heuristic. The results are given in Table 14.6, where the best lower bound is obtained using the 4/8/6 reformulation and the best upper bound using the relax-and-fix heuristic without reformulation. The overall duality gap obtained is 8.4% .

Table 14.6. Results for Making and Packing with 10% Demand Increase

K_{U-B}^{bulk}	K_{U-B}^{end}	K_{CC-B}^{end}	R& F Fix	R& F Bin	Cons (Model Cuts)	Vars (Bin/ Int)	Best Lower Bound	Best Upper Bound	Number Nodes	Total Time (secs)
0	0	0	2	4	1,692 (188)	1,635	16,173	39,153	7,167	121
0	0	0	2	6	1,692 (282)	1,635	23,001	37,158	115,266	2,415
4	8	6	0	0	24,028 (4,672)	9,375 (705)	34,053	–	1,184	7,200

Table 14.7. Making and Packing Solution with 10% Demand Increase

Objective function value	37,158
Backlogging [units*days]	35,773
Intermediate Inventory [units*days]	11,080
End product Inventory [units*days]	104,939
Average making capacity utilization [%]	61
Average packing capacity utilization [%]	82
Average feeder capacity utilization [%]	50
Average robot capacity utilization [%]	60
Number of making changeovers [-]	22
Number of packing changeovers [-]	35

The information shown in Table 14.7 about the best solution obtained suggests the possible impact of this demand increase. The end product inventory levels are decreased, the intermediate inventory level is increased by a factor of five, and total inventory is decreased. There is enough capacity on average to satisfy a 10% demand increase, but the backlogging is almost doubled. However, testing the validity of such a conclusion would require considerably more analysis and computation.

14.2 Storage Rack Production

14.2.1 Problem Description

General Context

Many discrete manufacturing companies are confronted with multi-level production planning problems and have implemented integrated manufacturing planning and control systems, like ERP or MRP systems, in the past decades. In most cases, these systems (or the first versions of these) have been pure transactional IT systems able to record, update, and communicate the status of the production system (i.e., in progress and planned purchase and manufacturing orders, customer orders, demand forecasts, available and allocated inventory, etc.); see Chapter 2.

Unfortunately, these systems and their underlying production planning models are not detailed and powerful enough to build optimized production plans. In this respect, the stakes for the future of analytical planning IT systems will be the ability to model accurately and take into account the joint optimization of capacity utilization and customer demand satisfaction. This is illustrated in the following case study where the emphasis is put on modeling the joint resource utilization of the various products or components in an MRP like planning problem.

Description

We start our analysis from a very standard initial formulation. Before that, and in order to allow the reader to practice the modeling approach outlined in Chapter 1, we give a complete and verbal description of the planning problem to be solved.

- A plant produces storage racks of differing height and depth giving in total six end products. Production of each storage rack (end product) is organized so that each item or component (i.e., end product, semi-finished product or raw material) has only one successor item. The assembly product structure or bill of materials of a typical storage rack is illustrated in Figure 14.2, in which the material flow is from bottom to top. This pure assembly structure comes from the kitting of varied amounts of the same key components required by different end products. The six end product structures are independent of one another, and there are 78 items in total.

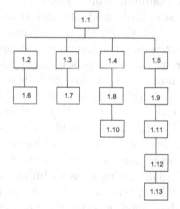

Figure 14.2. Multilevel BOM for storage rack production.

- Items (end products and intermediate items) are grouped into *item families*, and each item may belong to several families. A family is defined as a set of technologically similar items in the sense that the members of a family are grouped together for some production stages, and therefore share some resources or consume the same resources at these stages (machines and machine set-ups).

    Families are defined by the planning team in order to model capacity utilization more accurately. More precisely, they represent machines and set-ups in the following way.

    - A *C-family* models a set-up cost, which means that a set-up cost is incurred (once) in each period in which any member of the family is set-up or produced.

- A *T-family* models a set-up time, which means that a set-up time is incurred (once) in each period in which a member of the family is set up or produced. This set-up time consumes the available capacity of an associated machine (associated R-family).
- An *R-family* models a manufacturing machine or resource, which means that the family has a given available capacity (in hours per planning period), and family members are characterized by a unit production time (in hours per unit) on the resource. The capacity is also consumed by the set-up times from some associated T-families. The available capacity of a resource (in hours per period) may be period dependent.

Note that a given family may be simultaneously of the C-type, S-type, and/or R-type. For example, a C-family may consist of all items using some equipment, an R-family contains all items produced on the assembly machine, and a T-family contains all items of a certain depth and produced together (single preparation) on the R-family assembly machine.

Resource consumption rates vary greatly from product to product, but are constant over the planning time horizon.

- The plant produces standard products and is operated under a make-to-stock production policy. Given the limited number of bill of materials levels, and the automated material handling systems allowing the company to use small transfer batches, the production cycle is quite short, and a short-term planning horizon of 16 time periods is used for the MPS and the MRP. Each production activity is planned using zero lead-times.
- Demand forecasts for the six end products are available for each period of the planning horizon. To maintain a high level of customer service, the planning must fulfill demand without backlogging. There is no external demand – as spare parts – for the intermediate items.
- The objective of the planning team is to build, and maintain over time using a rolling schedule approach, a short term production plan meeting demand without backlogging, satisfying the capacity restrictions, and minimizing the inventory and set-up costs. Items vary greatly in echelon holding costs but these are constant over the time horizon.

Study Questions

The main objective of this project is to build a model and a MIP optimization algorithm to solve this planning problem. In particular, the following technical questions about the model and its mathematical formulation should be addressed.

- How well does the initial formulation perform?
- Are there ways in which the formulation can be improved?
- In using the extended formulations in a black box approach, how should one choose the approximation parameter TK?

- If one only has limited computation time, what approach gives the best feasible solution, or the smallest duality gap?
- Are there heuristics that work well for the problem, or should one run the MIP for all the computing time that is available?

In addition, and once an efficient optimization tool is available, the following modeling question should be tackled.

- Given the very short production cycle (lead-times can be assumed to be zero), and given the fact that the planning tool will be used in a rolling horizon manner, is it really necessary to optimize production over a planning horizon of 16 periods? What does one lose by considering a planning horizon of 8 periods ?

We analyze the technical questions in this section, and leave the modeling question as an exercise for Section 14.6.5.

14.2.2 Classification

This problem is a multi-level production planning problem classified as follows.

Multi-Level

- An assembly or convergent product structure $ML\text{-}A$, with six independent bills of materials (one for each end product) and 78 items in total.

Multi-Item

- There are no mode constraints in this problem because there is no limit on the number of items set up or produced on each resource per planning period.
- Structure $PQ = [PC\text{-}ST(C)]$ for the resource constraints because there is a capacity constraint limiting production for each R-family in each period. This constraint involves both production levels for the members of the R-family, and constant (over time) set-up times for members of an associated T-family.

  Nevertheless, observe that the set-up times are defined here for item families rather than for individual items. A set-up time is incurred in a period on a resource (i.e., a R-family) if any member of an associated T-family is set up for production. We thus need to adapt our usual or classical formulation to model this particular feature.
- All the machines are supposed to work in parallel, and there are no machine choice or machine assignment decisions in this problem. Each R-family models a machine or resource, and each member of the R-family represents an item that is produced on that resource.

Single-Item

- Structure $WW\text{-}U$ for each individual item because the objective function involves only inventory and set-up costs, and therefore satisfies the nonspeculative WW cost condition. There are no individual capacity constraints.

  This structure $WW\text{-}U$ is part of the initial formulation for end products, and can be built for each intermediate item by using the classical echelon stock reformulation.

  Again, observe that the set-up costs are defined here for item families rather than for individual items. The set-up cost of a C-family is incurred in a period if any member of the C-family is set up for production. We need to adapt our classical formulation to this feature.

- We could as well classify the individual items as $LS\text{-}U$ because the joint capacity restrictions impose implicit and time-dependent unit production costs on the items, and optimal solutions may not be stock minimal solutions.

14.2.3 Initial Formulation

The main feature of this problem is the definition of item families to model joint set-up costs and times. The rest of the problem is a classical multi-level MRP planning problem with resource capacity constraints. Here we transform the verbal problem description into a mathematical formulation using our systematic modeling approach outlined in Chapter 1. The resulting initial formulation is the one proposed in the original publication describing this case.

Objects and Indices	Mathematical Notation
Items or components	Object: set of (all) items I : $\|I\| = NI = 78$
	Index: $i \in I = \{1, \ldots, NI\}$
Time periods	Object: set of time periods T : $\|T\| = NT = 16$
	Index: $t \in T = \{1, \ldots, NT\}$
Set-up cost families	Object: set of C-families CF
	Index: $f, g \in CF$
Set-up time families	Object: set of T-families TF
	Index: $f, g \in TF$
Resource families	Object: set of R-families RF
	Index: $f, g \in RF$
Families	Object: set of families $F = CF \cup TF \cup RF$
	Index: $f, g \in F$

Remarks:

- Note that each family is defined by a subset of items.

Data	Mathematical Notation
Assembly product structure:	
Immediate succ. of an item	$\forall i \in I$ [-]: $\sigma(i) \in I \cup \{0\}$
Independent demand	$\forall \, i \in I$ with $\sigma(i) = 0$, $t \in T$ [units of i]: d_t^i
Definition of item families	$\forall \, f \in F$ [-]: $IF(f) \subseteq I$
Unit production time	$\forall \, f \in RF$, $i \in IF(f)$,
	[hours of f/unit of i]: α^{if}
Resource capacity	$\forall \, f \in RF$, $t \in T$ [hours]: L_t^f
Associated resources	$\forall \, g \in TF$ [-]: $ARF(g) \subseteq RF$
Resource set-up times	$\forall \, g \in TF$, $f \in ARF(g)$ [hours]: β^{gf}
Family set-up cost	$\forall \, f \in CF$ [hours]: c^f
Inventory cost	$\forall \, i \in I$ [euro/unit of i, period]: h^i

Remarks:

- In the assembly product structure, end products have no successor item. We model this by defining the successor of item i as $\sigma(i) = 0$ when i is an end product, and $\sigma(i) \in I$ otherwise. Note that there is no external or independent demand d_t^i for intermediate items (i.e., items i with $\sigma(i) \in I$). Each family $f \in F$ is defined as a subset $IF(f)$ of items.
- For each R-family $f \in RF$, the unit production time is defined for all items belonging to the resource family (i.e., all $i \in IF(f)$) and the production capacity in period t is denoted by L_t^f. For each T-family $g \in TF$, a set-up time is incurred on the subset $ARF(g)$ of the resource families when at least one member of the T-family is set up for production in a period. This set-up time for family g on resource $f \in ARF(g)$ is defined by β^{gf}. For the C-family $f \in CF$, the set-up cost is denoted by c^f.

Variables	Mathematical Notation
Production lot size	$\forall \, i \in I$, $t \in T$ [units of i]: $x_t^i \geq 0$
Production set-up	$\forall \, i \in I$, $t \in T$ [-]: $y_t^i \in \{0,1\}$
Inventory level	$\forall \, i \in I$, $t \in T$ [units of i]: $s_t^i \geq 0$
Family set-up	$\forall \, f \in CF \cup TF$, $t \in T$ [-]: $\eta_t^f \in \{0,1\}$

Remarks:

- The variables in this model are classical production, set-up, and inventory variables defined for all items in every time period.
- As usual, inventory variables represent the inventory at the end of each time period.
- In addition, set-up variables must be defined for C- and T-families in every period to model the family set-up costs and times, respectively.

Constraints	Mathematical Notation
ML-A/LS-U	
Demand satisfaction	$\forall\, i \in I, t \in T$ [units of i]: $dem\_sat_t^i$
Item set-up enforcement	$\forall\, i \in I, t \in T$ [units of i]: vub_t^i
PQ = [PC-ST(C)]	
Resource capacity	$\forall\, f \in RF, t \in T$ [res. hours]: $capa_t^f$
Family set-up enforcement	$\forall\, f \in CF \cup TF, i \in IF(f), t \in T$ [-]: fam_t^{if}

Remarks:

- We have listed first the set of constraints that can be identified from the problem description and derived from the classification $ML\text{-}A/PQ = [PC\text{-}ST(C)]/LS\text{-}U$. Next we will formulate each constraint separately.
- The only nonclassical constraints are the constraints fam_t^{if} required to model the family set-ups.

Objective function	Mathematical Notation
Minimize sum of	
inventory and set-up costs	[euro]: $cost$

The Complete Formulation

The complete initial formulation of the model, called INI, is the following.

$$(INI) \quad cost \quad := \min \sum_{i \in I} \sum_{t \in T} h^i s_t^i + \sum_{f \in CF} \sum_{t \in T} c^f \eta_t^f \qquad (14.22)$$

$$dem\_sat_t^i \quad := s_{t-1}^i + x_t^i = d_t^i + x_t^{\sigma(i)} + s_t^i$$
$$\text{for all } i \in I, t \in T \;\; (s_0^i = 0) \quad (14.23)$$

$$vub_t^i \quad := x_t^i \le M y_t^i$$
$$\text{for all } i \in I, t \in T \qquad (14.24)$$

$$capa_t^f \quad := \sum_{i \in IF(f)} \alpha^{if} x_t^i + \sum_{g \in TF: f \in ARF(g)} \beta^{gf} \eta_t^g \le L_t^f$$
$$\text{for all } f \in RF, t \in T \qquad (14.25)$$

$$fam_t^{if} \quad := y_t^i \le \eta_t^f$$
$$\text{for all } f \in CF \cup TF, i \in IF(f), t \in T \qquad (14.26)$$

$$s_t^i, x_t^i \in \mathbb{R}_+^1, \; y_t^i \in \{0,1\}$$
$$\text{for all } i \in I, t \in T \qquad (14.27)$$

$$\eta_t^f \in \{0,1\}$$
$$\text{for all } f \in CF \cup TF, t \in T, \qquad (14.28)$$

where in constraint (14.23), variable $x_t^{\sigma(i)}$ represents the dependent demand of item i and is assumed to be zero (or does not exist) when i is an end product, that is, when $\sigma(i) = 0$. On the contrary, d_t^i represents the independent demand and only exists for end products (i.e., $d_t^i = 0$ for all i with $\sigma(i) > 0$). In constraint (14.24), M is a large positive value because there is no individual capacity constraint. Constraint (14.25) is the classical big bucket capacity constraint with family set-up times. Finally, constraint (14.26) defines the link between the item set-up and the family set-up. A family f is forced to be set up in period t (i.e., $\eta_t^f = 1$) when any of its members $i \in IF(f)$ is set up in period t (i.e., $y_t^i = 1$).

14.2.4 Improving the Initial Formulation

We now make several observations relative to the initial formulation INI.

Observation 14.1 *i. The integrality constraints on the item set-up variables y_t^i can be relaxed, thereby significantly decreasing the number of integer variables.*

ii. As there are no item set-up costs, the y_t^i variables can be eliminated. To obtain a valid formulation, we can replace the constraints (14.24) and (14.26) by

$$x_t^i \le M\eta_t^f \quad \text{for all } f \in CF \cup TF, \ i \in IF(f), \ t \in T$$

The formulation obtained by relaxing the integrality requirements on the y variables is denoted $INI - yreal$. The formulation obtained by removing the y variables, denoted $INI - red$, is

$$(INI{-}red) \quad \min \sum_{i\in I}\sum_{t\in T} h^i s_t^i + \sum_{f\in CF}\sum_{t\in T} c^f \eta_t^f$$

$$s_{t-1}^i + x_t^i = d_t^i + x_t^{\sigma(i)} + s_t^i$$
$$\text{for all } i \in I, \ t \in T$$

$$x_t^i \le M\eta_t^f \qquad \text{for all } f \in CF \cup TF, \ i \in IF(f), \ t \in T$$

$$\sum_{i\in IF(f)} \alpha^{if} x_t^i + \sum_{g\in TF: f\in ARF(g)} \beta^{gf} \eta_t^g \le L_t^f$$
$$\text{for all } f \in RF, \ t \in T$$

$$x_t^i, s_t^i \in \mathbb{R}_+^1 \qquad \text{for all } i \in I, \ t \in T$$

$$\eta_t^f \in \{0,1\} \qquad \text{for all } f \in CF \cup TF, \ t \in T .$$

As we are dealing with a multi-level assembly problem, we can also rewrite the two formulations using echelon stock variables, as described in Chapter 13. We need to introduce the following notation.

- The echelon stock variable e_t^i, for all $i \in I$ and $t \in T$, is defined (for an assembly product structure) recursively as $e_t^i = s_t^i + e_t^{\sigma(i)}$. In other words, by expanding the recursion, the echelon stock e_t^i represents the sum of the inventory levels of item i and of all its successors up to the single end product obtained from i.
- In the assembly product structure, we denote by $q(i)$ the single end product obtained from item i, for all i.
- We define the echelon holding cost v^i, that is, the holding cost associated with the echelon stock, as $v^i = h^i - \sum_{j \in I : \sigma(j)=i} h^j$.

Using the equality $s_t^i = e_t^i - e_t^{\sigma(i)}$ to replace the inventory variables s_t^i by the echelon stock variables e_t^i, we obtain the echelon reformulation of the reduced formulation, denoted by $INI\text{-}red\text{-}ech$, see Chapter 13 for further or more detailed explanations:

$$(INI\text{-}red\text{-}ech) \quad \min \sum_{i \in I} \sum_{t \in T} v^i e_t^i + \sum_{f \in CF} \sum_{t \in T} c^f \eta_t^f$$

$$e_{t-1}^i + x_t^i = d_t^{q(i)} + e_t^i$$
$$\text{for all } i \in I, \, t \in T$$

$$e_t^i \geq e_t^{\sigma(i)} \quad \text{for all } i \in I : \sigma(i) > 0, \, t \in T$$

$$x_t^i \leq M \eta_t^f \quad \text{for all } f \in CF \cup TF, \, i \in IF(f), \, t \in T$$

$$\sum_{i \in IF(f)} \alpha^{if} x_t^i + \sum_{g \in TF : f \in ARF(g)} \beta^{gf} \eta_t^g \leq L_t^f$$
$$\text{for all } f \in RF, \, t \in T$$

$$x_t^i, e_t^i \in \mathbb{R}_+^1 \quad \text{for all } i \in I, \, t \in T$$

$$\eta_t^f \in \{0, 1\} \quad \text{for all } f \in CF \cup TF, \, t \in T.$$

14.2.5 Choosing the Appropriate Extended Reformulations

The next possibility is to tighten the two echelon stock formulations using the $LS\text{-}U$ classification of single-item subproblems.

Rather than using the extended reformulations for $LS\text{-}U$ involving $O(n)$ or $O(n^2)$ constraints (with $n = NT$) and many ($O(n^2)$) new variables, we have used the reformulation $WW\text{-}U$ by moving to the right in Table 4.4; see Section 4.4.4.

This $WW\text{-}U$ reformulation does not require any new variables, and for $INI\text{-}red\text{-}ech$ and each item $i \in I$, these additional constraints take the form

$$e_{t-1}^i + \sum_{u=t}^{l} d_{ul}^{q(i)} \eta_u^f \geq d_{tl}^{q(i)} \quad \text{for all } t, l \in T, \text{ and } l \geq t,$$

where f is any family containing item i; that is, $f \in CF \cup TF$ and $i \in IF(f)$. The resulting formulation is denoted $INI\text{-}red\text{-}ech\text{-}WWU$.

Clearly, these inequalities are only unique when each item belongs to just one family. If item i belongs to several families, but forming a nested set of families, then the above inequalities have to be added only for the smallest family. In other words, if $i \in IF(f_1) \subseteq IF(f_2) \subseteq IF(f_3) \subseteq \cdots$, then the $WW\text{-}U$ inequalities can be added for f_1 only, and the following constraints added to the formulation to link the family set-ups:

$$\eta_t^{f_1} \leq \eta_t^{f_2} \leq \eta_t^{f_3} \leq \cdots.$$

If item i belongs to several nonnested families of items, then the $WW\text{-}U$ inequalities must be added for these multiple families.

Note that the trade-off between the size and the quality of these reformulations can also be controlled by the approximation parameter TK introduced in Chapter 5 and used in our library of extended reformulations LS–LIB.

14.2.6 Results with Extended Reformulations

To analyze the best way of using the extended reformulations, we start from the initial problem, formulation, and data proposed in the literature.

We know that there are 6 end products, and that 6 independent assembly product structures are used to manufacture these end items. There are 78 items in total, and 16 planning periods. There are 20 nested families of items (i.e., any two families are either disjoint, or one is included in the other). All the data used here correspond to the low level of resource availability ($C1$) and low level of fixed costs ($S1$) from the original publication (see Section 14.6.5). Only the demand set has been generated randomly based on the distribution proposed.

We first ran each of the different formulations in default MIP mode for 450 seconds. The results are shown in Table 14.8. Column "Int" contains the number of integer variables, "LP" is the value of the first linear relaxation, "XLP" is the value of the first node relaxation after addition of the Xpress-MP cuts. Columns "BUB" and "BLB" give, respectively, the value of the best feasible solution and the best lower bound after 450 seconds. In the last column we show the best solution found after 60 and 150 seconds, respectively. The duality gap is computed as usual by $\text{Gap} = \frac{BUB - BLB}{BUB} \times 100$.

The results in Table 14.8 show that:

- It seems very important to eliminate the y variables from the model, and to express the variable upper bound constraints using directly the η set-up variables; see model $INI\text{-}red$. This improves both the initial lower bound (through some system preprocessing) and the root node lower bound after the addition of Xpress-MP cuts. So, this substitution improves the effectiveness of the system cuts, and allows one to reduce the gap substantially.

Table 14.8. Storage Rack Model: Reformulation Results in 450 Seconds

| Formulation | Cons | Vars | LP | BLB | Gap % | BUB-60 secs |
	Int		XLP	BUB	Nodes	BUB-150 secs
INI	3,312	3,648	2,111.3	6,493.2	48.2	∞
		1,152	5,532.9	12,531.0	27,524	∞
INI − yreal	3,312	3,648	997.5	6,683.8	45.1	12,217.0
		192	4,900.0	12,166.2	41,390	12,217.0
INI − red	2,352	2,688	4,033.1	10,421.4	12.6	12,219.7
		192	9,998.1	11,928.9	18,245	12,066.3
INI − red − ech	3504	2,688	4,035.4	9,792.9	18.0	12,083.9
		192	9,202.3	11,937.5	31,935	12,017.3
INI − red − ech − WWU	6,528	2,688	10,777.0	11,166.2	3.6	11,810.6
		192	10,841.5	11,587.9	4,074	11,587.9

- The echelon stock reformulation by itself does not improve or tighten the formulation of the model; see *INI-red-ech*.
- The echelon stock reformulation exhibits *LS-U* structure for each item. Using an extended reformulation *WW-U* for all the items is crucial in reducing the gap down to less than 4% in 450 seconds; see *INI-red-ech-WWU*.
- Using a tight reformulation allows one to obtain good solutions quickly. It takes less than 1 minute with formulation *INI-red-ech-WWU* to obtain a solution with a gap of 5.5%, and less than 2.5 minutes for a gap of 3.8% (it takes 450 seconds to prove that the actual gap is 3.8%, after 150 seconds the best lower bound is only 11065.7, and the provable gap is only 4.5%).

14.2.7 Results with Primal Heuristics

Next, to answer the study questions, we consider the effect of certain simple MIP-based heuristics described in Chapter 3, Section 3.6. In particular, we check whether it is possible to

- reduce the duality gap,
- improve the quality of the feasible solutions obtained, or
- reduce the time needed to obtain good solutions

by using these heuristics. The heuristics were applied to our best reformulation *INI-red-ech-WWU*.

The heuristics tried were the following:

i. Default MIP search truncated after 150 seconds of running time;
ii. Relax-and-fix with $R = 2$ iterations, time intervals $Q^1 = [1, 11]$; $U^1 = \emptyset$; $Q^2 = [12, 16]$, and 75 seconds of running time for each subproblem;
iii. Relax-and-fix with $R = 2$ iterations, time intervals $Q^1 = [1, 8]$; $U^1 = [9, 11]$; $Q^2 = [9, 16]$, and 75 seconds of running time for each subproblem;

iv. Relax-and-fix with $R = 2$ iterations, time intervals $Q^1 = [1, 9]$; $U^1 = \emptyset$; $Q^2 = [10, 16]$, and 75 seconds of running time for each subproblem;

v. Relax-and-fix with $R = 2$ iterations, time intervals $Q^1 = [1, 6]$; $U^1 = [7, 9]$; $Q^2 = [7, 16]$, and 75 seconds of running time for each subproblem;

vi. Relax-and-fix with $R = 2$ iterations, time intervals $Q^1 = [1, 6]$; $U^1 = [7, 9]$; $Q^2 = [7, 16]$, and 50 seconds of running time for each subproblem, followed by one iteration of 50 seconds of the RINS heuristic;

vii. Default MIP search truncated after 75 seconds of running time, followed by one iteration of 75 seconds of the RINS heuristic.

The time intervals in the relax-and-fix heuristic define the status of the η_t^f variables; see Section 3.6 for details. When applied, relax-and-fix (R&F) has been applied in two iterations, where

- η_t^f variables with $t \in Q^1 \cup U^1$ were binary at iteration 1,
- η_t^f variables with $t \in Q^1$ were fixed at the end of iteration 1,
- η_t^f variables with $t \in Q^2$ were binary at iteration 2.

The relaxation-induced neighborhood search (RINS) heuristic consists of

- fixing the η_t^f variables that have the same value in the linear relaxation solution of formulation $INI\text{-}red\text{-}ech\text{-}WWU$ and in the current best integer feasible solution (obtained by MIP or by R&F), and
- then solving the restricted MIP problem using the default MIP solver; see Section 3.6 for details.

Table 14.9. Storage Rack Model: Heuristic Results in 150 Seconds

Heuristic	Run T. (secs)	R&F Time Intervals (secs)	BLB	BUB	Gap (%)
MIP	150		11,065.7	11,587.9	4.5
R&F	75, 75	$Q^1 = [1, 11]; U^1 = \emptyset; Q^2 = [12, 16]$	11,019.1	11,535.1	4.5
R&F	75, 75	$Q^1 = [1, 8]; U^1 = [9, 11]; Q^2 = [9, 16]$	11,019.1	11,518.4	4.3
R&F	75, 75	$Q^1 = [1, 9]; U^1 = \emptyset; Q^2 = [10, 16]$	11,011.9	11,589.9	5.0
R&F	75, 75	$Q^1 = [1, 6]; U^1 = [7, 9]; Q^2 = [7, 16]$	11,011.9	11,589.9	5.0
R&F-RINS	50, 50, 50	$Q^1 = [1, 6]; U^1 = [7, 9]; Q^2 = [7, 16]$	10,975.2	11,654.4	5.8
MIP-RINS	75, 75	−	10,979.8	11,656.4	5.8

In Table 14.9, we show the running times and parameters of the various heuristics tested, the value of the best valid lower bound ("BLB") obtained (if any), the best feasible solution ("BUB"), and the gap after a maximum of 150 seconds. These results have not been obtained with LS–LIB, but with a direct Mosel implementation of the heuristics. In all these heuristics, the best valid lower bound is the best lower bound obtained before any heuristic variable fixing has been applied. For instance, in the R&F heuristic, this corresponds to the best lower bound obtained at the end of iteration 1.

In Table 14.9, we compare the default MIP branch-and-bound search with the heuristics, and we observe that the heuristics can typically improve the quality of the best feasible solution found in 150 seconds, but provide weaker lower bounds. The best heuristic solution is obtained here by relax-and-fix where enough variables keep their binary status during iteration 1 (i.e., $Q^1 \cup U^1$ is large), but not all of these variables are fixed at the end of iteration 1 (i.e., $U^1 \neq \emptyset$).

In conclusion, by using MIP-based heuristics and tight reformulations it seems possible to obtain good solutions quickly, better than those obtained from a pure MIP approach, but at the price of weaker lower bounds.

14.3 Insulating Board Extrusion

14.3.1 Problem Description

General Context

This case is an example of planning and scheduling in the process industry, where planning and scheduling cannot be as easily separated as in the case of discrete manufacturing. Here the distinction between planning and scheduling is not solely based on the production cycle length and on the production policy, but should also depend or be based on production process characteristics. For instance, when production runs take many days, the scheduling problem must be defined over a longer horizon. When set-ups are time-consuming, lot-sizing and product processing frequency decisions become crucial in optimizing capacity utilization and have to be taken into account in medium-term planning models. When set-ups are sequence-dependent, which is very often the case in the process industry, sequencing decisions have an impact on the processing capacity. As a consequence, planning and scheduling decisions have to be coordinated or jointly optimized.

We illustrate this required coordination on a real case taken from the literature. In this case, production planning decisions (i.e., the quantity to be produced in the coming month for each product) have already been made based on existing inventory, global available capacity, short-term customer orders, and sales forecasts. The objectives of the short-term (monthly) integrated batching and scheduling problem are the following:

- Define the production batches of each product to be processed during the month on each machine, in such a way that the global production quantities decided at the higher planning level are met for each product;
- Assign the batches to the different machines, taking into account the capability and relative speed of machines, as well as their available capacity;
- Sequence the batches on the machines in order to optimize capacity utilization.

We start our analysis by describing the initial MIP formulation proposed in the original publication. Then we reformulate the model by using the standard formulations proposed in Chapter 12 to represent a sequence of set-ups with sequence-dependent changeover times, and compare the numerical results obtained with these different reformulations. Finally, we analyze the multi-period extension, as suggested in the problem description.

Description of the Short-Term Batching and Scheduling Problem

- A plant produces insulating boards by extrusion. Several ingredients and additives are mixed in an extruding machine; this mix is conducted through the machine by screws and heated by friction. The resulting paste expands at the end of the extruder through a die. The die controls the final width and thickness, and is thus specific to each product. The production quantity is measured in $[m^3]$.

- There are several extruders working in parallel (four in our test instances). The production rates ($[m^3/h]$) are known for each feasible assignment of a product to a machine. A given machine is only able to process a subset of the products. There are sequence-dependent changeover times between the products corresponding to die removal, die installation, and process stabilization times. These production rates and changeover times are machine dependent.

- A die has a limited lifetime, and if the extrusion time of a given product exceeds this lifetime then the die has to be replaced. This corresponds to a changeover time of a product to itself. A batch is defined as the processing of a product on a machine without any interruption. The duration of a batch is therefore limited by the die-lifetime.

- There is at most one *campaign* (i.e., one sequence of consecutive batches of a same product) of each product per machine per month. This restriction is imposed to simplify the plant organization. There is some flexibility in the duration of a campaign. The end of a campaign can be delayed by at most 24 hours. This means that the length of the campaign, and the total lifetime of all dies used in the campaign, can be increased by 24 hours.

- The status of each machine at the beginning of the scheduling horizon is known and described by the product being processed and by the age of the die, that is, the duration in the past of the current batch.

- The products are grouped into families of products (seven in our tests). Each group corresponds to a specific die, and the changeover time between products of the same group, as well as the differences in production rates among products of a group, are neglected. A global quantity to be produced during the scheduling horizon is given for each product group (in $[m^3]$). This quantity is typically the output of the master planning system. For the rest of this case a *product* refers to a product family.

- The capacity available on each machine over the horizon is also given (in [hours]). It is machine-dependent because of planned shifts, maintenance, holidays, and so on.
- The objective of the integrated batching and scheduling problem is to be able to produce the planned quantities and to minimize the total machine utilization time required. If the machines are under-utilized in the optimal solution, then the master planning system will be allowed to increase the production requirements and a new iteration will be performed. If there is no feasible solution, then the master planning system will be asked to reduce the production requirements.

Study Questions

The first objective in the development of this model is the integration or common optimization of the batching and scheduling decisions over a single-period horizon of one month. A secondary or future objective would be to go one step further by integrating also the medium-term planning decisions over a horizon of three to four months. These objectives suggest the study of the following questions.

- Does the integrated approach for batching and scheduling provide globally better solutions than the traditional procedure? The current procedure first creates batches of products for the different machines, with only a rough estimate of capacity available for extrusion (changeover times and sequencing decisions are totally ignored at this stage). This estimate is equal to total capacity available minus estimated changeover times. Then the sequence is optimized for each machine individually in order to minimize real or effective capacity utilization. Finally, several iterations of successive batching and scheduling are performed, and the capacity available for extrusion in a batching iteration is updated to reflect the solution of the previous scheduling iteration. This process is repeated as long as global (i.e., over all machines) capacity utilization is decreased.
- Is it possible to formulate and solve the multi-period master planning and scheduling problem that integrates the production quantity decisions over several months with the batching and scheduling decisions within each month? Such an integration allows one to plan the whole plant in a single step, instead of using several iterations of separated planning and batching/scheduling steps. Here again, several iterations are needed because exact capacity utilization depends on the batches and on the sequence of products, and is not known initially when solving the planning subproblem.
- Does this integrated planning and batching/scheduling approach give significantly better solutions than the traditional iterative approach?
- The objective function used for the batching and scheduling has always been machine capacity utilization. This is because the main objective of

batching and scheduling was to offer the largest possible capacity to the planning module. When these planning, batching, and scheduling decisions are integrated into a single module, are there other objective functions to use than global capacity utilization over all machines and over the whole planning horizon?

The feasibility of the multi-period extension is treated later in this section. The problems/exercises posed in Section 14.6.5 provide a way to answer the other questions.

14.3.2 Classification

This problem is a single-level, single-period, multi-item production planning problem classified as follows.

Multi-Item

This problem is mainly a sequencing problem, with no limit on the number of items produced per period, and with sequence-dependent changeover times.

- For each machine k, the production mode classification is $PM = [M_\infty\text{-}SQ]$, where sequence-dependent changeover variables are required to model the capacity constraints and the objective function.
- There is an additional mode-type restriction that there exists at most one campaign of each product on each machine. Our basic formulations for $M_\infty\text{-}SQ$ proposed in Section 12.3 need to be adapted to take this into consideration.
- For each machine k, the production quantity classification for capacity constraints is $PQ = [PC\text{-}SQ]$ in order to model the capacity or time consumed during the changeover operations.

Single-Item

The single-item classification is degenerate in this case because there is a single time period.

- For each product, there is a global net requirement or demand to be satisfied from production on different machines. The initial inventory has already been deducted from the gross requirements.
- All the machines are supposed to work in parallel, and the machines used to satisfy the demand, as well as the lot sizes on these machines, have to be decided upon.
- For each product on each machine, a number of batches can be produced, and the capacity or maximum size of a single batch is limited by the die life time.
- Therefore, the single-item subproblem exhibits all the characteristics of $LS\text{-}C$, but with multiple machines, and any number of batches per machine.

14.3.3 Initial Formulation

We start with the description of the formulation proposed in the literature.

Objects and Indices	Mathematical Notation
Product (families)	Object: set of products I : $\|I\| = NI = 7$ Index: $i \in I = \{1, \ldots, NI\}$
Extruders (machines)	Object: set of machines K : $\|K\| = NK = 4$ Index: $k \in K = \{1, \ldots, NK\}$

Remarks:

- The products considered here are the families of products, because there is no need to distinguish the individual products within a family. We also denote by I_0 the set of products $I \cup \{0\}$, where the additional dummy product 0 is used to represent the idle state of a machine.

Data	Mathematical Notation
Feasible assignments: Products made on each machine Machines used for each product	 $\forall k \in K$ [-]: $I(k) \subseteq I$ $\forall i \in I$ [-]: $K(i) \subseteq K$
Demand or required production	$\forall i \in I$ $[m^3]$: D^i
Production rates	$\forall i \in I, k \in K(i)$ $[m^3/h]$: α_k^i
Changeover times	$\forall i, j \in I(k) \cup \{0\}$ $[h]$: γ_k^{ij}
Production capacity	$\forall k \in K$ $[h]$: L_k
Die lifetime	$\forall i \in I, k \in K(i)$ $[h]$: β_k^i
Extra die lifetime per campaign	$[h]$: ν
Starting product	$\forall k \in K$ [-]: $P_k \in I(k) \cup \{0\}$
Starting age of die	$\forall k \in K$ $[h]$: $A_k \in [0, \beta_k^{P_k}]$

Remarks:

- For each machine k, the die lifetime is the maximum duration of a batch.
- The extra die lifetime ν (24 hours in our tests) is the additional duration of a campaign allowed in excess of the normal die lifetime.
- The starting age of die A_k is the past duration of the batch of product P_k which is in process at the beginning of the planning horizon. Note that when machine k is initially idle, we take $P_k = 0$ and $A_k = 0$.

Variables	Mathematical Notation
Production or extrusion times	$\forall i \in I, k \in K(i)$ $[h]$: $x_k^i \in \mathbb{R}_+$
Campaign changeovers	$\forall k \in K, i, j \in I(k) \cup \{0\},$ $\quad i \neq j$ [-]: $\chi_k^{ij} \in \{0, 1\}$
Batch changeovers in a campaign	$\forall k \in K, i \in I(k)$ [-]: $z_k^i \in \mathbb{Z}_+$
Campaign sequence number	$\forall i \in I, k \in K(i)$ $[h]$: $u_k^i \in \mathbb{R}_+$

Remarks:

- The variables in this model are classical production and changeover variables, plus some additional variables to take into account the restrictions on campaigns and on die lifetimes.
- We define two types of die changes or changeovers. The campaign changeover variable χ_k^{ij} represents a transition from a campaign of product i to a campaign of a different product j. The batch changeover variable z_k^i represents the number of die changes within the campaign of product i (i.e., a changeover from product i to itself). Therefore, if there is a campaign of product i on machine k, there are $z_k^i + 1$ batches in this campaign.
- The continuous variable u_k^i is used to model the elimination of subtours, and represents the sequence number of the campaign of product i on machine k, if such a campaign exists.

Constraints	Mathematical Notation
$PM = [M_\infty\text{-}SQ]$	
Single predecessor	$\forall\ k \subset K,\ i \in I(k) \cup \{0\}$ [-]: $pred_k^i$
Single successor	$\forall\ k \in K,\ i \in I(k) \cup \{0\}$ [-]: $succ_k^i$
Subtour elimination	$\forall\ k \in K,\ i,j \in I(k),\ i \neq j$ [-]: $subtour_k^{ij}$
Continuation starting campaign	$\forall\ k \in K$ [-]: $start_k$
$PQ = [PC\text{-}SQ]$	
Machine capacity	$\forall\ k \in K$ [h]: $capa_k$
$LS\text{-}C$	
Demand satisfaction	$\forall\ i \in I$ [m^3]: dem_sat^i
Campaign capacity	$\forall\ k \in K,\ i \in I(k)$ [m^3]: $vub\_pred_k^i$ and $vub\_succ_k^i$
Batch capacity	$\forall\ k \in K,\ i \in I(k)$ [m^3]: $vub\_batch_k^i$

Remarks:

- We have listed the set of constraints that can be identified from the problem description. Although this formulation has been derived independently of the classification, to facilitate the analysis, we associate each constraint with the structures identified in the classification. Then we will formulate each constraint separately.

Objective function	Mathematical Notation
Minimize total capacity utilization	[h]: $used\_capa$

The Complete Formulation

The complete initial formulation of the model, called INI, is the following.

$$(INI) \ used\_capa \quad := \min \sum_{k \in K} \Big[\sum_{i \in I(k)} (x_k^i + \gamma_k^{ii} z_k^i) +$$

$$\sum_{i,j \in I(k) \cup \{0\}: j \neq i} \gamma_k^{ij} \chi_k^{ij} - \gamma_k^{0,P_k} \Big] \qquad (14.29)$$

$$pred_k^i \quad := \sum_{j \in I(k) \cup \{0\}} \chi_k^{ji} \leq 1 \ \text{ for } k \in K, \ i \in I(k) \cup \{0\} \quad (14.30)$$

$$succ_k^i \quad := \sum_{j \in I(k) \cup \{0\}} \chi_k^{ij} \leq 1 \ \text{ for } k \in K, \ i \in I(k) \cup \{0\} \quad (14.31)$$

$$subtour_k^{ij} \quad := u_k^j \geq u_k^i - |I(k)| + (|I(k)| + 1)\chi_k^{ij}$$
$$\text{for } k \in K, \ i,j \in I(k) : i \neq j \qquad (14.32)$$

$$start_k \quad := 1 \ = \ \chi_k^{0,P_k} \quad \text{ for } k \in K \qquad (14.33)$$

$$capa_k \quad := \sum_{i \in I(k)} (x_k^i + \gamma_k^{ii} z_k^i) + \sum_{i,j \in I(k) \cup \{0\}: j \neq i} \gamma_k^{ij} \chi_k^{ij} -$$
$$\gamma_k^{0,P_k} \ \leq \ L_k \quad \text{ for } k \in K \qquad (14.34)$$

$$dem\_sat^i \quad := \sum_{k \in K(i)} \alpha_k^i x_k^i \geq D^i \quad \text{ for } i \in I \qquad (14.35)$$

$$vub\_pred_k^i \quad := x_k^i \leq L_k \sum_{j \in I(k) \cup \{0\}} \chi_k^{ji} \quad \text{ for } k \in K, \ i \in I(k) \quad (14.36)$$

$$vub\_succ_k^i \quad := x_k^i \leq L_k \sum_{j \in I(k) \cup \{0\}} \chi_k^{ij} \quad \text{ for } k \in K, \ i \in I(k) \quad (14.37)$$

$$vub\_batch_k^i \quad := x_k^i + A_k \leq \beta_k^i z_k^i + \beta_k^i + \nu$$
$$\text{for } k \in K, \ i \in I(k) : i = P_k \qquad (14.38)$$

$$vub\_batch_k^i \quad := x_k^i \leq \beta_k^i z_k^i + \beta_k^i + \nu$$
$$\text{for } k \in K, \ i \in I(k) : i \neq P_k, \qquad (14.39)$$

where constraints (14.30) and (14.31) express the condition that each product has at most one predecessor and one successor in the campaign sequence relative to each machine. Constraint (14.32) eliminates subtours, which are defined as campaign sequences leaving a product i and coming back to i after some transitions. Such subtours form infeasible campaign sequences and are eliminated by associating a label or sequence number u_k^i with each product i on each machine k, and by imposing $u_k^j \geq u_k^i + 1$ when there is transition from i to j. Constraint (14.33) treats the initial transition from 0 to P_k. Constraint (14.34) is the usual single-machine capacity constraint with changeover times, taking all transition types into account. The objective (14.29) is the total capacity utilization over all machines. Constraint (14.35) models demand satisfaction. Constraints (14.36) and (14.37) ensure that the production of product i on machine k is zero when there is no campaign of i on k. Finally, constraints (14.38) and (14.39) ensure that there are enough batches ($z_k^i + 1$

of them) in a campaign of product i on machine k, in the case when i is the starting product on machine k and when it is not.

14.3.4 Improving the Initial Formulation

We now make several observations based on the classification in order to improve the initial formulation INI.

- For each machine k, we may use the vehicle routing formulation (12.10)–(12.14) of $PM = [M_\infty\text{-}SQ]$ to get a tight formulation for the sequence of campaigns.
- In this formulation, we restrict variables y_k^i, the number of set-ups of product i on machine k, to be binary because there is at most one campaign of each product on each machine.
- In order to improve the formulation of the production quantity structure $PQ = [PC\text{-}SQ]$, we can, as suggested in Section 12.4, build single-node flow relaxations of our model, and add flow cover or reverse flow cover valid inequalities; see Section 8.9. As these flow cover inequalities, as well as other mixed integer rounding inequalities based on the same relaxations, are now generated automatically in the standard MIP solvers, the easiest and fastest way to test this idea is to provide these (or some) single-node flow relaxations as part of the initial formulation.
- These relaxations are defined by flow constraints – capacity and demand satisfaction constraints – and variable upper bound constraints. To obtain pure variable upper bound constraints in the formulation:
  - We decompose the extrusion time in a campaign into normal extrusion time (limited by the die lifetime) and additional extrusion time at the end of a campaign (limited to 24 hours).
  - We artificially fix the start of the planning horizon on each machine at the past date at which the last batch was started, which requires us to update the product demands and the production capacities (see below).
- Finally, in order to keep a compact formulation and because the number of items is small, we retain the weak initial formulation of the subtour elimination constraints. An alternative to be tested here would be to add the subtour constraints (12.16) with $q = 1$. This would allow us to avoid the creation of the u_k^i variables, but at the potential expense of many subtour inequalities.

VRP Variables

We redefine the variables as

- $x_k^i \in \mathbb{R}_+$ denotes the *normal* extrusion time of the campaign of product i on machine k, for $k \in K$ and $i \in I(k)$ (i.e., without the additional lifetime).

- $v_k^i \in \mathbb{R}_+$ denotes the *additional* extrusion time of the campaign of product i on machine k, for $k \in K$ and $i \in I(k)$.
- $y_k^i \in \{0,1\}$ takes the value 1 if there exists a campaign of product i on machine k, and 0 otherwise, for $k \in K$ and $i \in I(k)$.
- $z_k^i \in \mathbb{Z}_+$ denotes the *total* number of batches in the campaign of product i on machine k, for $k \in K$ and $i \in I(k)$.
- $\chi_k^{ij} \in \{0,1\}$ takes the value 1 when there is a changeover from a campaign of product i to a campaign of product j on machine k, and 0 otherwise, for $k \in K$ and $i,j \in I(k) \cup \{0\}$, $i \neq j$.
- $u_k^i \in \mathbb{R}_+$ is used to model the subtour elimination restrictions, for $k \in K$ and $i \in I(k)$.

VRP Formulation

The above observations lead to the following alternative formulation, called VRP.

(VRP)

$$used\_capa \quad := \min \sum_{k \in K} \Big[\sum_{i \in I(k)} (x_k^i + v_k^i + \gamma_k^{ii} z_k^i) - A_k - \gamma_k^{P_k,P_k} +$$

$$\sum_{i,j \in I(k):j \neq i} (\gamma_k^{ij} - \gamma_k^{jj})\chi_k^{ij} +$$

$$\sum_{j \in I(k)} (\gamma_k^{0j} - \gamma_k^{jj})\chi_k^{0j}\, \delta^{0,P_k} \Big] \tag{14.40}$$

$$pred_k^0 \quad := \sum_{j \in I(k) \cup \{0\}} \chi_k^{j0} = 1 \quad \text{for } k \in K \tag{14.41}$$

$$pred_k^i \quad := \sum_{j \in I(k) \cup \{0\}} \chi_k^{ji} = y_k^i \quad \text{for } k \in K,\ i \in I(k) \tag{14.42}$$

$$succ_k^0 \quad := \sum_{j \in I(k) \cup \{0\}} \chi_k^{0j} = 1 \quad \text{for } k \in K \tag{14.43}$$

$$succ_k^i \quad := \sum_{j \in I(k) \cup \{0\}} \chi_k^{ij} = y_k^i \quad \text{for } k \in K,\ i \in I(k) \tag{14.44}$$

$$subtour_k^{ij} \quad := u_k^j \geq u_k^i - |I(k)| + (|I(k)| + 1)\chi_k^{ij}$$
$$\text{for } k \in K,\ i,j \in I(k) : i \neq j \tag{14.45}$$

$$start_k \quad := \chi_k^{0,P_k} = 1 \quad \text{for } k \in K \text{ with } P_k \neq 0 \tag{14.46}$$

$$capa_k \quad := \sum_{i \in I(k)} (x_k^i + v_k^i + \gamma_k^{ii} z_k^i) +$$

$$\sum_{i,j \in I(k): j \neq i} (\gamma_k^{ij} - \gamma_k^{jj}) \chi_k^{ij} +$$

$$\sum_{j \in I(k)} (\gamma_k^{0j} - \gamma_k^{jj}) \chi_k^{0j} \, \delta^{0,P_k} \leq$$

$$L_k + \gamma_k^{P_k, P_k} + A_k \quad \text{for } k \in K \qquad (14.47)$$

$$dem\_sat^i \quad := \sum_{k \in K(i)} \alpha_k^i (x_k^i + v_k^i) \geq$$

$$D^i + \sum_{k \in K: P_k = i} \alpha_k^i A_k \quad \text{for } i \in I \qquad (14.48)$$

$$past_k \quad := x_k^i \geq A_k \quad \text{for } k \in K, \; i \subset I(k) : i = P_k \qquad (14.49)$$

$$vub\_x_k^i \quad := x_k^i \leq \beta_k^i z_k^i \quad \text{for } k \in K, \; i \in I(k) \qquad (14.50)$$

$$vub\_v_k^i \quad := v_k^i \leq \nu y_k^i \quad \text{for } k \in K, \; i \in I(k) \qquad (14.51)$$

$$vlb\_z_k^i \quad := z_k^i \geq y_k^i \quad \text{for } k \in K, \; i \in I(k) \qquad (14.52)$$

$$vub\_z_k^i \quad := z_k^i \leq \lceil \frac{D^i}{\alpha_k^i \beta_k^i} \rceil y_k^i \quad \text{for } k \in K, \; i \in I(k) \qquad (14.53)$$

$$flow1_k^i \quad := \sum_{k \in K(i)} \alpha_k^i \beta_k^i z_k^i + \sum_{k in K(i)} \alpha_k^i v_k^i \geq$$

$$D^i + \sum_{k \in K: P_k = i} \alpha_k^i A_k \quad \text{for } i \in I \qquad (14.54)$$

$$flow2_k^i \quad := \sum_{k \in K(i)} \alpha_k^i \beta_k^i z_k^i + \sum_{k in K(i)} \alpha_k^i \nu \, y_k^i \geq$$

$$D^i + \sum_{k \in K: P_k = i} \alpha_k^i A_k \quad \text{for } i \in I, \qquad (14.55)$$

where constraints (14.41)–(14.44) model the sequence of campaigns, and constraint (14.46) fixes the starting product, on each machine. Note that all campaign sequences start from the dummy product, and end with the dummy product. In the capacity constraint (14.47), each batch of product i on machine k requires a batch start-up time γ_k^{ii} to change the die, and an additional changeover time of $\gamma_k^{ji} - \gamma_k^{ii}$ when the batch start-up coincides with the beginning of a new campaign of product i, coming from product j. The only case where the transition from the idle state 0 consumes capacity is when the starting product is the idle product, with the notation $\delta^{0,P_k} = 1$ when $P_k = 0$ and 0 otherwise. Because of the redefinition of the beginning of the planning horizon, the capacity in the rhs of (14.47) is increased by the past production time A_k and the past changeover time $\gamma_k^{P_k, P_k}$. The objective function (14.40) is again the total machine utilization time, without past utilization. Similarly, in the demand satisfaction constraint (14.48), the required production of product i is increased by its past production $\sum_{k \in K : P_k = i} \alpha_k^i A_k$. This past production level is imposed as a constraint in (14.49). The modification of the planning horizon allows us to write the variable lower and upper bound constraints simply as (14.50)–(14.53). Finally, the constraints (14.47)–(14.48),(14.54), and (14.55) define single-node flow and integer continuous knapsack relaxations, together with the variable upper bounds (14.50)–(14.51). The latter two relaxations are obtained from (14.48) by replacing the extrusion times using the variable upper bound constraints.

UF Variables

Finally, an alternative formulation to VRP is to formulate explicitly the sequence of campaigns on each machine by adapting the unit flow formulation (12.1)–(12.4) from Chapter 12.

We keep the variables x_k^i, y_k^i, z_k^i, v_k^i as in formulation VRP, and introduce the following new variables to represent the sequence of campaigns.

- $u_{kc}^i \in \{0,1\}$ takes the value 1 if the cth campaign on machine k is of product i, and 0 otherwise, for $k \in K$, and either $i \in I(k), c \in \{1,\ldots,|I_0|\}$, or $P_k = i = 0, c = 1$.
- $\chi_{kc}^{ij} \in \{0,1\}$ takes the value 1 when there is a changeover on machine k from the $(c-1)$th campaign which is of product i to the cth campaign which is of product j, and 0 otherwise, for $k \in K$, and either $i,j \in I(k)$, $i \neq j, c \in \{2,\ldots,|I_0|\}$, or $P_k = i = 0, j \in I(k), c = 2$.

Note that the only case in which we need to use the dummy product 0 is when it is the starting product or campaign on a machine.

UF Formulation

This leads to the following alternative formulation called UF.

(UF)

$$used\_capa \quad := \min \sum_{k \in K} \Big[\sum_{i \in I(k)} (x_k^i + v_k^i + \gamma_k^{ii} z_k^i) - A_k - \gamma_k^{P_k,P_k} +$$

$$\sum_{i,j \in I(k) \cup \{0\} : j \neq i} \sum_{c \geq 2} (\gamma_k^{ij} - \gamma_k^{jj}) \chi_{kc}^{ij} \Big] \tag{14.56}$$

$$pred_{k1}^i \quad := u_{k1}^{P_k} = 1 \quad \text{for } k \in K \tag{14.57}$$

$$pred_{k1}^i \quad := u_{k1}^i = 0 \quad \text{for } k \in K, i \in I(k) : i \neq P_k \tag{14.58}$$

$$pred_{kc}^i \quad := u_{kc}^i = \sum_{j \in I(k) \cup \{0\} : j \neq i} \chi_{kc}^{ji}$$

$$\text{for } k \in K, i \in I(k), c \geq 2 \tag{14.59}$$

$$succ_{kc}^i \quad := u_{k,c-1}^i \geq \sum_{j \in I(k) \cup \{0\}: j \neq i} \chi_{kc}^{ij}$$

$$\text{for } k \in K, i \in I(k), c \geq 2 \tag{14.60}$$

$$campaign_k^i \quad := y_k^i = \sum_{c \geq 1} u_{kc}^i \quad \text{for } k \in K, i \in I(k) \tag{14.61}$$

$$capa_k \quad := \sum_{i \in I(k)} (x_k^i + v_k^i + \gamma_k^{ii} z_k^i) +$$

$$\sum_{i,j \in I(k) \cup \{0\}: j \neq i} \sum_{c \geq 2} (\gamma_k^{ij} - \gamma_k^{jj}) \chi_{kc}^{ij} \leq$$

$$L_k + \gamma_k^{P_k, P_k} + A_k \quad \text{for } k \in K \tag{14.62}$$

$$(x, y, z, v) \text{ satisfy } (14.48)\text{--}(14.55), \tag{14.63}$$

where constraints (14.57)–(14.58) define the first campaign (entering flow), constraint (14.59) defines the cth campaign as the result of a transition from the previous $(c-1)$th campaign, whereas constraint (14.60) allows a transition from the $(c-1)$th campaign to the next cth campaign. Observe that the inequality $\geq$ constraint permits termination of the sequence after any campaign. Constraint (14.61) defines whether there exists a campaign of product i on machine k ($y_k^i = 1$), or not ($y_k^i = 0$). Finally, the capacity constraint (14.62) and the objective function (14.56) are similar to those in the VRP formulation.

14.3.5 Results with Reformulations

We have solved a test instance of the problem involving seven products and four machines, for which we ran each of the different formulations in default MIP mode with Xpress-MP. All three formulations INI, VRP, and UF solved this single-period batching and scheduling problem to optimality. The results are shown in Table 14.10. Columns "Cons," "Vars," and "Int" contain, respectively, the number of constraints, variables, and integer or binary variables. "LP" is the value of the first linear relaxation; "XLP" is the value of the first node relaxation after addition of the Xpress-MP cuts. Column "OPT" gives the common value of the optimal (integer) solution. Columns "Nodes" and "Time" give the total number of nodes and the running time to prove optimality.

The results show the effectiveness of the proposed VRP and UF reformulations in solving this problem. This is due to a much better quality lower bound. Although the linear relaxation of UF gives a slightly better bound, it takes more nodes than the VRP formulation to solve the problem to optimality.

Table 14.10. Insulating Boards: Reformulation Results

Formulation	Cons	Vars	Int	LP	XLP	OPT	Nodes	Time
INI	206	186	146	2,003.44	2,054.33	2,077.09	36,553	67
VRP	257	222	162	2,046.26	2,066.48	2,077.09	611	1
UF	447	814	774	2,046.55	2,066.21	2,077.09	729	2

14.3.6 The Multi-Period Planning and Scheduling Extension

We now consider the multi-period master planning, batching, and scheduling problem that optimizes all such decisions over a horizon of several months. We propose and test a formulation based on the single-period reformulations studied before. Given the results in Table 14.10, we use formulation VRP as a starting point. A similar model can easily be built from formulation UF.

As the extension is straightforward, we only comment on the modifications with respect to formulation VRP. Except for the demands, all sets and data remain the same as in the single-period model. An additional index t is used for time periods $t \in T = \{1, \dots, NT\}$. The demand for item i in time period t is denoted D_t^i, and the capacity of machine k in period t is denoted L_{kt}.

$MVRP$ Variables

The following VRP variables are retained and have the same interpretation. The only difference is that all these decisions have to be taken in each time period: $y_{kt}^i \in \{0,1\}$, $\chi_{kt}^{ij} \in \{0,1\}$, $u_{kt}^i \in \mathbb{R}_+$.

The other variables, defined for $t \in T$, $k \in K$, and $i \in I(k)$, are new or have a slightly different interpretation.

- $x_{kt}^i \in \mathbb{R}_+$ denotes the normal extrusion time of the campaign of product i on machine k in period t, without the additional lifetime at the end of a campaign (variable v) and without the remaining extrusion time performed in period t from the last batch of period $t-1$ (variable r).
- $v_{kt}^i \in \mathbb{R}_+$ denotes the additional extrusion time at the end of the campaign of product i on machine k in period t.
- $r_{kt}^i \in \mathbb{R}_+$ denotes the remaining extrusion time in period t from the last batch of period $t-1$ on machine k, if this last batch is of product i, and 0 otherwise.
- $z_{kt}^i \in \mathbb{Z}_+$ denotes the number of batches of product i started on machine k in period t.
- $s_t^i \in \mathbb{R}_+$ denotes the inventory of product i at the end of period t.

So, the main difference with respect to VRP is the decomposition of the extrusion times into three distinct sources: remaining time r from the previous period, normal time x, and additional time v.

$MVRP$ Formulation

The formulation of the multi-period extension is called $MVRP$.

$(MVRP)$

$$used\_capa \quad := \min \sum_{t \in T} \sum_{k \in K} \rho^t \Big[\sum_{i \in I(k)} (r_{kt}^i + x_{kt}^i + v_{kt}^i + \gamma_k^{ii} z_{kt}^i) +$$

$$\sum_{i,j \in I(k): j \neq i} (\gamma_k^{ij} - \gamma_k^{jj}) \chi_{kt}^{ij} \Big] +$$

$$\sum_{k \in K} \sum_{j \in I(k)} \rho (\gamma_k^{0,j} - \gamma_k^{jj}) \chi_{k1}^{0,j} \, \delta^{0,P_k} \tag{14.64}$$

$$pred_{kt}^0 \quad := \sum_{j \in I(k) \cup \{0\}} \chi_{kt}^{j,0} = 1 \quad \text{for } k \in K, \, t \in T \tag{14.65}$$

$$pred_{kt}^i \quad := \sum_{j \in I(k) \cup \{0\}} \chi_{kt}^{ji} = y_{kt}^i \quad \text{for } k \in K, \, i \in I(k), \, t \in T \tag{14.66}$$

$$succ_{kt}^0 \quad := \sum_{j \in I(k) \cup \{0\}} \chi_{kt}^{0,j} = 1 \quad \text{for } k \in K, \, t \in T \tag{14.67}$$

$$succ_{kt}^i \quad := \sum_{j \in I(k) \cup \{0\}} \chi_{kt}^{ij} = y_{kt}^i \quad \text{for } k \in K, \, i \in I(k), \, t \in T \tag{14.68}$$

$$subtour_{kt}^{ij} \quad := u_{kt}^j \geq u_{kt}^i - |I(k)| + (|I(k)| + 1) \chi_{kt}^{ij}$$

$$\text{for } k \in K, \, i,j \in I(k) : i \neq j, \, t \in T \tag{14.69}$$

$$start_{k,1}^i \quad := \chi_{k1}^{0,i} = \delta^{i,P_k} \quad \text{for } k \in K, \, i \in I(k) \tag{14.70}$$

$$start_{kt}^i \quad := \chi_{kt}^{0,i} = \chi_{k,t-1}^{i,0} \quad \text{for } k \in K, \, i \in I(k), \, 2 \leq t \leq NT \tag{14.71}$$

$$capa_{k,1} \quad := \sum_{i \in I(k)} (r_{k,1}^i + x_{k,1}^i + v_{k,1}^i + \gamma_k^{ii} z_{k,1}^i) +$$

$$\sum_{i,j \in I(k): j \neq i} (\gamma_k^{ij} - \gamma_k^{jj}) \chi_{k,1}^{ij} +$$

$$\sum_{j \in I(k)} (\gamma_k^{0,j} - \gamma_k^{jj}) \chi_{k1}^{0,j} \, \delta^{0,P_k} \leq L_{k,1} \quad \text{for } k \in K \tag{14.72}$$

$$capa_{kt} \quad := \sum_{i \in I(k)} (r_{kt}^i + x_{kt}^i + v_{kt}^i + \gamma_k^{ii} z_{kt}^i) +$$

$$\sum_{i,j \in I(k): j \neq i} (\gamma_k^{ij} - \gamma_k^{jj}) \chi_{kt}^{ij} \leq L_{kt}$$

$$\text{for } k \in K, \, 2 \leq t \leq NT \tag{14.73}$$

$$dem\_sat_t^i \quad := s_{t-1}^i + \sum_{k \in K(i)} \alpha_k^i (r_{kt}^i + x_{kt}^i + v_{kt}^i) =$$

$$D_t^i + s_t^i \quad \text{for } i \in I,\ t \in T \tag{14.74}$$

$$ub\_r_{k,1}^i \quad := r_{k,1}^i \le (\beta_k^i - A_k)\,\delta^{i,P_k} \quad \text{for } k \in K,\ i \in I(k) \tag{14.75}$$

$$vub\_r_{kt}^i \quad := r_{kt}^i \le \beta_k^i \chi_{kt}^{0i} \quad \text{for } k \in K,\ i \in I(k),\ 2 \le t \le NT \tag{14.76}$$

$$vub\_x_{kt}^i \quad := x_{kt}^i + r_{k,t+1}^i \le \beta_k^i z_{kt}^i \quad \text{for } k \in K,\ i \in I(k),\ t \in T \tag{14.77}$$

$$vub\_v_{kt}^i \quad := v_{kt}^i \le \nu\,(y_{kt}^i - \chi_{kt}^{i,0}) \quad \text{for } k \in K,\ i \in I(k),\ t \in T \tag{14.78}$$

$$vub\_z_{kt}^i \quad := z_{kt}^i \le \lceil \frac{L_{kt}}{\beta_k^i + \gamma_k^{ii}} \rceil y_{kt}^i \quad \text{for } k \in K,\ i \in I(k),\ t \in T \tag{14.79}$$

$$vlb\_z_{kt}^i \quad := z_{kt}^i \ge y_{kt}^i - \chi_{kt}^{0,i} \quad \text{for } k \in K,\ i \in I(k),\ t \in T, \tag{14.80}$$

where constraints (14.65)–(14.69) model the sequence of campaigns on each machine in each time period, constraints (14.70)–(14.71) impose the initial conditions and transitions between periods; that is, the ending batch of period $t-1$ is the starting batch of period t. Constraints (14.72)–(14.74) model the capacity restrictions and demand satisfaction, taking into account the decomposition of the extrusion times from three sources. The objective function (14.64) is the sum of machine utilization times, with a discount factor $\rho < 1$ to penalize capacity utilization in earlier periods, or to minimize stocks expressed in terms of production hours.

Constraint (14.75) limits the remaining production from the last batch prior to the planning horizon, and constraint (14.76) prevents any remaining production in period t if the production campaign is not the first in period t, that is, the last in period $t-1$. Constraint (14.77) ensures in each period a number z_{kt}^i of batches that is large enough to cover the normal production x_{kt}^i in period t, plus the remaining production $r_{k,t+1}^i$ postponed till period $t+1$. Constraint (14.78) restricts the extra production to campaigns that are finished in period t. Constraint (14.79) limits the number of batches in a period, and prevents the creation of batches in a period without an active campaign in that period. Finally, constraint (14.80) ensures that a new campaign contains at least one batch.

Improving Formulation $MVRP$

As for VRP, formulation $MVRP$ can be improved by adding low-level relaxations in order to help the MIP solver to generate mixed integer rounding and flow covers cuts. To take into account the integer variables involved in the formulation, we have added the following integer continuous knapsack sets (see Section 8.7) obtained by first summing the demand satisfaction constraints over periods t up to l, $1 \le t \le l \le NT$, and then replacing the continuous variables using their variable upper bounds:

$$s_{t-1}^i + \sum_{k \in K(i)} \alpha_k^i \beta_k^i \chi_{kt}^{0,i} + \sum_{k \in K(i)} \sum_{u=t}^l \alpha_k^i \beta_k^i z_{ku}^i$$

$$+ \sum_{k \in K(i)} \sum_{u=t}^l \alpha_k^i \nu (y_{ku}^i - \chi_{ku}^{i,0}) \geq \sum_{u=t}^l D_u^i \quad \text{for } i \in I, \ 2 \leq t \leq l \leq NT$$

$$(14.81)$$

$$\sum_{k \in K(i)} \alpha_k^i (\beta_k^i - A_k) \, \delta^{i,P_k} + \sum_{k \in K(i)} \sum_{u=1}^l \alpha_k^i \beta_k^i z_{ku}^i$$

$$+ \sum_{k \in K(i)} \sum_{u=1}^l \alpha_k^i \nu (y_{ku}^i - \chi_{ku}^{i,0}) \geq \sum_{u=1}^l D_u^i \quad \text{for } i \in I, \ 1 \leq l \leq NT.$$

$$(14.82)$$

The formulation $MVRP$ augmented by constraints (14.81)–(14.82) is called $MVRP^+$.

Another way to improve formulation $MVRP$ is to add extended formulations for higher-level relaxations of the model. We have identified three such relaxations for the multi-period single-item subproblems.

First, a $LS\text{-}U$ relaxation is defined for each product i. If (s, r, v, x, y, z, χ) is feasible for $MVRP$, then $(S, X, Y) \in \text{conv}(X^{LS-U}(DD))$ with

$$\begin{aligned}
S(t) \quad &:= \quad s_t^i & \text{for all } t, \\
X(t) \quad &:= \quad \sum_{k \in K(i)} \alpha_k^i \left(r_{kt}^i + x_{kt}^i + v_{kt}^i \right) & \text{for all } t, \\
Y(t) \quad &:= \quad \sum_{k \in K(i)} y_{kt}^i & \text{for all } t, \\
DD(t) \quad &:= \quad D_t^i & \text{for all } t,
\end{aligned}$$

where $X^{LS-U}(DD)$ denotes the set of feasible solutions of $LS\text{-}U$ with demand vector DD. This is a valid relaxation because for all (s, r, v, x, y, z, χ) feasible for $MVRP$, we have $S(t-1) + X(t) = DD(t) + S(t)$ and $Y(t) = 0$ implies $X(t) = 0$, for all t.

Next, we have identified two relaxations $WW\text{-}U$ and $WW\text{-}CC$ for each product i.

If (s, r, v, x, y, z, χ) is feasible for $MVRP$, this implies that $(S, Y) \in \text{conv}(X^{WW-U}(DD))$ and $(S, Y) \in \text{conv}(X^{WW-CC}(DD, CC))$ with

$$S(t) \quad := \quad s_t^i + \sum_{k \in K(i)} \alpha_k^i \left(r_{k,t+1}^i + \nu \chi_{kt}^{i,0} \right) \qquad \text{for all } t,$$

$$Y(t) \quad := \quad \sum_{k \in K(i)} z_{kt}^i \qquad \qquad \qquad \text{for all } t,$$

$$DD(t) \quad := \quad D_t^i \qquad \qquad \qquad \qquad \text{for all } t,$$

$$CC \quad := \quad \max_{k \in K(i)} \alpha_k^i \left(\beta_k^i + \nu \right),$$

where $X^{WW-U}(DD)$ denotes the set of feasible solutions of WW-U with demand vector DD, and $X^{WW-CC}(DD, CC)$ denotes the set of feasible solutions of WW-CC with demand vector DD, and constant capacity CC. These are valid relaxations because $Y(t)$ is the number of new batches started in period t, CC is an upper bound on the capacity of a single new batch (the term ν is needed because campaigns of a single batch are allowed, and therefore a single batch can give as much capacity as $\alpha_k^i (\beta_k^i + \nu)$), and $S(t)$ is an upper bound on the demand after period t satisfied from batches started before or in period t. In other words, these are valid relaxations because $S(t-1) + CC \sum_{u=t}^{l} Y(u) \geq \sum_{u=t}^{l} DD(u)$, for $1 \leq t \leq l \leq NT$. Observe that we would obtain a tighter relaxation by taking into account the varying capacities among the machines, but no appropriate (approximate) extended formulation is known.

These reformulations can be easily implemented using the LS–LIB library, either using the extended reformulations from the $XForm$ procedures, or the cutting planes from the $Xcut$ procedures. Formulation $MVRP$ augmented with the three extended formulations is denoted by $MVRP$-F, whereas formulation $MVRP$ tightened by the three cutting plane procedures is denoted by $MVRP$-C.

Results for the Multi-Period Extension

Here we compare different versions of formulation $MVRP$, namely

$$MVRP, \ MVRP^+, \ MVRP{-}F, \ MVRP^+{-}C, \ MVRP^+{-}F \ ,$$

for problem instances with $NT = 3$ and 4 time periods, with $NI = 7$ products, $NK = 4$ machines, and with a discounting factor $\rho = 0.98$. In our implementation of $MVRP$-C, we generate up to 20 rounds of cuts at the root node, and in the branch-and-bound tree with a cut frequency of 3, and a maximum depth of 30.

The results obtained with a maximum run-time of 600 seconds for each problem instance are given in Table 14.11. Column "LP" gives the value of the linear relaxation, before any cut. Column "XLP" reports on the objective value at the root node, after the addition of Xpress-MP and LS–LIB cuts. Columns "BUB" and "BLB" contain the best solution and the best lower bound at the end of the computing time (optimum proved or time limit reached). Column "Gap" gives the final duality gap at the end of the computation. Columns "Nodes" and "Time" contain the final information at the end of the computation. A * indicates that the time limit has been reached and the optimization stopped.

Table 14.11. Insulating Boards: Multi-Period Results

Formulation-NT	Cons	Vars	Int	LP	XLP	BUB	BLB	Nodes	Time (secs)	Gap (%)
$MVRP$-3	851	747	486	4,886	4,899	4,922	4,922	53,314	358	0
$MVRP^+$-3	899	749	486	4,886	4,900	4,922	4,922	29,859	241	0
$MVRP^+$-C-3	899	749	486	4,886	4,902	4,922	4,922	24,054	294	0
$MVRP$-F-3	1,103	894	486	4,895	4,904	4,922	4,922	18,372	213	0
$MVRP^+$-F-3	1,151	896	486	4,895	4,903	4,922	4,922	15,271	185	0
$MVRP$-4	1,144	996	648	7,035	7,048	7,091	7,061	51,286*	600*	0.42
$MVRP^+$-4	1,224	999	648	7,035	7,049	7,094	7,063	43,472*	600*	0.44
$MVRP^+$-C-4	1,224	999	648	7,035	7,053	7,091	7,063	28,467*	600*	0.39
$MVRP$-F-4	1,536	1,234	648	7,043	7,052	7,086	7,065	30,859*	600*	0.30
$MVRP^+$-F-4	1,616	1,237	648	7,043	7,052	7,088	7,066	26,872*	600*	0.31

The instance with three time periods can be solved easily by all formulations, but the instance with four time periods cannot be solved to optimality in less than ten minutes. However, in all cases the VRP-based formulations give very good quality solutions very quickly. In less than one minute, these formulations suggest solutions where capacity utilization is less than 1% away from optimality.

There is little difference between the various formulations, but it seems that formulation $MVRP$-F (or $MVRP^+$-F) outperforms the others. The performance of $MVRP$-C is a little worse although, in theory, the bounds it provides are as good as those obtained with $MVRP$-F. This is partly due to the fact that, in the current version of Mosel/Xpress-MP, the preprocessing has to be turned off when LS–LIB cuts are generated (bounds are not strengthened, redundant variables and constraints are not removed). Nevertheless, observe that more nodes are inspected in 600 seconds in $MVRP^+$-C-4 than in $MVRP^+$-F-4, which means that the cut generation and cut management times are compensated by the reduced optimization time for a smaller-size formulation.

These results are very preliminary, and this problem deserves further analysis. In particular, new formulations and specific valid inequalities for the structure $PQ = [PC\text{-}SQ]$ (a way to achieve this is described in Section 12.4) seem necessary, as well as some better formulations for the single-item LS-C and WW-C structures identified above with varying capacities and several machines.

14.4 Pigment Sequencing

This is a multi-item lot-sizing problem with capacity constant over time and production at full capacity. There are storage costs and sequence-dependent changeover costs , and at most one item is produced per period. Backlogging is not allowed.

Below we first discuss this basic problem and how to improve its formulation. In a second stage we consider how to formulate correctly several natural ways of assigning costs when there are idle or "no production" periods.

14.4.1 Initial Formulation

The basic problem we start from, with m items, n time periods, and no initial stock, is defined by the formulation

$$\min \sum_{i=1}^{m}\sum_{t=1}^{n} h^i s_t^i + \sum_{i,j=1}^{m}\sum_{t=1}^{n} c^{ij} \chi_t^{i,j} \tag{14.83}$$

$$s_{t-1}^i + x_t^i = d_t^i + s_t^i \qquad \text{for all } i,t \tag{14.84}$$

$$s_0^i = 0 \qquad \text{for all } i \tag{14.85}$$

$$x_t^i \le y_t^i \qquad \text{for all } i,t \tag{14.86}$$

$$\sum_{i=1}^{m} y_t^i = 1 \qquad \text{for all } t \tag{14.87}$$

$$\chi_t^{ij} \ge y_{t-1}^i + y_t^j - 1 \qquad \text{for all } i,j,t \tag{14.88}$$

$$x,y \subset \{0,1\}^{mn}, \; s \in \mathbb{R}_+^{mn}, \; \chi \in \mathbb{R}_+^{m^2 n}. \tag{14.89}$$

where constraint (14.84) is the classical flow balance constraint, with zero initial inventory by (14.85). Constraint (14.86) is the set-up forcing constraint, where we suppose without loss of generality that the demands have been normalized so that $d_t^i \in \{0,1\}$; see Section 10.5.1. Constraint (14.87) is the production mode restriction, and (14.88) defines the changeover variables. The objective (14.83) is the sum of inventory and changeover costs. We denote this basic problem *PIG-A*, and *PIG-A*-1 is the specific formulation (14.83)–(14.89).

14.4.2 Classification

This model has the following classification:

- Single machine;
- Single level;
- Multi-item with production mode $PM = [M_1\text{-}SQ]$, and production quantity $PQ = [PC\text{-}U]$ (i.e., there is no joint production quantity restriction);

- Each single item has classification *DLS-CC-SC*.

Note that the sequence-dependent changeover costs necessarily call for information about start-ups and switch-offs of single items, which explains the *SC* in the single-item classification.

14.4.3 Reformulations

Reformulation of the Changeover Variables

The weak formulation (14.88) of the changeover variables can be replaced by the unit flow reformulation presented in Section 12.2.2. This leads to the constraints

$$\sum_i \chi_t^{ij} = y_t^j \qquad \text{for all } j,t$$

$$\sum_j \chi_t^{ij} = y_{t-1}^i \qquad \text{for all } i,t$$

$$\sum_i y_0^i = 1$$

$$y \in \{0,1\}^{mn}, \ \chi \in \mathbb{R}_+^{m^2 n},$$

representing the flow of a single unit passing from item set-up to item set-up over time, plus the constraints linking the changeover variables to the start-up and switch-off variables

$$y_t^j - z_t^j = y_{t-1}^j - w_{t-1}^j = \chi_t^{jj} \quad \text{for all } j,t.$$

The formulation in which the constraints (14.88) are replaced in this way is denoted *PIG-A-2*.

Reformulation of the Single-Item Subproblems

Next we see in Table 4.6 that there is a tight reformulation of *DLS-CC-SC* with $O(n^2)$ constraints and $O(n)$ variables described in Section 10.5.1; see Theorem 10.18. Adding this to formulation *PIG-A-2* gives a final formulation *PIG-A-3*.

14.4.4 Computational Results with Reformulations

We consider an instance with $n = NT = 100, m = NI = 10$; the storage costs $h^i = 10$ are item-independent. The changeover costs c^{ij} are nonnegative, with $c^{ii} = 0$ for all i. They do not satisfy the triangle inequality. Total demand for the instance is 83, whereas total production capacity is 100, the number of periods. In Table 14.12 we present computational results showing the effects of the reformulations. These results have been obtained using the default version of Xpress-MP and LS–LIB to implement the single item reformulation. For the final formulation PIG-A-3, we set $TK = 31$, with model cuts switched on ($MC = 1$). The time limit has been fixed to 1800 seconds. As usual, the first four columns represent the formulation used, the number of constraints, variables, and integer variables. The next four columns represent the value of the linear programming relaxation at the root node before cuts, and after both the model cuts from DLS-CC-SC and system cuts from Xpress-MP, and the best upper and lower bounds at the end of the computation. The last three columns give the total time in seconds, the total number of nodes, and the final duality gap. A * indicates that the time limit has been reached and the optimization stopped.

Table 14.12. Results with Different Formulations for Problem PIG–A

Formulat.	Cons	Vars	Int	LP	XLP	BUB	BLB	Time (secs)	Nodes	Gap (%)
PIG–A–1	31,800	12,900	1,000	1,661.6	2,251.2	13,178	2,603.7	1,800*	17,800*	80.2
PIG–A–2	6,060	14,880	1,990	3,282.6	3,839.2	12,213	4,061.5	1,800*	17,300*	66.7
PIG–A–3	8,498	14,880	1,990	3,282.6	8,282.5	8,312	8,312	133	48	0

Note that with formulations PIG-A-1 and PIG-A-2, the instance of problem PIG-A is unsolved after 1800 seconds. Note also that most of the gap closed at the root node in formulation PIG-A-3 is due to the extended reformulation for DLS-CC-SC. In fact the value of the LP relaxation with the model cuts from DLS-CC-SC but without the Xpress-MP cuts is 8270.

14.4.5 Modeling Periods with No Production

Consider a time interval $[\tau, \tau+1, \ldots, \tau+p]$ with $p \geq 2$, in which some item i is produced in period τ ($x_\tau^i = 1$), nothing is produced in periods $\tau+1, \ldots, \tau+p-1$ ($x_k^l = 0$ for $k = \tau+1, \ldots, \tau+p-1$ and all items l) and then item j is produced in period $\tau + p$ ($x_{\tau+p}^j = 1$). There are different ways of modeling such idle time intervals, and of computing the corresponding changeover costs, such as:

a. *Default Changeover Costs.* This corresponds to the basic model PIG-A represented by the formulation (14.83)–(14.89). The changeover costs associated with such a sequence will be $\min_{i_1, \ldots, i_{p-1}} (c^{i,i_1} + c^{i_1,i_2} + \cdots + c^{i_{p-1},j})$ because the set-up sequence $y_\tau^i = y_{\tau+1}^{i_1} = \cdots = y_{\tau+p-1}^{i_{p-1}} = y_{\tau+p}^j = 1$ is feasible for any selection of items $i_1, \ldots, i_{p-1}$.
b. *Direct Changeover Costs.* The changeover cost of the sequence is c^{ij} (which may be costlier than the earlier solution if the triangle inequality is not satisfied). The corresponding model is denoted PIG-B.

c. *Switch-Off and Switch-On Costs.* Here we assume that there is a cost cf^i of passing from production of i to idle, and a cost cn^j of passing from idle to production of j. The changeover cost of the sequence is thus $cf^i + cn^j$. The corresponding model is denoted *PIG-C*.

d. *No Idle Periods.* In this case some item must be produced in every period, and sequences with idle periods are not feasible. So there is a build up of stock if the capacity exceeds the demand. The corresponding model is denoted *PIG-D*.

Now we consider how to modify our basic formulation $PIG\text{-}A\text{-}\{1,2,3\}$ so as to treat cases b, c and d.

Modeling case b: Assuming that $c^{ii} = 0$ for all i, it suffices to add $z_t^i \leq x_t^i$ for all i,t. This ensures that the set-up sequence for the idle time interval $[\tau, \tau+1, \ldots, \tau+p]$ will be $y_\tau^i = y_{\tau+1}^i = \cdots = y_{\tau+p-1}^i = 1, y_{\tau+p}^j = 1$ giving the required changeover cost c^{ij}.

Modeling case c: Here it is natural to introduce a dummy item $i = 0$ to correspond to "no production". The changeover formulation is extended to include this new item. We set $c^{i0} = cf^i$ and $c^{0i} = cn^i$ for all $i \neq 0$ and $c^{00} = 0$, and add $x_t^i = y_t^i$ for all i,t to the original $PIG\text{-}A\text{-}\{1,2,3\}$ formulation.

Modeling case d: Here it suffices to add $x_t^i = y_t^i$ for all i,t to the original $PIG\text{-}A\text{-}\{1,2,3\}$ formulation. Note that when the storage costs are item-independent $h^i = h$ for all i, the overall storage cost is constant, and can be dropped.

In Table 14.13, we show results using the tight formulation on the four different problem variants $PIG\text{-}\{A, B, C, D\}$.

Table 14.13. Various Idle Time Cost Models: Tightened Formulations $PIG\text{-}\{\star\}\text{-}3$

Formulat.	Cons	Vars	Int	LP	XLP	BUB	BLB	Time (secs)	Nodes	Gap (%)
PIG-A-3	8,498	14,880	1,990	3,282.6	8,282.5	8,312	8,312	133	48	0
PIG-B-3	9,488	14,880	1,990	3,285.9	8,300.8	8,360	8,360	133	41	0
PIG-C-3	7,795	16,367	2,189	3,567.4	8,513.9	8,627	8,267	266	190	0
PIG-D-3	7,498	13,880	1,990	13,890.4	16,084.8	16,831	16,115.5	1,800*	1,300*	4.3
PIG-D-3-r	7,498	13,880	1,990	13,889.1	16,289.8	16,392	16,392.0	6,700	5,382	0

We observe that the variants $PIG\text{-}A, PIG\text{-}B$, and $PIG\text{-}C$ have solution values that are fairly close to one another. On the other hand, due to the forced production in each period, problem $PIG\text{-}D$ is significantly different from the others both in cost and in solution difficulty, leaving a gap of of 4.3% after 1800 seconds. To obtain this relatively small gap, we removed the Gomory cuts (set *gomcuts* and *treegomcuts* to 0) of Xpress-MP. To solve this instance to optimality, we used the observation from above that the solution was independent of the storage costs. Thus we removed the constant storage cost (equal to 1381 times the unit storage cost) and increased the approximation parameter to $TK = 61$ to get a stronger lower bound. The results for this revised formulation PIG-D-3-r are shown in the last line of Table 14.13. Note that it took just over one hour to prove optimality.

14.5 Process Manufacturing

This is a multi-item, multi-level, multi-machine production planning problem with backlogging and lower bounds on stocks for finished products, and upper bounds on stocks for each product.

There are two production levels treating the intermediate items IP and the end items EP, respectively. For each item i, the machines have different production rates. The number of intermediate products is small, with a few machines dedicated to them. The coefficients RML^{ij} represent the number of units of intermediate product $i \in IP$ required to produce one unit of final product $j \in EP$. Note that the production structure is of distribution type: the set of end items is partitioned according to the unique intermediate item that each uses as input.

It is a small bucket model with at most two set-ups allowed per period. The special feature of the problem is the requirement of lower bounds MB^i (in units of item i) on production runs for each end-item $i \in EP$. A typical production run lasts for several periods, and full capacity production in all but the first and last periods of a production run is imposed. As opposed to a model with only one set-up per period, this two set-up model is used to allow some flexibility at the beginning and end of a production run. Finally, there are item dependent cleaning times CLT^i at the end of each production run.

14.5.1 Classification

This model has the classification:

- Multi-machine, where several machines can produce the same item but at different production rates;
- Two-level, with intermediate items IP, end items EP, and an ML-D structure;
- Multi-item with production mode $PM = [M_2\text{-}SC]$, and production quantity $PQ = [PC\text{-}CLT]$ (the same as $PC\text{-}ST$ but cleaning times instead of start-up times);
- Each single item has classification LS-C-B, SC, CLT, AFC, MR, RLS, SS, SUB.

Note that there is no pure lower bound LB in the single-item classification because lower bounds are imposed only on the total size of each production run. Note also that the restricted (minimum) length sequences appear as a consequence of mode M_2 combined with AFC and MR.

14.5.2 Initial Formulation

We define first some additional notation:

- ρ^{ik} is the amount of item i produced on machine k per unit of time.
- L^k is the time available for production on machine k in each period.
- $\beta^{ik} = \lceil \frac{MB^i}{L^k} \rceil$, where $\beta^{ik} + 1$ is the maximum number of periods required to produce the amount MB^i if i is produced on k.
- $\alpha^{ik} = \lceil \frac{MB^i + CLT^i - L^k}{L^k} \rceil$ where $\alpha^{ik} + 1$ is the minimum number of periods in a production run or set-up sequence.

Before writing an initial formulation, we remove the lower bounds on stocks by using net stock variables (i.e., s_t^i will be the stock level above the minimum stock) and adapt the demands for end items accordingly, as in Section 4.5. Using lot-size variables x_t^{ik} representing the production time of item i on machine k in period t, an initial and natural formulation of this problem is

$$\min \sum_{i,k,t} (p^i \rho^{ik} x_t^{ik} + c^i w_t^{ik}) + \sum_{i,t} (h^i s_t^i + b^i r_t^i) \tag{14.90}$$

$$s_{t-1}^i + \sum_{k} \rho^{ik} x_t^{ik} = \sum_{j,k:j \neq i} RML^{ij} \rho^{jk} x_t^{jk} + s_t^i \quad \text{for } i \in IP, \text{ all } t$$
$$\tag{14.91}$$

$$s_{t-1}^i - r_{t-1}^i + \sum_{k} \rho^{ik} x_t^{ik} = d_t^i + s_t^i - r_t^i \quad \text{for } i \in EP, \text{ all } t$$
$$\tag{14.92}$$

$$w_{t-1}^{ik} \geq y_{t-1}^{ik} - y_t^{ik} \qquad \text{for all } i, k, t \tag{14.93}$$

$$x_t^{ik} + CLT^i w_t^{ik} \leq L^k y_t^{ik} \qquad \text{for all } i, k, t \tag{14.94}$$

$$\sum_{i} x_t^{ik} + \sum_{i} CLT^i w_t^{ik} \leq L^k \qquad \text{for all } k, t \tag{14.95}$$

$$w_t^{ik} + w_{t+1}^{ik} \leq y_t^{ik} \qquad \text{for all } i, k, t \tag{14.96}$$

$$\sum_{i} y_t^{ik} - \sum_{i} w_t^{ik} \leq 1 \qquad \text{for all } k, t \tag{14.97}$$

$$\sum_{i} w_t^{ik} \leq 1 \qquad \text{for all } k, t \tag{14.98}$$

$$x_t^{ik} \geq L^k (y_{t-1}^{ik} + y_t^{ik} - w_t^{ik} - w_{t-1}^{ik} - 1) \qquad \text{for all } i, k, t \tag{14.99}$$

$$w_l^{ik} \leq y_t^{ik} \qquad \text{for } i \in EP, \text{ all } k, t, l :$$
$$t \leq l \leq t + \alpha^{ik}$$
$$\tag{14.100}$$

$$\sum_{l=t-\beta^{ik}}^{t} \rho^{ik} x_l^{ik} \geq MB^i w_t^{ik} \qquad\qquad \text{for } i \in EP, \text{ all } k, t \quad (14.101)$$

$$s_0^i = S_0^i, r_0^i = 0, \ s_t^i \leq \overline{S_t^i} \qquad\qquad \text{for all } i, t \qquad\qquad (14.102)$$

$$s_t^i, r_t^i \in \mathbb{R}_+^1, \ x_t^{ik} \in \mathbb{R}_+^1, \ y_t^{ik}, w_t^{ik} \in \{0,1\} \qquad \text{for all } i, k, t, \qquad (14.103)$$

where constraints (14.91)–(14.92) are the flow balance constraints, (14.93) defines the switch-off variables, (14.94) the individual capacity constraints with cleaning times, and (14.95) the machine capacity constraints with cleaning times. Constraints (14.96)–(14.98) model the restriction of two set-ups per period, (14.100) the minimum run length constraint expressing the minimum number of periods in a production run, (14.101) the minimum run constraint in units of products, and (14.99) the almost full capacity production constraint.

14.5.3 Reformulation

Below we reformulate the problem. From Section 12.3, we obtain (14.110)–(14.111) to represent the two set-ups per period model Y^{M_2-SC}. From Section 11.3, we use the stronger constraint (14.112) to describe the almost full capacity constraint, and we add the minimum run valid inequalities (14.115) to strengthen the minimum production run constraint (14.114). From Section 11.4 we adapt constraint (11.10) to get (14.113) representing a minimum run-time of $\alpha^{ik}+1$ time periods. Note that in each case the version of the constraint is expressed with the switch-off variables w_t^{ik} rather than the start-up variables z_t^{ik}. Finally each end item has as a compact relaxation WW-U-B, SC, for which its extended formulation can be easily added using LS–LIB.

$$\min \sum_{i,k,t} (p^i \rho^{ik} x_t^{ik} + c^i w_t^{ik}) + \sum_{i,t} (h^i s_t^i + e^i r_t^i) \qquad\qquad (14.104)$$

$$s_{t-1}^i + \sum_k \rho^{ik} x_t^{ik} = \sum_{j,k:j \neq i} RML^{ij} \rho^{jk} x_t^{jk} + s_t^i \quad \text{for } i \in IP, \text{ all } t$$

$$(14.105)$$

$$s_{t-1}^i - r_{t-1}^i + \sum_k \rho^{ik} x_t^{ik} = d_t^i + s_t^i - r_t^i \qquad\qquad \text{for } i \in EP, \text{ all } t$$

$$(14.106)$$

$$w_{t-1}^{ik} \geq y_{t-1}^{ik} - y_t^{ik} \qquad\qquad\qquad \text{for all } i, k, t \quad (14.107)$$

$$x_t^{ik} + CLT^i w_t^{ik} \leq L^k y_t^{ik} \qquad\qquad \text{for all } i, k, t \quad (14.108)$$

$$\sum_i x_t^{ik} + \sum_i CLT^i w_t^{ik} \leq L^k \qquad\qquad \text{for all } k, t \quad (14.109)$$

$$\sum_i y_t^{ik} - \sum_i w_t^{ik} \leq 1 \qquad\qquad\qquad \text{for all } k, t \quad (14.110)$$

$$\sum_i w_t^{ik} \leq 1 \qquad \text{for all } k, t \qquad (14.111)$$

$$x_t^{ik} \geq L^k(y_{t-1}^{ik} - w_t^{ik} - w_{t-1}^{ik}) \qquad \text{for all } i, k, t \cdot \qquad (14.112)$$

$$\sum_{l=t}^{t+\alpha^{ik}} w_l^{ik} \leq y_t^{ik} \qquad \text{for } i \in EP, \text{ all } k, t \qquad (14.113)$$

$$\sum_{l=t-\beta^{ik}}^{t} \rho^{ik} x_l^{ik} \geq MB^i w_t^{ik} \qquad \text{for } i \in EP, \text{ all } k, t \qquad (14.114)$$

$$r_{t-1}^i + s_l^i \geq (MB^i - d_{tl}^i)^+ \sum_k \sum_{\tau=t+\beta^{ik}}^{l} w_\tau^{ik} \qquad \text{for all } i, k, l, t$$

$$\text{with } t + \beta^{ik} \leq l \qquad (14.115)$$

$$s_0^i = S_0^i, r_0^i = 0, \ s_t^i \leq \overline{S_t^i} \qquad \text{for all } i, t \qquad (14.116)$$

$$s_t^i, r_t^i \in \mathbb{R}_+^1, \ x_t^{ik} \in \mathbb{R}_+^1, \ y_t^{ik}, w_t^{ik} \in \{0,1\} \qquad \text{for all } i, k, t. \qquad (14.117)$$

14.5.4 Computational Results

Below we provide results for the original formulation (14.90)–(14.103), denoted *pm-NT-a*, and the tightened formulation (14.104)–(14.117) to which we have added the reformulation for $WW\text{-}U\text{-}B, SC$ for each end-product with $TK = NT$, denoted *pm-NT-b*. The formulations have been adapted to reflect the initial conditions of the facility (machines set-up status and past production for in-process batches). Results are for instances with $NT = 12$ periods and $NT = 20$ periods with a time limit of 1800 seconds. After a run of several hours, the optimal value of pm-20-b has been proven to be 23,358.2.

Table 14.14. Results for Process Manufacturing

Instance	Cons	Vars	Int	LP	XLP	BUB	BLB	Time (secs)	Nodes	Gap (%)
pm-12-a	1,911	1,416	804	1,421.5	15,002.2	16,103.9	16,075.4	1,800*	225,300	0.2
pm-12-b	4,268	1,459	785	15,499.6	15,740.3	16,103.9	16,103.9	61	1,763	0
pm-20-a	3,544	2,640	1,520	19,841.7	20,837.2	23,710.4	22,321.2	1,800*	73,500	5.9
pm-20-b	6,507	2,730	1,498	22,005.3	22,332.2	23,407.9	23,168.6	1,800*	9,600	1.0

14.6 Powder Production

This problem is a simplified version of a laundry powder production/packing problem. There are 60 types of packed powder, which are the end products.

There is a known demand to be met for each of these 60 end products over 30 periods. Backlogging is allowed. The set of end products is partitioned into seven different groups sharing a common production line (resource) and, on the other hand, into two distinct groups sharing a common manpower resource.

The production of each end product consumes one given type among 17 available powders (bulk products). Planning the production of the bulk products is also part of the problem. There is a common resource shared by all bulks. Other complicating constraints at the bulk level are perishability (the powder must be packed at the latest one period after production), and a maximum total stock level for the powders.

None of the resources in the problem involve set-up times. However, there is a time-independent minimum production quantity for all end and bulk products. The objective is to minimize the sum of the stocks of intermediate and end products and the sum of backlogs of end products, with a higher cost/weight for the backlogs.

14.6.1 Classification

The model is classified as:

- Two-level with a distribution structure ML-D;
- Multi-machine with 10 resources (7 specialized machines for end products, 2 specialized manpower resources for end products, 1 machine for bulks), no mode restrictions $PM = [M_\infty]$ and big bucket resource constraints without set-up times $PQ = [PC]$;
- Multi-item with 60 end products denoted I_{end} and 17 bulks denoted I_{bulk};
- End products are classified as WW-U-B, LB with constant lower bounds on production;
- Bulk products are classified as WW-U-LB, PER with constant lower bounds on production and perishability restrictions.

14.6.2 Initial Formulation

An initial MIP formulation is as follows.

$$\min \sum_{i \in I_{end}} \sum_t r_t^i + \sum_{i \in I_{end}} \sum_t 0.5 s_t^i + \sum_{i \in I_{bulk}} \sum_t 0.25 s_t^i \qquad (14.118)$$

$$s_{t-1}^i + r_t^i + x_t^i = d_t^i + r_{t-1}^i + s_t^i \qquad \text{for } i \in I_{end}, 1 \le t \le NT \qquad (14.119)$$

$$s_{t-1}^i + x_t^i = \sum_{j \in succ(i)} x_t^j + s_t^i \qquad \text{for } i \in I_{Bulk}, 1 \le t \le NT \qquad (14.120)$$

$$LB^i y_t^i \leq x_t^i \leq d_{1,NT}^i y_t^i \qquad \text{for } i \in I_{end} \cup I_{Bulk}, 1 \leq t \leq NT \qquad (14.121)$$

$$\sum_{i \in I_f} x_t^i \leq L_f \qquad \text{for } 1 \leq f \leq NF, 1 \leq t \leq NT \qquad (14.122)$$

$$s_{t-1}^i \leq \sum_{j \in succ(i)} x_t^j \qquad \text{for } i \in I_{Bulk}, 1 \leq t \leq NT \qquad (14.123)$$

$$\sum_{i \in I_{bulk}} s_t^i \leq \overline{S} \qquad \text{for } 1 \leq t \leq NT \qquad (14.124)$$

$$r_0^i = s_0^i = 0 \qquad \text{for } i \in I_{end} \cup I_{Bulk} \qquad (14.125)$$

$$r_t^i, s_t^i \in \mathbb{R}_+^1, \ y_t^i \in \{0,1\} \qquad \text{for } i \in I_{end} \cup I_{Bulk}, 1 \leq t \leq NT, \qquad (14.126)$$

where Equations (14.119)–(14.120) are the balance constraints for a two-level lot-sizing problem, where $succ(i)$ is the set of end items using bulk i. Equations (14.121) are the set-up forcing constraints. Note that each item is uncapacitated and involves a constant lower bound on production. There are 10 joint resources in this problem, modeled as (14.122) where the set I_f is the set of items using resource f. As resources are specialized, variables s, r, x and demand d are expressed in capacity units (production hours). Equation (14.123) models the perishability at the bulk product level: the stock of i in t must come directly off the production line (i.e., cannot originate from the stock in period $t-1$). Finally, Equation (14.124) imposes a global limit on the amount of bulk product that can be in stock at any time.

14.6.3 Testing the Initial Formulation and Reformulations

We first carry out several runs to see whether the problem is easy to solve. First we run the above formulation, denoted *pp-a* in default branch-and-cut mode with MAXTIME of 900 seconds. The results are shown in the row *pp-a* of Table 14.15.

We then make some important observations concerning the initial formulation *pp-a*. Note first that the perishability constraints (14.123) are written more naturally, but equivalently, as $s_t^i \leq x_t^i$ by using the flow constraints (14.120). In addition there are several end items for which the total demand is less than the minimum production quantity. Given the objective to minimize stocks and backlogs, it follows that these items are not produced more than once in an optimal solution. Thus for the items $i \in I_{end}$ with $d_{1,NT}^i \leq LB^i$, if we introduce an additional 0–1 variable z^i where $z^i = 1$ indicates that end item i is not produced at all, we can write

$$\sum_{t=1}^{NT} y_t^i + z^i = 1$$

$$x_t^i = LB^i y_t^i \qquad\qquad \text{for all } t$$

$$s_t^i = (LB^i - d_{1t}^i) \sum_{\tau=1}^{t} y_\tau^i \qquad\qquad \text{for all } t$$

$$r_t^i = d_{1t}^i \Big(\sum_{\tau=t+1}^{NT} y_t^i + z^i \Big) \qquad\qquad \text{for all } t \ .$$

Furthermore, this reformulation can be added also for each bulk product $i \in I_{bulk}$ by using the echelon stock transformation from Section 13.3, where we define the echelon stock variable as $es_t^i = s_t^i + \sum_{j \in succ(i)} s_t^j$, the echelon backlog variable as $er_t^i = \sum_{j \in succ(i)} r_t^j$ and the echelon demand as $ed_t^i = \sum_{j \in succ(i)} d_t^j$. So, for $i \in I_{bulk}$ with $ed_{1,NT}^i \leq LB^i$, we add also the formulation

$$\sum_{t=1}^{NT} y_t^i + z^i = 1$$

$$x_t^i = LB^i y_t^i \qquad\qquad \text{for all } t$$

$$es_t^i = (LB^i - ed_{1t}^i) \sum_{\tau=1}^{t} y_\tau^i \qquad\qquad \text{for all } t$$

$$er_t^i = ed_{1t}^i \Big(\sum_{\tau=t+1}^{NT} y_t^i + z^i \Big) \qquad\qquad \text{for all } t \ .$$

After making these changes to the model, we denote the new formulation *pp-b* and obtain the results shown in Table 14.15.

For the third test run, we use the fact that the reformulation *WW-U-B* is compact and can be added for all items. We add this reformulation with $TK = 5$ to *pp-b* for each bulk item i for which $ed_{1,NT}^i > LB^i$ using variables es, er and demand vector ed, and for each end item i for which $d_{1,NT}^i > LB^i$ using s, r, and d. Results for the resulting formulation, denoted *pp-c*, and using model cuts (MC) are given in Table 14.15. Here it takes 114 seconds to obtain the LP–MC value corresponding to the initial linear relaxation plus the model cuts from *WW-U-B*, 193 seconds in total to obtain the XLP value with automatically generated system cuts of Xpress-MP, using *covercuts* = 20, *Gomcuts* = 2. The lower bound is drastically improved, but no feasible solution is found within 900 seconds. We then reduce the parameter to $TK = 3$, but still no solution is found within 900 seconds.

To keep the size of the reformulation as small as possible, another option is to add cutting planes for *WW-U-B* in place of the extended formulation. This corresponds to formulation *pp-cc* in Table 14.15. The XLP bound is obtained by switching off the preprocessing, and then adding system cuts followed by

Table 14.15. Powder Production: Initial Runs with MAXTIME $= 900$

Formulation	Cons	Vars	Int	LP(-MC)	XLP	BUB	BLB	Gap (%)
pp-a	7,693	8,867	2,310	217.1	822.4	2,531.4	868.2	66
pp-b	8,308	8,878	2,321	471.4	1,029.8	2,545.8	1,081.0	58
pp-c	27,545	12,598	2,321	1,548.3	1,587.8	∞	1,602.3	∞
pp-cc	8,308	8,878	2,201	470.1	1,710.7	∞	1,711.3	∞

$WW\text{-}U\text{-}B$ cuts. The observation that the XLP bound is better than that of instance $pp\text{-}c$ can be explained by the fact that there is no approximation parameter TK in the separation routine, so essentially $TK = NT = 30$ here.

Our tentative conclusion from these initial runs is that proving optimality for this problem is probably out of the question. So we decide to address the question of finding a feasible solution guaranteed within 25% of optimality or better within say 30 minutes, and a solution within 10% of optimality within several hours.

14.6.4 Finding Lower Bounds for Powder Production

Here we have little choice but to use the linear programming bounds provided by the extended formulations, as improving the lower bounds through partial enumeration only will be very slow.

In Table 14.16 we show the results of runs for $pp\text{-}c$ with four different values of TK in reformulation $WW\text{-}U\text{-}B$. In a fifth run we combine the formulation of $WW\text{-}U\text{-}B$ with $TK = 8$ with the less compact reformulation $WW\text{-}U\text{-}B, LB$ for all items with significant demands (i.e., $ed^i_{1,NT} > LB^i$ for $i \in I_{bulk}$ and $d^i_{1,NT} > LB^i$ for $i \in I_{end}$) using $TK = 4$. The latter formulation is called $pp\text{-}d$.

Table 14.16. Powder Production: Lower Bounds by Reformulation and LP

Formulation	$U - B$ TK	$U - B, LB$ TK	Cons	Vars	LP	XLP	Time (secs)
pp-c	3	0	20,423	12,598	1,399.9	1,456.3	115
pp-c	5	0	26,653	12,598	1,547.2	1,601.2	178
pp-c	8	0	35,089	12,598	1,657.3	1,716.0	287
pp-c	15	0	52,948	12,598	1,692.1	1,728.9	403
pp-d	8	4	134,106	76,094	1,682.2	1,717.3	3915

14.6.5 Finding Upper Bounds for Powder Production

Given our preliminary tests, it appears likely that we can apply our full range of heuristics to the weak formulations $pp\text{-}a$ and $pp\text{-}b$. However, if we want to use the tightened formulation *pp-c* to find feasible solutions, we will need to start with a decomposition heuristic such as relax-and-fix or exchange, which divides the problem up into smaller subproblems.

In Table 14.17 we show our results starting first with the weak formulation *pp-b*, and then with the tightened formulation *pp-c*. In both cases we have applied relax-and-fix followed by two rounds of exchange. The parameter "MAXT" is the maximum time given to the subproblems, more precisely the running time for a subproblem is the maximum of "MAXT" and the run-time to find the first feasible solution. As usual "FIX" is the number of variables fixed in each subproblem, and here BIN = FIX for all the relax-and-fix runs. The columns "R&F", "EXCH1", and "EXCH2" contain, respectively, the value of the feasible solution after relax-and-fix, the first and second exchange rounds. On the tightened formulation *pp-c*, the time to find a feasible solution even for the smaller subproblems is long and somewhat unpredictable.

Table 14.17. Powder Production: Primal Heuristics – Upper Bounds

Formulation	MAXT	FIX	R&F	EXCH1	EXCH2	Time (secs)	Time (secs)	Time (secs)
pp-b	60	5	2,239.4	2,136.0	2,128.3	< 360	< 360	< 360
pp-b	30	5	2,254.7	2,191.7.0	2,153.5	< 180	< 180	< 180
pp-c(TK=8)	60	5	2,126.2	2,053.2	2,051.0	300	301	303
pp-c(TK=5)	60	5	2,093.3	2,049.1	2,037.4	1,510	302	302

Thus using reformulation *pp-c* with $TK = 8$ we can obtain a lower bound of 1716.0 in 287 secs, and using formulation *pp-b* an upper bound of 2136.0 in less than 360 secs. This gives a gap of 19.67%. Using the best possible bounds, a lower bound of 1728.9 in 403 secs and an upper bound of 2037.4 in about 2100 secs give a gap of 15.1%.

To get a further improvement, we probably need to run RINS or Local branching using a tight formulation, starting from our best feasible solution. Such runs will almost certainly take a long time.

Exercises

Making and Packing

Exercise 14.1 Because the new technology allows more flexibility in running the lines, one would like to analyze whether the assignment restrictions have strong implications on the global line productivity.

Formulate and solve the problem in which it is possible to change the product assignment at each shift rather than at each day, for all machines, feeders, and robots. This can be done simply by using the same formulations but with reduced length time periods. Backlogging and inventory costs are per period costs; consequently they have to be divided by three to keep comparable units in the objective function.

Try to obtain good lower and upper bounds, using reformulations and heuristics. What is the impact of this more flexible organization on customer service level, and on inventory costs?

Exercise 14.2 Given the current product mix, what is the bottleneck stage ? To respond to a uniform market increase, does it make sense to invest in new feeders and robots?

Exercise 14.3 To find out whether the relative backlogging cost has a strong impact on the solutions of the planning model, answer the following questions. Would a higher relative backlogging cost substantially improve customer service level, and would a lower backlogging cost improve line productivity? Run some tests to answer these questions.

Storage Rack Production

Exercise 14.4 Given the very short production cycle (lead-times can be assumed to be zero), and the fact that the planning system will be used in a rolling horizon manner, analyze whether the planning horizon can be reduced to eight periods, without affecting the quality of solutions.

Run tests to simulate the rolling horizon procedure, and compare the results obtained using different planning horizons.

Insulating Board Extrusion

Exercise 14.5 For the single-period batching and scheduling model, implement the current [first batching – then scheduling] iterative procedure, and compare its performance with that of the integrated [batching–scheduling] approach using formulation VRP or UF. Use the data available on the book Web site.

Exercise 14.6 Compare the performance (quality of solutions versus run time) of the integrated [planning–batching–scheduling] approach using formulation $MVRP$ with the performance of the current [first planning – then [batching–scheduling]] iterative procedure.

In order to assess the specific contribution of the multi-period extension, use formulation VRP for the [batching–scheduling] subproblem in the implementation of the current procedure. The data D_t^i of model VRP is the output of the planning subproblem of the current procedure.

This planning subproblem is a single-level multi-item problem with a single global capacity constraint per time period, that is, $LS\text{-}U/PM = [M_\infty]/PQ = [PC, PC\text{-}SU]^1$, where the capacity is an estimate (supposed to become more accurate from iteration to iteration) of the total amount of extrusion time available in each time period. It does not contain any decision variable for the number of batches (batching) or changeover times (scheduling).

Compare also the design and performance of the current procedure with that of a time decomposition relax-and-fix heuristic for $MVRP$.

Exercise 14.7 For formulation $MVRP$, suggest different objective functions, test them on the data available on the Web site, and compare the solutions obtained. In particular, is it necessary to introduce inventory holding costs, and how should they be computed?

Process Manufacturing

Exercise 14.8 The results given in Table 14.14 show that it is possible to obtain good solutions for the instance of the process manufacturing problem with $NT = 20$ time periods by using appropriate reformulations. However. it takes about 30 minutes to obtain a solution with a gap of 1%.

In order to obtain good solutions quickly, develop construction and improvement heuristics based on the initial formulation and on the reformulations; see Section 3.6. Analyze the performance of these heuristics on the same problem instance.

Exercise 14.9 In the process manufacturing problem, there is an upper bound on the stock level of each item at the end of each time period. Therefore, there exists a relaxation WW-U-B, SUB for each end product and a relaxation WW-U-SUB for each intermediate product.

Use the reformulation results and valid inequalities from Section 11.7 to improve the formulation of the problem to see if you can obtain better lower and upper bounds. Test your reformulations on the two instances with $NT = 12$ and $NT = 20$ time periods.

Notes

Section 14.1. This case is inspired by a real case study, but its data are disguised.

Section 14.2. This case, its mathematical programming formulation, and data are taken from Simpson and Erenguc [152]. Results have been given earlier in Wolsey [194], Danna et al. [52], and Belvaux and Wolsey [25].

Section 14.3. This real case of the required coordination between planning and scheduling, as well as the initial mathematical programming formulation, are taken from Batta and Teghem [24]. The multi-period extension presented here is original.

Section 14.4. The problem description and its mathematical programming formulation are from Fleischmann [68].

Section 14.5. This industrial application has been described and tackled earlier in Belvaux and Wolsey [26]. It is inspired by a real case study, but its data are disguised.

Section 14.6. This industrial application has been described and tackled earlier in Van Vyve [178] and Van Vyve and Wolsey [181].

References

1. P. Afentakis and B. Gavish. Optimal lot-sizing algorithms for complex product structures. *Operations Research*, 34:237–249, 1986.
2. P. Afentakis, B. Gavish, and U. Karmarkar. Computationally efficient optimal solutions to the lot-sizing problem in multistage assembly systems. *Management Science*, 30:222–239, 1984.
3. A. Aggarwal and J. Park. Improved algorithms for economic lot-size problems. *Operations Research*, 41:549–571, 1993.
4. E.H. Aghezzaf and L.A. Wolsey. Lot-sizing polyhedra with a cardinality constraint. *Operations Research Letters*, 11:13–18, 1992.
5. E.H. Aghezzaf and L.A. Wolsey. Modelling piecewise linear concave costs in a tree partitioning problem. *Discrete Applied Mathematics*, 50:101–109, 1994.
6. A. Agra and M. Constantino. Lotsizing with backlogging and start-ups: The case of Wagner–Whitin costs. *Operations Research Letters*, 25:81–88, 1999.
7. R.K. Ahuja, T.L. Magnanti, and J.B. Orlin. *Network Flows: Theory, Algorithms, and Applications*. Prentice-Hall, 1993.
8. D. Aksen, K. Altinkemer, and S. Chand. The single-item lot-sizing problem with immediate lost sales. *European Journal of Operational Research*, 147:558–566, 2003.
9. S. Anily and M. Tzur. Shipping multiple-items by capacitated vehicles - an optimal dynamic programming approach. *Transportation Science*, 39:233–248, 2005.
10. S. Anily, M. Tzur, and L.A. Wolsey. Lot-sizing with family set-ups. CORE DP 2005/70, Université catholique de Louvain, 2005.
11. E. Arkin, D. Joneja, and R. Roundy. Computational complexity of uncapacitated multi-echelon production planning problems. *Operations Research Letters*, 8:61–66, 1989.
12. B.C. Arntzen, G.G. Brown, T.P. Harrison, and L.L. Trafton. Global supply chain management at Digital Equipment Corporation. *Interfaces*, 25:69–93, 1995.
13. A. Atamtürk. Strong formulations of robust mixed 0–1 programming. Technical report, Dept. of IE and OR, Univ. of California, Berkeley, 2003, revised 2005.
14. A. Atamtürk and S. Küçükyavuz. Lot sizing with inventory bounds and fixed costs. *Operations Research*, 53:711–730, 2005.

15. A. Atamtürk and J.C. Munoz. A study of the lot-sizing polytope. *Mathematical Programming*, 99:443–466, 2004.
16. A. Atamtürk and D. Rajan. On splittable and unsplittable flow capacitated network design arc–set polyhedra. *Mathematical Programming*, 92:315–333, 2002.
17. T.E. Baker and J.A. Muckstadt. The CHES problems. Technical report, Chesapeake Decision Sciences Inc., New Providence, NJ 07974, 1989.
18. E. Balas. Disjunctive programs: Cutting planes from logical conditions. In O.L. Mangasarian et al., editor, *Nonlinear Programming*, volume 2, pages 279–312. Academic Press, 1975.
19. E. Balas. Facets of the knapsack polytope. *Mathematical Programming*, 8:146–164, 1975.
20. E. Balas. The prize collecting traveling salesman problem. *Networks*, 19:621–636, 1989.
21. E. Balas. Disjunctive programming: Properties of the convex hull of feasible points. Invited paper with foreword by G. Cornuéjols and W.R. Pulleyblank. *Discrete Applied Mathematics*, 89:1–44, 1998.
22. I. Barany, J. Edmonds, and L.A. Wolsey. Packing and covering a tree by subtrees. *Combinatorica*, 6:245–257, 1986.
23. I. Barany, T.J. Van Roy, and L.A. Wolsey. Uncapacitated lot sizing: The convex hull of solutions. *Mathematical Programming*, 22:32–43, 1984.
24. C. Batta and J. Teghem. Optimization of production scheduling in plastics processing industry. *JORBEL*, 34:55–78, 1994.
25. G. Belvaux and L.A. Wolsey. Bc-prod: A specialized branch-and-cut system for lot-sizing problems. *Management Science*, 46:724–738, 2000.
26. G. Belvaux and L.A. Wolsey. Modelling practical lot-sizing problems as mixed integer programs. *Management Science*, 47:993–1007, 2001.
27. P.J. Billington, J.O. McClain, and L.J. Thomas. Heuristics for multi-level lot-sizing with a bottleneck. *Management Science*, 32:989–1006, 1986.
28. G.R. Bitran and H.H. Yanasse. Computational complexity of the capacitated lot size problem. *Management Science*, 28:1174–1186, 1982.
29. N. Brahimi. *Planification de la production: modèles et algorithmes pour les problemes de dimensionnement de lots*. PhD thesis, Université de Nantes, 2004.
30. N. Brahimi, S. Dauzère-Pérès, and N.M. Najid. Capacitated multi-item lot-sizing problems with time windows. Technical report, Ecole des Mines de Nantes, 2005.
31. G. Brown, J. Keegan, B. Vigus, and K. Wood. The Kellogg Company optimizes production, inventory, and distribution. *Interfaces*, 31 (6):1–15, 2001.
32. J. Browne, J. Harhen, and J. Shivnan. *Production Management Systems: An Integrated Perspective*. Addison-Wesley, 1996.
33. M.R. Bussieck, A. Fink, and M.E. Lübbeke. Yet another note on "an efficient zero-one formulation of the multilevel lot-sizing problem". Technical report, Braunschweig University of Technology, 1998.
34. D. Cattrysse, J. Maes, and L.N. van Wassenhove. Set partitioning and column generation heuristics for capacitated dynamic lot-sizing. *European Journal of Operational Research*, 46:38–47, 1990.
35. D. Cattrysse, M. Salomon, R. Kuik, and L.N. van Wassenhove. A dual ascent and column generation heuristic for the DLSP with set-up times. *Management Science*, 39:477–486, 1993.

36. W.-H. Chen and J.-M. Thizy. Analysis of relaxations for the multi-item capacitated lot-sizing problem. *Annals of Operations Research*, 26:29–72, 1990.

37. S. Chopra and P. Meindl. *Supply Chain Management: Strategy, Planning and Operation.* Prentice-Hall, 2001.

38. M. Christopher. *Logistics and Supply Chain Management: Strategies for Reducing Cost and Improving Service.* Pitman, 1998.

39. S. Chubanov, M.Y. Kovalyov, and E. Pesch. A FPTAS for a single-item capacitated economic lot-sizing problem with monotone cost structure. *Mathematical Programming*, (Online First. DOI: 10.1007/s10107-005-0641-0), 2006.

40. C. Chung and M. Lin. An $O(T^2)$ algorithm for the $NI/G/NI/ND$ capacitated single item lot size problem. *Management Science*, 34:420–426, 1988.

41. A.J. Clark and H. Scarf. Optimal policies for multi-echelon inventory problems. *Management Science*, 6:475–490, 1960.

42. M. Conforti, M. di Summa, and L.A. Wolsey. The mixing set with flows. DP 2005/92, CORE,Université catholique de Louvain, 2005.

43. M. Conforti, F. Eisenbrand, and L.A. Wolsey. Extended formulation and projections for generalized mixing sets. Technical report, CORE, Université catholique de Louvain, 2005.

44. M. Conforti and L.A. Wolsey. Compact formulations as a union of polyhedra. DP 2005/62, CORE,Université catholique de Louvain, 2005.

45. M. Constantino. *A polyhedral approach to production planning models: Start-up costs and times, and lower bounds on production.* PhD thesis, Université catholique de Louvain, 1995.

46. M. Constantino. A cutting plane approach to capacitiated lot-sizing with start-up costs. *Mathematical Programming*, 75:353–376, 1996.

47. M. Constantino. Lower bounds in lot-sizing models: A polyhedral study. *Mathematics of Operations Research*, 23:101–118, 1998.

48. M. Constantino, A. Miller, Y. Pochet, M. Van Vyve, B. Verweij, and L.A. Wolsey. New MIP cuts for supply chain structures: Final report on MIP cuts. Technical Report LISCOS: Large Scale Integrated Supply Chain Optimisation Software based upon Branch-and-Cut and Constraint Programming, GROWTH Project G1RD-1999-00034, DR2.3.1/U, CORE, 2002.

49. W.J. Cook, R. Kannan, and A. Schrijver. Chvátal closures for mixed integer programming problems. *Mathematical Programming*, 47:155–174, 1990.

50. H. Crowder, E.L. Johnson, and M.W. Padberg. Solving large-scale 0–1 linear programming problems. *Operations Research*, 31:803–834, 1983.

51. W.B. Crowston and M.H. Wagner. Dynamic lot size models for multi-stage assembly systems. *Management Science*, 20:14–21, 1973.

52. E. Danna, E. Rothberg, and C. Le Pape. Exploring relaxation induced neighborhoods to improve MIP solutions. *Mathematical Programming*, 102:71–90, 2005.

53. G.B. Dantzig and P. Wolfe. Decomposition principle for linear programs. *Operations Research*, 8:101–111, 1960.

54. S. Dauzère-Pérès, N. Brahimi, N.M. Najid, and A. Nordli. Uncapacitated lot-sizing problems with time windows. Technical report, Ecole des Mines de Saint-Etienne, 2005.

55. N.P. Dellaert and J. Jeunet. Randomized multi-level lot-sizing heuristics for general product structures. *European Journal of Operational Research*, 148:211–228, 2003.

56. M. Diaby, H.C. Bahl, M.H. Karwan, and S. Zionts. A Lagrangean relaxation approach for very large-scale capacitated lot-sizing. *Management Science*, 38:1329–1340, 1992.

57. P.S. Dixon and E.A. Silver. A heuristic solution procedure for the multi-item, single-level, limited capacity, lot-sizing problem. *J. of Operations Management*, 2:23–39, 1981.

58. A. Drexl and A. Kimms. Lot sizing and scheduling — survey and extensions. *European Journal of Operational Research*, 99:221–235, 1997.

59. B. Dzielinski and R. Gomory. Optimal programming of lot-sizes, inventories and labor allocations. *Management Science*, 11:874–890, 1965.

60. J. Edmonds. Maximum matching and a polyhedron with 0–1 vertices. *Journal of Research of the National Bureau of Standards*, 71B:125–130, 1966.

61. G.D. Eppen and R.K. Martin. Solving multi-item lot-sizing problems using variable definition. *Operations Research*, 35:832–848, 1987.

62. A. Federgrün, J. Meissner, and M. Tzur. Progressive interval heuristics for multi-item capacitated lot-sizing problems. Technical report, Columbia University, 2002.

63. A. Federgrün and M. Tzur. A simple forward algorithm to solve general dynamic lot-size models with n periods in $O(n \log n)$ or $O(n)$ time. *Management Science*, 37:909–925, 1991.

64. A. Federgrün and M. Tzur. The joint replenishment problem with time-varying parameters: Efficient asymptotic and epsilon-optimal solutions. *Operations Research*, 42:1067–1087, 1994.

65. A. Federgrün and M. Tzur. Time-partitioning heuristics: Application to one warehouse, multi-item, multi-retailer lot-sizing problems. *Naval Research Logistics Quarterly*, 46:463–486, 1999.

66. M. Fischetti and A. Lodi. Local branching. *Mathematical Programming*, 98:23–48, 2003.

67. B. Fleischmann. The discrete lot-sizing and scheduling problem. *European Journal of Operational Research*, 44:337–348, 1990.

68. B. Fleischmann. The discrete lot-sizing and scheduling problem with sequence-dependent setup costs. *European Journal of Operational Research*, 75:395–404, 1994.

69. B. Fleischmann and H. Meyr. Planning hierarchy, modeling and advanced planning systems. In A.G. de Kok and S.C. Graves, editors, *Supply Chain Management: Design, Coordination and Operation*, volume 11 of *Handbooks in Operations Research and Management Science*. North Holland, 2003.

70. B. Fleischmann, H. Meyr, and M. Wagner. Advanced planning. In H. Stadtler and C. Kilger, editors, *Supply Chain Management and Advanced Planning: Concepts, Models, Software and Case Studies*, pages 71–96. Springer-Verlag, 2002.

71. M. Florian and M. Klein. Deterministic production planning with concave costs and capacity constraints. *Management Science*, 18:12–20, 1971.

72. R. Fukasawa, H. Longo, J. Lysgaard, M. Poggi de Arago, M. Reis, E. Uchoa, and R.F. Werneck. Robust branch-and-cut-and-price for the capacitated vehicle routing problem. *Mathematical Programming*, (Online First. DOI: 10.1007/s10107-005-0644-x), 2006.

73. A.M. Geoffrion. Lagrangean relaxation for integer programming. *Mathematical Programming Study*, 2:82–114, 1974.

74. A. Ghoula-Houri. Caractérisation des matrices totalement unimodulaire. *C.R. Academy of Sciences of Paris*, 254:1192–1194, 1962.

75. P.C. Gilmore and R.E. Gomory. A linear programming approach to the cutting stock problem. *Operations Research*, 9:849–859, 1961.

76. M. Goetschalckx. Strategic network planning. In *Supply Chain Management and Advanced Planning: Concepts, Models, Software and Case Studies* [155], pages 105–121.

77. R.E. Gomory. An algorithm for the mixed integer problem. Technical Report RM-2597, The RAND Corporation, 1960.

78. S.C. Graves, A.H.G. Rinnooy Kan, and P.H. Zipkin (Eds). *Logistics of Production and Inventory*, volume 4 of *Handbooks in Operations Research and Management Science*. North Holland, 1993.

79. M. Grötschel, L. Lovász, and A. Schrijver. The ellipsoid method and its consequences in combinatorial optimization. *Combinatorica*, 1:169–197, 1981.

80. M. Grötschel, L. Lovász, and A. Schrijver. *Geometric Algorithms and Combinatorial Optimization*. Springer, Berlin, 1988.

81. Z. Gu, G.L. Nemhauser, and M.W.P. Savelsbergh. Lifted flow cover inequalities for mixed 0–1 integer programs. *Mathematical Programming*, 85:439–467, 1998.

82. Y. Guan, S. Ahmed, A.J. Miller, and G.L. Nemhauser. On formulations of the stochastic uncapacitated lot-sizing problem. *Operations Research Letters*, to appear, 2006.

83. Y. Guan, S. Ahmed, G.L. Nemhauser, and A.J. Miller. A branch-and-cut algorithm for the stochastic uncapacitated lot-sizing problem. *Mathematical Programming*, 105:55–84, 2005.

84. M. Guignard and S. Kim. Lagrangean decomposition for integer programming: Theory and applications. *RAIRO*, 21:307–323, 1987.

85. O. Günlük and Y. Pochet. Mixing mixed integer inequalities. *Mathematical Programming*, 90:429–457, 2001.

86. P.L. Hammer, E.L. Johnson, and U.N. Peled. Facets of regular 0–1 polytopes. *Mathematical Programming*, 8:179–206, 1975.

87. A.C. Hax and H.C. Meal. Hierarchical integration of production planning and scheduling. In M. Geisler, editor, *TIMS studies in Management Science*, chapter 1. North Holland, 1975.

88. S. Heipcke. *Applications of Optimization with Xpress*. Dash Optimization Ltd, 2002.

89. M. Held and R.M. Karp. The traveling salesman problem and minimum spanning trees: Part II. *Mathematical Programming*, 1:6–25, 1971.

90. D.S. Hochbaum and J.G. Shantikumar. Convex separable optimization is not much harder than linear optimization. *Journal of ACM*, 37:843–862, 1990.

91. W.J. Hopp and M.L. Spearman. *Factory Physics: Foundations of Manufacturing Management*. McGraw-Hill, 2000.

92. V.H. Hsu. Dynamic economic lot size model with perishable inventory. *Management Science*, 46:1159–1169, 2000.

93. L.A. Johnson and D.C. Montgomery. *Operations Research in Production Planning, Scheduling and Inventory Control*. John Wiley and Sons, 1974.

94. K. Jornsten and M. Nasberg. A new Lagrangian relaxation approach to the generalized assignment problem. *European Journal of Operational Research*, 27:313–323, 1986.

95. S. Kang, K. Malik, and L.J. Thomas. Lotsizing and scheduling in parallel machines with sequence dependent setup costs. *Management Science*, 45 (2):273–289, 1999.

96. E.P.C. Kao. A multi-product dynamic lot-size model with individual and joint set-up costs. *Operations Research*, 27:279–289, 1979.

97. S. Karlin. The application of renewal theory to the study of inventory policies. In K.J. Arrow, S. Karlin, and H. Scarf, editors, *Studies in the Mathematical Theory of Inventory and Production*, pages 270–297. Stanford University Press, 1958.

98. A. Kolen. Lovász proof of $LS - U$. Personal communication, Limburg University, 1985.

99. E. Kondili, C.C. Pantelides, and R.W.H. Sargent. A general algorithm for short-term scheduling of batch operations – I: MILP formulation. *Computers Chem. Engng*, 17:211–227, 1993.

100. J. Krarup and O. Bilde. Plant location, set covering and economic lot sizes: An $O(mn)$ algorithm for structured problems. In L. Collatz et al., editors, *Optimierung bei Graphentheoretischen und Ganzzahligen Probleme*, pages 155–180. Birkhauser Verlag, Basel, 1977.

101. R. Kuik, M. Salomon, and L.N. van Wassenhove. Batching decisions: Structure and models. *European Journal of Operational Research*, 75:243–263, 1994.

102. C-Y Lee, S. Cetinkaya, and A.P.M. Wagelmans. A dynamic lot-sizing model with demand time windows. *Management Science*, 47:1384–1395, 2001.

103. J. Lee, J. Leung, and F. Margot. Min-up/min-down polytopes. *Discrete Optimization*, 1:77–85, 2004.

104. J. Leung, T.M. Magnanti, and R. Vachani. Facets and algorithms for capacitated lot-sizing. *Mathematical Programming*, 45:331–359, 1989.

105. R. Levi, R. Roundy, and D.B. Shmoys. Primal-dual algorithms for deterministic inventory problems. In *Proceedings of STOC '04*. ACM, 2004.

106. M. Loparic, H. Marchand, and L.A. Wolsey. Dynamic knapsack sets and capacitated lot-sizing. *Mathematical Programming*, 95:53–69, 2003.

107. M. Loparic, Y. Pochet, and L.A. Wolsey. The uncapacitated lot-sizing problem with sales and safety stocks. *Mathematical Programming*, 89:487–504, 2001.

108. Q. Louveaux and L.A. Wolsey. Lifting, superadditivity, mixed integer rounding and single node flow sets revisited. *4OR*, 1:173–208, 2003.

109. L. Lovász. Graph theory and integer programming. *Annals of Discrete Mathematics*, 4:141–158, 1979.

110. S.F. Love. A facilities in series inventory model with nested schedules. *Management Science*, 18:327–338, 1972.

111. S.F. Love. Bounded production and inventory models with piecewise concave costs. *Management Science*, 20:313–318, 1973.

112. J. Maes and L. van Wassenhove. Multi-item single level capacitated dynamic lot-sizing heuristics: a general review. *J. of the Operational Research Society*, 39:991–1004, 1988.

113. T.L. Magnanti, P. Mirchandani, and R. Vachani. The convex hull of two core capacitated network design problems. *Mathematical Programming*, 60:233–250, 1993.

114. P. Malkin. Minimum runtime and stoptime polyhedra. Report, CORE, Université catholique de Louvain, 2003.

115. A.S. Manne. Programming of economic lot sizes. *Management Science*, 4:115–135, 1958.

116. A.S. Manne. Capacity expansion and probabilistic growth. *Econometrica*, 29:632–649, 1961.
117. H. Marchand and L.A. Wolsey. The 0–1 knapsack problem with a single continous variable. *Mathematical Programming*, 85:15–33, 1999.
118. H. Marchand and L.A. Wolsey. Aggregation and mixed integer rounding to solve MIPs. *Operations Research*, 49:363–371, 2001.
119. R.K. Martin. *Large Scale Linear and Integer Optimization*. Kluwer, 1999.
120. H. Meyr, J. Rohde, H. Stadtler, and C. Sürie. Supply chain analysis. In *Supply Chain Management and Advanced Planning: Concepts, Models, Software and Case Studies* [155], pages 29–43.
121. A. Miller, G.L. Nemhauser, and M.W.P. Savelsbergh. On the capacitated lot-sizing and continuous 0-1 knapsack polyhedra. *European Journal of Operational Research*, 125:298–315, 2000.
122. A. Miller, G.L. Nemhauser, and M.W.P. Savelsbergh. A multi-item production planning model with setup times: Algorithms, reformulations, and polyhedral characterizations for a special case. *Mathematical Programming B*, 95:71–90, 2003.
123. A. Miller, G.L. Nemhauser, and M.W.P. Savelsbergh. On the polyhedral structure of a multi-item production planning model with setup times. *Mathematical Programming B*, 94:375–406, 2003.
124. A. Miller and L.A. Wolsey. Tight formulations for some simple MIPs and convex objective IPs. *Mathematical Programming B*, 98:73–88, 2003.
125. A. Miller and L.A. Wolsey. Tight MIP formulations for multi-item discrete lot-sizing problems. *Operations Research*, 51:557–565, 2003.
126. G.L. Nemhauser and L.A. Wolsey. *Integer and Combinatorial Optimization*. John Wiley and Sons, 1988.
127. J. Orlicky. *Material Requirements Planning*. McGraw-Hill, 1975.
128. F. Ortega. *Formulations and algorithms for fixed charge networks and lot-sizing problems*. PhD thesis, Université catholique de Louvain, 2001.
129. F. Ortega and M. Van Vyve. Lot-sizing with fixed charges on stocks: The convex hull. *Discrete Optimization*, 1:189–203, 2004.
130. M.W. Padberg, T.J. Van Roy, and L.A. Wolsey. Valid inequalities for fixed charge problems. *Mathematical Programming*, 33:842–861, 1985.
131. G. Parker and R. Rardin. *Discrete Optimization*. Academic Press, 1988.
132. O. Pereira and L.A. Wolsey. On the Wagner–Whitin lot-sizing polyhedron. *Mathematics of Operations Research*, 26:591–600, 2001.
133. Y. Pochet. *Lot-sizing problems: Reformulations and cutting plane algorithms*. PhD thesis, Université catholique de Louvain, 1987.
134. Y. Pochet. Valid inequalities and separation for capacitated economic lot-sizing. *Operations Research Letters*, 7:109–116, 1988.
135. Y. Pochet, M. Van Vyve, and L.A. Wolsey. LS-LIB: A library of reformulations, cut separation algorithms and primal heuristics in a high-level modeling language for solving MIP production planning problems. DP 2005/47, CORE, Université catholique de Louvain, 2005.
136. Y. Pochet and R. Weismantel. The sequential knapsack polytope. *SIAM Journal on Optimization*, 8:248–264, 1998.
137. Y. Pochet and L.A. Wolsey. Lot-size models with backlogging: Strong formulations and cutting planes. *Mathematical Programming*, 40:317–335, 1988.
138. Y. Pochet and L.A. Wolsey. Solving multi-item lot sizing problems using strong cutting planes. *Management Science*, 37:53–67, 1991.

139. Y. Pochet and L.A. Wolsey. Lot-sizing with constant batches: Formulation and valid inequalities. *Mathematics of Operations Research*, 18:767–785, 1993.

140. Y. Pochet and L.A. Wolsey. Polyhedra for lot-sizing with Wagner–Whitin costs. *Mathematical Programming*, 67:297–324, 1994.

141. Y. Pochet and L.A. Wolsey. Algorithms and reformulations for lot sizing problems. In W. Cook, L. Lovász, and P. Seymour, editors, *Combinatorial Optimization*, volume 20 of *DIMACS Series in Discrete Mathematics and Theoretical Computer Science*. American Mathematical Society, 1995.

142. Y. Pochet and L.A. Wolsey. Integer knapsacks and flow covers with divisible coefficients: Polyhedra, optimization and separation. *Discrete Applied Mathematics*, 59:57–74, 1995.

143. M. Poggi de Arago and E. Uchoa. Integer program formulation for robust branch-and-cut-and-price algorithms. PUC Rio de Janeiro, November 2003.

144. R. Rardin and L.A. Wolsey. Valid inequalities and projecting the multicommodity extended formulation for uncapacitated fixed charge network flow problems. *European Journal of Operational Research*, 71:95–109, 1993.

145. R.L. Rardin and U. Choe. Tighter relaxations of fixed charge network flow problems. Technical Report report J-79-18, School of Industrial and Systems Engineering, Georgia Institute of Technology, 1979.

146. J. Rohde and M. Wagner. Master planning. In *Supply Chain Management and Advanced Planning: Concepts, Models, Software and Case Studies* [155], pages 143–160.

147. M. Salomon. *Deterministic Lot-Sizing Models for Production Planning*. Number 355 in Lecture Notes in Economics and Mathematical Systems. Springer-Verlag, 1991.

148. A. Schrijver. *Theory of Linear and Integer Programming*. John Wiley and Sons, 1986.

149. J.F. Shapiro. *Modeling the Supply Chain*. Duxbury, 2001.

150. D.X. Shaw and A.P.M. Wagelmans. An algorithm for single-item capacitated economic lot sizing with piecewise linear production and general holding costs. *Management Science*, 44:831–838, 1998.

151. E.A. Silver, D.F. Pyke, and R. Peterson. *Inventory Management and Production Planning and Scheduling*. John Wiley and Sons, 1998.

152. N.C. Simpson and S.S. Erenguc. Production planning in multiple stage manufacturing environments with joint costs, limited resources and set-up times. Technical report, Department of Management Science and Systems, University of Buffalo, 1998.

153. H. Stadtler. Reformulations of the shortest-route model for dynamic multi-item multi-level capacitated lot-sizing. *OR Spektrum*, 19:87–96, 1997.

154. H. Stadtler. Multilevel lot sizing with setup times and multiple constrained resources: Internally rolling schedules with lot-sizing windows. *Operations Research*, 51:487–502, 2003.

155. H. Stadtler and C. Kilger (Eds.). *Supply Chain Management and Advanced Planning: Concepts, Models, Software and Case Studies*. Springer-Verlag, 2002.

156. J.I.A. Stallaert. The complementary class of generalized flow cover *Discrete Applied Mathematics*, 77:73–80, 1997.

157. C. Surie and H. Stadtler. The capacitated lot-sizing problem with linked lot-sizes. *Management Science*, 49(8):1039–1054, 2003.

158. H. Tempelmeier and M. Derstoff. A Lagrangean-based heuristic for dynamic multi-level multi-item constrained lot-sizing with set-up times. *Management Science*, 42:738–757, 1996.

159. H. Tempelmeier and S. Helber. A heuristic for dynamic multi-item multi-level capacitated lot-sizing for general product structures. *European Journal of Operational Research*, 75:296–311, 1994.

160. J.M. Thizy and L.N. Van Wassenhove. Lagrangean relaxation for the multi-item capacitated lot-sizing problem: A heuristic implementation. *IIE Transactions*, 17:308–313, 1985.

161. W. Trigeiro, L.J. Thomas, and J.O. McClain. Capacitated lot-sizing with set-up times. *Management Science*, 35:353–366, 1989.

162. C.A. van Eijl. *A polyhedral approach to the discrete lot-sizing and scheduling problem*. PhD thesis, Eindhoven University of Technology, 1996.

163. C.A. van Eijl and C.P.M. van Hoesel. On the discrete lot-sizing and scheduling problem with Wagner–Whitin costs. *Operations Research Letters*, 20:7–13, 1997.

164. C.P.M. van Hoesel. *Models and algorithms for single-item lot sizing problems*. PhD thesis, Erasmus University, Rotterdam, 1991.

165. C.P.M. van Hoesel and A.W.J. Kolen. A linear description of the discrete lot-sizing and scheduling problem. *European Journal of Operational Research*, 75:342–353, 1994.

166. C.P.M. van Hoesel, R. Kuik, M. Salomon, and L.N. van Wassenhove. The single item discrete lot-sizing and scheduling problem: Optimization by linear and dynamic programming. *Discrete Applied Mathematics*, 48:289–303, 1994.

167. C.P.M. van Hoesel, H.E. Romeijn, D. Romero Morales, and A.P.M. Wagelmans. Polynomial time algorithms for some multi-level lot-sizing problems with production capacities. Research Memorandum RM/02/018, Universiteit Maastricht, 2002.

168. C.P.M. van Hoesel, H.E. Romijn, D. Romero Morales, and A. Wagelmans. Integrated lot-sizing in serial supply chains with production capacities. *Management Science*, 51:1706–1719, 2005.

169. C.P.M. van Hoesel, A. Wagelmans, and A. Kolen. A dual algorithm for the economic lot-sizing problem. *European Journal of Operational Research*, 52:315–325, 1991.

170. C.P.M. van Hoesel, A. Wagelmans, and L.A. Wolsey. Polyhedral characterization of the economic lot-sizing problem with start-up costs. *SIAM Journal of Discrete Mathematics*, 7:141–151, 1994.

171. C.P.M. van Hoesel and A.P.M. Wagelmans. An $O(T^3)$ algorithm for the economic lot-sizing problem with constant capacities. *Management Science*, 42:142–150, 1996.

172. C.P.M. van Hoesel and A.P.M. Wagelmans. Fully polynomial approximation schemes for single-item capacitated economic lot-sizing problems. *Mathematics of Operations Research*, 26:339–357, 2001.

173. T.J. Van Roy and L.A. Wolsey. Valid inequalities and separation for uncapacitated fixed charge networks. *Operations Research Letters*, 4:105–112, 1985.

174. T.J. Van Roy and L.A. Wolsey. Valid inequalities for mixed 0–1 programs. *Discrete Applied Mathematics*, 14:199–213, 1986.

175. T.J. Van Roy and L.A. Wolsey. Solving mixed 0–1 programs by automatic reformulation. *Operations Research*, 35:45–57, 1987.

176. M. Van Vyve. Algorithms for single item constant capacity lotsizing problems. DP 2003/07, CORE, Université catholique de Louvain, 2003.

177. M. Van Vyve. Lot-sizing with constant lower bounds on production. draft 2, CORE, Université catholique de Louvain, 2003.

178. M. Van Vyve. *A solution approach of production planning problems based on compact formulations for single-item lot-sizing models*. PhD thesis, Université catholique de Louvain, 2003.

179. M. Van Vyve. The continuous mixing polyhedron. *Mathematics of Operations Research*, 30:441–452, 2005.

180. M. Van Vyve. Formulations for the single item lot-sizing problem with backlogging and constant capacity. *Mathematical Programming*, to appear, 2006.

181. M. Van Vyve and L.A. Wolsey. Approximate extended formulations. *Mathematical Programming B*, 105:501–522, 2006.

182. F. Vanderbeck. Lot-sizing with start-up times. *Management Science*, 44:1409–1425, 1998.

183. A.F. Veinott. Minimum concave cost solution of Leontief substitution systems of multifacility inventory systems. *Operations Research*, 17:262–291, 1969.

184. B. Verweij and L.A. Wolsey. Uncapacitated lot-sizing with buying, sales and backlogging. *Optimization Methods and Software*, 19:427–436, 2004.

185. T.E. Vollman, W.L. Berry, and D.C. Whybark. *Manufacturing Planning and Control Systems*. Richard D. Irwin, 1997.

186. S. Voss and D.L. Woodruff. *Introduction to Computational Optimization Models for Production Planning in a Supply Chain*. Springer-Verlag, 2003.

187. A.P.M. Wagelmans, C.P.M. van Hoesel, and A.W.J. Kolen. Economic lotsizing: an $O(n \log n)$ algorithm that runs in linear time in the Wagner–Whitin case. *Operations Research*, 40:145–156, 1992.

188. H.M. Wagner and T.M. Whitin. Dynamic version of the economic lot size model. *Management Science*, 5:89–96, 1958.

189. H. Waterer. Trigeiro with startups and changeovers. Report, CORE, Université catholique de Louvain, 2002.

190. L.A. Wolsey. Faces for linear inequalities in 0–1 variables. *Mathematical Programming*, 8:165–178, 1975.

191. L.A. Wolsey. Submodularity and valid inequalities in capacitated fixed charge networks. *Operations Research Letters*, 8:119–124, 1989.

192. L.A. Wolsey. Uncapacitated lot-sizing problems with start-up costs. *Operations Research*, 37:741–747, 1989.

193. L.A. Wolsey. *Integer Programming*. John Wiley and Sons, 1998.

194. L.A. Wolsey. Solving multi-item lot-sizing problems with an MIP solver using classification and reformulation. *Management Science*, 48:1587–1602, 2002.

195. L.A. Wolsey. Lot-sizing with production and delivery time windows. *Mathematical Programming*, (Online First. DOI: 10.1007/s10107-005-0675-3), 2006.

196. W.I. Zangwill. A deterministic multi-period production scheduling model with backlogging. *Management Science*, 13:105–119, 1966.

197. W.I. Zangwill. Minimum concave cost flows in certain networks. *Management Science*, 14:429–450, 1968.

198. W.I. Zangwill. A backlogging model and a multi-echelon model of a dynamic economic lot size production system – A network approach. *Management Science*, 15:506–526, 1969.

Index